Barriers and Channels Formed by Tight Junction Proteins

Barriers and Channels Formed by Tight Junction Proteins

Edited by Gwen Johnston

Larsen & Keller
www.larsen-keller.com

Barriers and Channels Formed by Tight Junction Proteins
Edited by Gwen Johnston
ISBN: 979-8-88836-045-3 (Hardback)

Larsen & Keller
Published by Larsen and Keller Education,
5 Penn Plaza,
19th Floor,
New York, NY 10001, USA

Cataloging-in-Publication Data

Barriers and channels formed by tight junction proteins / edited by Gwen Johnston.
p. cm.
Includes bibliographical references and index.
ISBN 979-8-88836-045-3
1. Tight junctions (Cell biology). 2. Cell junctions. 3. Membrane proteins. 4. Claudins. I. Johnston, Gwen.
QH603.C4 B372 2023
571.6--dc23

Table of Contents

reface

ht junctions (TJ) are cell adhesion structures or junction structures that are located at the uppermost portion of the acent epithelial or endothelial cells in organs and tissues. TJs comprise a branched network of sealing strands, where ry strand acts independently. The main function of TJs is maintaining the relative stability of intracellular substances. It ctions by closing or opening the intercellular pathways and by coordinating the entry and exit of molecules of different es and charges. TJs are also responsible for regulating the permeability of the paracellular barrier, preventing microbial asion, regulating cell proliferation, and maintaining cell polarity. TJ proteins connect endothelial and epithelial cells, d are also responsible for creating specific barriers and channels for paracellular transport. This book contains some th-breaking studies related to barriers and channels formed by tight junction proteins. It will serve as a valuable source reference for graduate and postgraduate students.

is book has been the outcome of endless efforts put in by authors and researchers on various issues and topics within e field. The book is a comprehensive collection of significant researches that are addressed in a variety of chapters. It ll surely enhance the knowledge of the field among readers across the globe.

gives us an immense pleasure to thank our researchers and authors for their efforts to submit their piece of writing fore the deadlines. Finally in the end, I would like to thank my family and colleagues who have been a great source of spiration and support.

Editor

1

Computational Nanoscopy of Tight Junctions at the Blood–Brain Barrier Interface

Nandhini Rajagopal, Flaviyan Jerome Irudayanathan and Shikha Nangia *

Department of Biomedical and Chemical Engineering, Syracuse University, Syracuse, NY 13244, USA; nrajagop@syr.edu (N.R.); firudaya@syr.edu (F.J.I.)
* Correspondence: snangia@syr.edu

Abstract: The selectivity of the blood–brain barrier (BBB) is primarily maintained by tight junctions (TJs), which act as gatekeepers of the paracellular space by blocking blood-borne toxins, drugs, and pathogens from entering the brain. The BBB presents a significant challenge in designing neurotherapeutics, so a comprehensive understanding of the TJ architecture can aid in the design of novel therapeutics. Unraveling the intricacies of TJs with conventional experimental techniques alone is challenging, but recently developed computational tools can provide a valuable molecular-level understanding of TJ architecture. We employed the computational methods toolkit to investigate claudin-5, a highly expressed TJ protein at the BBB interface. Our approach started with the prediction of claudin-5 structure, evaluation of stable dimer conformations and nanoscale assemblies, followed by the impact of lipid environments, and posttranslational modifications on these claudin-5 assemblies. These led to the study of TJ pores and barriers and finally understanding of ion and small molecule transport through the TJs. Some of these in silico, molecular-level findings, will need to be corroborated by future experiments. The resulting understanding can be advantageous towards the eventual goal of drug delivery across the BBB. This review provides key insights gleaned from a series of state-of-the-art nanoscale simulations (or computational nanoscopy studies) performed on the TJ architecture.

Keywords: claudin; tight junctions; blood–brain barrier; in silico; drug discovery; membrane proteins; protein interactions; molecular dynamics

1. The Blood–Brain Barrier

The blood–brain barrier (BBB) is a selective permeability barrier that impedes the influx of potentially harmful blood-borne chemicals into the brain [1–4]. Although discovered by Ehrlich in 1885, the term blood–brain barrier was coined almost four decades later in the 1920s, after a series of reports by Lewandowsky, Goldmann, Stern, and Gautier [5–22]. Their collective work in vertebrate brain showed that tracers and toxins were unable to permeate the blood–brain interface; for example, trypan blue dye injected into the brain of a mouse remained confined to the brain, whereas the same dye injected outside the brain stained the entire body except the brain (Figure 1a). This permeability barrier is considered an evolutionary adaptation in vertebrates, and is vital for maintaining homeostasis in the CNS [1,5,21–23]. Unlike other organs, the brain lacks energy reserves [24–26]; therefore, for normal function, it requires a constant supply of nutrients from the bloodstream [27,28]. The brain's microvasculature regulates the transfer of essential nutrients while protecting the brain against small chemical imbalances, which can lead to serious disease pathologies [29–33].

A critical component of the BBB is a specialized physical barrier comprised of tight junctions (TJs) present along the apical side of endothelial cells [1,23,34]. These cell-to-cell adhesion structures act as gatekeepers of the paracellular pathway (Figure 1b) by regulating the passive diffusion of molecules and ions into the brain [23,35,36]. The establishment of TJs at the BBB leads to a high electric resistance

(>1000–3000 Ω cm^2) across the interface [37,38]. The presence of TJs also leads to cell polarization whereby the apical and basolateral sides of the endothelia are isolated [2,35]. The polarized endothelia control transcellular permeability by specific expression of transporters and receptor proteins in the apical side of the membrane (Figure 1b). In the event of disease pathologies, TJs are compromised, leading to the loss of barrier selectivity in both the transcellular and paracellular pathways [33,35].

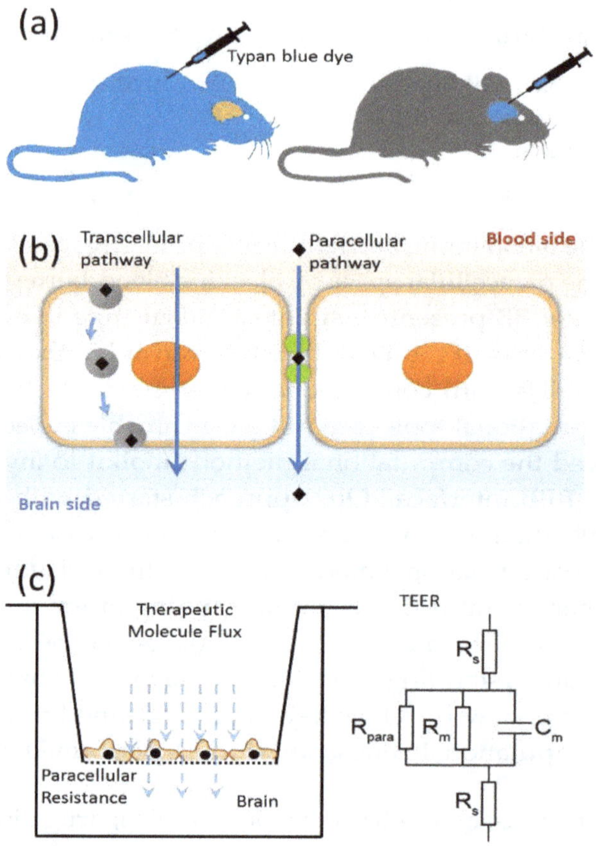

Figure 1. The blood–brain barrier. (**a**) Discovery of the blood–brain barrier (BBB), (**b**) pathways of molecular transport across the neurovascular endothelial barrier: transcellular (arrow through the cell) and paracellular (arrow through space between the two cells) and (**c**) in vitro characterization of the paracellular barrier strength of movement of therapeutic molecules (downward arrows) using transepithelial electric resistance (TEER) model.

Although the BBB is vital for neurological functions, its selective permeability is a deterrent to drug delivery into the brain [31,39–41]. The BBB obstructs effective treatment of neurodegenerative diseases such as Alzheimer's and Parkinson's, because therapeutics are unable to cross the barrier [39]. To develop new therapeutics, in vivo animal models provide an experimental environment that mimics the intricacies of human physiology. However, high cost and lack of complete translation of disease pathologies from animal models to humans make in vivo studies less effective for drug development. In vitro models are therefore being used early in a drug's development to investigate its permeability across the BBB [42–44]. Drug permeability is typically studied using cell cultures grown on permeable membranes in a transwell, where the transendothelial/epithelial electrical resistance (TEER) is used to measure the change in the electrical resistance across the cellular monolayer (Figure 1c) [37,45]. The TEER measurements evaluate the permeability and integrity of the endothelial cell layer [45,46], but the static cell culture cannot mimic the fluid characteristics of the BBB. More recently, microfluidic organ-on-chip in vitro models have been employed for the assessment of TJ barrier properties. Efforts are underway to develop BBB-on-a-chip that can recreate the human BBB using cells derived from the brain endothelial cells [47,48]. Additionally, in silico models of the BBB TJs have been developed to reliably compute the permeability of ions and small molecules across the paracellular barrier [49–52]. With advances in computer algorithms and hardware, strategies are being developed to design novel therapeutics that can cross the BBB. A concerted in vivo, in vitro, and in silico effort is underway to unravel the complexities of the BBB.

2. Tight Junctions: Role of Claudin-5 in the BBB

Claudin family of membrane proteins establishes TJs in endothelial and epithelial cells [53–55]. Claudins are expressed in vascular endothelial cells and all known epithelial cells throughout the body [56–58]. To date, 27 members of the claudin family are known to be functionally expressed in mammals [59,60]. They are classified into classic claudins (1–10, 14, 15, 17, and 19) and non-classic claudins (11–13, 16, 18, and 20–27) based on their sequence similarity and structural homology [59,61,62]. Most tissues express multiple members of the claudin family at the TJs [56,63–67]. Claudin proteins (~20–30 kDa) fold into a four-transmembrane helix bundle (TM1–4) with two extracellular loops (ECL1–2) and an intracellular loop (ICL; Figure 2) [60,62,68]. The lateral interactions mediated by TM and ECL domains of claudins within the membrane of the same cell are termed cis interactions. Cis interactions between multiple claudins occur in membranes of expressing cells. Subsequently, the claudins undergo trans assembly via head-on interactions of their ECL loops and form the macromolecular TJ assembly with adjacent claudin expressing cells. The TJ strands (often observable under freeze-fracture microscopy) are multi-component structures with claudins as the primary component alongside other TJ-associated proteins such as occludin, tricellulin, junctional adhesion molecules (JAMs), and zonula occludens (ZO-1,2) [53,60,61].

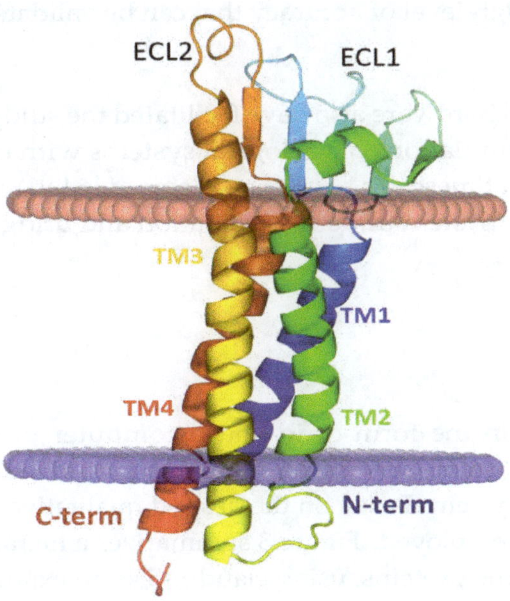

Figure 2. Homology modeled claudin-5 structure shows four transmembrane helices (TM1–4), extracellular loops (ECL1 and ECL2), and intracellular loop (ICL) in cartoon representation colored from N-terminal (blue) to truncated C-terminal (red). The membrane spanning region lies within the horizontal (red and blue) planes.

Claudin-5 is the key TJ protein at the BBB with an expression level that is much higher than claudin-1, -3, and -12 [69,70]. Knockout experiments have shown that claudin-5 is responsible for controlling the paracellular permeability of molecules up to ~800 Da [37,38,71]. In vitro biochemical analyses have suggested that claudin-5's ECL2 loop strongly contributes to the trans interactions [72,73]. Mutations to conserved residues in ECL2 lead to a marked increase in paracellular permeability because of compromised size selectivity [59,74]. Studies also show that claudin-5 knockout mice die within 10 h of birth, although their TJs remain impermeable to molecules >800 Da in size [71]. In contrast a double knockdown study of both claudin-5 and occludin showed enhanced permeability of 3–10 kDa tracers indicating that occludin serves a structural role in the BBB TJs [75,76].

3. Need for Computational Modeling

In silico models complement in vitro and in vivo approaches used to investigate biological systems. Computer models augment our ability to analyze aspects of biological systems that are too complex to be investigated by conventional experiments alone [77–80]. Computational approaches provide a molecular-level basis for experimentally observed biomacromolecular assembly, selectivity, dynamics, and macroscopic properties.

For example, X-ray crystallography of a membrane protein provides essential structural information about its conformation at low temperatures [81,82]. However, proteins participate in dynamic processes and have thermal motions at physiological temperatures that cannot be resolved by current experimental techniques [83,84]. In silico methods can readily build upon the crystal structure data to provide information about molecular displacements, conformational dynamics, and reorientation frequencies of proteins [80]. The resulting models can also provide insights into the intramolecular and intermolecular interactions of a protein in its native environment, including other proteins, lipids, ions, and solvents. Another area where computational methods have been transformative is in predicting the three-dimensional structure of membrane proteins. The hydrophobic nature of membrane proteins makes them experimentally unwieldy for crystallization and X-ray structure determination [83–85]. In recent years, advances in protein sequencing have enabled the utilization of sequence co-evolution information to infer and predict the native fold(s) of a protein to a high-level of accuracy that can be validated by experimental biochemical assays [86–90].

Advances in computational hardware also have facilitated the study of complex biological systems. In the last decade, molecular simulations of biological systems with millions of interacting particles over microsecond to millisecond timescales have been reported [64,91–116]. These simulations provide unprecedented insights into the system under investigation and bridge nanoscale data to macroscale experimental results.

4. Computational Toolkit

Computational methods, in the form of in-house computer programs, freely-distributed and commercial software suites, and online servers, offer an extensive toolkit that can be used to investigate different aspects of a biological system. Based on the desired qualitative or quantitative output, an array of computational models can be employed. Figure 3 summarizes a hierarchical computational approach relevant to the study of membrane proteins, using claudin-5 as an example.

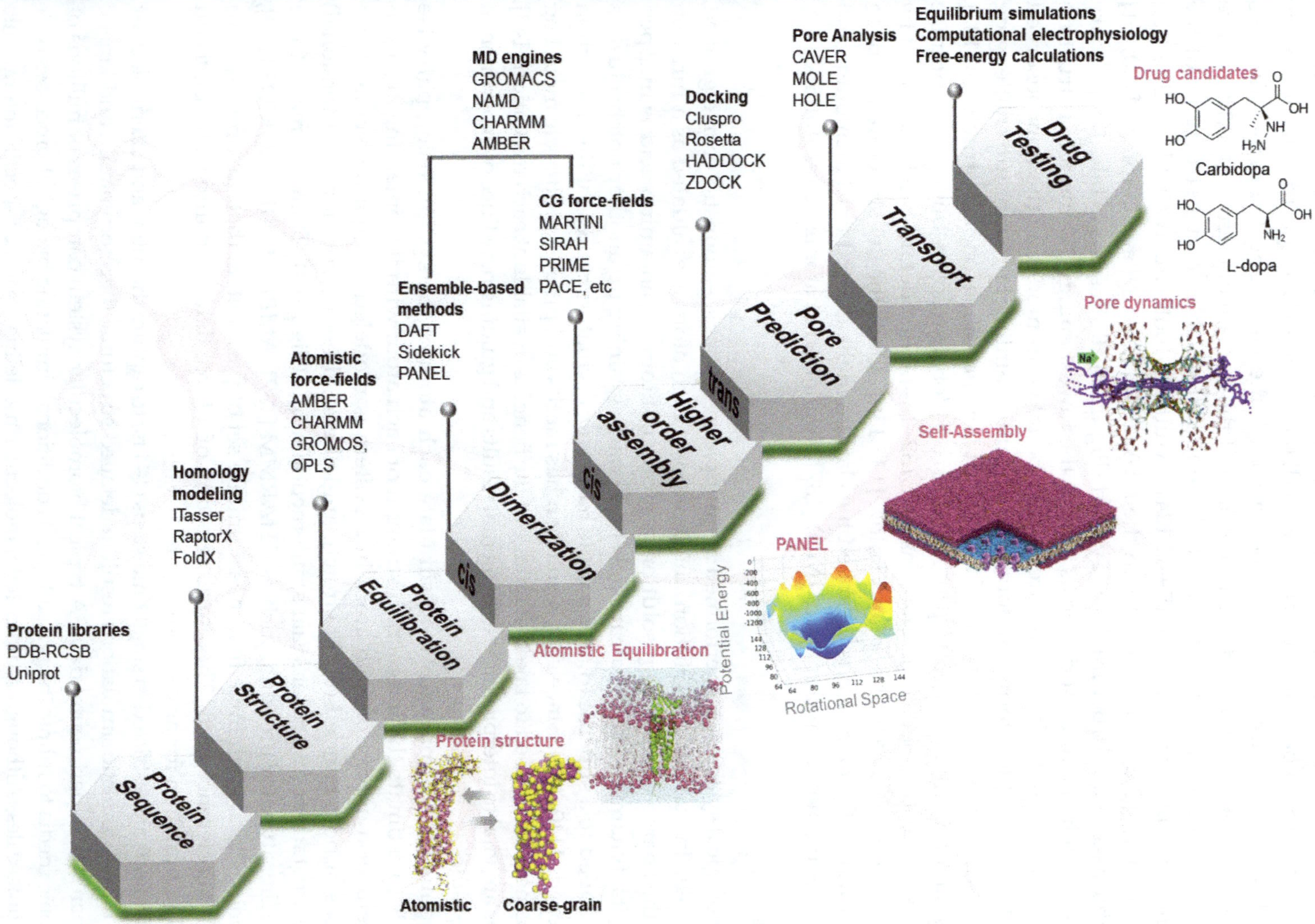

Figure 3. An overview of computational approaches and tools used to study the tight junctions at the blood–brain barrier. The stepwise research strategy included: obtaining protein sequence from online libraries; determining protein structure; equilibrating the protein in membrane lipids; investigating cis assembly (dimer and higher-order); investigating trans assembly for predicting tight junction barrier and pores; evaluating transport properties on ions and molecules; and testing tight junction modulators for enhanced drug delivery. The computational approaches (homology modeling, atomistic and coarse-grained molecular dynamics self-assembly simulations, PANEL, etc.) and tools (online libraries, webservers, force-fields, and software suites) associated with each step of our research strategy are also provided.

4.1. *Protein Structure Prediction*

Crystal structures of murine and human claudins mClaudin-15 (4P79) [68], mClaudin-19 (3X29) [117], hClaudin-4 (5B2G) [118], and mClaudin-3 (6AKE) [119] are available. Most of the currently available claudin structures have missing loop and c-terminal domains, due to crystallographic artifacts. To study a claudin that has not yet been structurally resolved, the first step is to construct its homology model [120–122]. Reliable homology models can be derived if the target and template structures have at least 30% residue identity spanning the entire length of the domain of interest [120,123]. Since most claudins share ~30% sequence identity, this technique can reliably model structures of the other members of the claudin protein family [59,62,124]. Multiple tools are available online that can build homology models; [125,126] among them the I-TASSER [127–129] and RaptorX [130,131] servers are well established in the structural biology field (Figure 3). Both servers take the primary sequence of the target protein as input and give as output the lowest energy 3D-modeled structure of the protein. In building homology models care should be taken in modeling the nonconserved domains. For example, the flexible c-terminal domains of claudin family of proteins are challenging to model because they are intrinsically disordered and adopt nonspecific secondary structures [132]. Homology modeling methods are designed to provide a correct global structure but do not optimize side-chain orientations or flexible-loop conformations [121,132–134]. Such details can be obtained by loop modeling techniques such as ModLoop and FoldX [135,136]. Once the 3D-structure has been computationally determined, the next step is to embed the claudin in the desired lipid membrane and optimize the structure using molecular dynamics (MD) [137] simulation at atomistic resolution.

4.2. *Protein Dynamics*

Simulations of biological systems are performed at multiple resolutions—atomistic, coarse-grained, and mesoscale [137–139]. In atomistic resolution, the motion of each atom is evaluated as a function of time, whereas, in the coarse-grain (CG) resolution, a group of atoms in an amino acid is mapped to a bead to evaluate its dynamics. Mesoscale models take coarse graining up another notch; clusters of amino acids are mapped to a superbead, and simple harmonic oscillators represent the interaction between those superbeads [140–142]. Since each resolution has merits and pitfalls, multiscale modeling approaches are increasingly being used to take advantage of the benefits available at each resolution [143]. Multiscale simulations are vital to interpreting experimental results and guiding new lines of experimental investigation [80,144–148].

The classical atomistic MD approach is to simulate each atom of the system using pairwise interatomic potentials to estimate the interaction strength of each atom with its remaining atoms in the system. The interaction energy of each atom in a system is modeled using a molecular force field, which comprises of both bonded (bond stretch, angle bend, dihedral rotation, and torsional) and non-bonded (electrostatic and van der Waal's) interaction terms (Figure 4). There are many force fields available in the literature for proteins, such as CHARMM [137], AMBER [149], GROMOS [150], and OPLS [151] among others. Each force field has a unique set of parameters that are optimized to reproduce experimental reference data. Therefore, selection of a force field is an important decision before starting an MD simulation for any system.

The all-atom simulations are restricted to small time-steps (1–4 fs) to capture the intra- and intermolecular motion of atoms over time. Atomistic simulations result in detailed structural, thermodynamic, and kinetic data that can be compared to experimental data. A typical biomolecular system comprises of millions of atoms, embedded in ionic media to mimic the physiological conditions, simulating biological processes over micro- or millisecond timescales. Although atomic- level molecular modeling provides specific interatomic interactions, its practical application is limited by computational resources.

Setting up membrane protein systems for molecular simulations is challenging when it comes to embedding a protein in multicomponent lipid membranes. To streamline the process, the recent development of publicly hosted interactive web-based servers (such as charmm-gui) [152] is enabling the generation of input files required for membrane protein simulations. Among the existing simulation packages are popular MD engines such as GROMACS [153,154], NAMD [155], AMBER [156,157], CHARMM [152,158,159], and visualization tools such as VMD [160], Pymol [161], and YASARA (Figure 3) [162–164].

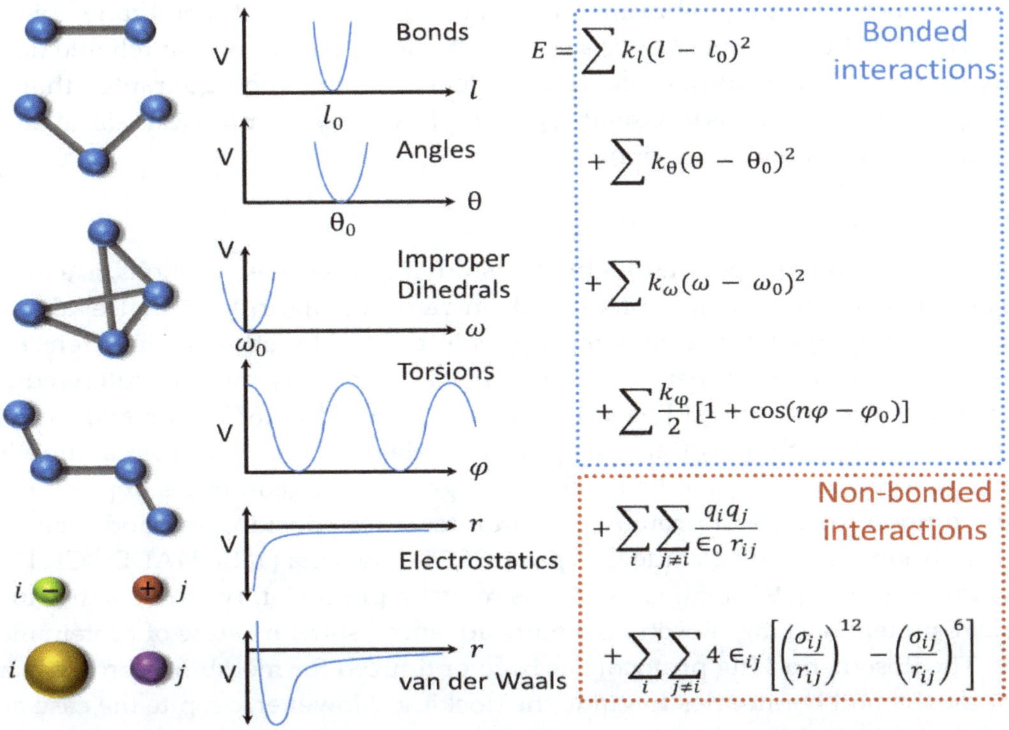

Figure 4. Typical form of a molecular force-field constituted by bonded (bond, angle, dihedral, and torsion) interactions and nonbonded (electrostatic and van der Waal's) interaction energy terms. The force filed parameter set include: equilibrium bond length (l_0), bond angle (θ_0), dihedral angle (ω_0), torsion angle (φ_0), and their respective force constants, k_l, k_θ, k_ω, and k_φ, as well as charge (q_i) on each atom i, dielectric constant (ϵ_0), strength of dispersion interactions (ϵ_{ij}), and contact distance (σ_{ij}) between the atoms i and j. The interatomic distance is represented by r_{ij}.

The use of CG simulations has increased in the last 10 years with the development of several force fields for a variety of biomolecules. Popular CG force fields include MARTINI [165–169], SIRAH [170,171], PRIME [172,173], and PACE [174,175]. Among these, MARTINI is the most popular for membrane protein simulations [165,166,176]. The MARTINI force field maps ~4 nonhydrogenic atoms into one bead; for example, a glycine amino acid residue is mapped as one bead [165,166,168]. By mapping a small number of atoms into a bead, CG resolution reduces the number of particles in a simulation and allows the system to be sampled less frequently with a larger time-step (20–40 fs). This approach enables simulation timescales that are 2 to 3 orders of magnitude larger than atomistic simulations. Thus, the advantages of CG simulations have spurred interest in simulations of larger sizes (hundreds of nanometers) and longer timescales (tens of microseconds) [177–180]. Others and we have rigorously demonstrated the ability of MD simulations to capture residue-level precision in membrane protein interactions [91,94,97,108,110,115].

4.3. Protein–Protein Interactions

4.3.1. Self-Assembly Simulations

Molecular self-assembly simulation is a powerful technique that mimics interactions among molecules over time [94,98,145,181]. These simulations capture the spontaneous organization of molecules dictated by intra- and intermolecular nonbonded interactions (van der Waals forces, electrostatics, hydrogen bonds, hydrophobic, and hydrophilic interactions) under near-equilibrium conditions. Self-assembly simulations are routinely used to study complex membrane protein–protein interactions. These simulations, however, have several limitations including high computational cost. Membrane protein assembly is particularly computationally expensive in part because protein diffusion requires much longer equilibration time than the relaxation of membrane lipids. Moreover, the simulations need to be performed over a sufficiently long time to obtain reliable data, although ensuring convergence of such simulations is a challenge. There is no guarantee that a 10–100 μs timescale is long enough for proteins to assemble [182,183]. Employing multiscale simulations mitigates the challenges of long timescale simulation.

4.3.2. Molecular Docking

Molecular docking, another popular method of studying protein associations, is a highly effective method to examine and rank interactions based on rigorous algorithms that evaluate empirical energetics data. Docking approaches are simple geometry-based methods where tens of thousands of protein–protein interfaces are generated via geometric rigid body motions followed by a scoring scheme where interfaces are ranked based on the energetics [184,185]. The energy function that is used in the scoring algorithm is often empirically modeled or derived from a knowledge-based method. Docking approaches tend to be fast because generating geometries of proteins is a simple task. Depending upon the type of protein–protein interaction, docking methods can give realistic results. Popular docking methods include Cluspro [186,187], Rosetta [188], HADDOCK [189,190], and ZDOCK [191] among others [192–195]. Cluspro has robust reproducibility, and it is the current state of the art in protein–protein docking. Rosetta is a more advanced software suite of protein modeling and analysis tools. The Rosetta docking protocol has been optimized for membrane proteins, and one can define membrane domain boundaries to guide the docking. However, despite the ease and speed of docking methods, static docking is not optimal for macromolecular assemblies that involve multiple proteins [196]. Moreover, for membrane proteins, the presence of the lipid environment significantly influences proteins' association patterns, but none of the currently available docking methods offers a realistic lipid environment for protein docking [197–202].

4.3.3. Protein Association Energy Landscape

An alternate strategy to study membrane protein association is to employ ensemble-based sampling methods. These methods use hundreds to a few thousand starting structures that each run independently to yield an unbiased sampling of the protein–protein association landscape. One such ensemble-based method, docking assay for transmembrane (DAFT) components [203] uses the population density of the resulting protein association conformations as a proxy for stability. Other ensemble-based methods use amino acid contact-based stability as a metric [204,205]; however, none of the available methods provides the association energies needed to determine the relative stability of protein–protein conformations obtained in the ensemble.

In our work, we used a combined stochastic sampling and MD approach to sample pairwise protein–protein interactions [206]. Taking advantage of in-plane diffusion of membrane proteins relative to each other in the lipid bilayer, we defined a two-dimensional rotational space of the interacting proteins. We adopted a stochastic sampling technique to generate thousands of seed geometries that uniformly sample the rotational space around a pair of interacting proteins. These seed geometries were then propagated independently to equilibrium using standard MD. The interaction energy associated with each conformation visited in the dynamics was computed to generate the protein association energy landscape (PANEL) [206]. By placing the proteins within the interacting distance of the seed geometries, we eliminated the need for protein diffusion, thereby reducing the simulation times for each system and boosting the PANEL performance by two orders of magnitude over classic self-assembly simulations. The PANEL method also provides an absolute measure of interaction energy for any pair of interacting proteins in dimeric or even in trimeric association [206]. We used the PANEL method to report the landscape of claudin-5 dimers (Figure 3), corroborating our previous self-assembly simulation results [115]. Furthermore, the robustness, computational affordability, and user-friendly implementation of the PANEL method makes it attractive for application to any membrane protein.

4.4. Pore Characterization and Paracellular Transport

The molecular architecture of the TJ pores and barriers are the determinants of the charge and size-selective transport of ions and molecules. Analysis of the pore's structural features is important to understand TJ selectivity. The structural characteristics of TJ pores (such as length, diameter, volume, and cross-sectional area) provide valuable insight into the size selectivity of the pore. Additionally, identification of pore-lining residues in the channel walls and the bottleneck regions of the pore offer significant clues about their charge selectivity. Several software tools, such as CAVER [207], MOLE [208], HOLE [209], MOLAXIS [210,211], and POREWALKER [212], (each based on slightly different algorithms) provide structural analysis of protein pores (Figure 3).

Paracellular transport involves calculation of the free energy profiles of the translocation of a substrate through the pore. This calculated energy directly correlates to the permeability of the substrate across the pore [213,214]. The free energy profile can be obtained from the advanced sampling techniques of umbrella sampling [215] or metadynamics [216–218]. In the umbrella sampling technique, where multiple frames or conformations of a solute are generated along the length of the pore, an external force-constant is applied on the solute to restrain its position. Based on the amount of force applied, the free energy profile along the transport path can be evaluated [215,219]. An alternative method to determine the free energy profile is the recently introduced metadynamics approach where the system is biased gradually such that the ion translocation is made feasible within the affordable timescale of MD simulation. Metadynamics is available in GROMACS through the PLUMED plugin [220].

A simplistic equilibrium MD simulation can also be set up with different ion concentrations to observe the number of ion translocation events that occur during the simulation. This method may provide insights about preferred ion transport under unbiased, equilibrium conditions; however, the limitation of the method is the timesale in which an ion transport event will occur may not be within the simulaion time frame. To accelerate the ion transport event, computational electrophysiology has emerged as a method that involves biasing the molecular system by creating localized charge imbalances in system, which induces ion transport [221,222]. The small charge imbalances provide biologically realistic electrochemical gradients across the pore and facilitate ion transport.

5. Computational Nanoscopy of Claudin-5 in BBB Tight Junction Architecture

Claudin proteins embedded in the lipid bilayer undergo cis assembly via their transmembrane helical domains, or by the extracellular loops, or by both [94,115]. In general, protein–protein cis interactions are mediated by a multicomponent mixture of saturated and unsaturated lipids with variety of charged head groups and lipid tail lengths [91,106,201,202,223]. The helical bundle of membrane proteins has a rigid tertiary structure that provides stability to the protein to reside in the membrane environment. This rigidity in turn affects the local lipid environment and manifests as a hydrophobic mismatch between the protein and the lipids [198,224,225]. Hydrophobic mismatch is energetically unfavorable and serves as a driving force for the embedded protein to find interactions that would minimize its hydrophobic mismatch with the lipids and undergo cis assembly [103,226–230].

There are multiple aspects of claudin cis and trans assembly that influence the TJ architecture. The cis-assembled claudins can provide the template for the trans interactions [97,231,232]; therefore, the study of cis interfaces is vital to understand the TJ architecture. Besides, claudins are reported to be post-translationally modified (PTM) with phosphorylation and palmitoylation. Of these PTMs, studies suggest that palmitoylation alters recruitment of claudin proteins in the TJs and TJ permeability [91,233,234]. Since most aspartate–histidine–histidine–cysteine (DHHC) enzymes that assist in palmitoylation [235] are thought to be localized to the endoplasmic reticulum (ER) [198,236], it is likely that claudin cis interactions that are influenced by palmitoylation are localized in the ER. However, DHHC enzymes have been reported in other organelles and in the plasma membrane [237], which makes both the location of claudin-5 palmitoylation in the secretory pathway and the cis dimerization ambiguous. It is not surprising that the complexity of the BBB TJ architecture has been a deterrent in unraveling the mystery of the size and charge selectivity of the TJ. A list of basic questions that have confounded experts in the TJ research field include:

(1) How do claudin proteins form contiguous strands? Do the claudin proteins have preferred cis conformations?
(2) Where along the secretory pathway do claudin-5 proteins undergo cis assembly?
(3) How do posttranslational modifications of claudin-5 influence the cis interactions?
(4) How are TJ pores and barriers formed?
(5) How do BBB TJs get their size and change selectivity?

To provide a new perspective to these long-standing challenges, we developed tools for in silico investigation of the TJ claudin-5 proteins. In a series of publications over the past few years [91,94,97,115,206], we have reported on the various aspects of claudin-5 BBB TJ architecture. Here, we reviewed the outcomes of these five lines of inquiry (Figure 5).

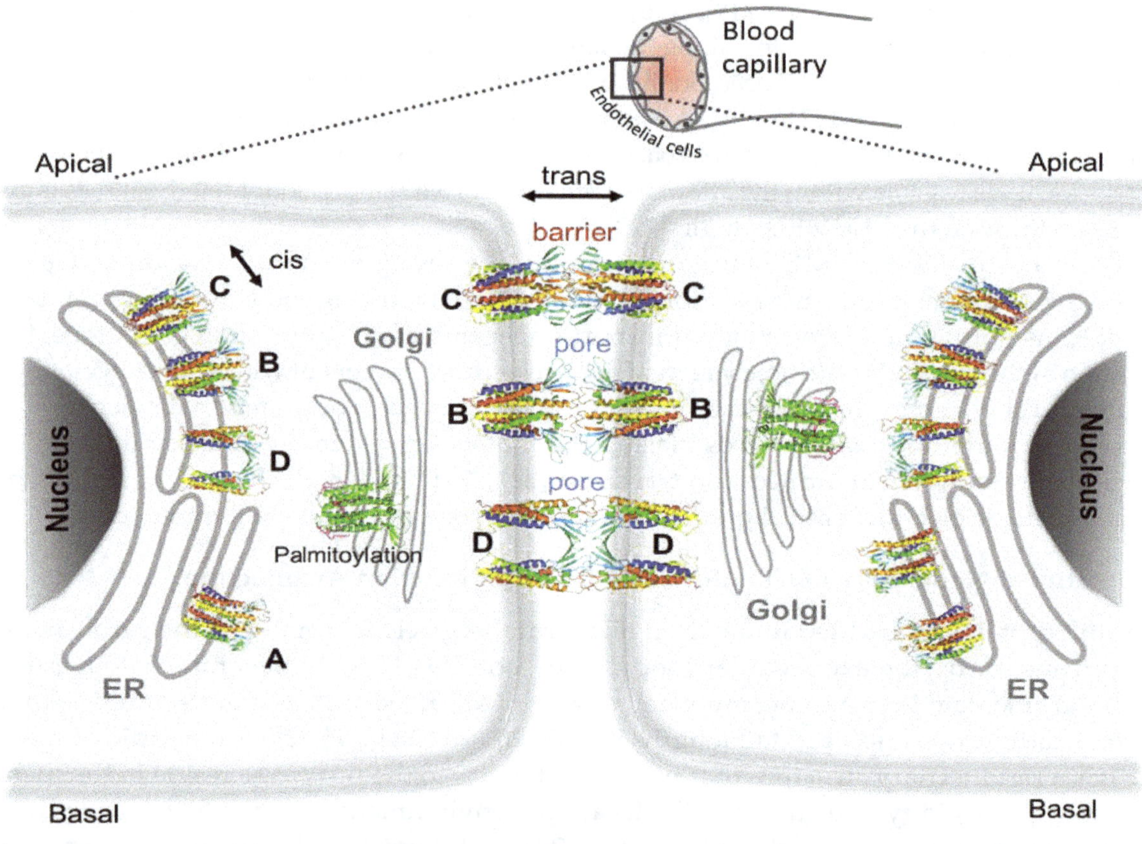

Figure 5. A schematic view of claudin-5's cis and trans assembly along the secretory pathway. The cis dimerization of claudin-5 proteins in endoplasmic membrane (ER) and Golgi complex membrane can result in multiple conformations that are categorized as dimers A, B, C, and D [94,97,115]; posttranslational modification can influence the relative stability of these dimers [91]; the trans interaction of C dimer pair results in a paracellular barrier, whereas trans interaction of B–B or D–D dimer pairs result in paracellular pore.

5.1. *Claudin-5 Forms Symmetric and Asymmetric Dimer Conformations that Form Contiguous Strands*

The cis interactions among claudins, determined by computational methods, corroborate results of biochemical assays [115]. The claudin-5 cis interactions were studied using both the self-assembly and the PANEL approaches [115,206].

The self-assembly simulations, performed in CG resolution, suggest that claudin-5 monomers embedded in a simple mixture of saturated and unsaturated lipids readily form dimers (within 400–500 ns) [115]. These dimers subsequently (in 5–10 μs) form contiguous strands and high-order assemblies (Figure 6a). For example, in a system composed of 64 claudin-5 monomers, a contiguous strand of 36 monomers was observed with a ring-like pentamer (Figure 6b), which match the size description of ~10 nm particles observed under freeze fracture [238–240]. The claudin strands present at the end of the 10 μs simulation were extracted and reverse mapped to atomistic resolution for conformational and structural analysis (Figure 6c). A pair-wise conformational analysis of the self-assembled claudins revealed four, most prevalent conformations (labeled dimer A–D) that concurrently exist to form the strands. Several repeat simulations have been performed that all show formation of dimers A–D although in varying populations. Since it is practically impossible to achieve convergence in self-assembly simulations due to finite timescales, we were unable to provide converged relative populations of each of the four dimers. However, these simulations indicate that a dimer is the smallest stable unit of the self-assembled strand and that the cis assembly can occur in the absence of trans interacting partners from the adjacent cell. In an experimental study, claudin-2 cis dimerization was also observed independent of cell-surface expression and intercellular cross-linking [241].

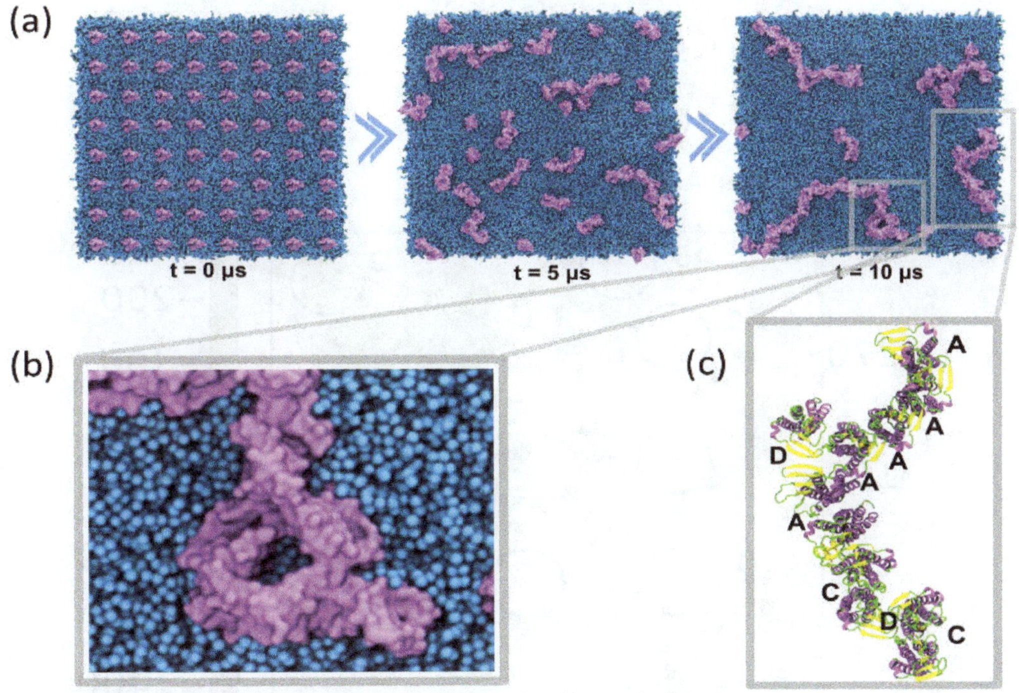

Figure 6. Claudin-5 strand formation in coarse-grain molecular dynamics (CG MD) self-assembly simulations. (**a**) Snapshots (top view) of claudin-5 (purple beads) assembly in lipid (cyan beads) membrane at $t = 0$, 5, and 10 μs, (**b**) close up view of a self-assembled claudin-5 pentamer, and (**c**) reversed-mapped strand showing dimer A, C, and D conformations.

To delve deeper into the stability of the claudin-5 dimers, we developed the PANEL approach [206]. The PANEL plot for claudin-5 provided a comprehensive sampling of dimer conformations and a clear visualization of the entire energy landscape as a function of the orientation angles (Figure 7). The claudin-5 dimer interaction energies range from 0 (least stable) to approximately -1500 kJ mol^{-1} (most stable) [206]. The PANEL contour plot shows multiple low energy conformations. Conformations along the diagonal (Figure 7b) represent symmetric dimers, while those off the diagonal are asymmetric dimers. It was reassuring that the key symmetric Dimers B, C, and D, also observed in self-assembly and in experiments, are shown in the minimum energy PANEL plot. Dimer B is a TM3:TM3-mediated dimer, while Dimer C is mediated by TM4:TM4, and Dimer D, the symmetric ECL1-mediated interaction, also proposed by Suzuki et al. [242]. There are also several asymmetric dimers (Dimer A), which are marked as Dimers

A1, A2, and A3 in the PANEL plot. Dimer A1 corresponds to the linear arrangement observed in the claudin-15 crystal structure [242]. The relative orientation of the claudins in each of the Dimers A to D (Figure 7b) is described in our earlier work [115,206]. Although the presence of claudin dimers and higher order complexes has been reported in several experimental studies [73,240,241,243,244], isolating intact dimers or higher order complexes still remains a challenge. Use of computational approach provides a direct look into the structural details of these claudin interfaces.

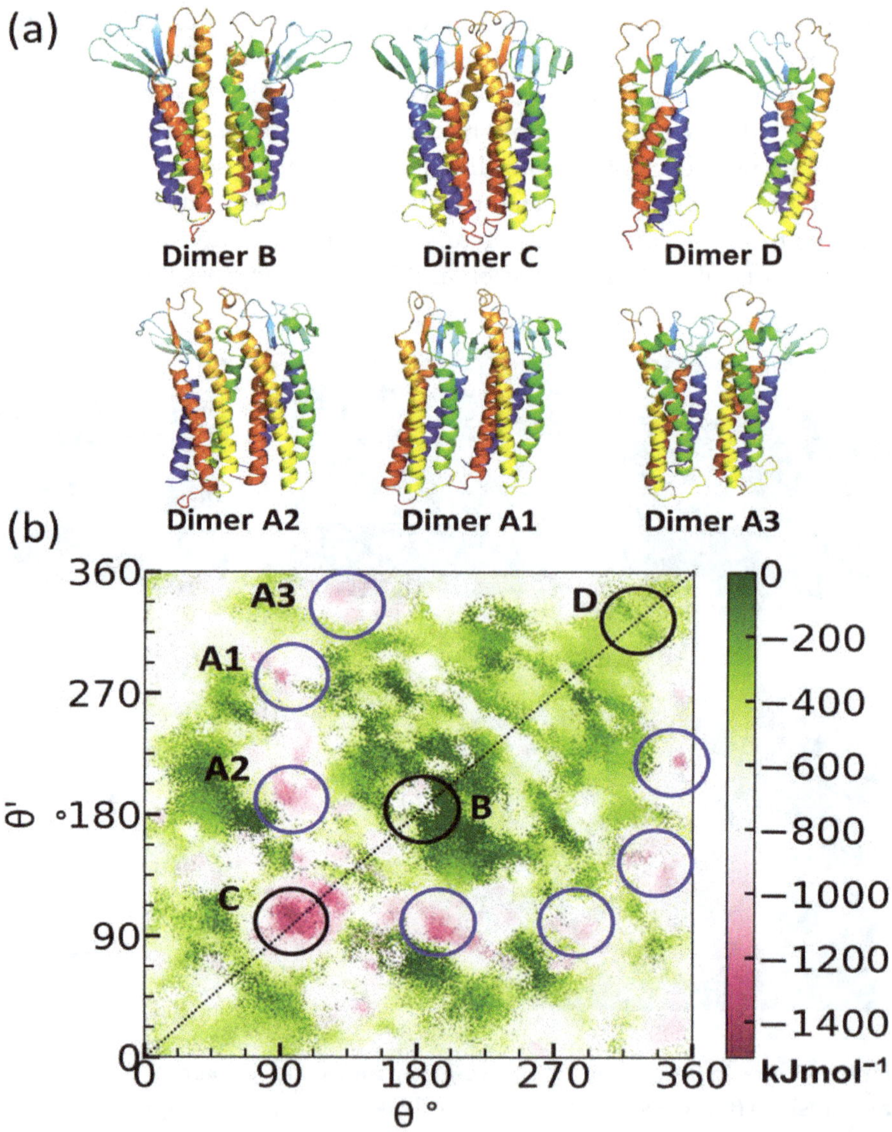

Figure 7. Claudin-5 cis dimerization. (**a**) Key cis dimers formed in claudin-5, reverse mapped to atomistic resolutions, and (**b**) PANEL plot showing location of low energy dimers A1, A2, and A3 (blue circles); and dimers B, C, and D (black circles).

5.2. Primary Cis Interactions Occur Early in the Secretory Pathway

Like other TJ proteins, claudins are secreted in the endoplasmic reticulum (ER) and reach the plasma membrane via the Golgi complex [231] Along this secretory pathway, the membrane composition changes drastically [245–247] To mimic the complexity of cell organelles encountered by a typical membrane protein during its secretory pathway, we performed claudin-5 self-assembly in three model membrane compositions: ER, cholesterol-enriched endoplasmic reticulum (ERc), and plasma membrane (PM) [91]. The ER membrane model constituted a binary mixture of saturated and unsaturated lipids in 1:1 ratio; the ERc was a 2:2:1 ternary mixture of saturated lipid, unsaturated lipid, and cholesterol; and the PM model comprised ten lipid components including gangliosides, sphingomyelin, and cholesterol [91].

Our findings indicate that claudin-5 monomers are more likely to form dimers in simple membranes like the ER where saturated and unsaturated phospholipids are the principal components. As claudins move to the Golgi complex, trans-Golgi vesicles, and the plasma membrane, the higher cholesterol concentrations in these membranes limit claudin-5 assembly. We observed that an increase in membrane complexity was negatively correlated with the length of claudin-5 strands (Figure 8a) [91].

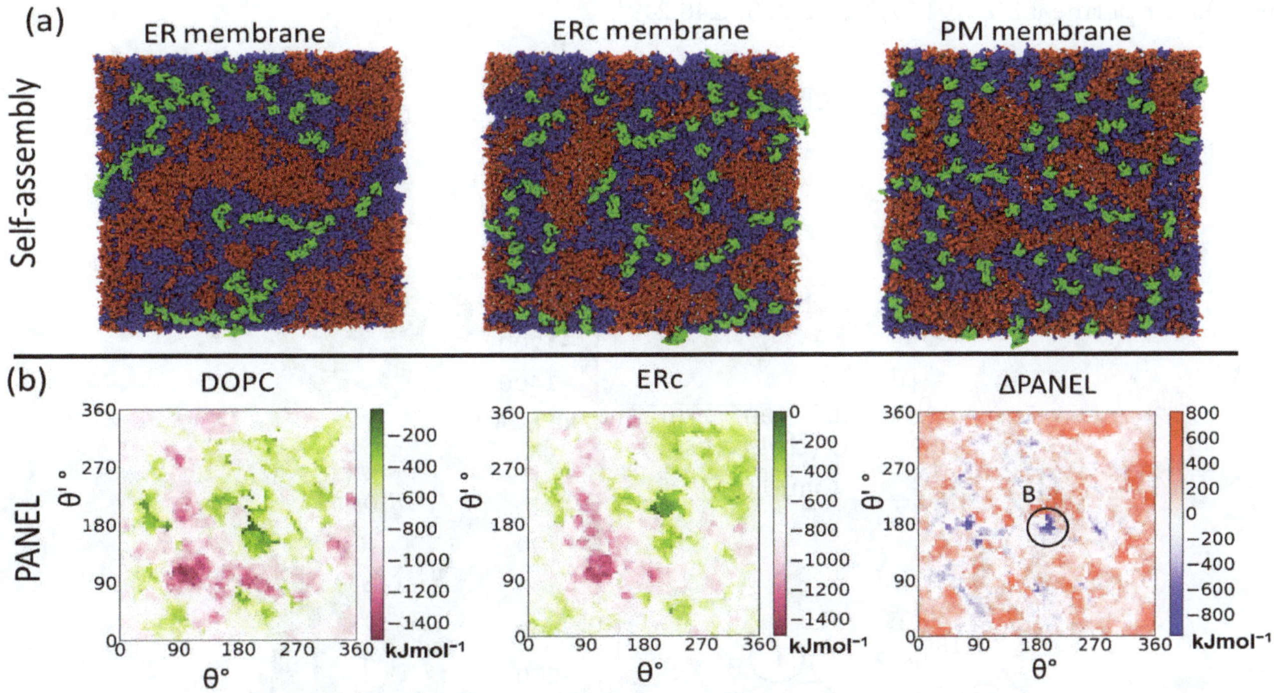

Figure 8. Effects of lipid complexity on claudin-5 dimerization. (**a**) Snapshots (top view) of self-assembled claudins (green) embedded in ER, cholesterol-enriched endoplasmic reticulum (ERc), and plasma membrane (PM) membranes with saturated lipids (red), unsaturated (blue), and cholesterol (white). (**b**) PANEL plots generated for claudin-5 dimers embedded in symmetric DOPC bilayer and ERc membrane. The ΔPANEL plot shows stabilization (blue) and destabilization (red) of the dimer geometries due to change in lipid environment DOPC to ERc. The stabilized dimer B location is marked (circle, black) on the ΔPANEL plot.

A comparison of PANEL plots for claudin-5 dimers in a simple unsaturated di-oleoyl-phosphatidylcholine (DOPC) lipid membrane and ERc membrane shows the impact of the lipid composition on dimer stability (Figure 8a). The effect is quantified by taking the difference in interaction energies (ΔPANEL) between corresponding geometries (θ, θ') on the two plots. A vast majority of the dimer interactions were destabilized in ERc membrane, although a few conformations did show more favorable interactions. In particular, Dimer B conformation (mediated via TM3:TM3 interaction) is stabilized. Overall, our results suggest that dimerization is more favorable in conditions existing in the early stages of the secretory pathway.

5.3. Posttranslational Modifications Alter Relative Stability of cis Dimers

Several studies have indicated the significance of palmitoylation in the recruitment of specific claudins to TJ and the consequent effect on the TJ permeability [233,234]. Since palmitoylation and depalmitoylation are highly labile reactions, the evaluation of these modifications on claudin assembly is experimentally challenging [237]. However, in silico methods provide the necessary spatial and temporal resolutions required to study these posttranslational modifications.

The presence of palmitoyl chains at the protein–lipid interface introduces variations in interfacial properties, effecting both the extent of dimerization as well as the distribution of specific dimer interfaces. Our results show that palmitoylation increases cholesterol and saturated lipid accumulation around the claudins and enhances cis interactions to the Dimer B interface while disrupting Dimer C largely due to steric factors (Figure 9a,b) [91]. The impact of palmitoylation on claudin-5 cis dimers is evident from the ΔPANEL landscape plotted as the difference between palmitoylated claudin (claudin-5P) and non-palmitoylated claudin-5 PANEL plots (Figure 9c). The ΔPANEL plot shows the stabilization of Dimer B and the destabilization of Dimer C regions. The change in dimer stability upon palmitoylation indicates that posttranslational modification can be a potential strategy to control paracellular permeability [91,97,233,234,237,248,249].

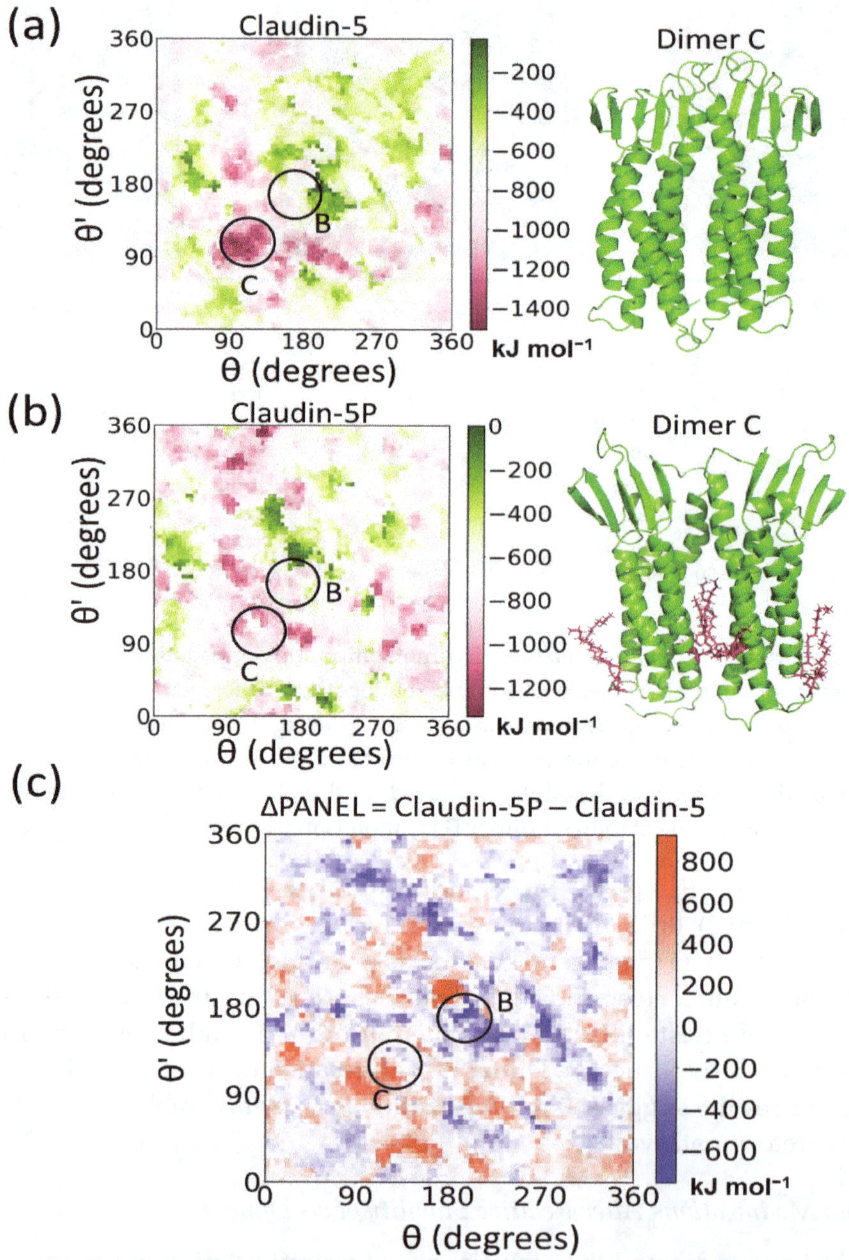

Figure 9. Effects of palmitoylation on claudin-5 cis dimerization. (**a**) PANEL plot Figure claudin. and dimer C conformation (green ribbon), (**b**) PANEL plot for claudin-5P and dimer C conformation (green ribbon) with palmitoylation (pink sticks), and (**c**) ΔPANEL showing the difference in stability upon palmitoylation. The location of dimers B and C are marked by circles (black) on all plots.

5.4. *Cis Dimers Act as Precursors to Form Trans Interfaces*

Trans interactions are formed when claudins on adjacent cells interact head-on to form the macromolecular TJ assembly. We adopted both CG- and atomistic-level simulations to decipher the nature of trans interfaces. Numerous challenges were posed both on conceptual and computational resource fronts while designing the trans assembly simulations.

To mimic paracellular space, our initial studies of trans interactions used two membrane patches with multiple (64) equidistant claudin-5 proteins [97]. They were placed head-on, such that the ECL loops from either membrane could interact within the gap between the two membranes (Figure 10). Sampling 20 μs of trans self-assembly took about 12 days on a supercomputer using 20 high performance computational nodes. Despite the intensive computation, the simulations represented only a small membrane patch that contained 128 claudin monomers across the two membranes. In contrast to a single membrane simulation, trans assembly simulations were unable to capture a few cis dimeric interactions that were prevalent in the single membrane self-assembly simulations. The simulations were intentionally designed so that a diffusing claudin-5 monomer would encounter a trans interacting partner before encountering a cis interacting partner; this ensured generation of more trans interaction variants. The absence of cis dimers rendered this a monomer-driven, trans interaction assembly [97]. Multiple pore-like assemblies were observed in these simulations showing ECL contacts stabilized by hydrogen-bonds and salt bridges. These trans pairs also strongly resembled pore models suggested in previous studies [72–74,250]. However, this model represented a very low fraction of the actual trans interactions, which span the entire perimeter of a cell and involve tens of millions of protein monomers.

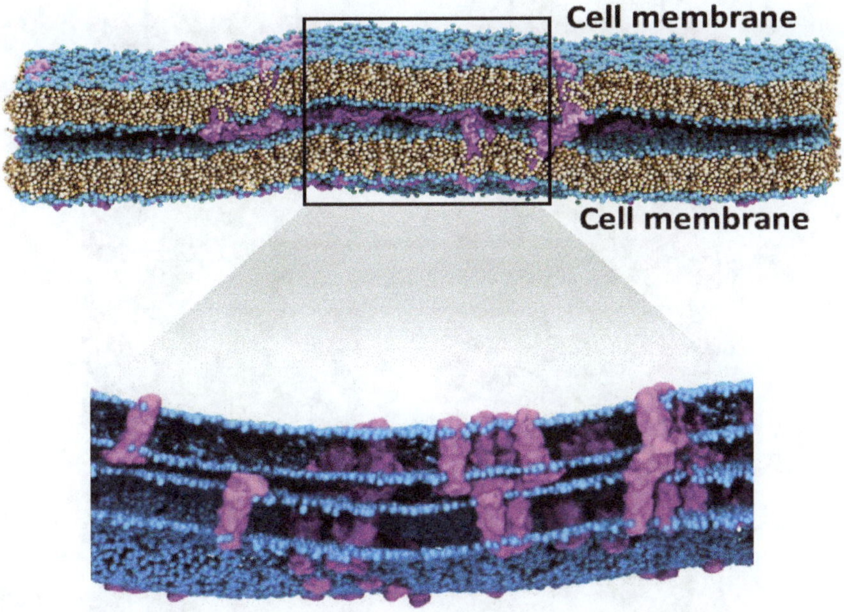

Figure 10. Claudin-5 assembly in two adjacent membranes. A 20 μs snapshot of claudin-5 (purple) trans and cis assembly in a coarse gained MD simulation. The zoomed-in view shows the interacting claudins embedded in the membrane (lipid head groups (blue); lipid tails (light brown). Water and ions in the system are not shown for clarity.

From the insights obtained from the cis and trans self-assembly simulations a dimer-guided trans interaction study was carried out using molecular docking approaches. This method yielded highly promising pore-forming trans interfaces. Moreover, the formation of cis dimers before trans interactions has been suggested in several findings [231,241,251,252]. Based on our work, we suggest that the cis dimers act as precursors for forming trans interfaces [91,97,115]. The fact that cis assembly simulation also self-assembles into a dimer in shorter timescales, lends support to that cis interactions may act as the template to guide the trans assembly.

The findings from these simulations were able to fill significant gaps in our understanding of the structure–function relationships between claudin monomers. Although several questions still need to be answered, and further investigations need to be performed to achieve a comprehensive understanding of the TJ architecture, our current understanding has been made possible by the an array of techniques in the computational toolkit.

5.5. Sub-Nanometer Wide Claudin-5 TJ Pores, Lined by Charged Residue Determine Pore Selectivity

For claudin-5, two pore models, mediated by different cis dimers, have been reported based on trans interaction simulations and mutation analyses of cis dimers [97]. The pore I model is a face-to-face stacking of the dimer D interface. This pore resembles the proposed claudin-15 pore [242]. The pore II model was based on the Dimer B interactions observed in claudin-5. These models were generated using docking methods in combination with symmetric refinement to arrive at the pore structures based on Dimers D and B (Figure 11). Fundamental pore features, such as diameter, length of the pore, pore bottleneck diameter, and the pore lining residues, were characterized.

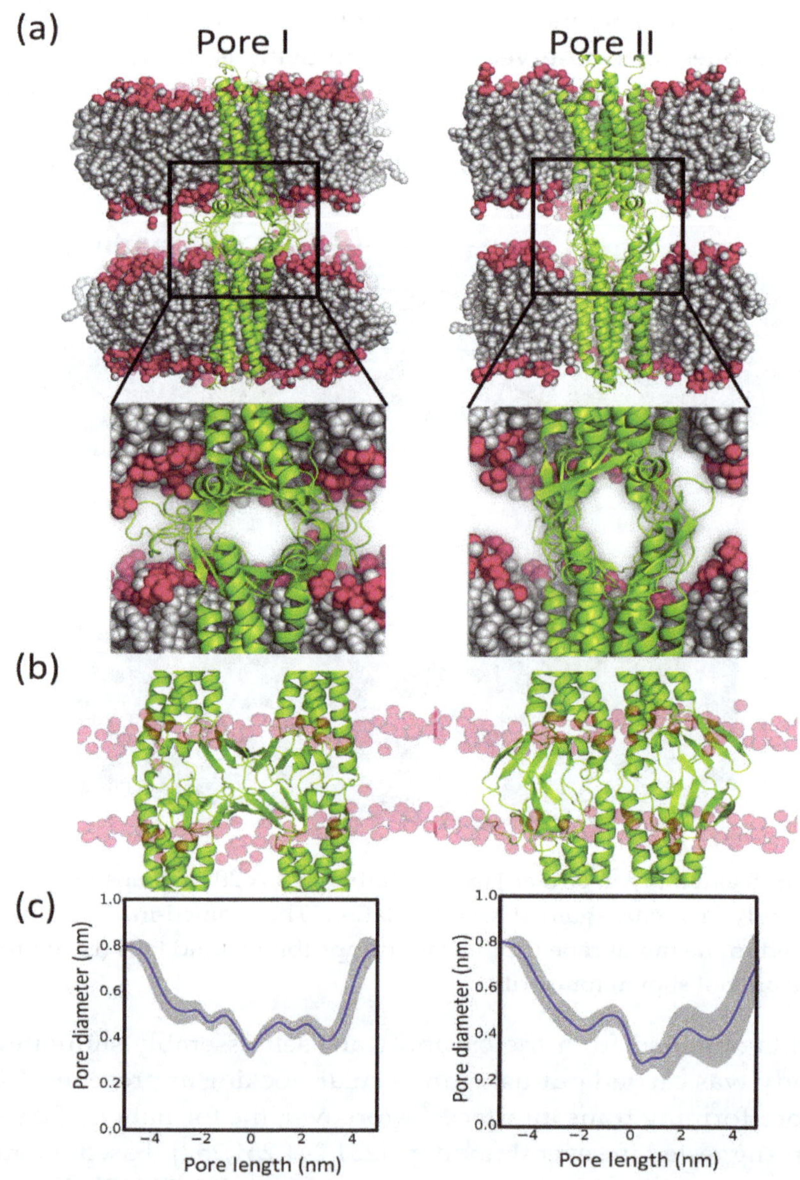

Figure 11. Model for claudin-5 pore I (left) and pore II (right). **(a)** Front-view, **(b)** side-view, and **(c)** pore diameter along the length of the pore.

Pore characteristics provide a general idea of the nature of the pore. However, evaluating the energetics of molecular transport along these pores is essential to further enhance our understanding of paracellular selectivity of the BBB's TJs. Using advanced sampling methods, translocation of water and glucose was reported across claudin-5 pore model formed by Dimer B. These simulations showed that the claudin-5 Dimer B pore allows water to translocate without energy barriers but had a higher energy barrier for glucose translocation compared with the GLUT1 glucose transporter [97,253,254]. Further extension of this work revealed that claudin-2 is selective to sodium ions with multiple observations of ion translocation under equilibrium conditions. Similar approaches are gaining attention within the field (e.g., claudin-15 pores probed using MD simulations) [255–257]. The more comprehensive undertaking of looking at ion transport in pores formed by different claudins is still in its infancy, but several computational groups are developing the essential methods needed to pursue this line of inquiry. In the meantime, the proposed pore structures present opportunities for further refinement to obtain better insights about their structural as well as functional aspects.

6. Additional Factors that Influence Tight Junctions

Computational methods can efficiently study cis and trans interactions of claudins at the molecular level. Claudins and other TJ proteins exist in a chemical environment that varies in type and amount of proteins and lipids, hydrophobicity, protein lipidation states, and other aspects [63,200,225,231,246,247]. These physiological factors contribute to the final TJ architecture. Moreover, TJs are known to be dynamic entities that undergo constant structural change [258,259]. Thus, to understand paracellular transport in its entirety, other factors need to be considered. A few key aspects important to TJ formation are discussed below.

6.1. Higher Order Claudin Aggregates

Several studies have reported the presence of claudin aggregates (in addition to strands) [61,118], suggesting that higher-order aggregates such as trimers, tetramers, or hexamers may be formed. *Clostridium perfringens* enterotoxin, which causes food poisoning, has been reported to form higher-order aggregates upon interacting with claudins. We proposed that the trimeric form of the enterotoxin could potentially bind to the trimeric form of claudins in the intestinal epithelium [94,117]. In another recent report of claudin-11 TJs, the carbon replica method revealed an interesting view of consistent double-stranded structures [260]. These reports emphasize that in order to fully understand the architecture of the TJ strands, it is important to consider the higher-order interfaces in addition to dimers.

6.2. Involvement of Other TJ Proteins: Key to TJ Structural Stability

In addition to claudins, proteins from approximately 80 other families are known to be involved in the TJ protein complex assembly including occludins and JAMs [53,54,60]. Occludins and JAMs contribute to the structural stability of TJs. Although these proteins are not known to affect directly TJs' selective permeability or barrier function, their role in maintaining the structural integrity of the TJ strongly suggests they form interfaces with claudins in order to support the strand. This may lead to masking of some claudins' interfacing surfaces while binding to others, thereby redistributing the relative populations of dimer interfaces. The study of occludins and JAMs is crucial to provide a complete perspective of TJs due to their significant presence within the TJ assembly.

6.3. Cytoskeletal Involvement

Claudins and other TJ membrane proteins are dynamically regulated by scaffolding proteins (zonula occludens-1/2/3) and actin cytoskeleton [53,231]. Although studying claudins individually provides insights about their interface forming behavior, thereby helping us to predict the architecture and nature of the corresponding TJ strands, continuous regulation by the cytoskeleton may induce several factors such as localized redistribution of claudin populations, or regulation of biophysical properties such as protein tilting, membrane anchoring, and so on. Since the cytoskeletal regulation tends to be highly dynamic and specific to the needs of the cell, deciphering their roles remains one of the greatest challenges.

7. Future Work

Future work on the BBB research will be directed towards the development of strategies to enable and enhance drug delivery to the brain. A fundamental understanding of the molecular construct that guards the blood–brain interface has been obtained through computational tools, which have enabled the design of in silico models to explore and the understanding of the architecture of this interface. One future research goal is to identify small molecules that can act as TJ modulators for use in screening claudin-5 trans interactions, reducing specificity, increasing pore size, and enhancing pore permeability in a controlled and reversible manner. Having the molecular-level pore structure is foundational to this endeavor. We envisioned that in vitro studies would be designed in the future to validate the predictions of the computational models. Other future work includes pursuing an integrated computational and in vitro/in vivo study of multiple drug pathways for neurotherapeutics, evaluating molecular drug interactions, obtaining the dynamics of transport of potential drug molecules with the vision of quantifying drug permeability through the BBB.

Author Contributions: N.R., F.J.I., and S.N.; conceptualization, S.N., N.R., and F.J.I.; methodology, S.N., N.R., and F.J.I.; software, N.R. and F.J.I.; validation, N.R. and F.J.I.; formal analysis, N.R. and F.J.I.; investigation, N.R. and F.J.I.; data curation, N.R. and F.J.I.; writing—original draft preparation, N.R. and F.J.I.; writing—review and editing, S.N., N.R., and F.J.I.; visualization, S.N., N.R., and F.J.I.; supervision, S.N.; project administration, S.N.; funding acquisition, S.N.

Abbreviations

BBB	Blood–brain barrier
CG	Coarse grained
Claudin-5P	Palmitoylated claudin
DAFT	Docking assay for transmembrane
DHHC	Aspartate-Histidine-Histidine-Cysteine
DOPC	1,2-dioleoyl-sn-glycero-3-phosphocholine
ECL	Extracellular loop
ER	Endoplasmic Reticulum
ERc	Cholesterol enriched Endoplasmic Reticulum
ICL	Intracellular loop
JAMS	Junctional adhesion molecules
MD	Molecular dynamics
PANEL	Protein Association Energy Landscape
TEER	Transendothelial/epithelial electrical resistance
TJ	Tight junction
TM	Transmembrane loop
ZO	Zonula Occuldens

References

1. Abbott, N.J. Comparative Physiology of the Blood-Brain Barrier. In *Physiology and Pharmacology of the Blood-Brain Barrier*; Bradbury, M.W.B., Ed.; Springer: Berlin/Heidelberg, Germany, 1992; pp. 371–396.
2. Bauer, H.C.; Krizbai, I.A.; Bauer, H.; Traweger, A. "You Shall Not Pass"-tight junctions of the blood brain barrier. *Front. Neurosci.* **2014**, *8*, 392. [CrossRef] [PubMed]
3. Begley, D.J.; Brightman, M.W. Structural and functional aspects of the blood-brain barrier. In *Peptide Transport and Delivery into the Central Nervous System*; Prokai, L., Prokai-Tatrai, K., Eds.; Birkhäuser Basel: Basel, Switzerland, 2003; pp. 39–78.

4. Daneman, R.; Prat, A. The blood-brain barrier. *Cold Spring Harb. Perspect. Biol.* **2015**, *7*, a020412. [CrossRef] [PubMed]
5. Saunders, N.R.; Dreifuss, J.-J.; Dziegielewska, K.M.; Johansson, P.A.; Habgood, M.D.; Møllgård, K.; Bauer, H.-C. The rights and wrongs of blood-brain barrier permeability studies: A walk through 100 years of history. *Front. Neurosci.* **2014**, *8*, 404. [CrossRef] [PubMed]
6. Lewandowsky, M. Zur Lehre von der Cerebrospinalflüssgkeit. *Z. Clin. Med.* **1900**, *40*, 480–494.
7. Goldmann, E.E. Die äussere und innere Sekretion des gesunden und kranken Organismus im Lichte der 'vitalen Färbung. *Beiträg Klin. Chir.* **1909**, *64*, 192–265.
8. Goldmann, E.E. Vitalfärbung am Zentralnervensyatem. Beitrag zur Physio-Pathologie des plexus chorioideus und der Hirnhäute. *Abh. Preuss Akad. Wiss. Phys. Math. Kl.* **1913**, *1*, 1–60.
9. Ehrlich, P. *The Relations Existing between Chemical Constitution, Distribution and Pharmacological Action*; Collected Studies on Immunity New York: John Wiley & Sons, translated by C Bolduana from Ch XXXIV of Gesammelte Arbeiten zur Immunitätsforschung; Ehrlich, P., Ed.; Hirschwald: Berlin, Germany, 1906.
10. Ehrlich, P. Das Sauerstoffbedürfnis des Organismus. In *Eine Farbenanalytische Studie*; Hirschwald: Berlin, Germany, 1885.
11. Stern, L.; Gautier, R. II.—Les Rapports Entre Le Liquide Céphalo-Rachidien Et Les éléments Nerveux De L'axe Cerebrospinal. *Arch. Int. Physiol.* **1922**, *17*, 391–448. [CrossRef]
12. Stern, L.; Gautier, R. III.—Rapports Entre Le Liquide Céphalo-Rachidien Des Espaces Ventriculaires Et Celui Des Espaces Sous-Arachnoïdiens. *Arch. Int. Physiol.* **1923**, *20*, 403–436. [CrossRef]
13. Stern, L.; Gautier, R. Passage simultané des substances dans le liquide céphalo-rachidien et dans les centres nerveux. *Rcr D. Ia Soc. De Phys. Et D'hist. Natur. De Genève* **1918**, *35*, 58–60.
14. Stern, L.; Gautier, R. Le passage dans le liquide céphalo-rachidien de substances introduites dans la circulation et leur action sur le système nerveux central chez les différentes espèces animales. *Rcr D. Ia Soc. De Phys. Et D'hist. Natur. De Genève* **1918**, *35*, 91–94.
15. Stern, L.; Gautier, R. Recherches sur le liquide céphalo-rachidien. 1. Les rapports entre le liquide céphalo-rachidien et la circulation sanguine. *Arch. Int. Physiol.* **1921**, *17*, 138–192. [CrossRef]
16. Roux, E.; Borrel, A. Tétanos cérébral et immunité contre le tétanus. *Ann. Inst. Pasteur* **1898**, *12*, 225–239.
17. Bouffard, G. Injection des couleurs de benzidine aux animaux normaux. *Ann. D. L'lnst. Pasteur. Paris* **1906**, *20*, 539–548.
18. Weed, L.H. An anatomical consideration of the cerebro-spinal fluid. *Anat. Res.* **1917**, 461–496. [CrossRef]
19. Weed, L.H. The development of the cerebrospinal fluid spaces in pig and in man. *Contrib. Embryol. Carnegie Inst.* **1917**, *5*, 3–116.
20. Wislocki, G.B. Experimental studies on fetal absorption. I. The vitality stained fetus. *Embryol. Carnegie Inst.* **1920**, *11*, 45–60.
21. Cohen, H.; Davies, S. The Morphology and Permeability of the Roof of the Fourth Ventricle in some Mammalian Embryos. *J. Anat.* **1938**, *72*, 430–455.
22. Friedemann, U. Blood-brain barrier. *Physiol. Rev.* **1942**, *22*, 125–145. [CrossRef]
23. Wolburg, H.; Lippoldt, A. Tight junctions of the blood–brain barrier: Development, composition and regulation. *Vasc. Pharmacol.* **2002**, *38*, 323–337. [CrossRef]
24. Duelli, R.; Kuschinsky, W. Brain Glucose Transporters: Relationship to Local Energy Demand. *Physiology* **2001**, *16*, 71–76. [CrossRef]
25. Howarth, C.; Gleeson, P.; Attwell, D. Updated energy budgets for neural computation in the neocortex and cerebellum. *J. Cereb. Blood Flow Metab.* **2012**, *32*, 1222–1232. [CrossRef] [PubMed]
26. Harris, J.; Jolivet, R.; Attwell, D. Synaptic Energy Use and Supply. *Neuron* **2012**, *75*, 762–777. [CrossRef] [PubMed]
27. Mergenthaler, P.; Lindauer, U.; Dienel, G.A.; Meisel, A. Sugar for the brain: The role of glucose in physiological and pathological brain function. *Trends Neurosci.* **2013**, *36*, 587–597. [CrossRef] [PubMed]
28. Patching, S.G. Glucose Transporters at the Blood-Brain Barrier: Function, Regulation and Gateways for Drug Delivery. *Mol. Neurobiol.* **2017**, *54*, 1046–1077. [CrossRef]
29. Abbott, N.J. Inflammatory Mediators and Modulation of Blood–Brain Barrier Permeability. *Cell. Mol. Neurobiol.* **2000**, *20*, 131–147. [CrossRef]
30. Huber, J.D.; Egleton, R.D.; Davis, T.P. Molecular physiology and pathophysiology of tight junctions in the blood–brain barrier. *Trends Neurosci.* **2001**, *24*, 719–725. [CrossRef]

31. Pardridge, W.M. Drug and Gene Delivery to the Brain: The Vascular Route. *Neuron* **2002**, *36*, 555–558. [CrossRef]
32. Ballabh, P.; Braun, A.; Nedergaard, M. The blood–brain barrier: An overview: Structure, regulation, and clinical implications. *Neurobiol. Dis.* **2004**, *16*, 1–13. [CrossRef]
33. Hawkins, B.T.; Davis, T.P. The Blood-Brain Barrier/Neurovascular Unit in Health and Disease. *Pharmacol. Rev.* **2005**, *57*, 173. [CrossRef]
34. Kniesel, U.; Wolburg, H. Tight Junctions of the Blood–Brain Barrier. *Cell. Mol. Neurobiol.* **2000**, *20*, 57–76. [CrossRef]
35. Luissint, A.-C.; Artus, C.; Glacial, F.; Ganeshamoorthy, K.; Couraud, P.-O. Tight junctions at the blood brain barrier: Physiological architecture and disease-associated dysregulation. *Fluids Barriers CNS* **2012**, *9*, 23. [CrossRef] [PubMed]
36. Greene, C.; Campbell, M. Tight junction modulation of the blood brain barrier: CNS delivery of small molecules. *Tissue Barriers* **2016**, *4*, e1138017. [CrossRef] [PubMed]
37. Butt, A.M.; Jones, H.C.; Abbott, N.J. Electrical resistance across the blood-brain barrier in anaesthetized rats: A developmental study. *J. Physiol.* **1990**, *429*, 47–62. [CrossRef] [PubMed]
38. Abbott, N.J.; Patabendige, A.A.K.; Dolman, D.E.M.; Yusof, S.R.; Begley, D.J. Structure and function of the blood–brain barrier. *Neurobiol. Dis.* **2010**, *37*, 13–25. [CrossRef] [PubMed]
39. Pardridge, W.M. The blood-brain barrier: Bottleneck in brain drug development. *NeuroRX* **2005**, *2*, 3–14. [CrossRef]
40. Pardridge, W.M. Blood–brain barrier delivery. *Drug Discov. Today* **2007**, *12*, 54–61. [CrossRef]
41. Pardridge, W.M. Drug Transport across the Blood–Brain Barrier. *J. Cereb. Blood Flow Metab.* **2012**, *32*, 1959–1972. [CrossRef]
42. Cucullo, L.; Bennani-Baiti, B.; Rapp, E.; Janigro, D. Drug delivery and in vitro models of the blood–brain barrier. *Curr. Opin. Drug Discov. Dev.* **2005**, *8*, 89–99.
43. Wilhelm, I.; Krizbai, I. In Vitro Models of the Blood-Brain Barrier for the Study of Drug Delivery to the Brain. *Mol. Pharm.* **2014**, *11*, 1949–1963. [CrossRef]
44. Heymans, M.; Sevin, E.; Gosselet, F.; Lundquist, S.; Culot, M. Mimicking brain tissue binding in an in vitro model of the blood-brain barrier illustrates differences between in vitro and in vivo methods for assessing the rate of brain penetration. *Eur. J. Pharm. Biopharm.* **2018**, *127*, 453–461. [CrossRef]
45. Srinivasan, B.; Kolli, A.R.; Esch, M.B.; Abaci, H.E.; Shuler, M.L.; Hickman, J.J. TEER Measurement Techniques for In Vitro Barrier Model Systems. *J. Lab. Autom.* **2015**, *20*, 107–126. [CrossRef] [PubMed]
46. Santaguida, S.; Janigro, D.; Hossain, M.; Oby, E.; Rapp, E.; Cucullo, L. Side by side comparison between dynamic versus static models of blood–brain barrier in vitro: A permeability study. *Brain Res.* **2006**, *1109*, 1–13. [CrossRef] [PubMed]
47. Achyuta, A.K.H.; Conway, A.J.; Crouse, R.B.; Bannister, E.C.; Lee, R.N.; Katnik, C.P.; Behensky, A.A.; Cuevas, J.; Sundaram, S.S. A modular approach to create a neurovascular unit-on-a-chip. *Lab. A Chip* **2013**, *13*, 542–553. [CrossRef] [PubMed]
48. Wang, Y.I.; Abaci, H.E.; Shuler, M.L. Microfluidic blood-brain barrier model provides in vivo-like barrier properties for drug permeability screening. *Biotechnol. Bioeng.* **2017**, *114*, 184–194. [CrossRef] [PubMed]
49. Bennion, B.J.; Be, N.A.; McNerney, M.W.; Lao, V.; Carlson, E.M.; Valdez, C.A.; Malfatti, M.A.; Enright, H.A.; Nguyen, T.H.; Lightstone, F.C.; et al. Predicting a Drug's Membrane Permeability: A Computational Model Validated With in Vitro Permeability Assay Data. *J. Phys. Chem. B* **2017**, *121*, 5228–5237. [CrossRef] [PubMed]
50. Dickson, C.J.; Hornak, V.; Pearlstein, R.A.; Duca, J.S. Structure–Kinetic Relationships of Passive Membrane Permeation from Multiscale Modeling. *J. Am. Chem. Soc.* **2017**, *139*, 442–452. [CrossRef]
51. Carpenter, T.S.; Kirshner, D.A.; Lau, E.Y.; Wong, S.E.; Nilmeier, J.P.; Lightstone, F.C. A method to predict blood-brain barrier permeability of drug-like compounds using molecular dynamics simulations. *Biophys. J.* **2014**, *107*, 630–641. [CrossRef]
52. Geldenhuys, W.J.; Mohammad, A.S.; Adkins, C.E.; Lockman, P.R. Molecular determinants of blood-brain barrier permeation. *Ther. Deliv.* **2015**, *6*, 961–971. [CrossRef]
53. González-Mariscal, L.; Betanzos, A.; Nava, P.; Jaramillo, B.E. Tight junction proteins. *Prog. Biophys. Mol. Biol.* **2003**, *81*, 1–44. [CrossRef]
54. Tsukita, S.; Furuse, M.; Itoh, M. Multifunctional strands in tight junctions. *Nat. Rev. Mol. Cell Biol.* **2001**, 2 285–293. [CrossRef]

55. Morita, K.; Furuse, M.; Fujimoto, K.; Tsukita, S. Claudin multigene family encoding four-transmembrane domain protein components of tight junction strands. *Proc. Natl. Acad. Sci. USA* **1999**, *96*, 511. [CrossRef] [PubMed]
56. Schulzke, J.D.; Ploeger, S.; Amasheh, M.; Fromm, A.; Zeissig, S.; Troeger, H.; Richter, J.; Bojarski, C.; Schumann, M.; Fromm, M. Epithelial Tight Junctions in Intestinal Inflammation. *Ann. N. Y. Acad. Sci.* **2009**, *1165*, 294–300. [CrossRef] [PubMed]
57. Morita, K.; Sasaki, H.; Furuse, M.; Tsukita, S. Endothelial Claudin. *J. Cell Biol.* **1999**, *147*, 185. [CrossRef] [PubMed]
58. Furuse, M.; Hata, M.; Furuse, K.; Yoshida, Y.; Haratake, A.; Sugitani, Y.; Noda, T.; Kubo, A.; Tsukita, S. Claudin-based tight junctions are crucial for the mammalian epidermal barrier. *J. Cell Biol.* **2002**, *156*, 1099. [CrossRef]
59. Günzel, D.; Yu, A.S.L. Claudins and the Modulation of Tight Junction Permeability. *Physiol. Rev.* **2013**, *93*, 525–569. [CrossRef]
60. Günzel, D.; Fromm, M. Claudins and Other Tight Junction Proteins. *Compr. Physiol.* **2012**, *2*, 1819–1852.
61. Van Itallie, C.M.; Anderson, J.M. Claudin interactions in and out of the tight junction. *Tissue Barriers* **2013**, *1*, e25247. [CrossRef]
62. Krause, G.; Winkler, L.; Mueller, S.L.; Haseloff, R.F.; Piontek, J.; Blasig, I.E. Structure and function of claudins. *Biochim. Biophys. Acta (BBA) Biomembr.* **2008**, *1778*, 631–645. [CrossRef]
63. Buckley, A.; Turner, J.R. Cell Biology of Tight Junction Barrier Regulation and Mucosal Disease. *Cold Spring Harb. Perspect. Biol.* **2018**, *10*, a029314. [CrossRef]
64. Garcia-Hernandez, V.; Quiros, M.; Nusrat, A. Intestinal epithelial claudins: Expression and regulation in homeostasis and inflammation. *Ann. N. Y. Acad. Sci.* **2017**, *1397*, 66–79. [CrossRef]
65. Hou, J.; Rajagopal, M.; Yu, A.S.L. Claudins and the kidney. *Annu. Rev. Physiol.* **2013**, *75*, 479–501. [CrossRef] [PubMed]
66. Hewitt, K.J.; Agarwal, R.; Morin, P.J. The claudin gene family: Expression in normal and neoplastic tissues. *BMC Cancer* **2006**, *6*, 186. [CrossRef] [PubMed]
67. Wang, H.; Yang, X. The expression patterns of tight junction protein claudin-1, -3, and -4 in human gastric neoplasms and adjacent non-neoplastic tissues. *Int. J. Clin. Exp. Pathol.* **2015**, *8*, 881–887. [PubMed]
68. Suzuki, H.; Nishizawa, T.; Tani, K.; Yamazaki, Y.; Tamura, A.; Ishitani, R.; Dohmae, N.; Tsukita, S.; Nureki, O.; Fujiyoshi, Y. Crystal Structure of a Claudin Provides Insight into the Architecture of Tight Junctions. *Science* **2014**, *344*, 304. [CrossRef] [PubMed]
69. Ohtsuki, S.; Yamaguchi, H.; Katsukura, Y.; Asashima, T.; Terasaki, T. mRNA expression levels of tight junction protein genes in mouse brain capillary endothelial cells highly purified by magnetic cell sorting. *J. Neurochem.* **2008**, *104*, 147–154. [CrossRef]
70. Daneman, R.; Zhou, L.; Agalliu, D.; Cahoy, J.D.; Kaushal, A.; Barres, B.A. The Mouse Blood-Brain Barrier Transcriptome: A New Resource for Understanding the Development and Function of Brain Endothelial Cells. *PLoS ONE* **2010**, *5*, e13741. [CrossRef]
71. Nitta, T.; Hata, M.; Gotoh, S.; Seo, Y.; Sasaki, H.; Hashimoto, N.; Furuse, M.; Tsukita, S. Size-selective loosening of the blood-brain barrier in claudin-5-deficient mice. *J. Cell Biol.* **2003**, *161*, 653–660. [CrossRef]
72. Krause, G.; Winkler, L.; Piehl, C.; Blasig, I.; Piontek, J.; Müller, S.L. Structure and Function of Extracellular Claudin Domains. *Mol. Biol. Cell* **2009**, *1165*, 4333–4346. [CrossRef]
73. Rossa, J.; Ploeger, C.; Vorreiter, F.; Saleh, T.; Protze, J.; Günzel, D.; Wolburg, H.; Krause, G.; Piontek, J. Claudin-3 and Claudin-5 Protein Folding and Assembly into the Tight Junction Are Controlled by Non-conserved Residues in the Transmembrane 3 (TM3) and Extracellular Loop 2 (ECL2) Segments. *J. Biol. Chem.* **2014**, *289*, 7641–7653. [CrossRef]
74. Piehl, C.; Piontek, J.; Cording, J.; Wolburg, H.; Blasig, I.E. Participation of the second extracellular loop of claudin-5 in paracellular tightening against ions, small and large molecules. *Cell. Mol. Life Sci.* **2010**, *67*, 2131–2140. [CrossRef]
75. Keaney, J.; Walsh, D.M.; O'Malley, T.; Hudson, N.; Crosbie, D.E.; Loftus, T.; Sheehan, F.; McDaid, J.; Humphries, M.M.; Callanan, J.J.; et al. Autoregulated paracellular clearance of amyloid-β across the blood-brain barrier. *Sci. Adv.* **2015**, *1*, e1500472. [CrossRef] [PubMed]
76. Greene, C.; Hanley, N.; Campbell, M. Claudin-5: Gatekeeper of neurological function. *Fluids Barriers CNS* **2019**, *16*, 3. [CrossRef] [PubMed]

77. Di Ventura, B.; Lemerle, C.; Michalodimitrakis, K.; Serrano, L. From in vivo to in silico biology and back. *Nature* **2006**, *443*, 527–533. [CrossRef] [PubMed]
78. Ekins, S.; Mestres, J.; Testa, B. In silico pharmacology for drug discovery: Methods for virtual ligand screening and profiling. *Br. J. Pharmacol.* **2007**, *152*, 9–20. [CrossRef]
79. McInnes, C. Virtual screening strategies in drug discovery. *Curr. Opin. Chem. Biol.* **2007**, *11*, 494–502. [CrossRef]
80. Huggins, D.J.; Biggin, P.C.; Dämgen, M.A.; Essex, J.W.; Harris, S.A.; Henchman, R.H.; Khalid, S.; Kuzmanic, A.; Laughton, C.A.; Michel, J.; et al. Biomolecular simulations: From dynamics and mechanisms to computational assays of biological activity. *Wiley Interdiscip. Rev. Comput. Mol. Sci.* **2019**, *9*, e1393. [CrossRef]
81. Hope, H. Crystallography of Biological Macromolecules at Ultra-Low Temperature. *Annu. Rev. Biophys. Biophys. Chem.* **1990**, *19*, 107–126. [CrossRef]
82. Smyth, M.S.; Martin, J.H. x ray crystallography. *Mol. Pathol.* **2000**, *53*, 8–14. [CrossRef]
83. Carpenter, E.P.; Beis, K.; Cameron, A.D.; Iwata, S. Overcoming the challenges of membrane protein crystallography. *Curr. Opin. Struct. Biol.* **2008**, *18*, 581–586. [CrossRef]
84. Caffrey, M. Crystallizing Membrane Proteins for Structure Determination: Use of Lipidic Mesophases. *Annu. Rev. Biophys.* **2009**, *38*, 29–51. [CrossRef]
85. Seddon, A.M.; Curnow, P.; Booth, P.J. Membrane proteins, lipids and detergents: Not just a soap opera. *Biochim. Biophys. Acta (BBA) Biomembr.* **2004**, *1666*, 105–117. [CrossRef] [PubMed]
86. Dimmic, M.W.; Hubisz, M.J.; Bustamante, C.D.; Nielsen, R. Detecting coevolving amino acid sites using Bayesian mutational mapping. *Bioinformatics* **2005**, *21*, i126–i135. [CrossRef] [PubMed]
87. Lui, S.; Tiana, G. The network of stabilizing contacts in proteins studied by coevolutionary data. *J. Chem. Phys.* **2013**, *139*, 155103. [CrossRef] [PubMed]
88. Jana, B.; Morcos, F.; Onuchic, J.N. From structure to function: The convergence of structure based models and co-evolutionary information. *Phys. Chem. Chem. Phys.* **2014**, *16*, 6496–6507. [CrossRef] [PubMed]
89. Wang, Y.; Barth, P. Evolutionary-guided de novo structure prediction of self-associated transmembrane helical proteins with near-atomic accuracy. *Nat. Commun.* **2015**, *6*, 7196. [CrossRef] [PubMed]
90. Teixeira, P.L.; Mendenhall, J.L.; Heinze, S.; Weiner, B.; Skwark, M.J.; Meiler, J. Membrane protein contact and structure prediction using co-evolution in conjunction with machine learning. *PLoS ONE* **2017**, *12*, e0177866. [CrossRef]
91. Rajagopal, N.; Irudayanathan, F.J.; Nangia, S. Palmitoylation of Claudin-5 Proteins Influences Their Lipid Domain Affinity and Tight Junction Assembly at the Blood–Brain Barrier Interface. *J. Phys. Chem. B* **2019**, *123*, 983–993. [CrossRef]
92. Murthy, A.C.; Dignon, G.L.; Kan, Y.; Zerze, G.H.; Parekh, S.H.; Mittal, J.; Fawzi, N.L. Molecular interactions underlying liquid−liquid phase separation of the FUS low-complexity domain. *Nat. Struct. Mol. Biol.* **2019**, *26*, 637–648. [CrossRef]
93. Zheng, M.; Zhao, J.; Cui, C.; Fu, Z.; Li, X.; Liu, X.; Ding, X.; Tan, X.; Li, F.; Luo, X.; et al. Computational chemical biology and drug design: Facilitating protein structure, function, and modulation studies. *Med. Res. Rev.* **2018**, *38*, 914–950. [CrossRef]
94. Irudayanathan, F.J.; Wang, X.; Wang, N.; Willsey, S.R.; Seddon, I.A.; Nangia, S. Self-Assembly Simulations of Classic Claudins—Insights into the Pore Structure, Selectivity, and Higher Order Complexes. *J. Phys. Chem. B* **2018**, *122*, 7463–7474. [CrossRef]
95. Ibsen, K.N.; Ma, H.; Banerjee, A.; Tanner, E.E.L.; Nangia, S.; Mitragotri, S. Mechanism of Antibacterial Activity of Choline-Based Ionic Liquids (CAGE). *ACS Biomater. Sci. Eng.* **2018**, *4*, 2370–2379. [CrossRef]
96. Sengupta, D.; Prasanna, X.; Mohole, M.; Chattopadhyay, A. Exploring GPCR–Lipid Interactions by Molecular Dynamics Simulations: Excitements, Challenges, and the Way Forward. *J. Phys. Chem. B* **2018**, *122*, 5727–5737. [CrossRef] [PubMed]
97. Irudayanathan, F.J.; Wang, N.; Wang, X.; Nangia, S. Architecture of the paracellular channels formed by claudins of the blood–brain barrier tight junctions. *Ann. N. Y. Acad. Sci.* **2017**, *1405*, 131–146. [CrossRef] [PubMed]
98. Grime, J.M.A.; Dama, J.F.; Ganser-Pornillos, B.K.; Woodward, C.L.; Jensen, G.J.; Yeager, M.; Voth, G.A. Coarse-grained simulation reveals key features of HIV-1 capsid self-assembly. *Nat. Commun.* **2016**, *7*, 11568. [CrossRef] [PubMed]

99. Ingólfsson, H.I.; Arnarez, C.; Periole, X.; Marrink, S.J. Computational 'microscopy' of cellular membranes. *J. Cell Sci.* **2016**, *129*, 257. [CrossRef] [PubMed]
100. Reddy, T.; Sansom, M.S. The Role of the Membrane in the Structure and Biophysical Robustness of the Dengue Virion Envelope. *Structure* **2016**, *24*, 375–382. [CrossRef]
101. Arnarez, C.; Marrink, S.J.; Periole, X. Molecular mechanism of cardiolipin-mediated assembly of respiratory chain supercomplexes. *Chem. Sci.* **2016**, *7*, 4435–4443. [CrossRef]
102. Holdbrook, D.A.; Huber, R.G.; Piggot, T.J.; Bond, P.J.; Khalid, S. Dynamics of Crowded Vesicles: Local and Global Responses to Membrane Composition. *PLoS ONE* **2016**, *11*, e0156963. [CrossRef]
103. Castillo, N.; Monticelli, L.; Barnoud, J.; Tieleman, D.P. Free energy of WALP23 dimer association in DMPC, DPPC, and DOPC bilayers. *Chem. Phys. Lipids* **2013**, *169*, 95–105. [CrossRef]
104. Bennett, W.F.D.; MacCallum, J.L.; Hinner, M.J.; Marrink, S.J.; Tieleman, D.P. Molecular View of Cholesterol Flip-Flop and Chemical Potential in Different Membrane Environments. *J. Am. Chem. Soc.* **2009**, *131*, 12714–12720. [CrossRef]
105. Periole, X.; Knepp, A.M.; Sakmar, T.P.; Marrink, S.J.; Huber, T. Structural Determinants of the Supramolecular Organization of G Protein-Coupled Receptors in Bilayers. *J. Am. Chem. Soc.* **2012**, *134*, 10959–10965. [CrossRef] [PubMed]
106. Schäfer, L.V.; de Jong, D.H.; Holt, A.; Rzepiela, A.J.; de Vries, A.H.; Poolman, B.; Killian, J.A.; Marrink, S.J. Lipid packing drives the segregation of transmembrane helices into disordered lipid domains in model membranes. *Proc. Natl. Acad. Sci. USA* **2011**, *108*, 1343. [CrossRef] [PubMed]
107. Duncan, A.L.; Reddy, T.; Koldsø, H.; Hélie, J.; Fowler, P.W.; Chavent, M.; Sansom, M.S.P. Protein crowding and lipid complexity influence the nanoscale dynamic organization of ion channels in cell membranes. *Sci. Rep.* **2017**, *7*, 16647. [CrossRef] [PubMed]
108. Koldsø, H.; Sansom, M.S.P. Organization and Dynamics of Receptor Proteins in a Plasma Membrane. *J. Am. Chem. Soc.* **2015**, *137*, 14694–14704. [CrossRef]
109. Shorthouse, D.; Hedger, G.; Koldsø, H.; Sansom, M.S.P. Molecular simulations of glycolipids: Towards mammalian cell membrane models. *Biochimie* **2016**, *120*, 105–109. [CrossRef]
110. Koldsø, H.; Reddy, T.; Fowler, P.W.; Duncan, A.L.; Sansom, M.S.P. Membrane Compartmentalization Reducing the Mobility of Lipids and Proteins within a Model Plasma Membrane. *J. Phys. Chem. B* **2016**, *120*, 8873–8881. [CrossRef]
111. Duncan, A.L.; Song, W.; Sansom, M.S.P. Lipid-Dependent Regulation of Ion Channels and G Protein–Coupled Receptors: Insights from Structures and Simulations. *Annu. Rev. Pharmacol. Toxicol.* **2019**. [CrossRef]
112. Koldsø, H.; Shorthouse, D.; Hélie, J.; Sansom, M.S.P. Lipid Clustering Correlates with Membrane Curvature as Revealed by Molecular Simulations of Complex Lipid Bilayers. *PLoS Comput. Biol.* **2014**, *10*, e1003911. [CrossRef]
113. Duncan, A.L.; Bandurka, M.A.R.; Chavent, M.G.; Rassam, P.; Song, W.; Birkholz, O.; Helie, J.; Reddy, T.; Beliaev, D.; Hambly, B.; et al. How Nanoscale Protein Interactions Determine the Mesoscale Dynamic Organisation of Membrane Proteins. *Biophys. J.* **2019**, *116*, 365a. [CrossRef]
114. Lin, X.; Gorfe, A.A.; Levental, I. Protein Partitioning into Ordered Membrane Domains: Insights from Simulations. *Biophys. J.* **2018**, *114*, 1936–1944. [CrossRef]
115. Irudayanathan, F.J.; Trasatti, J.P.; Karande, P.; Nangia, S. Molecular Architecture of the Blood Brain Barrier Tight Junction Proteins–A Synergistic Computational and In Vitro Approach. *J. Phys. Chem. B* **2016**, *120*, 77–88. [CrossRef] [PubMed]
116. Ma, H.; Irudayanathan, F.J.; Jiang, W.; Nangia, S. Simulating Gram-Negative Bacterial Outer Membrane: A Coarse Grain Model. *J. Phys. Chem. B* **2015**, *119*, 14668–14682. [CrossRef] [PubMed]
117. Saitoh, Y.; Suzuki, H.; Tani, K.; Nishikawa, K.; Irie, K.; Ogura, Y.; Tamura, A.; Tsukita, S.; Fujiyoshi, Y. Structural insight into tight junction disassembly by Clostridium perfringens enterotoxin. *Science* **2015**, *347*, 775. [CrossRef] [PubMed]
118. Shinoda, T.; Shinya, N.; Ito, K.; Ohsawa, N.; Terada, T.; Hirata, K.; Kawano, Y.; Yamamoto, M.; Kimura-Someya, T.; Yokoyama, S.; et al. Structural basis for disruption of claudin assembly in tight junctions by an enterotoxin. *Sci. Rep.* **2016**, *6*, 33632. [CrossRef] [PubMed]
119. Nakamura, S.; Irie, K.; Tanaka, H.; Nishikawa, K.; Suzuki, H.; Saitoh, Y.; Tamura, A.; Tsukita, S.; Fujiyoshi, Y. Morphologic determinant of tight junctions revealed by claudin-3 structures. *Nat. Commun.* **2019**, *10*, 816. [CrossRef] [PubMed]

120. Xiang, Z. Advances in homology protein structure modeling. *Curr. Protein Pept. Sci.* **2006**, *7*, 217–227. [CrossRef]
121. Shen, M.-y.; Sali, A. Statistical potential for assessment and prediction of protein structures. *Protein Sci.* **2006**, *15*, 2507–2524. [CrossRef]
122. Kopp, J.; Schwede, T. Automated protein structure homology modeling: A progress report. *Pharmacogenomics* **2004**, *5*, 405–416. [CrossRef]
123. Fiser, A. Template-based protein structure modeling. *Methods Mol. Biol.* **2010**, *673*, 73–94.
124. Mineta, K.; Yamamoto, Y.; Yamazaki, Y.; Tanaka, H.; Tada, Y.; Saito, K.; Tamura, A.; Igarashi, M.; Endo, T.; Takeuchi, K.; et al. Predicted expansion of the claudin multigene family. *FEBS Lett.* **2011**, *585*, 606–612. [CrossRef]
125. Kelley, L.A.; Mezulis, S.; Yates, C.M.; Wass, M.N.; Sternberg, M.J.E. The Phyre2 web portal for protein modeling, prediction and analysis. *Nat. Protoc.* **2015**, *10*, 845. [CrossRef] [PubMed]
126. Schwede, T.; Kopp, J.; Guex, N.; Peitsch, M.C. SWISS-MODEL: An automated protein homology-modeling server. *Nucleic Acids Res.* **2003**, *31*, 3381–3385. [CrossRef] [PubMed]
127. Roy, A.; Kucukural, A.; Zhang, Y. I-TASSER: A unified platform for automated protein structure and function prediction. *Nat. Protoc.* **2010**, *5*, 725–738. [CrossRef] [PubMed]
128. Yang, J.; Yan, R.; Roy, A.; Xu, D.; Poisson, J.; Zhang, Y. The I-TASSER Suite: Protein structure and function prediction. *Nat. Methods* **2014**, *12*, 7. [CrossRef] [PubMed]
129. Yang, J.; Zhang, Y. I-TASSER server: New development for protein structure and function predictions. *Nucleic Acids Res.* **2015**, *43*, W174–W181. [CrossRef]
130. Källberg, M.; Wang, H.; Wang, S.; Peng, J.; Wang, Z.; Lu, H.; Xu, J. Template-based protein structure modeling using the RaptorX web server. *Nat. Protoc.* **2012**, *7*, 1511–1522. [CrossRef]
131. Peng, J.; Xu, J. RaptorX: Exploiting structure information for protein alignment by statistical inference. *Proteins* **2011**, *79* (Suppl. 10), 161–171. [CrossRef]
132. Fiser, A.; Do, R.K.G.; Šali, A. Modeling of loops in protein structures. *Protein Sci.* **2000**, *9*, 1753–1773. [CrossRef]
133. Baker, D.; Sali, A. Protein Structure Prediction and Structural Genomics. *Science* **2001**, *294*, 93. [CrossRef]
134. Wiltgen, M. Algorithms for Structure Comparison and Analysis: Homology Modelling of Proteins. In *Encyclopedia of Bioinformatics and Computational Biology*; Ranganathan, S., Gribskov, M., Nakai, K., Schönbach, C., Eds.; Academic Press: Oxford, UK, 2019; pp. 38–61. [CrossRef]
135. Schymkowitz, J.; Borg, J.; Stricher, F.; Nys, R.; Rousseau, F.; Serrano, L. The FoldX web server: An online force field. *Nucleic Acids Res.* **2005**, *33*, W382–W388. [CrossRef]
136. Fiser, A.; Sali, A. ModLoop: Automated modeling of loops in protein structures. *Bioinformatics* **2003**, *19*, 2500–2501. [CrossRef] [PubMed]
137. MacKerell, A.D.; Bashford, D.; Bellott, M.; Dunbrack, R.L.; Evanseck, J.D.; Field, M.J.; Fischer, S.; Gao, J.; Guo, H.; Ha, S.; et al. All-Atom Empirical Potential for Molecular Modeling and Dynamics Studies of Proteins. *J. Phys. Chem. B* **1998**, *102*, 3586–3616. [CrossRef] [PubMed]
138. Wang, H.; Junghans, C.; Kremer, K. Comparative atomistic and coarse-grained study of water: What do we lose by coarse-graining? *Eur. Phys. J. E* **2009**, *28*, 221–229. [CrossRef] [PubMed]
139. Kmiecik, S.; Gront, D.; Kolinski, M.; Wieteska, L.; Dawid, A.E.; Kolinski, A. Coarse-Grained Protein Models and Their Applications. *Chem. Rev.* **2016**, *116*, 7898–7936. [CrossRef] [PubMed]
140. Arkhipov, A.; Freddolino, P.L.; Imada, K.; Namba, K.; Schulten, K. Coarse-Grained Molecular Dynamics Simulations of a Rotating Bacterial Flagellum. *Biophys. J.* **2006**, *91*, 4589–4597. [CrossRef]
141. Shih, A.Y.; Arkhipov, A.; Freddolino, P.L.; Schulten, K. Coarse Grained Protein−Lipid Model with Application to Lipoprotein Particles. *J. Phys. Chem. B* **2006**, *110*, 3674–3684. [CrossRef]
142. Akhmatskaya, E.; Reich, S. Meso-GSHMC: A stochastic algorithm for meso-scale constant temperature simulations. *Procedia Comput. Sci.* **2011**, *4*, 1353–1362. [CrossRef]
143. de Jong, D.H.; Periole, X.; Marrink, S.J. Dimerization of Amino Acid Side Chains: Lessons from the Comparison of Different Force Fields. *J. Chem. Theory Comput.* **2012**, *8*, 1003–1014. [CrossRef]
144. Ayton, G.S.; Voth, G.A. Systematic multiscale simulation of membrane protein systems. *Curr. Opin. Struct. Biol.* **2009**, *19*, 138–144. [CrossRef]
145. Yuan, C.; Li, S.; Zou, Q.; Ren, Y.; Yan, X. Multiscale simulations for understanding the evolution and mechanism of hierarchical peptide self-assembly. *Phys. Chem. Chem. Phys.* **2017**, *19*, 23614–23631. [CrossRef]

146. Singh, N.; Li, W. Recent Advances in Coarse-Grained Models for Biomolecules and Their Applications. *Int. J. Mol. Sci.* **2019**, *20*, 3774. [CrossRef] [PubMed]
147. Jiang, W.; Luo, J.; Nangia, S. Multiscale Approach to Investigate Self-Assembly of Telodendrimer Based Nanocarriers for Anticancer Drug Delivery. *Langmuir* **2015**, *31*, 4270–4280. [CrossRef] [PubMed]
148. Jiang, W.; Wang, X.; Guo, D.; Luo, J.; Nangia, S. Drug-Specific Design of Telodendrimer Architecture for Effective Doxorubicin Encapsulation. *J. Phys. Chem. B* **2016**, *120*, 9766–9777. [CrossRef] [PubMed]
149. Cornell, W.D.; Cieplak, P.; Bayly, C.I.; Gould, I.R.; Merz, K.M.; Ferguson, D.M.; Spellmeyer, D.C.; Fox, T.; Caldwell, J.W.; Kollman, P.A. A Second Generation Force Field for the Simulation of Proteins, Nucleic Acids, and Organic Molecules. *J. Am. Chem. Soc.* **1995**, *117*, 5179–5197. [CrossRef]
150. Oostenbrink, C.; Villa, A.; Mark, A.E.; Van Gunsteren, W.F. A biomolecular force field based on the free enthalpy of hydration and solvation: The GROMOS force-field parameter sets 53A5 and 53A6. *J. Comput. Chem.* **2004**, *25*, 1656–1676. [CrossRef] [PubMed]
151. Jorgensen, W.L.; Maxwell, D.S.; Tirado-Rives, J. Development and Testing of the OPLS All-Atom Force Field on Conformational Energetics and Properties of Organic Liquids. *J. Am. Chem. Soc.* **1996**, *118*, 11225–11236. [CrossRef]
152. Jo, S.; Kim, T.; Iyer, V.G.; Im, W. CHARMM-GUI: A web-based graphical user interface for CHARMM. *J. Comput. Chem.* **2008**, *29*, 1859–1865. [CrossRef]
153. Abraham, M.J.; Murtola, T.; Schulz, R.; Páll, S.; Smith, J.C.; Hess, B.; Lindahl, E. GROMACS: High performance molecular simulations through multi-level parallelism from laptops to supercomputers. *SoftwareX* **2015**, *1–2*, 19–25. [CrossRef]
154. Van Der Spoel, D.; Lindahl, E.; Hess, B.; Groenhof, G.; Mark, A.E.; Berendsen, H.J.C. GROMACS: Fast, flexible, and free. *J. Comput. Chem.* **2005**, *26*, 1701–1718. [CrossRef]
155. Phillips, J.C.; Braun, R.; Wang, W.; Gumbart, J.; Tajkhorshid, E.; Villa, E.; Chipot, C.; Skeel, R.D.; Kalé, L.; Schulten, K. Scalable molecular dynamics with NAMD. *J. Comput. Chem.* **2005**, *26*, 1781–1802. [CrossRef]
156. Thompson, J.M.T.; Grindon, C.; Harris, S.; Evans, T.; Novik, K.; Coveney, P.; Laughton, C. Large-scale molecular dynamics simulation of DNA: Implementation and validation of the AMBER98 force field in LAMMPS. *Philos. Trans. R. Soc. Lond. Ser. A Math. Phys. Eng. Sci.* **2004**, *362*, 1373–1386.
157. Case, D.A.; Cheatham Iii, T.E.; Darden, T.; Gohlke, H.; Luo, R.; Merz, K.M., Jr.; Onufriev, A.; Simmerling, C.; Wang, B.; Woods, R.J. The Amber biomolecular simulation programs. *J. Comput. Chem.* **2005**, *26*, 1668–1688. [CrossRef] [PubMed]
158. Wu, E.L.; Cheng, X.; Jo, S.; Rui, H.; Song, K.C.; Dávila-Contreras, E.M.; Qi, Y.; Lee, J.; Monje-Galvan, V.; Venable, R.M.; et al. CHARMM-GUI Membrane Builder toward realistic biological membrane simulations. *J. Comput. Chem.* **2014**, *35*, 1997–2004. [CrossRef] [PubMed]
159. Brooks, B.R.; Brooks Iii, C.L.; Mackerell Jr, A.D.; Nilsson, L.; Petrella, R.J.; Roux, B.; Won, Y.; Archontis, G.; Bartels, C.; Boresch, S.; et al. CHARMM: The biomolecular simulation program. *J. Comput. Chem.* **2009**, *30*, 1545–1614. [CrossRef] [PubMed]
160. Humphrey, W.; Dalke, A.; Schulten, K. VMD: Visual molecular dynamics. *J. Mol. Graph.* **1996**, *14*, 33–38. [CrossRef]
161. *The PyMOL Molecular Graphics System*; Version 2.0; Schrödinger, LLC: New York, NY, USA, 2015.
162. Land, H.; Humble, M.S. YASARA: A Tool to Obtain Structural Guidance in Biocatalytic Investigations. In *Protein Engineering: Methods and Protocols*; Bornscheuer, U.T., Höhne, M., Eds.; Springer New York: New York, NY, USA, 2018; pp. 43–67. [CrossRef]
163. Krieger, E.; Vriend, G. YASARA View—molecular graphics for all devices—From smartphones to workstations. *Bioinformatics* **2014**, *30*, 2981–2982. [CrossRef] [PubMed]
164. Krieger, E.; Koraimann, G.; Vriend, G. Increasing the precision of comparative models with YASARA NOVA—A self-parameterizing force field. *Proteins Struct. Funct. Bioinform.* **2002**, *47*, 393–402. [CrossRef]
165. Marrink, S.J.; Risselada, H.J.; Yefimov, S.; Tieleman, D.P.; de Vries, A.H. The MARTINI Force Field: Coarse Grained Model for Biomolecular Simulations. *J. Phys. Chem. B* **2007**, *111*, 7812–7824. [CrossRef]
166. Monticelli, L.; Kandasamy, S.K.; Periole, X.; Larson, R.G.; Tieleman, D.P.; Marrink, S.-J. The MARTINI Coarse-Grained Force Field: Extension to Proteins. *J. Chem. Theory Comput.* **2008**, *4*, 819–834. [CrossRef]
167. Marrink, S.J.; Tieleman, D.P. Perspective on the Martini model. *Chem. Soc. Rev.* **2013**, *42*, 6801–6822. [CrossRef]
168. Periole, X.; Marrink, S.-J. The Martini Coarse-Grained Force Field. In *Biomolecular Simulations: Methods and*

Protocols; Monticelli, L., Salonen, E., Eds.; Humana Press: Totowa, NJ, USA, 2013; pp. 533–565.
169. de Jong, D.H.; Singh, G.; Bennett, W.F.D.; Arnarez, C.; Wassenaar, T.A.; Schäfer, L.V.; Periole, X.; Tieleman, D.P.; Marrink, S.J. Improved Parameters for the Martini Coarse-Grained Protein Force Field. *J. Chem. Theory Comput.* **2013**, *9*, 687–697. [CrossRef] [PubMed]
170. Darré, L.; Machado, M.R.; Brandner, A.F.; González, H.C.; Ferreira, S.; Pantano, S. SIRAH: A Structurally Unbiased Coarse-Grained Force Field for Proteins with Aqueous Solvation and Long-Range Electrostatics. *J. Chem. Theory Comput.* **2015**, *11*, 723–739. [CrossRef] [PubMed]
171. Machado, M.R.; Pantano, S. SIRAH tools: Mapping, backmapping and visualization of coarse-grained models. *Bioinformatics* **2016**, *32*, 1568–1570. [CrossRef] [PubMed]
172. Kar, P.; Gopal, S.M.; Cheng, Y.-M.; Predeus, A.; Feig, M. PRIMO: A Transferable Coarse-grained Force Field for Proteins. *J. Chem. Theory Comput.* **2013**, *9*, 3769–3788. [CrossRef] [PubMed]
173. Cheon, M.; Chang, I.; Hall, C.K. Extending the PRIME model for protein aggregation to all 20 amino acids. *Proteins* **2010**, *78*, 2950–2960. [CrossRef] [PubMed]
174. Han, W.; Wan, C.-K.; Wu, Y.-D. PACE Force Field for Protein Simulations. 2. Folding Simulations of Peptides. *J. Chem. Theory Comput.* **2010**, *6*, 3390–3402. [CrossRef]
175. Han, W.; Wan, C.-K.; Jiang, F.; Wu, Y.-D. PACE Force Field for Protein Simulations. 1. Full Parameterization of Version 1 and Verification. *J. Chem. Theory Comput.* **2010**, *6*, 3373–3389. [CrossRef]
176. Uusitalo, J.J.; Ingólfsson, H.I.; Marrink, S.J.; Faustino, I. Martini Coarse-Grained Force Field: Extension to RNA. *Biophys. J.* **2017**, *113*, 246–256. [CrossRef]
177. Bernardi, R.C.; Melo, M.C.R.; Schulten, K. Enhanced sampling techniques in molecular dynamics simulations of biological systems. *Biochim. Et Biophys. Acta (BBA) Gen. Subj.* **2015**, *1850*, 872–877. [CrossRef]
178. Perilla, J.R.; Goh, B.C.; Cassidy, C.K.; Liu, B.; Bernardi, R.C.; Rudack, T.; Yu, H.; Wu, Z.; Schulten, K. Molecular dynamics simulations of large macromolecular complexes. *Curr. Opin. Struct. Biol.* **2015**, *31*, 64–74. [CrossRef]
179. Perilla, J.R.; Schulten, K. Physical properties of the HIV-1 capsid from all-atom molecular dynamics simulations. *Nat. Commun.* **2017**, *8*, 15959. [CrossRef] [PubMed]
180. Xiao, G.; Ren, M.; Hong, H. 50 million atoms scale molecular dynamics modelling on a single consumer graphics card. *Adv. Eng. Softw.* **2018**, *124*, 66–72. [CrossRef]
181. Orsi, M. 15-Molecular simulation of self-assembly. In *Self-Assembling Biomaterials*; Azevedo, H.S., da Silva, R.M.P., Eds.; Woodhead Publishing: Cambridge, MA, USA, 2018; pp. 305–318. [CrossRef]
182. Ben-Nissan, G.; Sharon, M. Capturing protein structural kinetics by mass spectrometry. *Chem. Soc. Rev.* **2011**, *40*, 3627–3637. [CrossRef] [PubMed]
183. Sekhar, A.; Vallurupalli, P.; Kay, L.E. Defining a length scale for millisecond-timescale protein conformational exchange. *Proc. Natl. Acad. Sci. USA* **2013**, *110*, 11391. [CrossRef]
184. Agrawal, P.; Singh, H.; Srivastava, H.K.; Singh, S.; Kishore, G.; Raghava, G.P.S. Benchmarking of different molecular docking methods for protein-peptide docking. *BMC Bioinform.* **2019**, *19*, 426. [CrossRef]
185. Pagadala, N.S.; Syed, K.; Tuszynski, J. Software for molecular docking: A review. *Biophys. Rev.* **2017**, *9*, 91–102. [CrossRef]
186. Kozakov, D.; Hall, D.R.; Xia, B.; Porter, K.A.; Padhorny, D.; Yueh, C.; Beglov, D.; Vajda, S. The ClusPro web server for protein–protein docking. *Nat. Protoc.* **2017**, *12*, 255. [CrossRef]
187. Comeau, S.R.; Gatchell, D.W.; Vajda, S.; Camacho, C.J. ClusPro: A fully automated algorithm for protein–protein docking. *Nucleic Acids Res.* **2004**, *32*, W96–W99. [CrossRef]
188. Lyskov, S.; Gray, J.J. The RosettaDock server for local protein-protein docking. *Nucleic Acids Res.* **2008**, *36*, W233–W238. [CrossRef]
189. de Vries, S.J.; van Dijk, M.; Bonvin, A.M.J.J. The HADDOCK web server for data-driven biomolecular docking. *Nat. Protoc.* **2010**, *5*, 883. [CrossRef]
190. Dominguez, C.; Boelens, R.; Bonvin, A.M.J.J. HADDOCK: A Protein−Protein Docking Approach Based on Biochemical or Biophysical Information. *J. Am. Chem. Soc.* **2003**, *125*, 1731–1737. [CrossRef] [PubMed]
191. Chen, R.; Li, L.; Weng, Z. ZDOCK: An initial-stage protein-docking algorithm. *Proteins Struct. Funct. Bioinform.* **2003**, *52*, 80–87. [CrossRef] [PubMed]
192. Liu, Y.; Zhao, L.; Li, W.; Zhao, D.; Song, M.; Yang, Y. FIPSDock: A new molecular docking technique driven by fully informed swarm optimization algorithm. *J. Comput. Chem.* **2013**, *34*, 67–75. [CrossRef] [PubMed]
193. Gagnon, J.K.; Law, S.M.; Brooks Iii, C.L. Flexible CDOCKER: Development and application of a pseudo-explicit

structure-based docking method within CHARMM. *J. Comput. Chem.* **2016** *37*, 753–762. [CrossRef] [PubMed]
194. Pons, C.; Garzon, J.I.; Lopéz-Blanco, J.R.; Fernandez-Recio, J.; Kovacs, J.; Chacon, P.; Abagyan, R. FRODOCK: A new approach for fast rotational protein–protein docking. *Bioinformatics* **2009**, *25*, 2544–2551.
195. Liu, M.; Wang, S. MCDOCK: A Monte Carlo simulation approach to the molecular docking problem. *J. Comput. Aided Mol. Des.* **1999**, *13*, 435–451. [CrossRef]
196. Yellapu, N.K. Molecular Modelling, Dynamics, and Docking of Membrane Proteins: Still a Challenge. In *Applied Case Studies and Solutions in Molecular Docking-Based Drug Design*; IGI Global: Hershey, PA, USA, 2016; pp. 186–208.
197. Burke, K.A.; Yates, E.A.; Legleiter, J. Biophysical insights into how surfaces, including lipid membranes, modulate protein aggregation related to neurodegeneration. *Front. Neurol* **2013**, *4*, 17. [CrossRef]
198. Bogdanov, M.; Dowhan, W.; Vitrac, H. Lipids and topological rules governing membrane protein assembly. *Biochim. Biophys. Acta* **2014**, *1843*, 1475–1488. [CrossRef]
199. Bogdanov, M.; Mileykovskaya, E.; Dowhan, W. Lipids in the assembly of membrane proteins and organization of protein supercomplexes: Implications for lipid-linked disorders. *Subcell. Biochem.* **2008**, *49*, 197–239.
200. Ho, C.-C. Chapter 7—Membranes for Bioseparations. In *Bioprocessing for Value-Added Products from Renewable Resources*; Yang, S.-T., Ed.; Elsevier: Amsterdam, The Netherlands, 2007; pp. 163–183. [CrossRef]
201. Page, R.C.; Li, C.; Hu, J.; Gao, F.P.; Cross, T.A. Lipid bilayers: An essential environment for the understanding of membrane proteins. *Magn. Reson. Chem.* **2007**, *45*, S2–S11. [CrossRef]
202. Sengupta, D.; Marrink, S.J. Lipid-mediated interactions tune the association of glycophorin A helix and its disruptive mutants in membranes. *Phys. Chem. Chem. Phys.* **2010**, *12*, 12987–12996. [CrossRef] [PubMed]
203. Wassenaar, T.A.; Pluhackova, K.; Moussatova, A.; Sengupta, D.; Marrink, S.J.; Tieleman, D.P.; Böckmann, R.A. High-Throughput Simulations of Dimer and Trimer Assembly of Membrane Proteins. The DAFT Approach. *J. Chem. Theory Comput.* **2015**, *11*, 2278–2291. [CrossRef] [PubMed]
204. Altwaijry, N.A.; Baron, M.; Wright, D.W.; Coveney, P.V.; Townsend-Nicholson, A. An Ensemble-Based Protocol for the Computational Prediction of Helix–Helix Interactions in G Protein-Coupled Receptors using Coarse-Grained Molecular Dynamics. *J. Chem. Theory Comput.* **2017**, *13*, 2254–2270. [CrossRef] [PubMed]
205. Hall, B.A.; Halim, K.B.A.; Buyan, A.; Emmanouil, B.; Sansom, M.S.P. Sidekick for Membrane Simulations: Automated Ensemble Molecular Dynamics Simulations of Transmembrane Helices. *J. Chem. Theory Comput.* **2014**, *10*, 2165–2175. [CrossRef] [PubMed]
206. Rajagopal, N.; Nangia, S. Obtaining Protein Association Energy Landscape (PANEL) for Integral Membrane Proteins. *J. Chem. Theory Comput.* **2019**. [CrossRef] [PubMed]
207. Petrek, M.; Otyepka, M.; Banás, P.; Kosinová, P.; Koca, J.; Damborský, J. CAVER: A new tool to explore routes from protein clefts, pockets and cavities. *BMC Bioinform.* **2006**, *7*, 316. [CrossRef]
208. Petřek, M.; Košinová, P.; Koča, J.; Otyepka, M. MOLE: A Voronoi Diagram-Based Explorer of Molecular Channels, Pores, and Tunnels. *Structure* **2007**, *15*, 1357–1363. [CrossRef]
209. Smart, O.S.; Neduvelil, J.G.; Wang, X.; Wallace, B.A.; Sansom, M.S. HOLE: A program for the analysis of the pore dimensions of ion channel structural models. *J. Mol Graph.* **1996**, *14*, 354–360. [CrossRef]
210. Yaffe, E.; Fishelovitch, D.; Wolfson, H.J.; Halperin, D.; Nussinov, R. MolAxis: A server for identification of channels in macromolecules. *Nucleic Acids Res.* **2008**, *36*, W210–W215. [CrossRef]
211. Yaffe, E.; Fishelovitch, D.; Wolfson, H.J.; Halperin, D.; Nussinov, R. MolAxis: Efficient and accurate identification of channels in macromolecules. *Proteins* **2008**, *73*, 72–86. [CrossRef]
212. Pellegrini-Calace, M.; Maiwald, T.; Thornton, J.M. PoreWalker: A Novel Tool for the Identification and Characterization of Channels in Transmembrane Proteins from Their Three-Dimensional Structure. *PLoS Comput. Biol.* **2009**, *5*, e1000440. [CrossRef] [PubMed]
213. Peter, C.; Hummer, G. Ion Transport through Membrane-Spanning Nanopores Studied by Molecular Dynamics Simulations and Continuum Electrostatics Calculations. *Biophys. J.* **2005**, *89*, 2222–2234. [CrossRef] [PubMed]
214. Hummer, G.; Pratt, L.R.; García, A.E. Molecular Theories and Simulation of Ions and Polar Molecules in Water. *J. Phys. Chem. A* **1998**, *102*, 7885–7895. [CrossRef]
215. Torrie, G.M.; Valleau, J.P. Nonphysical sampling distributions in Monte Carlo free-energy estimation: Umbrella sampling. *J. Comput. Phys.* **1977**, *23*, 187–199. [CrossRef]
216. Barducci, A.; Bonomi, M.; Parrinello, M. Metadynamics. *Wiley Interdiscip. Rev. Comput. Mol. Sci.* **2011**, *1*, 826–843. [CrossRef]

217. Laio, A.; Gervasio, F.L. Metadynamics: A method to simulate rare events and reconstruct the free energy in biophysics, chemistry and material science. *Rep. Prog. Phys.* **2008**, *71*, 126601. [CrossRef]
218. Barducci, A.; Bussi, G.; Parrinello, M. Well-Tempered Metadynamics: A Smoothly Converging and Tunable Free-Energy Method. *Phys. Rev. Lett.* **2008**, *100*, 020603. [CrossRef]
219. Kumar, S.; Rosenberg, J.M.; Bouzida, D.; Swendsen, R.H.; Kollman, P.A. THE weighted histogram analysis method for free-energy calculations on biomolecules. I. The method. *J. Comput. Chem.* **1992**, *13*, 1011–1021. [CrossRef]
220. Bonomi, M.; Branduardi, D.; Bussi, G.; Camilloni, C.; Provasi, D.; Raiteri, P.; Donadio, D.; Marinelli, F.; Pietrucci, F.; Broglia, R.A.; et al. PLUMED: A portable plugin for free-energy calculations with molecular dynamics. *Comput. Phys. Commun.* **2009**, *180*, 1961–1972. [CrossRef]
221. Kutzner, C.; Grubmüller, H.; de Groot, B.L.; Zachariae, U. Computational electrophysiology: The molecular dynamics of ion channel permeation and selectivity in atomistic detail. *Biophys. J.* **2011**, *101*, 809–817. [CrossRef]
222. Kutzner, C.; Köpfer, D.A.; Machtens, J.-P.; de Groot, B.L.; Song, C.; Zachariae, U. Insights into the function of ion channels by computational electrophysiology simulations. *Biochim. Biophys. Acta (BBA) Biomembr.* **2016**, *1858*, 1741–1752. [CrossRef]
223. Epand, R.M. Lipid polymorphism and protein–lipid interactions. *Biochim. Biophys. Acta (BBA) Rev. Biomembr.* **1998**, *1376*, 353–368. [CrossRef]
224. Killian, J.A. Hydrophobic mismatch between proteins and lipids in membranes. *Biochim. Biophys. Acta (BBA) Rev. Biomembr.* **1998**, *1376*, 401–416. [CrossRef]
225. de Jesus, A.J.; Allen, T.W. The determinants of hydrophobic mismatch response for transmembrane helices. *Biochim. Biophys. Acta (BBA) Biomembr.* **2013**, *1828*, 851–863. [CrossRef] [PubMed]
226. Parton, D.L.; Klingelhoefer, J.W.; Sansom, M.S.P. Aggregation of model membrane proteins, modulated by hydrophobic mismatch, membrane curvature, and protein class. *Biophys. J.* **2011**, *101*, 691–699. [CrossRef] [PubMed]
227. Bavi, O.; Vossoughi, M.; Naghdabadi, R.; Jamali, Y. The Combined Effect of Hydrophobic Mismatch and Bilayer Local Bending on the Regulation of Mechanosensitive Ion Channels. *PLoS ONE* **2016**, *11*, e0150578. [CrossRef] [PubMed]
228. Webb, R.J.; East, J.M.; Sharma, R.P.; Lee, A.G. Hydrophobic Mismatch and the Incorporation of Peptides into Lipid Bilayers: A Possible Mechanism for Retention in the Golgi. *Biochemistry* **1998**, *37*, 673–679. [CrossRef]
229. Milovanovic, D.; Honigmann, A.; Koike, S.; Göttfert, F.; Pähler, G.; Junius, M.; Müllar, S.; Diederichsen, U.; Janshoff, A.; Grubmüller, H.; et al. Hydrophobic mismatch sorts SNARE proteins into distinct membrane domains. *Nat. Commun.* **2015**, *6*, 5984. [CrossRef]
230. Fowler, P.W.; Williamson, J.J.; Sansom, M.S.P.; Olmsted, P.D. Roles of Interleaflet Coupling and Hydrophobic Mismatch in Lipid Membrane Phase-Separation Kinetics. *J. Am. Chem. Soc.* **2016**, *138*, 11633–11642. [CrossRef]
231. Koval, M. Differential pathways of claudin oligomerization and integration into tight junctions. *Tissue Barriers* **2013**, *1*, e24518. [CrossRef]
232. Koval, M. Claudins—Key Pieces in the Tight Junction Puzzle. *Cell Commun. Adhes.* **2006**, *13*, 127–138. [CrossRef] [PubMed]
233. Heiler, S.; Mu, W.; Zöller, M.; Thuma, F. The importance of claudin-7 palmitoylation on membrane subdomain localization and metastasis-promoting activities. *Cell Commun. Signal.* **2015**, 13. [CrossRef] [PubMed]
234. Van Itallie, C.M.; Gambling, T.M.; Carson, J.L.; Anderson, J.M. Palmitoylation of claudins is required for efficient tight-junction localization. *J. Cell Sci.* **2005**, *118*, 1427–1436. [CrossRef]
235. Fukata, Y.; Bredt, D.S.; Fukata, M. *Protein Palmitoylation by DHHC Protein Family, the Dynamic Synapse: Molecular Methods in Ionotropic Receptor Biology*; CRC Press/Taylor & Francis: Boca Raton, FL, USA, 2006.
236. Gorleku, O.A.; Barns, A.-M.; Prescott, G.R.; Greaves, J.; Chamberlain, L.H. Endoplasmic Reticulum Localization of DHHC Palmitoyltransferases Mediated by Lysine-based Sorting Signals. *J. Biol. Chem.* **2011**, *286*, 39573–39584. [CrossRef] [PubMed]
237. Chamberlain, L.H.; Shipston, M.J. The Physiology of Protein S-acylation. *Physiol. Rev.* **2015**, *95*, 341–376. [CrossRef]
238. Hou, J. Chapter 2—Paracellular Channel Formation. In *The Paracellular Channel*; Hou, J., Ed.; Academic Press: Cambridge, MA, USA, 2019; pp. 9–27. [CrossRef]

239. Hou, J.; Renigunta, A.; Konrad, M.; Gomes, A.S.; Schneeberger, E.E.; Paul, D.L.; Waldegger, S.; Goodenough, D.A. Claudin-16 and claudin-19 interact and form a cation-selective tight junction complex. *J. Clin. Investig.* **2008**, *118*, 619–628. [CrossRef]
240. Mitic, L.L.; Unger, V.M.; Anderson, J.M. Expression, solubilization, and biochemical characterization of the tight junction transmembrane protein claudin-4. *Protein Sci.* **2003**, *12*, 218–227. [CrossRef]
241. Van Itallie, C.M.; Mitic, L.L.; Anderson, J.M. Claudin-2 Forms Homodimers and Is a Component of a High Molecular Weight Protein Complex. *J. Biol. Chem.* **2011**, *286*, 3442–3450. [CrossRef]
242. Suzuki, H.; Tani, K.; Tamura, A.; Tsukita, S.; Fujiyoshi, Y. Model for the architecture of claudin-based paracellular ion channels through tight junctions. *J. Mol. Biol.* **2015**, *427*, 291–297. [CrossRef]
243. Coyne, C.B.; Gambling, T.M.; Boucher, R.C.; Carson, J.L.; Johnson, L.G. Role of claudin interactions in airway tight junctional permeability. *Am. J. Physiol. Lung Cell. Mol. Physiol.* **2003**, *285*, L1166–L1178. [CrossRef]
244. Angelow, S.; Yu, A.S.L. Structure-Function Studies of Claudin Extracellular Domains by Cysteine-scanning Mutagenesis. *J. Biol. Chem.* **2009**, *284*, 29205–29217. [CrossRef] [PubMed]
245. Anbazhagan, V.; Schneider, D. The membrane environment modulates self-association of the human GpA TM domain—Implications for membrane protein folding and transmembrane signaling. *Biochim. Biophys. Acta (BBA) Biomembr.* **2010**, *1798*, 1899–1907. [CrossRef] [PubMed]
246. van Meer, G.; Voelker, D.R.; Feigenson, G.W. Membrane lipids: Where they are and how they behave. *Nat. Rev. Mol. Cell Biol.* **2008**, *9*, 112. [CrossRef] [PubMed]
247. Von Heijne, G. The membrane protein universe: What's out there and why bother? *J. Intern. Med.* **2007**, *261*, 543–557. [CrossRef]
248. Blaskovic, S.; Blanc, M.; Goot, F.G. What does S-palmitoylation do to membrane proteins? *FEBS J.* **2013**, *280*, 2766–2774. [CrossRef]
249. Linder, M.E.; Deschenes, R.J. Palmitoylation: Policing protein stability and traffic. *Nat. Rev. Mol. Cell Biol.* **2007**, *8*, 74–84. [CrossRef]
250. Rossa, J.; Lorenz, D.; Ringling, M.; Veshnyakova, A.; Piontek, J. Overexpression of claudin-5 but not claudin-3 induces formation of trans-interaction–dependent multilamellar bodies. *Ann. N. Y. Acad. Sci.* **2012**, *1257*, 59–66. [CrossRef]
251. Piontek, J.; Winkler, L.; Wolburg, H.; Müller, S.L.; Zuleger, N.; Piehl, C.; Wiesner, B.; Krause, G.; Blasig, I.E. Formation of tight junction: Determinants of homophilic interaction between classic claudins. *FASEB J.* **2007**, *22*, 146–158. [CrossRef]
252. Kausalya, P.J.; Amasheh, S.; Günzel, D.; Wurps, H.; Müller, D.; Fromm, M.; Hunziker, W. Disease-associated mutations affect intracellular traffic and paracellular Mg^{2+} transport function of Claudin-16. *J. Clin Investig.* **2006**, *116*, 878–891. [CrossRef]
253. Galochkina, T.; Ng Fuk Chong, M.; Challali, L.; Abbar, S.; Etchebest, C. New insights into GluT1 mechanics during glucose transfer. *Sci. Rep.* **2019**, *9*, 998. [CrossRef]
254. Pan, Y.; Zhang, Y.; Gongpan, P.; Zhang, Q.; Huang, S.; Wang, B.; Xu, B.; Shan, Y.; Xiong, W.; Li, G.; et al. Single glucose molecule transport process revealed by force tracing and molecular dynamics simulations. *Nanoscale Horiz.* **2018**, *3*, 517–524. [CrossRef]
255. Samanta, P.; Wang, Y.; Fuladi, S.; Zou, J.; Li, Y.; Shen, L.; Weber, C.; Khalili-Araghi, F. Molecular determination of claudin-15 organization and channel selectivity. *J. Gen. Physiol.* **2018**, *150*, 949. [CrossRef] [PubMed]
256. Alberini, G.; Benfenati, F.; Maragliano, L. Molecular Dynamics Simulations of Ion Selectivity in a Claudin-15 Paracellular Channel. *J. Phys. Chem. B* **2018**, *122*, 10783–10792. [CrossRef] [PubMed]
257. Alberini, G.; Benfenati, F.; Maragliano, L. A refined model of claudin-15 tight junction paracellular architecture by molecular dynamics simulations. *PLoS ONE* **2017**, *12*, e0184190. [CrossRef] [PubMed]
258. Weber, C.R.; Turner, J.R. Dynamic modeling of the tight junction pore pathway. *Ann. N. Y. Acad. Sci.* **2017**, *1397*, 209–218. [CrossRef]
259. Shen, L.; Weber, C.R.; Turner, J.R. The tight junction protein complex undergoes rapid and continuous molecular remodeling at steady state. *J. Cell Biol.* **2008**, *181*, 683. [CrossRef]
260. Krystofiak, E.S.; Heymann, J.B.; Kachar, B. Carbon replicas reveal double stranded structure of tight junctions in phase-contrast electron microscopy. *Commun. Biol.* **2019**, *2*, 98. [CrossRef]

2

Claudins in the Renal Collecting Duct

Janna Leiz [1,2] and Kai M. Schmidt-Ott [1,2,3,*]

1 Department of Nephrology and Intensive Care Medicine, Charité-Universitätsmedizin Berlin, 12203 Berlin, Germany; janna.leiz@charite.de
2 Molecular and Translational Kidney Research, Max-Delbrück-Center for Molecular Medicine in the Helmholtz Association (MDC), 13125 Berlin, Germany
3 Berlin Institute of Health (BIH), 10178 Berlin, Germany
* Correspondence: kai.schmidt-ott@charite.de.

Abstract: The renal collecting duct fine-tunes urinary composition, and thereby, coordinates key physiological processes, such as volume/blood pressure regulation, electrolyte-free water reabsorption, and acid-base homeostasis. The collecting duct epithelium is comprised of a tight epithelial barrier resulting in a strict separation of intraluminal urine and the interstitium. Tight junctions are key players in enforcing this barrier and in regulating paracellular transport of solutes across the epithelium. The features of tight junctions across different epithelia are strongly determined by their molecular composition. Claudins are particularly important structural components of tight junctions because they confer barrier and transport properties. In the collecting duct, a specific set of claudins (Cldn-3, Cldn-4, Cldn-7, Cldn-8) is expressed, and each of these claudins has been implicated in mediating aspects of the specific properties of its tight junction. The functional disruption of individual claudins or of the overall barrier function results in defects of blood pressure and water homeostasis. In this concise review, we provide an overview of the current knowledge on the role of the collecting duct epithelial barrier and of claudins in collecting duct function and pathophysiology.

Keywords: epithelial barrier; barrier formation; collecting duct; tight junction; claudin

---

1. Introduction

Selective barriers formed by epithelial monolayers are vital for many critical physiological processes [1]. These barriers are composed of intercellular multiprotein complexes along the apical-basal axis of epithelial cells creating direct cell-cell interactions and forming apical junctional complexes [2]. These complexes include tight junctions (TJs), adherens junctions, and desmosomes.

TJs are the most apical and diverse of these junctional complexes. They are formed by transmembrane as well as cytoplasmic proteins and are linked to the cytoskeleton. The protein composition of TJs varies greatly among different epithelia and determines the barrier properties and permeability [3].

Overall, TJs serve a dual function in epithelial layers acting as (a) a fence, separating membrane proteins between the apical and basolateral membrane and (b) as a gate by regulating size- and charge-selective movements of ions, solutes, and small molecules via the paracellular route [4–7]. Consequently, TJs are not only critical to establish and maintain cell adhesion and epithelial polarity but are also necessary to create selective paracellular permeability between compartments.

Several studies show that the main protein family mediating TJ characteristics and functions are the membrane-spanning claudins. So far, more than 25 different claudins have been identified, and at least 10 of them are expressed in spatiotemporal patterns along the renal nephron (for a detailed review see [8]).

Structurally, the transmembrane proteins connect in both *cis* and *trans* to neighboring claudins of the same, as well as the opposing, cell membranes forming TJ strands [9]. Claudins are sufficient to arrange those strands when expressed in cells lacking endogenous TJ formation [10].

Here, we review the role of claudins in epithelial barrier formation with emphasis on the renal collecting duct as well as their implications in collecting duct physiology and function.

2. The Renal Collecting Duct

The renal collecting duct is the most distal part of the renal tubules. It connects renal nephrons with the renal pelvis and—together with the distal convoluted tubule and the connecting tubule—it contributes to the aldosterone-sensitive distal nephron. In general, the renal collecting duct plays important roles in fine-tuning urinary composition, extracellular fluid volume, electrolyte balance, blood pressure regulation, water homeostasis, and acid-base regulation [11,12].

Although most of the water and solute reabsorption in the kidney occurs in the more upstream segments of the nephron, transport variability in the renal collecting duct is significantly higher, and ion and water transport are under strict hormonal control [11]. This permits to adjust reabsorption and secretion to prevalent physiological conditions and to control the body's water and electrolyte balance closely.

The main solutes reabsorbed in the collecting duct are sodium (Na+) and chloride (Cl−), while potassium (K+) is secreted into the urine [13] (Figure 1). Transport of these ions occurs via the transcellular route—mediated by channels and membrane transporters—and via the paracellular route—mediated by TJs.

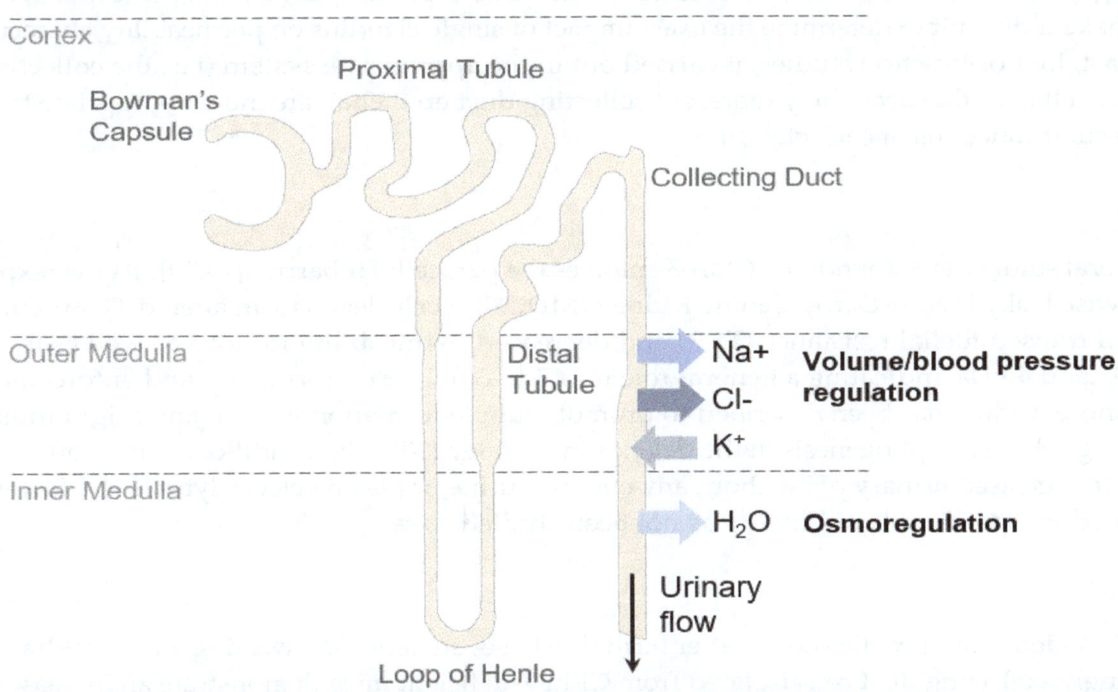

Figure 1. Transport in the renal collecting duct. In the renal collecting duct Na+ and Cl− are reabsorbed, while K+ is secreted into the urine. A steep osmolality gradient between the interstitium and lumen across the epithelial barrier is the driving force for water reabsorption.

3. Claudins and Their Roles in Collecting Duct Epithelial Barrier Formation and Paracellular Ion Transport

In general, claudins can be pore- or barrier-forming [14]. As in other segments of the renal nephron, it is believed that the expressed claudins are the main mediators of TJ permeability characteristics, and thus, control the paracellular pathway in the collecting duct [9,15]. Among the predominately expressed claudins in the renal collecting duct are claudin-3 (Cldn-3), -4 (Cldn-4), -7 (Cldn-7) and -8 (Cldn-8) [16,17]. Furthermore, a recently published study proposes a role for claudin-19 (Cldn-19) in the renal collecting duct [18]. Overall, the present knowledge strongly supports the theory that these claudins either enforce the epithelial barrier in general or build a paracellular Cl− channel, while other aspects of their functions are less clear (Table 1).

Table 1. Claudins expressed in the renal collecting duct and their proposed function in barrier formation and collecting duct tight junctions.

Claudin	Proposed Function in the Collecting Duct Tight Junction	Knockout Phenotype In Vivo
Cldn-3	General barrier function [19,20]	Increase in urinary pH, no electrolyte abnormalities [21]
Cldn-4	Cl− channel, cation barrier [22]	Hypotension, hypochloremia, metabolic alkalosis, renal wasting of Na+ and Cl− [23]
Cldn-7	Cl− and Na+ channel [24–27]	Severe renal salt wasting, dehydration [27]
Cldn-8	Cl− channel, cation barrier [28,29]	Hypotension, hypochloremia, metabolic alkalosis, renal wasting of Na+ and Cl− [29]
Cldn-19	Adaption of barrier selectivity to osmolality [18]	-

In the following sections, we will briefly review the published literature on the functions of these claudins in the collecting duct. As a general note, it should be acknowledged that gain-of-function studies utilizing overexpression of claudins in cell lines have a general limitation due to the influence of the endogenous components of the TJ in each cell line. The expressed claudins in the respective chosen model make it difficult to determine the exact impact of single claudins on permeability characteristics. In contrast, loss-of-function studies, if carried out in the appropriate system (i.e., the collecting duct in vivo or cell lines that accurately represent collecting duct epithelia), are more appropriate to outline the molecular functions of each claudin.

3.1. *Cldn-3*

Several studies have found that Cldn-3 enforces the paracellular barrier [19,20]. Its overexpression in otherwise leaky Madin-Darby Canine Kidney (MDCK) II cells leads to an altered TJ structure with increased transepithelial resistance (TER) and decreased permeability for ions as well as molecules of 332 Da and 4 kDa, indicating a general role for Cldn-3 in barrier formation and enforcement [19]. Furthermore, Cldn-3 has been described to promote tubule formation in vitro and might thus play a role during tubule morphogenesis in the developing kidney [30]. The conditional knockout of Cldn-3 resulted in increased urinary pH without any effect on urine or plasma electrolytes [21]. However, the collecting duct-specific role of Cldn-3 has not been studied so far.

3.2. *Cldn-7*

Cldn-7-deficient mice die shortly after birth due to severe renal salt wasting and dehydration [27]. Interestingly, collecting duct cells isolated from Cldn-7-deficient mice demonstrate an increase in TER and show decreased paracellular permeability for both Na+ and Cl− [31]. This suggests that Cldn-7 may form a non-selective paracellular channel facilitating Cl− and Na+ reabsorption in collecting ducts.

Different and partially conflicting roles of Cldn-7 have been suggested in other experimental systems, complicating the interpretation of Cldn-7 biology. Overexpression of Cldn-7 in LLC-PK1 cells induces an increase in TER accompanied by a reduced Cl− and elevated Na+ conductance [24]. In contrast, the knockdown of Cldn-7 in those cells leads to loss of anion selectivity and decreased Cl− permeation, whereas in cation-selective MDCK cells, knockdown of Cldn-7 leads to increased Na+ permeability [25]. Mutation experiments in vitro demonstrated the importance of specific negatively charged amino acids in the extracellular loop of Cldn-7 for paracellular permeability and charge selectivity [32]. These contrasting results could indicate different roles for Cldn-7, depending on the cellular background, and endogenously expressed TJ components.

3.3. *Cldn-4 and Cldn-8*

Both, Cldn-4 and Cldn-8, are thought to serve as a cation barrier [33–35] and an anion channel [22]. This hypothesis is supported by data from in vivo experiments. Collecting duct-specific knockout of either Cldn-4 or Cldn-8, causes hypotension, hypochloremia, metabolic alkalosis and renal wasting of Na+ and Cl− [23,29]. These phenotypes would be consistent with Cldn-4 and Cldn-8 acting as paracellular Cl− channels, which are necessary for a paracellular "chloride shunt" required for effective transcellular Na+ reabsorption (via the epithelial sodium channel).

Different effects of Cldn-4 overexpression on charge selectivity, depending on the used cell model, have been described. Overexpression of Cldn-4 leads to an increase in TER in cation-selective MDCK II and anion-selective LLC-PK1 cells, but a decrease in Na+ permeability is only observed in MDCK II cells [34].

Cldn-8 has been shown to be necessary to recruit Cldn-4 to the TJ and to implement the protein into the junctional complex. In the absence of Cldn-8, Cldn-4 is mainly found in the endoplasmic reticulum and the Golgi complex, but not in the apical cell membrane where TJs are located [22]. Thus, Cldn-8 knockout causes a functional double knockout on the TJ level due to its requirement for correct Cldn-4 localization [22].

3.4. *Cldn-19*

In addition to the longer known claudins of the renal collecting duct described above, a recently published study indicates a role for Cldn-19, which had previously been linked to thick ascending limb functions [36–38]. Cldn-19 is associated with tightness and cation selectivity of the epithelial barrier. Interestingly, the TJ localization of Cldn-19 is promoted by a low osmolality, whereas high osmolality favors an intracellular localization, suggesting that it may contribute to tonicity-induced changes in paracellular ion selectivity [18]. This role of a claudin in epithelial adaption to the changing osmolality along the corticomedullary axis provides an interesting aspect of TJ physiology but collecting duct-specific knockout models of Cldn-19 have not yet been generated.

4. The Collecting Duct Epithelial Barrier in Electrolyte-Free Water Reabsorption

Although water transport in the renal collecting duct is not directly facilitated by water channel-forming claudins (contrasting with observations of Cldn-2 in the proximal tubule [39]), TJs in the collecting duct contribute to water reabsorption indirectly.

The driving force for water reabsorption in the renal collecting duct is a steep osmolality gradient formed by high concentrations of osmolytes in the interstitium, which increases towards the renal medulla [40]. The tight collecting duct epithelial barrier is crucial to maintain the osmolality gradient between the tubular lumen and the interstitium.

Water transport in the renal collecting duct is mainly controlled by arginine vasopressin (AVP), also called antidiuretic hormone. If the water content in the body is low, AVP binds to its type 2 receptor (V2R) localized in the basal cell membrane of collecting duct principal cells and stimulates the expression of aquaporin-2 (AQP2) water channels. Furthermore, AVP triggers a signaling cascade leading to the accumulation of AQP2 in the apical membrane (for a detailed review see [11,41]). This mechanism enables reabsorption of electrolyte-free water from the intraluminal urine and forms the basis of urinary concentrating ability. Inactivation of either V2R or AQP2 leads to polyuria with massive excretion of electrolyte-free water, a condition called nephrogenic diabetes insipidus [42].

A recent study has demonstrated the importance of an intact epithelial barrier in the renal collecting duct for efficient water reabsorption [43]. Deletion of the transcriptional regulator Grainyhead-like 2 (Grhl2), an epithelial transcription factor that induces the expression of barrier-enforcing molecular TJ components including Cldn-4 [44], results in a leaky collecting duct epithelium and a decreased

TER across the collecting duct epithelium. Leakage of interstitial osmolytes across the Grhl2-deficient collecting duct epithelium is associated with defective retention of osmolytes in the interstitium of the inner medulla. Grhl2-deficient mice show signs of diabetes insipidus and fail to concentrate their urine adequately, although AQP-mediated water transport across the apical and basolateral membranes of Grhl2-deficient collecting ducts is intact [43]. This indicates that a tight collecting duct epithelial barrier is crucial for the maintenance of osmolality gradients and for effective collecting duct water reabsorption. Interestingly, Grhl2 deficiency (unlike deficiencies of individual claudins) was not associated with abnormalities of Cl− and Na+ reabsorption. It needs to be acknowledged that Grhl2 functions are not restricted to the effects on the TJ, which may explain the difference in phenotypes. In addition, it is possible that the barrier defect of Grhl2-deficient collecting duct cells is more profound, leading to non-ion-selective leakage of interstitial osmolytes into hypotonic urine.

5. Aldosterone and Its Role in Transcellular and Paracellular Transport Regulation

In the renal collecting duct, Na+ transport is separated from Cl− transport [45]. Na+ reabsorption from the intraluminal urine occurs transcellularly via the epithelial sodium channel (ENaC) that locates to the apical membrane of collecting duct principal cells. This generates a lumen-negative potential, providing a driving force for K+ secretion via the renal outer medullary potassium channel (ROMK). In contrast, Cl− reabsorption in the collecting duct occurs predominantly via the paracellular route. This "chloride shunt" is important to limit the built-up of a lumen-negative potential and, thereby, facilitates continued Na+ reabsorption and prevents excessive K+ secretion.

The key regulator of ENaC is aldosterone, a hormone secreted from the adrenal gland in response to hyperkalemia and hypovolemia [11,46]. Overall, aldosterone plays a central role in blood pressure regulation by controlling plasma Na+ and K+ levels and thus indirectly influences water retention or loss. However, growing evidence supports the hypothesis that aldosterone controls Na+ reabsorption and K+ secretion not only by regulating the abundance of ENaC and increasing transcellular transport but also by adjusting paracellular Cl− permeability in multiple ways:

For instance, aldosterone triggers the expression of Cldn-8 when ENaC is active, presumably to seal the paracellular route for Na+ back flux. Thereby net flux of Na+ can be increased [28].

Additionally, aldosterone regulates channel-activating protease 1 (Cap1) [47]. Cap1, in turn, stimulates ENaC and inhibits the Cl− conductivity by directly regulating Cldn-4 *trans*-interactions [23,48]. Thereby K+ secretion is favored over Cl− reabsorption (Figure 2a).

Aldosterone also induces the phosphorylation and activation of the with no lysine kinases 4 (WNK4) [49]. Expression of WNK4 in MDCK II cells has been shown to reduce the TER and increase Cl− permeability, without affecting TJ structure [50]. Activated WNK4 phosphorylates Cldn-4 on threonine residues decreasing the cells' TER and increasing apical to basal anion passage [51].

In the absence of aldosterone, WNK4 inhibits ENaC and ROMK activity and thus directly opposes Cap1 [52,53]. WNK4 phosphorylation suspends this inhibition (Figure 2b). Taken together, this indicates the possibility that aldosterone might regulate claudins through Cap1 and WNK4, coordinating Cl− reabsorption or K+ secretion, respectively.

Kahle and colleagues proposed WNK4 as the functional switch regulating Na+ and Cl− reabsorption independently from K+ secretion depending on the physiological conditions [53]. However, the factors facilitating different functional states of WNK4 haven't been provided, and to our knowledge, it is still unknown how the reverse actions of Cap1 and WNK4, that are both mediated by aldosterone, are regulated to decide in favor of K+ secretion or Cl− reabsorption, respectively.

In diabetes, the role of aberrant aldosterone signaling in the progression of renal disease has long been established [54,55]. Mediated by the divergent aldosterone levels, Cldn-4 and Cldn-8 are overexpressed in the distal nephron from type 1 diabetic rats, and the expression of WNK4 and its co-localization with Cldn-4 and Cldn-8 is also increased [56,57]. This might result in increased activation of Cldn-4 and Cldn-8 by WNK4 under diabetic conditions and could implicate disturbed paracellular transport in renal disease progression. However, additional experimental evidence verifying this hypothesis is needed.

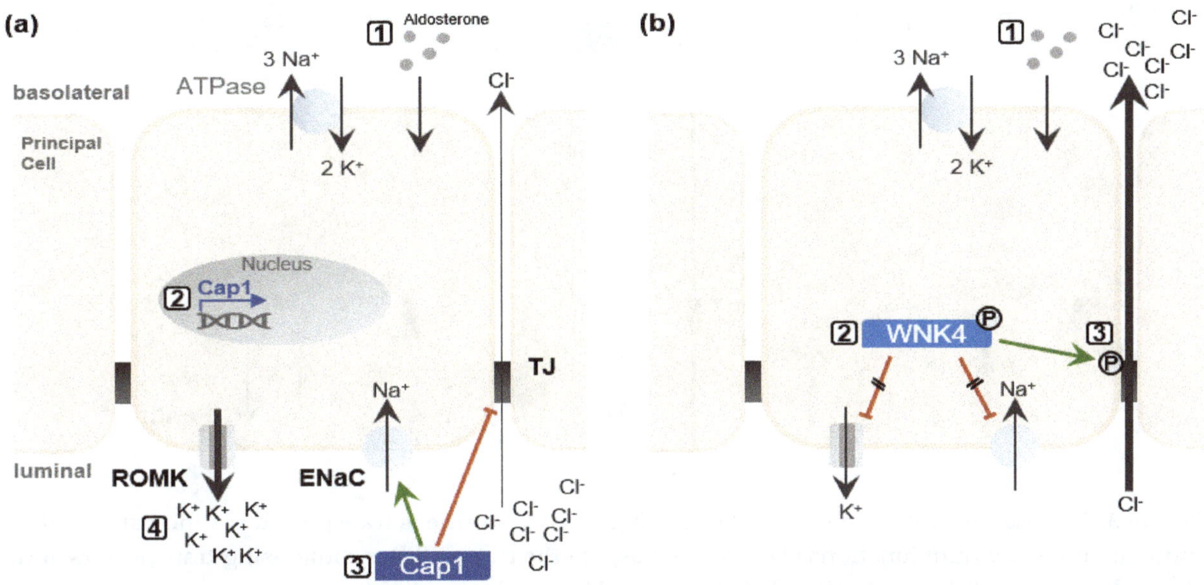

Figure 2. (**a**) Aldosterone-mediated induction of channel-activating protease 1 (Cap1) activates Na+ reabsorption, but simultaneously inhibits the paracellular "chloride shunt", resulting in excessive K+ excretion. When aldosterone is secreted (1), it triggers the expression of Cap1 in principal cells of the renal collecting duct (2). Cap1 inhibits Cl− reabsorption directly by disrupting *trans*-interactions of Cldn-4. Simultaneously, Cap1 activates apical Na+ channels (ENaC), and thus, increases the transcellular reabsorption of Na+ (3). Consequently, the growing luminal negative potential drives K+ secretion into the urine via renal outer medullary potassium (ROMK) channels (4). (**b**) Aldosterone activates ENaC, ROMK, and the paracellular "chloride shunt" via with no lysine kinases 4 (WNK4) phosphorylation. When aldosterone is secreted (1), it leads to the phosphorylation of WNK4. This suspends the tonic inhibition of ROMK and ENaC by WNK4 (2). Furthermore, WNK4 phosphorylates claudins located to the tight junctions (TJs) of the renal collecting duct and thereby increases Cl− reabsorption (3). Green arrows indicate positive regulation, red lines indicate inhibition.

6. Chloride Reabsorption in Renal Collecting Ducts and Potential Involvement in Disease

The aldosterone-sensitive distal nephron is the main side of Cl− reabsorption. It occurs via multiple ways, including paracellular transport via TJs as well as transcellular pathways in intercalated cells [12,23,29] and is driven by the lumen-negative transepithelial potential generated by the unilateral Na+ reabsorption to maintain electroneutrality (Figure 3). As in aldosterone signaling, it becomes increasingly evident that crosstalk between paracellular and transcellular transport occurs.

The importance of efficient Cl− reabsorption becomes obvious in claudin-deficient mouse models. For example, Cldn-7 deficiency in mice is lethal within 12 days after birth due to severe salt-wasting and subsequent chronic dehydration (see above) [27]. Cldn-7$^{-/-}$ mice show reduced ROMK and increased ENaC, Aqp2, and Na+ Cl− cotransporter mRNA. These changes in channel expressions are probably due to a compensatory mechanism to inhibit further urinary loss of salt and water [27].

It has long been established that accumulation of luminal Cl− depolarizes the membrane, and thereby, inhibits the apical Na+ channel ENaC [58]. Several studies also have demonstrated that pathological increases in Cl− reabsorption are associated with diseases, such as pseudo-hypoaldosteronism type II (PHA-II) or Gordon's syndrome.

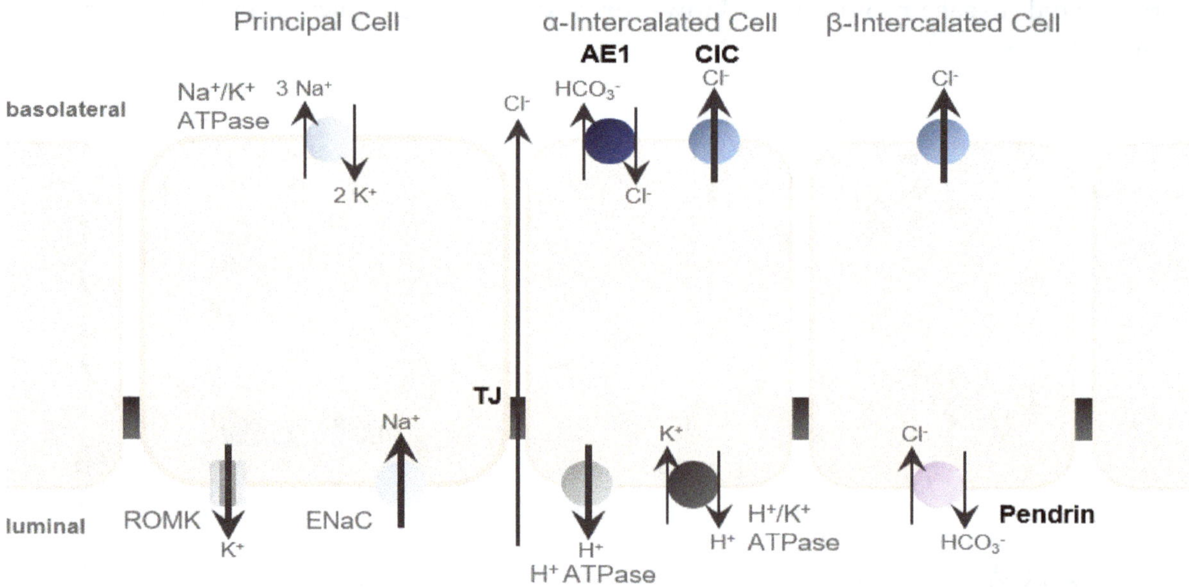

Figure 3. Chloride transport in the renal collecting duct. Chloride is transported via the paracellular route mediated by tight junctions (TJs), as well as, via the transcellular route using transporters and channels in α- and β-intercalated cells (depicted in bold).

PHA-II is a rare Mendelian syndrome leading to hypertension, hyperkalemia, and metabolic acidosis [59]. These symptoms are the exact opposite of the Cldn-4 and Cldn-8 knockout phenotypes in mice [23,29].

Interestingly, a long-known causative gene for PHA-II is WNK4 [60], and WNK4 can regulate Cl− conductance presumably by phosphorylation of Cldn-4 and Cldn-8 [50]. The PHA-II-causing mutation of WNK4 increases paracellular Cl− permeability in vitro [49,50]. Consistent with the proposed Cl− pore activity of Cldn-4 and Cldn-8, this is an additional indicator of an activating effect of claudin phosphorylation by WNK4. Furthermore, a recent study demonstrates that WNK4 is overexpressed in Cldn-7-deficient cultured, collecting duct cells [31]. Conversely, the PHA-II mutant of WNK4 is associated with increased Cldn-7 phosphorylation and enhanced paracellular Cl− conductivity [26]. This adds a new aspect and raises the possibility that claudins might act both up- and down-stream of WNK4 to regulate paracellular transport.

Another gene associated with PHA-II is KLHL3, encoding Kelch-like protein 3 [61,62]. Interestingly, KLHL3 normally induces ubiquitination and degradation of Cldn-8, while disease-associated mutations of KLHL3 abolish the interaction of KLHL3 with Cldn-8 [29]. In addition, KLHL3 leads to the ubiquitination of WNK4. In line with PHA-II symptoms, loss of KLHL3 increases Cl- permeability in vivo [29], which may contribute to the disease phenotype.

7. Conclusions

Collecting duct epithelia express claudins enforcing a high TER such as Cldn-3, -4, -7, -8, and -19, consistent with a demand for strong epithelial barrier function in the presence of steep transepithelial gradients. Based on our current knowledge, these claudins are thought to either support a tight barrier in general or to act as a cation barrier and/or an anion channel.

Consequently, the renal collecting duct is comprised of an especially tight epithelial barrier compared to other, more upstream segments of the nephron. As a result, the collecting duct lumen and interstitium are strictly separated. Nevertheless, the TJs of the collecting duct exhibit a regulated paracellular permeability for ions such as Cl-. Hence, the collecting duct TJ participates in a range of physiological functions:

It allows for controlled paracellular transport of Cl- to the interstitium, which is important in the setting of Na+ reabsorption via ENaC and secretion of K+ via ROMK.

It prevents paracellular diffusion and back flux of osmolytes into the urine and thus promotes the formation of steep gradients across the epithelium, which are necessary to drive electrolyte-free water reabsorption.

Future research will be needed to elucidate further the precise mechanisms that regulate TJ properties and that mediate crosstalk between paracellular and transcellular transport processes and how these processes relate to renal pathophysiology.

Author Contributions: Writing—original draft preparation: J.L., K.M.S.-O.; Writing—review and editing: J.L., K.M.S.-O. All authors have read and agreed to the published version of the manuscript.

Abbreviations

AQP	Aquaporin
AVP	Arginine vasopressin
Cap1	Channel-activating protease 1
Cldn	Claudin
Cl-	Chloride
ENaC	Epithelial sodium channel
Grhl2	Grainyhead-like 2
K+	Potassium
KLHL3	Kelch-like protein 3
MDCK	Madin-Darby Canine Kidney
mIMCD3	Mouse inner medullary collecting duct
Na+	Sodium
NCC	Na+ Cl− cotransporter
PHA-II	Pseudohypoaldosteronism type II (also known as Gordon's syndrome)
ROMK	Renal outer medullary potassium channel
TER	Transepithelial resistance
TJ	Tight junction
V2R	Vasopressin type 2 receptor
WNK4	With no lysine kinases 4

References

1. Powell, D.W. Barrier function of epithelia. *Am. J. Phys.* **1981**, *241*, G275–G288. [CrossRef] [PubMed]
2. Farquhar, M.G.; Palade, G.E. Junctional complexes in various epithelia. *J. Cell Biol.* **1963**, *17*, 375–412. [CrossRef] [PubMed]
3. Denker, B.M.; Sabath, E. The biology of epithelial cell tight junctions in the kidney. *J. Am. Soc. Nephrol.* **2011**, *22*, 622–625. [CrossRef] [PubMed]
4. Furuse, M.; Fujita, K.; Hiiragi, T.; Fujimoto, K.; Tsukita, S. Claudin-1 and -2: Novel integral membrane proteins localizing at tight junctions with no sequence similarity to occludin. *J. Cell Biol.* **1998**, *141*, 1539–1550. [CrossRef]
5. Colegio, O.R.; Van Itallie, C.M.; McCrea, H.J.; Rahner, C.; Anderson, J.M. Claudins create charge-selective channels in the paracellular pathway between epithelial cells. *Am. J. Physiol. Cell Physiol.* **2002**, *283*, C142–C147. [CrossRef]
6. Schneeberger, E.E.; Robert, D.L. The tight junction: A multifunctional complex. *Am. J. Physiol. Cell Physiol.* **2004**, *286*, C1213–C1228. [CrossRef]
7. Diamond, J.M. The epithelial junction: Bridge, gate, and fence. *Physiologist* **1977**, *20*, 10–18.
8. Yu, A.S. Claudins and the kidney. *J. Am. Soc. Nephrol.* **2015**, *26*, 11–19. [CrossRef]
9. Furuse, M.; Sasaki, H.; Tsukita, S. Manner of Interaction of Heterogeneous Claudin Species within and Between Tight Junction Strands. *J. Cell Biol.* **1999**, *147*, 891–903. [CrossRef]

10. Furuse, M.; Sasaki, H.; Fujimoto, K.; Tsukita, S. A Single Gene Product, Claudin-1 or -2, Reconstitutes Tight Junction Strands and Recruits Occludin in Fibroblasts. *J. Cell Biol.* **1998**, *143*, 391–401. [CrossRef]
11. Pearce, D.; Soundararajan, R.; Trimpert, C.; Kashlan, O.B.; Deen, P.M.T.; Kohan, D.E. Collecting duct principal cell transport processes and their regulation. *Clin. J. Am. Soc. Nephrol.* **2015**, *10*, 135–146. [CrossRef]
12. Rao, R.; Bhalla, V.; Pastor-Soler, N.M. Intercalated Cells of the Kidney Collecting Duct in Kidney Physiology. *Semin. Nephrol.* **2019**, *39*, 353–367. [CrossRef] [PubMed]
13. Wang, W.H.; Giebisch, G. Regulation of potassium (K) handling in the renal collecting duct. *Pflug. Arch.* **2009**, *458*, 157–168. [CrossRef] [PubMed]
14. Krause, G.; Winkler, L.; Mueller, S.L.; Haseloff, R.F.; Piontek, J.; Blasig, I.E. Structure and function of claudins. *Biochim. Biophys. Acta* **2008**, *1778*, 631–645. [CrossRef] [PubMed]
15. Simon, D.B.; Lu, Y.; Choate, K.A.; Velazquez, H.; Al-Sabban, E.; Praga, M.; Casari, G.; Bettinelli, A.; Colussi, G.; Rodriguez-Soriano, J.; et al. Paracellin-1, a Renal Tight Junction Protein Required for Paracellular Mg2+Resorption. *Science* **1999**, *285*, 103–106. [CrossRef] [PubMed]
16. Kiuchi-Saishin, Y.; Gotoh, S.; Furuse, M.; Takasuga, A.; Tano, Y.; Tsukita, S. Differential Expression Patterns of Claudins, Tight Junction Membrane Proteins, in Mouse Nephron Segments. *J. Am. Soc. Nephrol.* **2002**, *13*, 875–886. [PubMed]
17. Li, W.Y.; Huey, C.L.; Yu, A.S.L. Expression of claudin-7 and -8 along the mouse nephron. *Am. J. Physiol. Ren. Physiol.* **2004**, *286*, F1063–F1071. [CrossRef]
18. Ziemens, A.; Sonntag, S.R.; Wulfmeyer, V.C.; Edemir, B.; Bleich, M.; Himmerkus, N. Claudin 19 Is Regulated by Extracellular Osmolality in Rat Kidney Inner Medullary Collecting Duct Cells. *Int. J. Mol. Sci.* **2019**, *20*, 4401. [CrossRef]
19. Milatz, S.; Krug, S.M.; Rosenthal, R.; Gunzel, D.; Muller, D.; Schulzke, J.D.; Amasheh, S.; Fromm, M. Claudin-3 acts as a sealing component of the tight junction for ions of either charge and uncharged solutes. *Biochim. Biophys. Acta* **2010**, *1798*, 2048–2057. [CrossRef]
20. Coyne, C.B.; Gambling, T.M.; Boucher, R.C.; Carson, J.L.; Johnson, L.G. Role of claudin interactions in airway tight junctional permeability. *Am. J. Physiol. Lung Cell Mol. Physiol.* **2003**, *285*, L1166–L1178. [CrossRef]
21. Schröder, K. Generierung und Charakterisierung ei-nes Claudin-3-defizienten Mausmodells. *Humboldt-Universität zu Berlin, Mathematisch-Naturwissenschaftliche Fakultät I.* 2013. Available online: https://edoc.hu-berlin.de/handle/18452/17423 (accessed on 23 December 2019).
22. Hou, J.; Renigunta, A.; Yang, J.; Waldegger, S. Claudin-4 forms paracellular chloride channel in the kidney and requires claudin-8 for tight junction localization. *Proc. Natl. Acad. Sci. USA* **2010**, *107*, 18010–18015. [CrossRef] [PubMed]
23. Gong, Y.; Yu, M.; Yang, J.; Gonzales, E.; Perez, R.; Hou, M.; Tripathi, P.; Hering-Smith, K.S.; Hamm, L.L.; Hou, J. The Cap1-claudin-4 regulatory pathway is important for renal chloride reabsorption and blood pressure regulation. *Proc. Natl. Acad. Sci. USA* **2014**, *111*, E3766–E3774. [CrossRef] [PubMed]
24. Alexandre, M.D.; Lu, Q.; Chen, Y.H. Overexpression of claudin-7 decreases the paracellular Cl- conductance and increases the paracellular Na+ conductance in LLC-PK1 cells. *J. Cell Sci.* **2005**, *118*, 2683–2693. [CrossRef] [PubMed]
25. Hou, J.; Gomes, A.S.; Paul, D.L.; Goodenough, D.A. Study of claudin function by RNA interference. *J. Biol. Chem.* **2006**, *281*, 36117–36123. [CrossRef]
26. Tatum, R.; Zhang, Y.; Lu, Q.; Kim, K.; Jeansonne, B.G.; Chen, Y.H. WNK4 phosphorylates ser(206) of claudin-7 and promotes paracellular Cl(-) permeability. *FEBS Lett.* **2007**, *581*, 3887–3891. [CrossRef]
27. Tatum, R.; Zhang, Y.; Salleng, K.; Lu, Z.; Lin, J.J.; Lu, Q.; Jeansonne, B.G.; Ding, L.; Chen, Y.H. Renal salt wasting and chronic dehydration in claudin-7-deficient mice. *Am. J. Physiol. Ren. Physiol.* **2010**, *298*, F24–F34. [CrossRef]
28. Amasheh, S.; Milatz, S.; Krug, S.M.; Bergs, M.; Amasheh, M.; Schulzke, J.D.; Fromm, M. Na+ absorption defends from paracellular back-leakage by claudin-8 upregulation. *Biochem. Biophys. Res. Commun.* **2009**, *378*, 45–50. [CrossRef]
29. Gong, Y.; Wang, J.; Yang, J.; Gonzales, E.; Perez, R.; Hou, J. KLHL3 regulates paracellular chloride transport in the kidney by ubiquitination of claudin-8. *Proc. Natl. Acad. Sci. USA* **2015**, *112*, 4340–4345. [CrossRef]
30. Haddad, N.; El Andalousi, J.; Khairallah, H.; Yu, M.; Ryan, A.K.; Gupta, I.R. The tight junction protein claudin-3 shows conserved expression in the nephric duct and ureteric bud and promotes tubulogenesis in vitro. *Am. J. Physiol. Ren. Physiol.* **2011**, *301*, F1057–F1065. [CrossRef]

31. Fan, J.; Tatum, R.; Hoggard, J.; Chen, Y.H. Claudin-7 Modulates Cl(-) and Na(+) Homeostasis and WNK4 Expression in Renal Collecting Duct Cells. *Int. J. Mol. Sci.* **2019**, *20*, 3798. [CrossRef]
32. Alexandre, M.D.; Jeansonne, B.G.; Renegar, R.H.; Tatum, R.; Chen, Y.H. The first extracellular domain of claudin-7 affects paracellular Cl- permeability. *Biochem. Biophys. Res. Commun.* **2007**, *357*, 87–91. [CrossRef] [PubMed]
33. Van Itallie, C.; Rahner, C.; Anderson, J.M. Regulated expression of claudin-4 decreases paracellular conductance through a selective decrease in sodium permeability. *J. Clin. Investig.* **2001**, *107*, 1319–1327. [CrossRef] [PubMed]
34. Van Itallie, C.M.; Fanning, A.S.; Anderson, J.M. Reversal of charge selectivity in cation or anion-selective epithelial lines by expression of different claudins. *Am. J. Physiol. Ren. Physiol.* **2003**, *285*, F1078–F1084. [CrossRef] [PubMed]
35. Yu, A.S.; Enck, A.H.; Lencer, W.I.; Schneeberger, E.E. Claudin-8 expression in Madin-Darby canine kidney cells augments the paracellular barrier to cation permeation. *J. Biol. Chem.* **2003**, *278*, 17350–17359. [CrossRef]
36. Konrad, M.; Schaller, A.; Seelow, D.; Pandey, A.V.; Waldegger, S.; Lesslauer, A.; Vitzthum, H.; Suzuki, Y.; Luk, J.M.; Becker, C.; et al. Mutations in the Tight-Junction Gene Claudin 19 (CLDN19) Are Associated with Renal Magnesium Wasting, Renal Failure, and Severe Ocular Involvement. *Am. J. Hum. Genet.* **2006**, *79*, 949–957. [CrossRef]
37. Angelow, S.; El-Husseini, R.; Kanzawa, S.A.; Yu, A.S. Renal localization and function of the tight junction protein, claudin-19. *Am. J. Physiol. Ren. Physiol.* **2007**, *293*, F166–F177. [CrossRef]
38. Hou, J.; Renigunta, A.; Konrad, M.; Gomes, A.S.; Schneeberger, E.E.; Paul, D.L.; Waldegger, S.; Goodenough, D.A. Claudin-16 and claudin-19 interact and form a cation-selective tight junction complex. *J. Clin. Investig.* **2008**, *118*, 619–628. [CrossRef]
39. Rosenthal, R.; Milatz, S.; Krug, S.M.; Oelrich, B.; Schulzke, J.D.; Amasheh, S.; Gunzel, D.; Fromm, M. Claudin-2, a component of the tight junction, forms a paracellular water channel. *J. Cell Sci.* **2010**, *123*, 1913–1921. [CrossRef]
40. Knepper, M.A.; Kwon, T.H.; Nielsen, S. Molecular physiology of water balance. *N. Engl. J. Med.* **2015**, *372*, 1349–1358. [CrossRef]
41. Ranieri, M.; Di Mise, A.; Tamma, G.; Valenti, G. Vasopressin-aquaporin-2 pathway: Recent advances in understanding water balance disorders. *F1000Res* **2019**, *8*, F1000. [CrossRef]
42. Loonen, A.J.M.; Knoers, N.V.A.M.; van Os, C.H.; Deen, P.M.T. Aquaporin 2 Mutations in Nephrogenic Diabetes Insipidus. *Semin. Nephrol.* **2008**, *28*, 252–265. [CrossRef] [PubMed]
43. Hinze, C.; Ruffert, J.; Walentin, K.; Himmerkus, N.; Nikpey, E.; Tenstad, O.; Wiig, H.; Mutig, K.; Yurtdas, Z.Y.; Klein, J.D.; et al. GRHL2 Is Required for Collecting Duct Epithelial Barrier Function and Renal Osmoregulation. *J. Am. Soc. Nephrol.* **2018**, *29*, 857–868. [CrossRef] [PubMed]
44. Aue, A.; Hinze, C.; Walentin, K.; Ruffert, J.; Yurtdas, Y.; Werth, M.; Chen, W.; Rabien, A.; Kilic, E.; Schulzke, J.D.; et al. A Grainyhead-Like 2/Ovo-Like 2 Pathway Regulates Renal Epithelial Barrier Function and Lumen Expansion. *J. Am. Soc. Nephrol.* **2015**, *26*, 2704–2715. [CrossRef] [PubMed]
45. Sansom, S.C.; Weinman, E.J.; O'Neil, R.G. Microelectrode assessment of chloride-conductive properties of cortical collecting duct. *Am. J. Physiol. Cell Physiol.* **1984**, *274*, F291–F302. [CrossRef] [PubMed]
46. Fine, D.; Meiselas, L.E.; Auerbach, T. The effect of acute hypovolemia on the release of "aldosterone" and on the renal excretion of sodium. *J. Clin. Investig.* **1958**, *37*, 232–243. [CrossRef] [PubMed]
47. Narikiyo, T.; Kitamura, K.; Adachi, M.; Miyoshi, T.; Iwashita, K.; Shiraishi, N.; Nonoguchi, H.; Chen, L.-M.; Chai, K.X.; Chao, J.; et al. Regulation of prostasin by aldosterone in the kidney. *J. Clin. Investig.* **2002**, *109*, 401–408. [CrossRef]
48. Vallet, V.; Chraibi, A.; Gaeggeler, H.-P.; Horisberger, J.-D.; Rossier, B.C. An epithelial serine protease activates the amiloride-sensitive sodium channel. *Nature* **1997**, *389*, 607–610. [CrossRef]
49. Yamauchi, K.; Rai, T.K.K.; Sohara, E.; Suzuki, T.; Itoh, T.; Suda, S.; Hayama, A.; Sasaki, S.; Uchida, S. Disease-causing mutant WNK4 increases paracellular chloride permeability and phosphorylates claudins. *Proc. Natl. Acad. Sci. USA* **2004**, *101*, 4690–4694. [CrossRef]
50. Kahle, K.T.; MacGregor, G.G.; Wilson, F.H.; Van Hoek, A.N.; Brown, D.; Ardito, T.; Kashgarian, M.; Giebisch, G.; Hebert, S.C.; Boulpaep, E.L.; et al. Paracellular Cl- permeability is regulated by WNK4 kinase: Insight into normal physiology and hypertension. *Proc. Natl. Acad. Sci. USA* **2004**, *101*, 14877–14882. [CrossRef]

51. Le Moellic, C.; Boulkroun, S.; Gonzalez-Nunez, D.; Dublineau, I.; Cluzeaud, F.; Fay, M.; Blot-Chabaud, M.; Farman, N. Aldosterone and tight junctions: Modulation of claudin-4 phosphorylation in renal collecting duct cells. *Am. J. Physiol. Cell Physiol.* **2005**, *289*, C1513–C1521. [CrossRef]
52. Ring, A.M.; Cheng, S.X.; Leng, Q.; Kahle, K.T.; Rinehart, J.; Lalioti, M.D.; Volkman, H.M.; Wilson, F.H.; Hebert, S.C.; Lifton, R.P. WNK4 regulates activity of the epithelial Na+ channel in vitro and in vivo. *Proc. Natl. Acad. Sci. USA* **2007**, *104*, 4020–4024. [CrossRef]
53. Kahle, K.T.; Wilson, F.H.; Leng, Q.; Lalioti, M.D.; O'Connell, A.D.; Dong, K.; Rapson, A.K.; MacGregor, G.G.; Giebisch, G.; Hebert, S.C.; et al. WNK4 regulates the balance between renal NaCl reabsorption and K+ secretion. *Nat. Genet.* **2003**, *35*, 372–376. [CrossRef]
54. Siragy, H.M.; Awad, A.; Abadir, P.; Webb, R. The angiotensin II type 1 receptor mediates renal interstitial content of tumor necrosis factor-alpha in diabetic rats. *Endocrinology* **2003**, *144*, 2229–2233. [CrossRef]
55. Hollenberg, N.K.; Stevanovic, R.; Agarwal, A.; Lansang, M.C.; Price, D.A.; Laffel, L.M.; Williams, G.H.; Fisher, N.D. Plasma aldosterone concentration in the patient with diabetes mellitus. *Kidney Int.* **2004**, *65*, 1435–1439. [CrossRef]
56. Molina-Jijon, E.; Rodriguez-Munoz, R.; Namorado Mdel, C.; Pedraza-Chaverri, J.; Reyes, J.L. Oxidative stress induces claudin-2 nitration in experimental type 1 diabetic nephropathy. *Free Radic. Biol. Med.* **2014**, *72*, 162–175. [CrossRef]
57. Molina-Jijon, E.; Rodriguez-Munoz, R.; Gonzalez-Ramirez, R.; Namorado-Tonix, C.; Pedraza-Chaverri, J.; Reyes, J.L. Aldosterone signaling regulates the over-expression of claudin-4 and -8 at the distal nephron from type 1 diabetic rats. *PLoS ONE* **2017**, *12*, e0177362. [CrossRef]
58. Canessa, C.M.; Schild, L.; Buell, G.; Thorens, B.; Gautschi, I.; Horisberger, J.-D.; Rossier, B.C. Amiloride-sensitive epithelial Na+ channel is made of three homologous subunits. *Nature* **1994**, *367*, 463–467. [CrossRef]
59. Gordon, R.D. Syndrome of hypertension and hyperkalemia with normal glomerular filtration rate. *Hypertension* **1986**, *8*, 93–102. [CrossRef]
60. Wilson, F.H.; Disse-Nicodeme, S.; Choate, K.A.; Ishikawa, K.; Nelson-Williams, C.; Desitter, I.; Gunel, M.; Milford, D.V.; Lipkin, G.W.; Achard, J.-M.; et al. Human hypertension caused by mutations in WNK kinase. *Science* **2001**, *293*, 1107–1112. [CrossRef]
61. Boyden, L.M.; Choi, M.; Choate, K.A.; Nelson-Williams, C.J.; Farhi, A.; Toka, H.R.; Tikhonova, I.R.; Bjornson, R.; Mane, S.M.; Colussi, G.; et al. Mutations in kelch-like 3 and cullin 3 cause hypertension and electrolyte abnormalities. *Nature* **2012**, *482*, 98–102. [CrossRef]
62. Louis-Dit-Picard, H.; Barc, J.; Trujillano, D.; Miserey-Lenkei, S.; Bouatia-Naji, N.; Pylypenko, O.; Beaurain, G.; Bonnefond, A.; Sand, O.; Simian, C.; et al. KLHL3 mutations cause familial hyperkalemic hypertension by impairing ion transport in the distal nephron. *Nat. Genet.* **2012**, *44* (Suppl. S1–S3), 456–460. [CrossRef]

3

Super-Resolution Imaging of Tight and Adherens Junctions: Challenges and Open Questions

Hannes Gonschior [1], Volker Haucke [1,2] and Martin Lehmann [1,*]

1 Leibniz-Forschungsinstitut für Molekulare Pharmakologie (FMP), 13125 Berlin, Germany; gonschior@fmp-berlin.de (H.G.); haucke@fmp-berlin.de (V.H.)
2 Faculty of Biology, Chemistry, Pharmacy, Freie Universität Berlin, 14195 Berlin, Germany
* Correspondence: mlehmann@fmp-berlin.de

Abstract: The tight junction (TJ) and the adherens junction (AJ) bridge the paracellular cleft of epithelial and endothelial cells. In addition to their role as protective barriers against bacteria and their toxins they maintain ion homeostasis, cell polarity, and mechano-sensing. Their functional loss leads to pathological changes such as tissue inflammation, ion imbalance, and cancer. To better understand the consequences of such malfunctions, the junctional nanoarchitecture is of great importance since it remains so far largely unresolved, mainly because of major difficulties in dynamically imaging these structures at sufficient resolution and with molecular precision. The rapid development of super-resolution imaging techniques ranging from structured illumination microscopy (SIM), stimulated emission depletion (STED) microscopy, and single molecule localization microscopy (SMLM) has now enabled molecular imaging of biological specimens from cells to tissues with nanometer resolution. Here we summarize these techniques and their application to the dissection of the nanoscale molecular architecture of TJs and AJs. We propose that super-resolution imaging together with advances in genome engineering and functional analyses approaches will create a leap in our understanding of the composition, assembly, and function of TJs and AJs at the nanoscale and, thereby, enable a mechanistic understanding of their dysfunction in disease.

Keywords: super-resolution microscopy; structured illumination microscopy; stimulated emission depletion; single molecule localization microscopy; adherens junction; tight junction

1. Introduction

1.1. *Adherens Junctions and Tight Junctions*

Epithelial and endothelial cells line multiple organs to segregate internal and external compartments. Cell contacts are formed on the one hand by the adherens junction (AJ) which provides mechanical adhesion, and on the other by the tight junction (TJ) which primarily acts as a barrier to solutes and water and as a fence that separates the apical and basolateral plasma membrane domains. However, some TJ proteins form paracellular channels for small cations, anions, or water [1].

Both junctions are characterized by extracellularly interacting transmembrane proteins that bind via intracellular adapter proteins to the cytoskeleton [2].

TJs were first observed by transmission electron microscopy (TEM) as apical membrane contacts extending over 200–500 nm with electron-dense areas forming a continuous belt-like attachment by fusing the outer membrane leaflets of two neighboring cells into a ≈3 nm thick structure [3]. Later, by using freeze fracture electron microscopy (FFEM) their structure was described as an interconnecting strand network with a general distance between its single strands of 30–80 nm in epithelial cells and tissues [4]. The first protein components of TJs were identified as Zonula occludens (ZO-1) [5] and later the transmembrane proteins occludin [6], junctional adhesion molecule A (JAM-A) [7], and claudins [8].

Occludin, together with tricellulin and marvel-D3 form the tetraspan family of TJ-associated Marvel Domain Proteins (TAMPs) [9]. Claudins regulate interactions between the TAMPs and vice versa [10]. The mammalian genome codes for 27 claudin proteins with specific functions and tissue-specific expression patterns. Zonula occludens 1 and 2 (ZO-1, -2) turned out to be the major adapter proteins on the cytosolic side of TJs [5,11,12]. They act via various interaction domains as scaffolds for occludin, claudins, and other integral TJ proteins, providing their linkage to the actin cytoskeleton [13–16]. For JAM-A it could be recently shown that its assembly enables epithelial cell polarization. JAM-A is a single transmembrane protein that regulates membrane apposition. In contrast, tetra spanning proteins like claudins form via their two extracellular loops paracellular barriers or channels. So far it could be shown that claudin-2 [17] and claudin-15 [18] promote increased water permeability and both claudins, as well as claudin-10b, also promote increased cation permeability [19–21]. Whereas, claudin-10a [21] and claudin-17 [22] contribute to a higher anion permeability. It is also known that some specific claudin combinations can support paracellular ion permeability, e.g., claudin-16/-19 for cations [23] and claudin-4/-8 for small anions [24]. As recently described a knockout of multiple claudins results in increased ion permeability but does not affect macromolecular diffusion or polarity in epithelial cells indicating that other TJ-related proteins must be involved [25].

The initial strand and TJ-meshwork formation is highly dependent on claudins [26] which form via *cis*- and *trans*-interactions arrays of TJ strands which are about 10 nm in diameter. TJ strands assemble to a complex TJ meshwork at bicellular junctions. A specialized site of the TJ where macromolecules can pass has been identified to be the tricellular TJ (tTJ), the location where three cells meet [27]. The proteins of the tTJ are tricellulin [28], and three singe-transmembrane proteins, angulin-1 to -3 [29]. The tTJ is virtually impermeable at high tricellulin expression for macromolecules but becomes permeable for it at low expression as observed in the inflammatory bowel disease ulcerative colitis [30].

The AJ mediates mechanical cell–cell interactions during development and in differentiated tissues. Intercellular adhesion is mediated by *cis*- and *trans*-association of extracellular domains of cadherins coded by 20 different genes in vertebrates. Tissue specific cadherins are epithelial cadherins (E-cad), neuronal cadherin (N-cad), and vascular endothelial cadherins (VE-cad). The *trans*-interactions of E-cad extracellular domains are Ca^{2+} dependent. Intracellularly E-cad are linked via catenin [31–33], vinculin [34,35], and other actin-binding proteins to the actin cytoskeleton [36]. Two forms of AJ have been observed, first linear AJ linked to the circumferential actin belt in mature epithelial sheets and second punctuate AJ at free edges with direct actin linkage. Actin bundles in linear AJ can be contracted by actomyosins to produce planar-polarized apical constrictions during tissue remodeling. Rho GTPases, formins, and myosins regulate actin polymerization and contraction. Cadherin can be removed from the surface via clathrin mediated endocytosis (CME) that counterbalances actomyosin mediated constriction in neighboring cells [37]. Therefore, AJs notably integrate both cellular adhesion and cortical tension sensing [38] important for tissue homeostasis and remodeling during development.

In order to mediate cell adhesion and barrier formation, both AJ and TJ form dense and complex multiprotein structures. Some of their characteristics like formation of the TJ meshwork have been visualized by electron microscopy (EM) [4]. However, due to the inherent limitations of EM with respect to defining molecular and structural dynamics many key questions remain unsolved yet: for example, how are the individual protein components like channel and barrier forming claudins of TJs or the different components of AJs organized within the junction and how does their interaction contribute to the overall functionality? How are junctions remodeled, e.g., by strand reformation, endocytosis, during cell division, or upon injury? To answer these questions, it is indispensable to understand the nanoscale organization of the molecular components of TJs (claudins, TAMPs, JAMs, ZOs) and AJs (cadherins, catenins, vinculin, actin-binding proteins) in fixed cells and tissue and, most importantly, in living preparations. Such an approach requires multi-color imaging at nanoscale resolution by super-resolution microscopy (SRM). SRM offers the possibility to perform quantitative analysis of nanoclusters of AJs or measurements of the strand and meshwork organization of TJs.

Recent studies using SRM have begun to uncover the nanoscale organization and dynamics of TJs [13,39,40] and AJs and their components [41–43]. Here we provide an overview on recent advances and limitations in SRM, summarize important new insight in junction research from recent studies and provide an outlook for future research.

1.2. Resolution Limit

TJs were initially identified by electron microscopy analysis of ultrathin sections of resin embedded cells and tissues as µm-sized apical membrane contacts with characteristic electron dense areas. These membrane contacts appeared to act as a barrier to organic dyes and showed specific membrane kissing points [3]. Using freeze fracture sample preparation these close membrane appositions could be resolved as fence-like intramembrane strands with characteristic mesh sizes of ≈30–80 nm and strand morphology [4]. Using immunogold labelling on resin and freeze fracture samples the early biochemically identified components of TJs, namely claudins and occludin, were shown to be localized to these meshworks and specifically copolymerize in strands [6,8]. Since FFEM involves difficult sample preparation and the immunogold labelling of chemically fixed samples even with overexpressed proteins and with high affinity antibodies is often incomplete, a dense and specific nanoscale distribution of multiple TJ and AJ proteins was not achieved so far. Fluorescence microscopy on the other hand offers high density labelling and highly sensitive detection of AJ and TJ components in intact tissues and living cells. Generally, the resolution of fluorescence microscopes is limited by Abbe's law to 200–250 nm in the lateral and 500–700 nm in axial direction [44]. Therefore, TJs in polarized epithelia often appear as diffraction-limited spots in x-z scans. Several fluorescence microscopy techniques have overcome the resolution limit by either spatially or temporally controlling fluorescence emission of which structured illumination microscopy (SIM), time-gated stimulated emission depletion (gSTED), and single molecule localization microscopy (SMLM) have turned out to be the three main techniques to visualize biological specimens at the nanoscale (Figure 1).

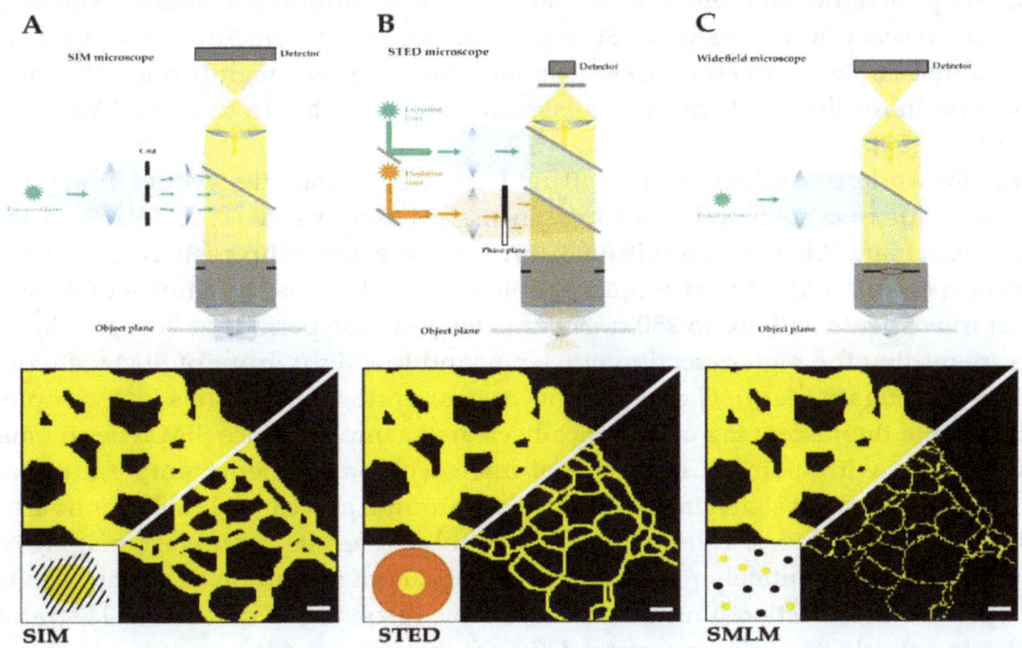

Figure 1. Schematic depiction of the main three SRM techniques. It is shown for each technique a detailed construction of the microscope and the light path as well as a schematic comparison of a reconstituted TJ meshwork resolved by structured illumination microscopy (SIM) (resolution limit: 100 nm) (**A**), time-gated stimulated emission depletion (gSTED) (resolution limit: 40 nm) (**B**), and single molecule localization microscopy (SMLM) (resolution limit: 20 nm) (**C**) in reference to confocal imaged TJ meshwork (resolution limit: 250 nm). Scale bars: 200 nm.

1.3. Structured Illumination Microscopy (SIM)

In structured illumination microscopy (SIM) spatially restricted fluorescence emissions are produced by stripe-patterned wide-field excitation (see Figure 1A) [45]. Multiple camera images are collected from an excitation pattern with different phases and orientations. The collected images contain high resolution information as moiré fringes are formed from the fluorescent sample illuminated by the stripe illumination. The high-resolution information from SIM images can be extracted from Fourier-transformed images in a process called deconvolution. Several illumination modes have been used to create SIM excitation patterns, e.g., in 3D [45], in a total internal reflection mode [46,47] at a grazing incidence angle [48] or in a lattice light sheet [49]. The final SIM images show a lateral resolution of typically 100–130 nm and an axial resolution of 100–250 nm that depends on the wavelength of emission light, the numerical aperture (NA) of the objective, the distance from the coverslip, and the illumination mode. Using an objective with the highest NA or nonlinear SIM with patterned activation resolutions of 84 and 48 nm, respectively, have been achieved [50].

SIM offers key advantages like the use of standard organic fluorophores and fluorescent proteins, low excitation power, and fast imaging speeds up to several hundred frames per second [48]. Since the image reconstruction relies heavily on mathematical processing the point-spread function (PSF) must be determined carefully, and low intensity raw images or high background can produce substantial reconstruction artefacts. Commercial wide-field SIM systems are offered by Zeiss (Elyra S1 and 7), GE Healthcare (Deltavision OMX), and Nikon (N-SIM), while lattice light sheet SIM is offered by 3i.

Point-scanning SIM or Re-Scan Microscopy uses a confocal microscope for scanning an excitation spot over the sample and collects the fluorescence emission directly on a multi-array detector in the absence of a pinhole element in the conjugate plane [51]. The signal at the central pixel and the signal distribution over multiple arrays represent accurately the PSF of the microscope and can be used to increase the resolution by a factor of 1.4 using linear deconvolution. The total signal intensity on the array is then reassigned to the scanned pixel. Since in the point-scanning SIM the pinhole of the confocal microscope is removed more light is collected, the resolution and signal-to-noise is increased by working with an accurately measured PSF and linear deconvolution. Alternatively, in a rescanning mode a camera detects the scanned emission signal with a higher magnification that also results in a 1.4-fold increase in resolution. Commercial systems are available from Zeiss (Airy Detector) and Confocal.nl (RCM) [52].

Standard SIM works best within the first 10 μm from the coverslip that would limit the observation of native TJs and AJs to certain cell types or thin tissues sections. If TJs and AJs form at flat cell contacts, e.g., punctuate AJs that are within 1 μm from the coverslip restricted excitation in total internal refection (TIRF) and GI-SIM would enable to visualize fast dynamics of TJ strands or AJ protein cluster movements with up to 250 fr/s over extended time periods at very low light doses [48]. Additionally, regarding the high detection efficiency and low light input of SIM we would see it as an ideal tool to investigate the AJ/TJ components at near endogenous levels for example by CrispR mediated knock-in of fluorescent tag or in optically cleared samples. Since SIM relies on mathematical image reconstruction with a strong influence of microscope alignment, temperature, and sample parameters, a high fluorescent signal intensity, low background, and careful calibrations are required to produce artefact-free SIM images. Overall SIM enables robust, sensitive, and fast imaging with highest contrast, but only achieves routinely resolution down to 100 nm resolution [53]. This resolution that is approximately half the diffraction limit would not enable resolving of individual TJ strands, small TJ meshes, cadherin nanoclusters, or the cortical actin organization at AJs.

1.4. Stimulated Emission Depletion (STED)

In stimulated emission depletion (STED) spatially restricted fluorescence emissions are produced in the center of a coaligned focused excitation spot surrounded by a ring-shaped depletion pattern (see Figure 1B) [54]. The emissions from the diffraction-limited excitation spot is depleted by the ring-shaped depletion focus except for the center region. By scanning the coaligned focused spots and detecting the emission in a confocal microscope, images with a lateral resolution of about 50 nm can be obtained by optimizing imaging conditions such as excitation and depletion energies. The confocal configuration eliminates out of focus light and offers both high contrast and optical sectioning. Pulsed excitation and time-gated detection have enabled resolutions below 50 nm [55]. Using a two-objective configuration or 3D STED depletion patterns a lateral and axially similar resolution of 50–70 nm has been demonstrated [56]. The overall STED resolution is mostly limited by depletion laser power, alignment quality and, most importantly, by labelling density and the availability of bright and photostable dyes with an optimal 5%–10% normalized emission intensity at the depletion laser wavelength. Commercial STED microscopes can be purchased from Leica, Picoquant, or Abberior Instruments.

Since both TJ and AJ are formed by oligomeric transmembrane proteins the lower signal to noise effect of STED can be outweighed by higher label density, e.g., by claudins forming strands by *cis*-and *trans*-interactions. Low abundant or under-labelled structures may not give enough signal for STED imaging. Multicolor STED is routinely performed with 2–3 channels. Therefore, STED imaging of TJs and AJs with optimal labelling [57,58] and optimized 2D and 3D cell culture conditions will show several TJ and AJ components as well as the actin cytoskeleton with <60 nm resolution deep inside cells [59]. Despite STED having the lowest signal to background ratio of all super-resolution techniques it produces raw images that usually do not require post-processing. During a practical comparison STED and SMLM gave similar resolutions on microtubules and vesicular structures [53]. Another advantage of STED is the possibility of imaging cellular structures in living cells and tissues with <50 nm resolution [58] and comparing the ultrastructural preservation after chemical fixation, permeabilization, and immune labelling [60]. Especially, the ultrastructure of tension-sensitive AJ, complex TJ strand morphologies, and the actin cytoskeleton could be altered by chemical fixation. Native AJs and TJs extend in the axial direction in polarized monolayers and are found several μm away from the glass surface. Here the optical sectioning of STED and the possibility to increase resolution in Z direction down to 70 nm would enable the visualization of TJ and AJ structures in a native context. For functional studies of nanoscale dynamics of AJ and TJ protein dynamics STED can be combined with fluorescence recovery after photobleaching (FRAP), fluorescence correlation spectroscopy (FCS), and fluorescence lifetime imaging microscopy (FLIM).

1.5. Single Molecule Localization Microscopy (SMLM)

Single Molecule Localization Microscopy (SMLM) increases resolution by exerting temporal control of fluorescence emission. Single molecule fluorescence emissions are detected from a sparse subset of fluorophores that show recovery from an inactive state (see Figure 1C) [61,62], photo activation [63], photo conversion, or short presence in the detection volume [64]. Depending on the brightness of the fluorescence signal the position of the single molecule can be determined with 2–50 nm precision [61,63,65]. The molecule positions are extracted from tens of thousands of images of sparse emitters, drift corrected, and plotted into a new super-resolution image. The SMLM image resolution is determined by both localization precision and localization density [66]. The limited photon output of single molecule emissions often requires strong laser excitation, sensitive detection by low noise cameras, and imaging in TIRF mode within 200 nm from the coverslip. Three-dimensional SMLM was achieved by astigmatic distortion of the PSF in z direction [67], by multi-plane detection [68] or an interferometric approach [69,70]. Both organic fluorophores and photoactivatable or blinking fluorescent proteins can be used as labels for SMLM. The movement of single biological molecules in cells could be analyzed by using tracking SMLM [71]. SMLM offers the highest lateral and axial resolution and can be performed with relatively simple setups when significant challenges of labelling density, sample drift, acquisition time, multicolor acquisition, and image analysis are overcome. A wide range of image analysis packages for 2D and 3D SMLM are available [72]. Commercial SMLM systems are offered by Leica (Leica GSD), Nikon (N-STORM), Zeiss (Zeiss Elyra P1) and Bruker (Vutara 350).

Similar to STED, the SMLM techniques offer diffraction unlimited resolution when high labelling densities can be achieved. Therefore, endogenous labelling of TJ components should be performed with polyclonal antibodies that can bind multiple epitopes. The maximal resolution of SMLM can be in the order of 10 nm that should be sufficient to resolve individual claudin strands or E-cad nanoclusters. Such high resolutions often require direct genetic labelling of proteins of interest with either photoactivatable proteins or self-labelling enzymatic tags. SMLM with photoactivatable fluorescent proteins especially in tissue is often limited by auto fluorescent and detection of rather dim fluorescence bursts and could therefore overestimate monomeric E-cad and underestimate clusters. Therefore, SMLM imaging should be performed close to the cover glass, e.g., in flat endothelial cells or in sparse cultures of epithelial cells that form flat cellular overlaps with TJ strands. Since flat portions of cellular contacts are often not within the typical depth of a TIRF, illumination of ≈200 nm SMLM should be performed with wide-field excitation and 3D SMLM. When 3D SMLM is used on AJ or TJ structures that are located several micrometers away from the glass coverslip the signal to noise ratio decreases that could result in more false-positive localization and a lower localization precision. The higher signal of SMLM in TIRF mode enables the analysis of individual AJ and TJ proteins formed in the lower plasma membrane, e.g., during early steps of polymerization, intracellular anchoring, or signal transduction outside of AJ and TJ. In order to reveal the nanoscale organization of multiple claudins in TJ meshworks or E-cad clusters together with intracellular AJ proteins multicolor SMLM is required. Major challenges for multicolor imaging are first to find spectrally different fluorophores with a similar performance as the widely used Alexa Fluor 647 dye with similar photon outputs, second harmonize buffer requirements of multiple fluorophores, and third register multicolor images after drift correction.

A general technical review [73] and comparison of SIM, STED, and SMLM can be found in [53]. In direct comparison both SMLM and STED gave the highest resolution of <50 nm compared to 110 nm for SIM, but SIM and SMLM offered the highest signal-to-noise ratio [53]. The principle and important considerations for TJ/AJ imaging are summarized in Figure 1, Figure 2 and Table 1.

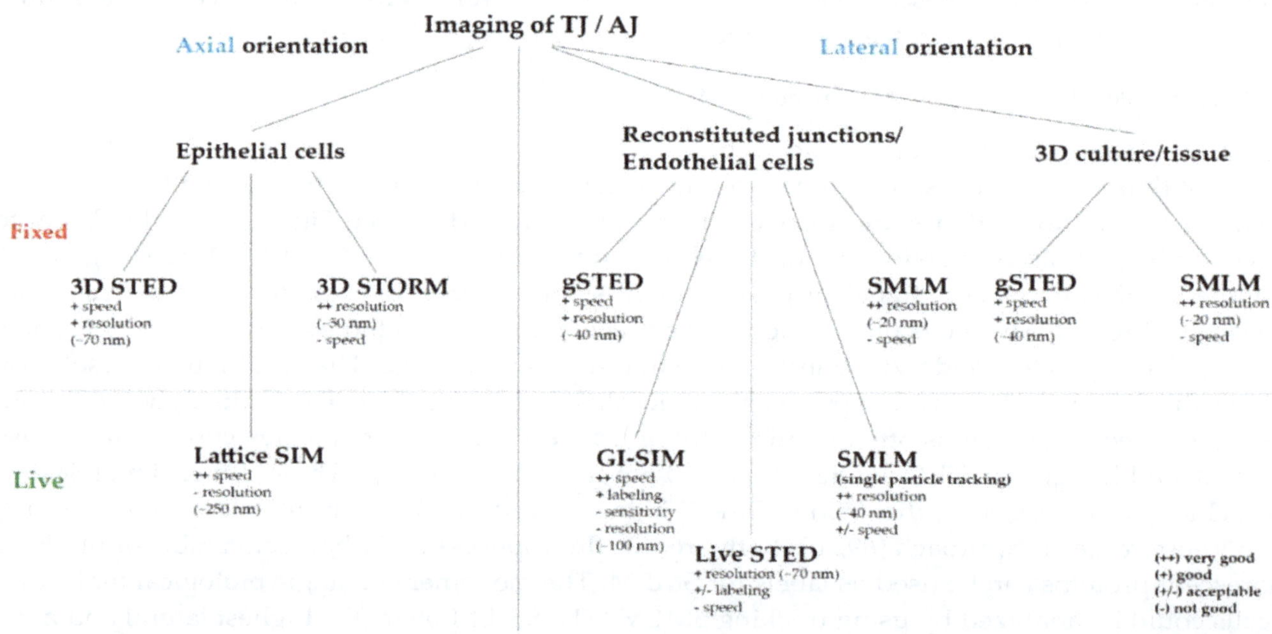

Figure 2. Practicable overview of the ideally used SRM technique for TJ and AJ imaging regarding the axial or lateral direction of the junction, the system (epi-, endothelial cells/fibroblasts/tissue), different labeling (fluorescent proteins/antibodies) and imaging (fixed/live) approaches.

Table 1. Overview and direct comparison of the key parameters of SIM, gSTED, and SMLM.

	SIM	gSTED	SMLM
Resolution limit	100 nm (lateral) 250 nm (axial)	40 nm (lateral) 70 nm (axial)	20 nm (lateral) 50 nm (axial)
Multi-color	4 colors	2–4 colors	2–4 colors
Sample preparation	Standard	Standard	Standard
Live imaging	Yes	Yes	Yes
Imaging speed	10 ms–10 s	10 s–5 min	10 min
Illumination power	10–100 W/cm^2	1–100 kW/cm^2	1–100 kW/cm^2

2. Results

2.1. *Super-Resolution Imaging of Adherens Junctions Reveals Nanoscale Clustering and Stratified Intracellular Organization*

AJs form at overlaps of endothelial cells and basolateral of TJs in epithelial cells and integrate tension sensing and adhesion remodeling [74]. In EM images AJs appear as close membrane appositions with an electron dense cytosolic plaque of actin [3] but freeze fracture images show no continuous strands of E-cad molecules. Instead, E-cad molecules were found in individual clusters using diffraction limited fluorescence microscopy [75].

At the nanoscale level endogenous E-cad in *Drosophila* embryos was found to form discrete nanometer-sized clusters by SMLM. The protein density estimations from fluorescence correlation spectroscopy calibrations and from single molecule counting were consistent with tightly packed E-cad molecules about eight nanometers apart from each other. Most E-cad was monomeric with varying cluster sizes that have been found to be regulated by dynamin-mediated endocytosis. Conversely, elimination of α-catenin or PAR3 reduced E-cad clustering, while interfering with action polymerization [41].

Similarly, when SMLM 3D-STORM was performed in epithelial cells with both bright organic dyes and photoactivatable proteins, highly homogenous cluster sizes of 50–60 nm with a fixed distance of ≈160 nm were identified [42]. E-cad clusters in epithelial cells contained six molecules similar to results in *Drosophila* embryos [41] and form independently of cadherin–cadherin interactions. Mutation in E-cad that interfered with *trans* or *cis*-interaction still produced clusters, albeit with a decreased molecular density. Coculture experiment found coexisting adhesive and non-adhesive E-cad clusters with similar size. Dual color SMLM revealed that adhesive and non-adhesive E-cad clusters are delimited by F-actin. Both the actin depolymerizing drug Latrunculin A and expression of a tail-less E-cad results in the formation of larger E-cad clusters. Therefore, SMLM revealed that E-cad forms a nanocluster surrounded by an actin fence that depends both on homophilic interactions and anchoring via the cytosolic tail. Even larger and more mature AJs are formed by groups of individual nanoclusters that can be linked to and stabilized by intracellular scaffold proteins [42]. Therefore, adhesion and tension sensing are mediated by small units of E-cad that are evenly spaced and can be finely regulated by incorporation into larger units and controlled by dynamin-dependent endocytosis.

How are the intracellular proteins of AJs organized? Using a combination of 3D PALM, surface-generated structured illumination together with biochemical perturbation on planarized biomimetic cadherin-based AJs the nanoscale architecture of AJ was analyzed [43]. Notably a plasma membrane-proximal cadherin-catenin compartment segregated from the actin cytoskeletal compartment, connected by an interface zone containing vinculin was found. The position of vinculin is determined by catenin and vinculin showed a tension and tyrosine phosphorylation dependent extended confirmation. The robust surface-generated structured illumination assay revealed the nanoscale changes of vinculin conformation in different cell types, under biochemical and pharmacological perturbations and could be verified by fluorescence resonance energy transfer

(FRET)-tension measurements. Therefore, vinculin integrates mechanical and chemical signals between cadherin and the actomyosin layer to control cell adhesion in cells and tissues [43].

Of note, a similar stratified nanoarchitecture and tension sensitive conformation of vinculin was previously found in focal adhesions (FA) [76]. Integrins form nanoclusters [77] while the tension sensing molecules paxillin and vinculin form linear and non-colocalizing areas within FA [78]. Vinculin FRET-based tension sensors revealed a high tension especially at small FA and that a general increase in cellular tension increases the size of FA [79].

2.2. Super-Resolution Imaging Shows Claudin Meshworks, Strand Dynamics, and Molecular Composition of the TJ

TJs are organized by claudin strands and meshworks that were revealed by FFEM of cell and tissue samples. Since EM requires strong fixation and often lacks molecular specificity SRM would be needed to confirm the nanoarchitecture of TJ in intact cells and tissues.

A first approach in this direction was done in 2003 when Sasaki et al. were able to image dynamics of claudin strands and whole TJ meshworks in claudin-1-GFP overexpressing L-fibroblasts using confocal microscopy. However, due to the previously described resolution limit a further analysis of multiple claudins and strand dynamics in denser not peripheral meshwork areas was not possible [80].

Van Itallie et al. reconstituted eGFP-claudin-2 strands in fibroblasts and imaged extended strands networks by live SIM with ≈120 nm resolution. Large meshes as well as single or double strand configurations and strand dynamics could be detected. Claudin-2 networks formed in the absence of claudin-2-ZO-1 binding and showed a higher mobility and failed to co-align with the actin cytoskeleton. Using multicolor SIM, a partial overlap between actin, ZO-1 and claudin-2 could be found, since actin and ZO-1 are also highly concentrated at AJ. Strand breaks and strand elongation occurred independently of ZO-1 binding, but preferentially at branch points. Branch points were also identified as polymerization sites of claudin and showed a weak enrichment of occludin. How other claudins copolymerize, associate with actin, ZO-1, and occludin are important questions for the nanostructure and the regulation of TJs. Since some claudins like claudin-3 form much smaller meshes in reconstituted TJ meshworks (e.g., Figure 3C) than claudin-2, super-resolution techniques with higher spatial resolution like SMLM and STED will help to resolve these questions [13].

Kaufmann et al. reconstituted TJs from C-terminally YFP tagged claudin-3 and claudin-5 in an epithelial cell line, devoid of endogenous claudin expression. SMLM with 50 nm structural resolution resolved meshworks and found two meshwork populations with ≈100/360 nm (claudin-3) and ≈60/480 nm (claudin-5) mesh sizes. Notably inhomogeneous claudin distribution along strands and in claudin proteins between strands were detected and could play a role in strand formation. Since SMLM requires >20,000 frames for image reconstruction, the dynamics of claudin meshworks were not resolved. Available multicolor SMLM could resolve mixtures of claudins or the nanoscale organization of TJ proteins in the future [39] (Figure 3).

An example for TJs resolved with STED microscopy has not been published yet. Just Cording et al. in 2017 were able to image changes of the Tricellular nanoarchitecture upon treatment with tricellulin derived peptides [81].

That STED microscopy in general is a tool to resolve TJ meshworks at nanoscale level without any further image processing is shown in Figure 4 by an example of a reconstituted meshwork of YFP-claudin-3 in Cos7 fibroblasts.

On an endogenous level, the group around Schlingmann and Koval studied the effect of alcohol on the nanostructure of alveolar epithelial tight junction using SMLM in combination with functional readouts. Alcohol increased claudin-5 levels, induced claudin-18 and actin containing spikes, and increased the paracellular flux. SMLM revealed mostly linear nanocluster arrangements of immunolabelled claudin-5, -18, and ZO1 in TJs. Under alcohol treatment, the specific nano-colocalization of claudin-18 and ZO1 decreased, while the colocalization of claudin-5 and claudin-18 increased. As molecular mechanism leading to alcohol induced paracellular flux the competition of claudin-5 and claudin-18 for interaction with ZO1 was proposed [40].

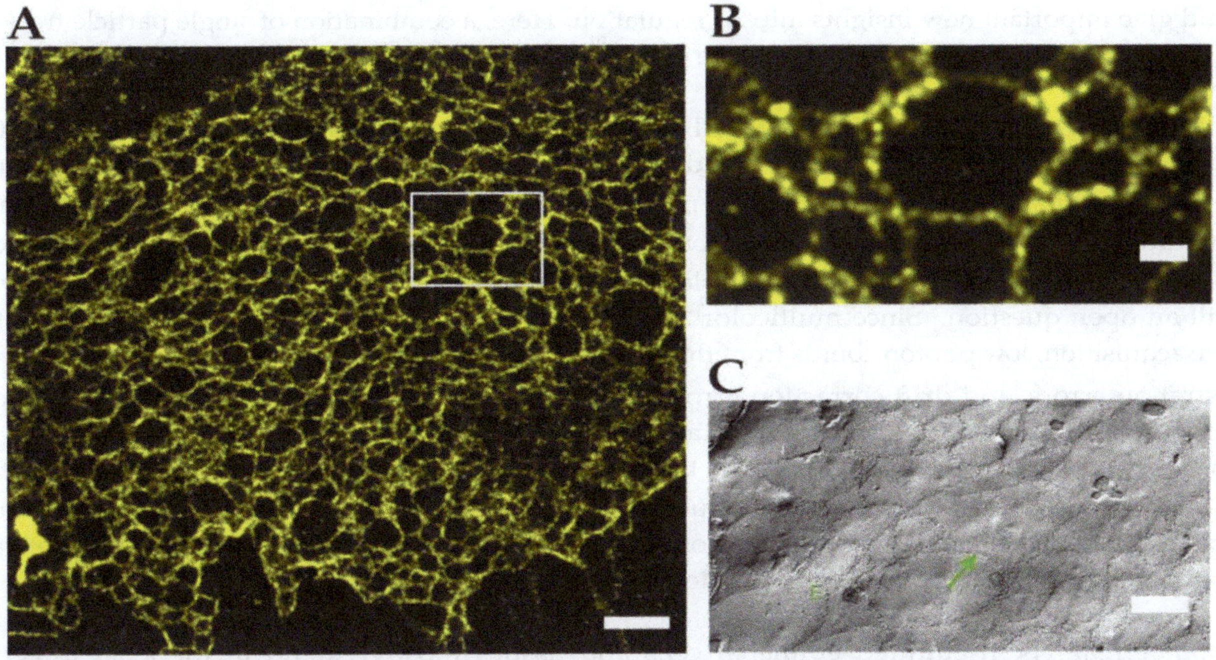

Figure 3. Reconstituted TJ networks formed by claudin3-YFP in HEK293 cells imaged by using SMLM (adapted from Kaufmann et al. 2012 [39]); (**A**) localization microscopy image of the region marked in the conventional wide-field fluorescence image with a mean effective optical resolution 48 nm. Scale bar: 1 µm; (**B**) magnification of claudin-3-YFP in HEK293 cells. Scale bar: 200 nm; (**C**) FFEM of claudin-5-YFP/claudin-3-YFP cotransfected HEK293 cells. Scale bar: 250 nm.

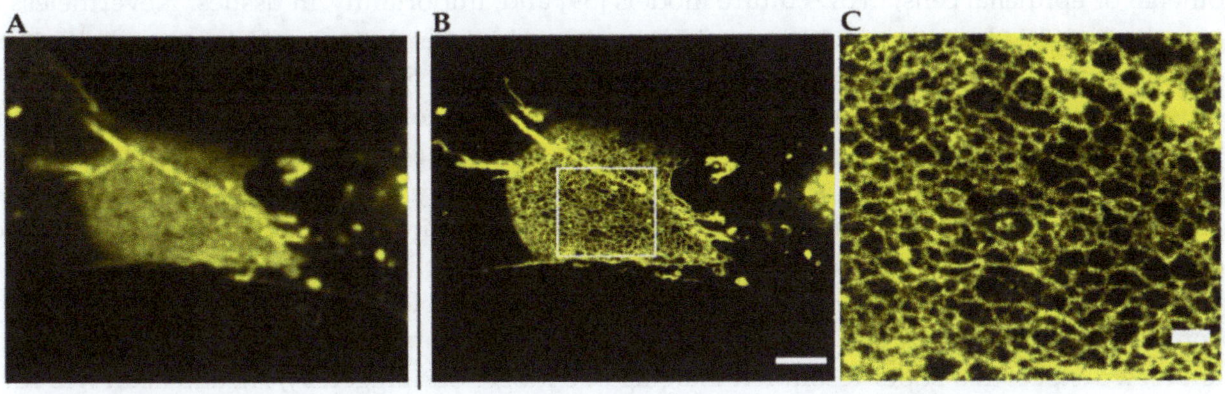

Figure 4. Reconstituted TJ networks formed by YFP-claudin-3 in Cos7 fibroblasts (unpublished data) imaged by using STED microscopy; (**A**) confocal overview image of a reconstituted TJ meshwork at the overlapping area of two transfected cells; (**B**) STED image of the same meshwork area from A. Scale bar: 2 µm; (**C**) magnification of the marked area in B. Scale bar: 500 nm.

3. Discussion and Outlook

Multicellular organisms and tissues rely on multiple cellular junction structures to mediate adherence, sense tension, and form a paracellular barrier. AJs and TJs form dense and complex multiprotein assemblies with a specific nanoscale architecture. While EM revealed claudin strands and membrane kissing points at the TJ or dense cytosolic plaques at the AJ, the intricate molecular nano-organization and dynamic remodeling of these cellular structures only begin to emerge. As shown by SMLM, E-cad forms in the AJ regularly spaced nanoclusters connected to a stratified intracellular layer of tension-sensing and actin interacting proteins at the AJ [41–43]. Revealing the dynamic nanoscale organization of AJs, especially movements of cadherins cluster, vinculin conformational changes, and actin remodeling under tension, during vascular development and endothelial tissue remodeling, should give important new insights into AJ regulation. Here, a combination of single particle tracking, and live STED could enable the study of the concerted action of multiple AJ components.

With recently developed, and now well matured, super-resolution microscopy techniques a first glimpse at claudin strand organization and dynamics at TJ was revealed. Since most TJs contain multiple claudins and other TJ associated proteins, e.g., occludin, or JAM, and various TJ-related scaffolding proteins that exchange on second timescales, fast multicolor imaging with nanoscale resolution would be required. How multiple claudins and occludin assemble into the TJ meshwork is still an open question. Since multicolor SMLM microscopy is still challenging because of slow image acquisition, low photon counts from fluorescent proteins, fluorophore incompatibilities, channel registration errors, and drift correction errors in the range of the resolution alternative multicolor techniques like STED and SIM are advantageous. STED imaging could also be applied to living cells and tissues [58]. Special care must be taken for the optimization of fixation, permeabilization, and labelling protocols as cellular nanostructures can be significantly altered during these sample preparation steps [60]. Ideally similar nanostructures should be visible in live STED of genetically tagged proteins, in fixed cells, and after endogenous labelling by antibodies.

Since real TJs in cultured epithelial cells and tissues mostly extend in the axial direction, the significantly lower z-resolution of most available super-resolution techniques sets a barrier to resolving the native TJ nano-architecture. Therefore, TJ components, namely claudins and occludin, were visualized in reconstituted flat cell culture systems that recapitulate early stages of epithelial formation. Here, claudin meshwork structures were visualized that were very similar to early FFEM studies. Importantly any observation in such a reconstituted system would have to be tested in endothelial or epithelial cells, in 3D culture models [59] and, importantly, in tissues. Nevertheless the improvement of SRM techniques which provide imaging at high resolution in thicker samples (>30 μm) and most importantly also with a sufficient z-resolution to image single strands at 50–100 nm spacing in tissue and epithelial cells will be one of the main and most important steps towards understanding the physiology and biology of TJs. One way to achieve this goal will be by using carefully aligned and refractive index matched 3D STED imaging of endogenously tagged abundant TJ proteins, like claudins, ZO-1, or occludin, providing novel insights about claudin localization within the meshwork but also its organization in the lateral membrane. Endogenous labelling still depends on the availability of specific antibodies or recently available genetic knock-in technology. Since SRM techniques resolve fluorophore positions down to a few nanometers, the sizes of antibodies become a limiting factor. Therefore, we expect that super-resolution imaging of endogenous tagged TJ, AJ, and actin will reveal the important nano-structural features of native junctions together with the associated cytoskeleton. Small, bright, and photostable fluorescent labels should be placed as close as possible to biological target structures that can be achieved via genetic tags like fluorescent proteins or self-labelling enzymes like SNAP or Halo tags [82]. Brighter fluorogenic dyes [83], more sensitive detectors, and microscopy systems like light sheet microscopes would enable faster and better resolved imaging of AJ and TJ structures. A better understanding of nano-structural details of TJs and AJs will ultimately enable discovery of new strategies to overcome or strengthen the intercellular barrier for drug delivery or during tissue damage, respectively.

Author Contributions: Conceptualization, H.G., M.L.; Funding Acquisition, V.H., M.L.; Visualization H.G. Writing—Original Draft Preparation H.G., M.L.; Writing—Review and Editing: H.G., V.H., M.L. All authors have read and agreed to the published version of the manuscript.

Abbreviations

AJ	Adherens junction
dSTORM	Direct stochastic optical reconstruction microscopy
E-cad	E-cad
eGFP	Enhanced green fluorescent protein
EM	Electron microscopy
FA	Focal adhesion
FCS	Fluorescence correlation spectroscopy
FFEM	Freeze fracture electron microscopy
fr	Frame
FRAP	Fluorescence recovery after photobleaching
FRET	Fluorescence resonance energy transfer
gSTED	Gated stimulated emission depletion
JAM-A	Junctional adhesion molecule A
kW	Kilowatt
min	Minute
ms	Millisecond
NA	Numerical aperture
nm	Nanometer
PALM	Photoactivated localization microscopy
PSF	Point spread function
s	Second
SIM	Structured illumination microscopy
SMLM	Single molecule localization microscopy
SRM	Super-resolution microscopy
TAMP	Tight junction-associated MARVEL protein
TEM	Transmission electron microscopy
TJ (bTJ, tTJ)	Tight junction (bicellular TJ, tricellular TJ)
YFP	Yellow fluorescent protein
ZO-1, -2, -3	Zonula occludens protein 1, 2, or 3

References

1. Günzel, D.; Fromm, M. Claudins and other tight junction proteins. *Compr. Physiol.* **2012**, *2*, 1819–1852. [PubMed]
2. Zihni, C.; Mills, C.; Matter, K.; Balda, M.S. Tight junctions: From simple barriers to multifunctional molecular gates. *Nat. Rev. Mol. Cell Biol.* **2016**, *17*, 564–580. [CrossRef] [PubMed]
3. Farquahr, M.G.; Palade, G.E. Junctional complexes in various epithelia. *J. Cell Biol.* **1963**, *17*, 375–412. [CrossRef] [PubMed]
4. Staehelin, L.A. Further observations on the fine structure of freeze-cleaved tight junctions. *J. Cell Sci.* **1973**, *13*, 763–786.
5. Stevenson, B.R.; Siliciano, J.D.; Mooseker, M.S.; Goodenough, D.A. Identification of ZO-1: A high molecular weight polypeptide associated with the tight junction (Zonula Occludens) in a variety of epithelia. *J. Cell Biol.* **1986**, *103*, 755–766. [CrossRef]
6. Furuse, M.; Hirase, T.; Itoh, M.; Nagafuchi, A.; Yonemura, S.; Tsukita, S.; Tsukita, S. Occludin: A novel integral membrane protein localizing at tight junctions. *J. Cell Biol.* **1993**, *123*, 1777–1788. [CrossRef]
7. Martìn-Padura, I.; Lostaglio, S.; Schneemann, M.; Williams, L.; Romano, M.; Fruscella, P.; Panzeri, C.; Stoppacciaro, A.; Ruco, L.; Villa, A.; et al. Junctional adhesion molecule, a novel member of the immunoglobulin superfamily that distributes at intercellular junctions and modulates monocyte transmigration. *J. Cell Biol.* **1998**, *142*, 117–127. [CrossRef]
8. Furuse, M.; Fujita, K.; Hiiragi, T.; Fujimoto, K.; Tsukita, S. Claudin-1 and -2: Novel integral membrane proteins localizing at tight junctions with no sequence similarity to occludin. *J. Cell Biol.* **1998**, *141*, 1539–1550. [CrossRef]

9. Raleigh, D.R.; Marchiando, A.M.; Zhang, Y.; Shen, L.; Sasaki, H.; Wang, Y.; Long, M.; Turner, J.R. Tight Junction–associated MARVEL Proteins MarvelD3, Tricellulin, and Occludin Have Distinct but Overlapping Functions. *Mol. Biol. Cell* **2010**, *21*, 1200–1213. [CrossRef]
10. Cording, J.; Berg, J.; Kading, N.; Bellmann, C.; Tscheik, C.; Westphal, J.K.; Milatz, S.; Gunzel, D.; Wolburg, H.; Piontek, J.; et al. In tight junctions, claudins regulate the interactions between occludin, tricellulin and marvelD3, which, inversely, modulate claudin oligomerization. *J. Cell Sci.* **2013**, *126*, 554–564. [CrossRef]
11. Jesaitis, L.A.; Goodenough, D.A. Molecular characterization and tissue distribution of ZO-2, a tight junction protein homologous to ZO-1 and the Drosophila discs-large tumor suppressor protein. *J. Cell Biol.* **1994**, *124*, 949–961. [CrossRef] [PubMed]
12. Balda, M.S.; Gonzalez-Mariscal, L.; Matter, K.; Cereijido, M.; Anderson, J.M. Assembly of the tight junction: The role of diacylglycerol. *J. Cell Biol.* **1993**, *123*, 293–302. [CrossRef] [PubMed]
13. Van Itallie, C.M.; Tietgens, A.J.; Anderson, J.M. Visualizing the dynamic coupling of claudin strands to the actin cytoskeleton through ZO-1. *Mol. Biol. Cell* **2017**, *28*, 524–534. [CrossRef] [PubMed]
14. Hirokawa, N.; Tilney, L.G. Interactions between actin filaments and between actin filaments and membranes in quick-frozen and deeply etched hair cells of the chick ear. *J. Cell Biol.* **1982**, *95*, 249–261. [CrossRef]
15. Madara, J.L.; Pappenheimer, J.R. Structural basis for physiological regulation of paracellular pathways in intestinal epithelia. *J. Membr. Biol.* **1987**, *100*, 149–164. [CrossRef]
16. Fanning, A.S.; Jameson, B.J.; Jesaitis, L.A.; Anderson, J.M. The tight junction protein ZO-1 establishes a link between the transmembrane protein occludin and the actin cytoskeleton. *J. Biol. Chem.* **1998**, *273*, 29745–29753. [CrossRef]
17. Rosenthal, R.; Milatz, S.; Krug, S.M.; Oelrich, B.; Schulzke, J.D.; Amasheh, S.; Günzel, D.; Fromm, M. Claudin-2, a component of the tight junction, forms a paracellular water channel. *J. Cell Sci.* **2010**, *123*, 1913–1921. [CrossRef]
18. Rosenthal, R.; Günzel, D.; Piontek, J.; Krug, S.M.; Ayala-Torres, C.; Hempel, C.; Theune, D.; Fromm, M. Claudin-15 forms a water channel through the tight junction with distinct function compared to claudin-2. *Acta Physiol.* **2019**, *2018*, 1–15. [CrossRef]
19. Amasheh, S.; Meiri, N.; Gitter, A.H.; Schöneberg, T.; Mankertz, J.; Schulzke, J.D.; Fromm, M. Claudin-2 expression induces cation-selective channels in tight junctions of epithelial cells. *J. Cell Sci.* **2002**, *115*, 4969–4976. [CrossRef]
20. Van Itallie, C.M.; Fanning, A.S.; Anderson, J.M. Reversal of charge selectivity in cation or anion-selective epithelial lines by expression of different claudins. *Am. J. Physiol. Ren. Physiol.* **2003**, *285*, 1078–1084. [CrossRef]
21. Van Itallie, C.M.; Rogan, S.; Yu, A.; Vidal, L.S.; Holmes, J.; Anderson, J.M. Two splice variants of claudin-10 in the kidney create paracellular pores with different ion selectivities. *Am. J. Physiol. Ren. Physiol.* **2006**, *291*, 1288–1299. [CrossRef] [PubMed]
22. Fromm, M.; Günzel, D.; Krug, S.M.; Amasheh, S.; Rosenthal, R.; Schulzke, J.D.; Fromm, A.; Conrad, M.P. Claudin-17 forms tight junction channels with distinct anion selectivity. *Cell. Mol. Life Sci.* **2012**, *69*, 2765–2778.
23. Hou, J.; Renigunta, A.; Konrad, M.; Gomes, A.S.; Schneeberger, E.E.; Paul, D.L.; Waldegger, S.; Goodenough, D.A. Claudin-16 and claudin-19 interact and form a cation-selective tight junction complex. *J. Clin. Investig.* **2008**, *118*, 619–628. [CrossRef] [PubMed]
24. Hou, J.; Renigunta, A.; Yang, J.; Waldegger, S. Claudin-4 forms paracellular chloride channel in the kidney and requires claudin-8 for tight junction localization. *Proc. Natl. Acad. Sci. USA* **2010**, *107*, 18010–18015. [CrossRef] [PubMed]
25. Otani, T.; Nguyen, T.P.; Tokuda, S.; Sugihara, K.; Sugawara, T.; Furuse, K.; Miura, T.; Ebnet, K.; Furuse, M. Claudins and JAM-A coordinately regulate tight junction formation and epithelial polarity. *J. Cell Biol.* **2019**, *218*, 3372–3396. [CrossRef] [PubMed]
26. Furuse, M.; Sasaki, H.; Fujimoto, K.; Tsukita, S. A single gene product, claudin-1 or -2, reconstitutes tight junction strands and recruits occludin in fibroblasts. *J. Cell Biol.* **1998**, *143*, 391–401. [CrossRef]
27. Krug, S.M.; Amasheh, S.; Richter, J.F.; Milatz, S.; Günzel, D.; Westphal, J.K.; Huber, O.; Schulzke, J.D.; Michael, F. Tricellulin Forms a Barrier to Macromolecules in Tricellular Tight Junctions without Affecting Ion Permeability. *Mol. Biol. Cell* **2009**, *20*, 3713–3724. [CrossRef]

28. Ikenouchi, J.; Furuse, M.; Furuse, K.; Sasaki, H.; Tsukita, S.; Tsukita, S. Tricellulin constitutes a novel barrier at tricellular contacts of epithelial cells. *J. Cell Biol.* **2005**, *171*, 939–945. [CrossRef]

29. Higashi, T.; Tokuda, S.; Kitajiri, S.I.; Masuda, S.; Nakamura, H.; Oda, Y.; Furuse, M. Analysis of the 'angulin' proteins LSR, ILDR1 and ILDR2—Tricellulin recruitment, epithelial barrier function and implication in deafness pathogenesis. *J. Cell Sci.* **2013**, *126*, 3797. [CrossRef]

30. Krug, S.M.; Bojarski, C.; Fromm, A.; Lee, I.M.; Dames, P.; Richter, J.F.; Turner, J.R.; Fromm, M.; Schulzke, J.D. Tricellulin is regulated via interleukin-13-receptor α2, affects macromolecule uptake, and is decreased in ulcerative colitis. *Mucosal Immunol.* **2018**, *11*, 345–356. [CrossRef]

31. Huber, A.H.; Weis, W.I.; West, C.D. The Structure of the ß-Catenin/E-Cadherin Complex and the Molecular Basis of Diverse Ligand Recognition by ß-Catenin. *Cell* **2001**, *105*, 391–402. [CrossRef]

32. Pokutta, S.; Weis, W.I. Structure of the Dimerization and ß-Catenin-Binding Region of a-Catenin. *Mol. Cell* **2000**, *5*, 533–543. [CrossRef]

33. Drees, F.; Pokutta, S.; Yamada, S.; Nelson, W.J.; Weis, W.I. α-Catenin Is a Molecular Switch that Binds E-Cadherin-β-Catenin and Regulates Actin-Filament Assembly. *Cell* **2005**, *123*, 903–915. [CrossRef] [PubMed]

34. Watabe-Uchida, M.; Uchida, N.; Imamura, Y.; Nagafuchi, A.; Fujimoto, K.; Uemura, T.; Vermeulen, S.; Van Roy, F.; Adamson, E.D.; Takeichi, M. α-Catenin-Vinculin Interaction Functions to Organize the Apical Junctional Complex in Epithelial Cells. *J. Cell Biol.* **1998**, *142*, 847–857. [CrossRef]

35. Weiss, E.E.; Kroemker, M.; Rüdiger, A.; Jockusch, B.M.; Rüdiger, M. Vinculin Is Part of the Cadherin–Catenin Junctional Complex: Complex Formation between α-Catenin and Vinculin. *J. Cell Biol.* **1998**, *141*, 755–764. [CrossRef]

36. Kobielak, A.; Fuchs, E.; Medical, H.H. α-Catenin: At the junction of intercellular adhesion and actin dynamics. *Nat. Rev. Mol. Cell Biol.* **2004**, *5*, 614–625. [CrossRef]

37. Nanes, B.A.; Chiasson-MacKenzie, C.; Lowery, A.M.; Ishiyama, N.; Faundez, V.; Ikura, M.; Vincent, P.A.; Kowalczyk, A.P. P120-Catenin Binding Masks an Endocytic Signal Conserved in Classical Cadherins. *J. Cell Biol.* **2012**, *199*, 365–380. [CrossRef]

38. Lecuit, T.; Yap, A.S. E-cadherin junctions as active mechanical integrators in tissue dynamics. *Nat. Cell Biol.* **2015**, *17*, 533–539. [CrossRef]

39. Kaufmann, R.; Piontek, J.; Grüll, F.; Kirchgessner, M.; Rossa, J.; Blasig, I.E.; Cremer, C. Visualization and Quantitative Analysis of Reconstituted Tight Junctions Using Localization Microscopy. *PLoS ONE* **2012**, *7*, 1–9. [CrossRef]

40. Schlingmann, B.; Overgaard, C.E.; Molina, S.A.; Lynn, K.S.; Mitchell, L.A.; Dorsainvil White, S.; Mattheyses, A.L.; Guidot, D.M.; Capaldo, C.T.; Koval, M. Regulation of claudin/zonula occludens-1 complexes by hetero-claudin interactions. *Nat. Commun.* **2016**, *7*, 12276. [CrossRef]

41. Quang, B.A.T.; Mani, M.; Markova, O.; Lecuit, T.; Lenne, P.F. Principles of E-cadherin supramolecular organization in vivo. *Curr. Biol.* **2013**, *23*, 2197–2207.

42. Wu, Y.; Kanchanawong, P.; Zaidel-Bar, R. Actin-Delimited Adhesion-Independent Clustering of E-Cadherin Forms the Nanoscale Building Blocks of Adherens Junctions. *Dev. Cell* **2015**, *32*, 139–154. [CrossRef] [PubMed]

43. Bertocchi, C.; Wang, Y.; Ravasio, A.; Hara, Y.; Wu, Y.; Sailov, T.; Baird, M.A.; Davidson, M.W.; Zaidel-bar, R.; Toyama, Y.; et al. Nanoscale architecture of cadherin-based cell adhesions. *Nat. Cell Biol.* **2017**, *19*, 28. [CrossRef] [PubMed]

44. Abbe, E. Beiträge zur Theorie des Mikroskops und der mikroskopischen Wahrnehmung. *Arch. Mikrosk. Anat.* **1873**, *9*, 413–418. [CrossRef]

45. Schermelleh, L.; Carlton, P.M.; Haase, S.; Shao, L.; Winoto, L.; Kner, P.; Burke, B.; Cardoso, M.C.; Agard, D.A.; Gustafsson, M.G.L.; et al. Subdiffraction multicolor imaging of the nuclear periphery with 3D structured illumination microscopy. *Science* **2008**, *320*, 1332–1336. [CrossRef] [PubMed]

46. Chung, E.; Kim, D.; So, P.T.C. Extended resolution wide-field optical imaging: Internal reflection fluorescence microscopy. *Opt. Lett.* **2006**, *31*, 945–947. [CrossRef]

47. Guo, M.; Chandris, P.; Giannini, J.P.; Trexler, A.J.; Fischer, R.; Chen, J.; Vishwasrao, H.D.; Rey-Suarez, I.; Wu, Y.; Wu, X.; et al. Single-shot super-resolution total internal reflection fluorescence microscopy. *Nat. Methods* **2018**, *15*, 425–428. [CrossRef]

48. Guo, Y.; Li, D.; Zhang, S.; Yang, Y.; Liu, J.J.; Wang, X.; Liu, C.; Milkie, D.E.; Moore, R.P.; Tulu, U.S.; et al. Visualizing Intracellular Organelle and Cytoskeletal Interactions at Nanoscale Resolution on Millisecond Timescales. *Cell* **2018**, *175*, 1430–1442. [CrossRef]
49. Chen, B.C.; Legant, W.R.; Wang, K.; Shao, L.; Milkie, D.E.; Davidson, M.W.; Janetopoulos, C.; Wu, X.S.; Hammer, J.A.; Liu, Z.; et al. Lattice light-sheet microscopy: Imaging molecules to embryos at high spatiotemporal resolution. *Science* **2014**, *346*, 1257998. [CrossRef]
50. Li, D.; Shao, L.; Chen, B.C.; Zhang, X.; Zhang, M.; Moses, B.; Milkie, D.E.; Beach, J.R.; Hammer, J.A.; Pasham, M.; et al. Extended-resolution structured illumination imaging of endocytic and cytoskeletal dynamics. *Science* **2015**, *349*, aab3500. [CrossRef]
51. De Luca, G.M.R.; Breedijk, R.M.P.; Brandt, R.A.J.; Zeelenberg, C.H.C.; de Jong, B.E.; Timmermans, W.; Azar, L.N.; Hoebe, R.A.; Stallinga, S.; Manders, E.M.M. Re-scan confocal microscopy: Scanning twice for better resolution. *Biomed. Opt. Express* **2013**, *4*, 2644. [CrossRef] [PubMed]
52. Huff, J. The Airyscan detector from ZEISS: Confocal imaging with improved signal-to-noise ratio and super-resolution. *Nat. Methods* **2015**, *12*, 1205. [CrossRef]
53. Wegel, E.; Göhler, A.; Lagerholm, B.C.; Wainman, A.; Uphoff, S.; Kaufmann, R.; Dobbie, I.M. Imaging cellular structures in super-resolution with SIM, STED and Localisation Microscopy: A practical comparison. *Sci. Rep.* **2015**, *6*, 1–13. [CrossRef] [PubMed]
54. Hell, S.W.; Wichmann, J. Breaking the diffraction resolution limit by stimulated emission: Stimulated-emission-depletion fluorescence microscopy. *Opt. Lett.* **1994**, *19*, 780. [CrossRef]
55. Vicidomini, G.; Moneron, G.; Han, K.Y.; Westphal, V.; Ta, H.; Reuss, M.; Engelhardt, J.; Eggeling, C.; Hell, S.W. Sharper low-power STED nanoscopy by time gating. *Nat. Methods* **2011**, *8*, 571–575. [CrossRef]
56. Curdt, F.; Herr, S.J.; Lutz, T.; Schmidt, R.; Engelhardt, J.; Sahl, S.J.; Hell, S.W. isoSTED nanoscopy with intrinsic beam alignment. *Opt. Express* **2015**, *23*, 30891–30903. [CrossRef]
57. Grimm, J.B.; Muthusamy, A.K.; Liang, Y.; Brown, T.A.; Lemon, W.C.; Patel, R.; Lu, R.; Macklin, J.J.; Keller, P.J.; Ji, N.; et al. A general method to fine-tune fluorophores for live-cell and in vivo imaging. *Nat. Methods* **2017**, *14*, 987. [CrossRef]
58. Bottanelli, F.; Kromann, E.B.; Allgeyer, E.S.; Erdmann, R.S.; Baguley, S.W.; Sirinakis, G.; Schepartz, A.; Baddeley, D.; Toomre, D.K.; Rothman, J.E.; et al. Two-colour live-cell nanoscale imaging of intracellular targets. *Nat. Commun.* **2015**, *7*, 1–5. [CrossRef]
59. Maraspini, R.; Wang, C.-H.; Honigmann, A. Optimization of 2D and 3D cell culture to study membrane organization with STED microscopy. *J. Phys. Appl. Phys.* **2020**, *53*, 014001. [CrossRef]
60. Schnell, U.; Dijk, F.; Sjollema, K.A.; Giepmans, B.N.G. Immunolabeling artifacts and the need for live-cell imaging. *Nat. Methods* **2012**, *9*, 152–158. [CrossRef]
61. Rust, M.J.; Bates, M.; Zhuang, X. Sub-diffraction-limit imaging by stochastic optical reconstruction microscopy (STORM). *Nat. Methods* **2006**, *3*, 793–795. [CrossRef] [PubMed]
62. Bates, M.; Huang, B.; Dempsey, G.T.; Zhuang, X. Multicolor super-resolution imaging with photo-switchable fluorescent probes. *Science* **2007**, *317*, 1749–1753. [CrossRef] [PubMed]
63. Betzig, E.; Patterson, G.H.; Sougrat, R.; Lindwasser, O.W.; Olenych, S.; Bonifacino, J.S.; Davidson, M.W.; Lippincott-Schwartz, J.; Hess, H.F. Imaging intracellular fluorescent proteins at nanometer resolution. *Science* **2006**, *313*, 1642–1645. [CrossRef] [PubMed]
64. Giannone, G.; Hosy, E.; Levet, F.; Constals, A.; Schulze, K.; Sobolevsky, A.I.; Rosconi, M.P.; Gouaux, E.; Tampe, R.; Choquet, D.; et al. Dynamic superresolution imaging of endogenous proteins on living cells at ultra-high density. *Biophys. J.* **2010**, *99*, 1303–1310. [CrossRef]
65. Vaughan, J.C.; Jia, S.; Zhuang, X. Ultrabright photoactivatable fluorophores created by reductive caging. *Nat. Methods* **2012**, *9*, 1181–1184. [CrossRef]
66. Jones, S.A.; Shim, S.H.; He, J.; Zhuang, X. Fast, three-dimensional super-resolution imaging of live cells. *Nat. Methods* **2011**, *8*, 499–505. [CrossRef]
67. Huang, B.; Wang, W.; Bates, M.; Zhuang, X. Three-Dimensional Super-Resolution Reconstruction Microscopy. *Science* **2008**, *319*, 810–813. [CrossRef]
68. Juette, M.F.; Gould, T.J.; Lessard, M.D.; Mlodzianoski, M.J.; Nagpure, B.S.; Bennett, B.T.; Hess, S.T.; Bewersdorf, J. Three-dimensional sub-100 nm resolution fluorescence microscopy of thick samples. *Nat. Methods* **2008**, *5*, 527–529. [CrossRef]

69. Shtengel, G.; Galbraith, J.A.; Galbraith, C.G.; Lippincott-Schwartz, J.; Gillette, J.M.; Manley, S.; Sougrat, R.; Waterman, C.M.; Kanchanawong, P.; Davidson, M.W.; et al. Interferometric fluorescent super-resolution microscopy resolves 3D cellular ultrastructure. *Proc. Natl. Acad. Sci. USA* **2009**, *106*, 3125–3130. [CrossRef]
70. Xu, K.; Babcock, H.P.; Zhuang, X. Dual-objective STORM reveals three-dimensional filament organization in the actin cytoskeleton. *Nat. Methods* **2012**, *9*, 185–188. [CrossRef]
71. Manley, S.; Gillette, J.M.; Patterson, G.H.; Shroff, H.; Hess, H.F.; Betzig, E.; Lippincott-Schwartz, J. High-density mapping of single-molecule trajectories with photoactivated localization microscopy. *Nat. Methods* **2008**, *5*, 155–157. [CrossRef] [PubMed]
72. Sage, D.; Pham, T.A.; Babcock, H.; Lukes, T.; Pengo, T.; Chao, J.; Velmurugan, R.; Herbert, A.; Agrawal, A.; Colabrese, S.; et al. Super-resolution fight club: Assessment of 2D and 3D single-molecule localization microscopy software. *Nat. Methods* **2019**, *16*, 387–395. [CrossRef] [PubMed]
73. Schermelleh, L.; Ferrand, A.; Huser, T.; Eggeling, C.; Sauer, M.; Biehlmaier, O.; Drummen, G.P.C. Super-resolution microscopy demystified. *Nat. Cell Biol.* **2019**, *21*, 72–84. [CrossRef] [PubMed]
74. Maître, J.L.; Berthoumieux, H.; Krens, S.F.G.; Salbreux, G.; Jülicher, F.; Paluch, E.; Heisenberg, C.P. Adhesion functions in cell sorting by mechanically coupling the cortices of adhering cells. *Science* **2012**, *338*, 253–256. [CrossRef] [PubMed]
75. Adams, C.L.; Nelson, W.J.; Smith, S.J. Quantitative analysis of cadherin-catenin-actin reorganization during development of cell-cell adhesion. *J. Cell Biol.* **1996**, *135*, 1899–1911. [CrossRef] [PubMed]
76. Case, L.B.; Baird, M.A.; Shtengel, G.; Campbell, S.L.; Hess, H.F.; Davidson, M.W.; Waterman, C.M. Molecular mechanism of vinculin activation and nanoscale spatial organization in focal adhesions. *Nat. Cell Biol.* **2015**, *17*, 880–892. [CrossRef] [PubMed]
77. Van Zanten, T.S.; Cambi, A.; Koopman, M.; Joosten, B.; Figdor, C.G.; Garcia-Parajo, M.F. Hotspots of GPI-anchored proteins and integrin nanoclusters function as nucleation sites for cell adhesion. *Proc. Natl. Acad. Sci. USA* **2009**, *106*, 18557–18562. [CrossRef]
78. Shroff, H.; Galbraith, C.G.; Galbraith, J.A.; White, H.; Gillette, J.; Olenych, S.; Davidson, M.W.; Betzig, E. Dual-color superresolution imaging of genetically expressed probes within individual adhesion complexes. *Proc. Natl. Acad. Sci. USA* **2007**, *104*, 20308–20313. [CrossRef]
79. Grashoff, C.; Hoffman, B.D.; Brenner, M.D.; Zhou, R.; Parsons, M.; Yang, M.T.; McLean, M.A.; Sligar, S.G.; Chen, C.S.; Ha, T.; et al. Measuring mechanical tension across vinculin reveals regulation of focal adhesion dynamics. *Nature* **2010**, *466*, 263–266. [CrossRef]
80. Sasaki, H.; Matsui, C.; Furuse, K.; Mimori-Kiyosue, Y.; Furuse, M.; Tsukita, S. Dynamic behavior of paired claudin strands within apposing plasma membranes. *Proc. Natl. Acad. Sci. USA* **2003**, *100*, 3971–3976. [CrossRef]
81. Cording, J.; Arslan, B.; Staat, C.; Dithmer, S.; Krug, S.M.; Krüger, A.; Berndt, P.; Günther, R.; Winkler, L.; Blasig, I.E.; et al. Trictide, a tricellulin-derived peptide to overcome cellular barriers. *Ann. N. Y. Acad. Sci.* **2017**, *1405*, 89–101. [CrossRef] [PubMed]
82. Liss, V.; Barlag, B.; Nietschke, M.; Hensel, M. Self-labelling enzymes as universal tags for fluorescence microscopy, super-resolution microscopy and electron microscopy. *Sci. Rep.* **2015**, *5*, 1–13. [CrossRef] [PubMed]
83. Lukinavičius, G.; Umezawa, K.; Olivier, N.; Honigmann, A.; Yang, G.; Plass, T.; Mueller, V.; Reymond, L.; Corrêa, I.R.; Luo, Z.G.; et al. A near-infrared fluorophore for live-cell super-resolution microscopy of cellular proteins. *Nat. Chem.* **2013**, *5*, 132–139. [CrossRef] [PubMed]

4

Claudin-12 Knockout Mice Demonstrate Reduced Proximal Tubule Calcium Permeability

Allein Plain [1], Wanling Pan [1], Deborah O'Neill [1], Megan Ure [1], Megan R. Beggs [1,2], Maikel Farhan [2,3], Henrik Dimke [4,5], Emmanuelle Cordat [1] and R. Todd Alexander [1,2,3,*]

1 Department of Physiology, The University of Alberta, Edmonton, AB T6J 2R7, Canada; aplain@gmail.com (A.P.); pan2@ualberta.ca (W.P.); oneilld@ualberta.ca (D.O.); ure@ualberta.ca (M.U.); mbeggs@ualberta.ca (M.R.B.); cordat@ualberta.ca (E.C.)
2 The Women's & Children's Health Research Institute, 11405-87 Avenue, Edmonton, AB T6G 1C9 Canada; maikel@ualberta.ca
3 Department of Pediatrics, The University of Alberta, Edmonton, AB T6J 2R7, Canada
4 Department of Cardiovascular and Renal Research, Institute of Molecular Medicine, University of Southern Denmark, 5230 Odense, Denmark; hdimke@health.sdu.dk
5 Department of Nephrology, Odense University Hospital, 5000 Odense, Denmark
* Correspondence: todd2@ualberta.ca.

Abstract: The renal proximal tubule (PT) is responsible for the reabsorption of approximately 65% of filtered calcium, primarily via a paracellular pathway. However, which protein(s) contribute this paracellular calcium pore is not known. The claudin family of tight junction proteins confers permeability properties to an epithelium. Claudin-12 is expressed in the kidney and when overexpressed in cell culture contributes paracellular calcium permeability (P_{Ca}). We therefore examined claudin-12 renal localization and its contribution to tubular paracellular calcium permeability. Claudin-12 null mice (KO) were generated by replacing the single coding exon with β-galactosidase from *Escherichia coli*. X-gal staining revealed that claudin-12 promoter activity colocalized with aquaporin-1, consistent with the expression in the PT. PTs were microperfused ex vivo and P_{Ca} was measured. P_{Ca} in PTs from KO mice was significantly reduced compared with WT mice. However, urinary calcium excretion was not different between genotypes, including those on different calcium containing diets. To assess downstream compensation, we examined renal mRNA expression. Claudin-14 expression, a blocker of P_{Ca} in the thick ascending limb (TAL), was reduced in the kidney of KO animals. Thus, claudin-12 is expressed in the PT, where it confers paracellular calcium permeability. In the absence of claudin-12, reduced claudin-14 expression in the TAL may compensate for reduced PT calcium reabsorption.

Keywords: proximal tubule; calcium permeability; claudin-12

1. Introduction

Calcium is essential for a myriad of physiological functions including intracellular signal transduction, blood clotting, and as a structural component of bone. It is therefore tightly maintained within a narrow physiologic range in serum. This is achieved through hormonal regulation by parathyroid hormone (PTH), vitamin D, and fibroblast growth factor 23 (FGF23). These hormones, in turn, mediate coordinated interactions between intestinal calcium absorption/secretion, bone resorption/deposition and filtration at the renal glomerulus, and consequent reabsorption along the nephron. The majority, approximately two-thirds of filtered calcium, is reabsorbed by the proximal tubule [1]. This occurs via a paracellular route, primarily driven by the reabsorption of water from the proximal tubule [2–6]. A failure to reabsorb calcium from the proximal tubule has been implicated in the pathogenesis of kidney stone formation [7]. Thus, understanding the molecular mediators of this process is a prerequisite to finding improved therapies for this disease.

Paracellular fluxes not only rely on a driving force, but are also dependent on the permeability of the epithelium to the ion being transported. The proximal tubule has significant permeability to Ca^{2+}, enabling paracellular Ca^{2+} flux [1,3–5]. Claudins are a family of four pass membrane proteins expressed in the tight junction that confer paracellular permeability to the tight junction [8]. Claudin-2 is expressed in the proximal tubule and claudin-2 knockout (KO) mice display increased urinary calcium excretion relative to their wild type littermates, consistent with claudin-2 conferring calcium permeability to the proximal tubule [8,9]. Claudin-2 and claudin-12 are expressed in the intestine, and are implicated in mediating calcium permeability [10,11]. Claudin-12 mRNA has been detected in the kidney [2,12]. We have previously reported claudin-12 expression in a renal proximal tubular cell culture model, opossum kidney (OK) cells [13], and mRNA expression has been identified in proximal tubules, where it could also contribute paracellular Ca^{2+} permeability [14].

To study claudin-12 renal expression and if it contributes Ca^{2+} permeability to the renal tubule, claudin-12 KO mice were generated by replacing the single coding exon with β-galactosidase from *E. coli*. The knockout mice grew and behaved similarly to their wild type (WT) littermates. We did not detect differences in plasma electrolytes nor in calciotropic hormone levels between KO and WT mice. We observed predominant X-gal staining in the renal cortex that colocalized with aquaporin-1, indicating expression of claudin-12 in the renal proximal tubule. We therefore perfused proximal tubules from claudin-12 KO mice and found reduced paracellular Ca^{2+} permeability. The KO mice did not however have increased urinary Ca^{2+} excretion. We thus examined the expression of genes that participate in tubular Ca^{2+} transport and found decreased expression of the paracellular Ca^{2+} blocker, claudin-14 [15], and propose that increased thick ascending limb (TAL) Ca^{2+} reabsorption compensates for reduced proximal tubular Ca^{2+} reabsorption in claudin-12 knockout mice.

2. Results

2.1. Generation of a Global Claudin-12 Knockout Model

In order to examine the renal localization and potential tubular transport role of claudin-12, we generated a global claudin-12 knockout mouse. The claudin-12 gene (*Cldn12*) was replaced by homologous recombination of exon 4, the only coding exon of the *Cld12* gene, with the β-galactosidase coding sequence from *E. coli* (Figure 1A). Specific PCR reactions failed to amplify the wild type sequence from KO animals. Similarly, PCR with primers specific for β-galactosidase did not amplify a product from wild type DNA. However, appropriate size PCR fragments could be amplified from both the wild type *Cld12* gene and β-galactosidase from heterozygous mice (Figure 1C). Moreover, quantitative real-time PCR performed on cDNA generated from RNA isolated from whole kidney of wild type mice detected *Cld12* that was not detectable in knockout mice (Figure 1B), consistent with known renal *Cld12* expression [2]. Unfortunately, we have been unable to identify a commercially available antibody or generate an antibody that was specific for claudin-12, as was previously reported by Professor Tsukita's group [16], and more recently by a group studying claudin-12 in nervous tissue [17]. We therefore performed a sequencing reaction on DNA from wild type and claudin-12 KO mice using a primer specific for a sequence approximately 50 bp 5′ to the start cogon in exon 4 of claudin-12. This confirmed that the claudin-12 coding sequence had been replaced with β-galactosidase (Figure 1D).

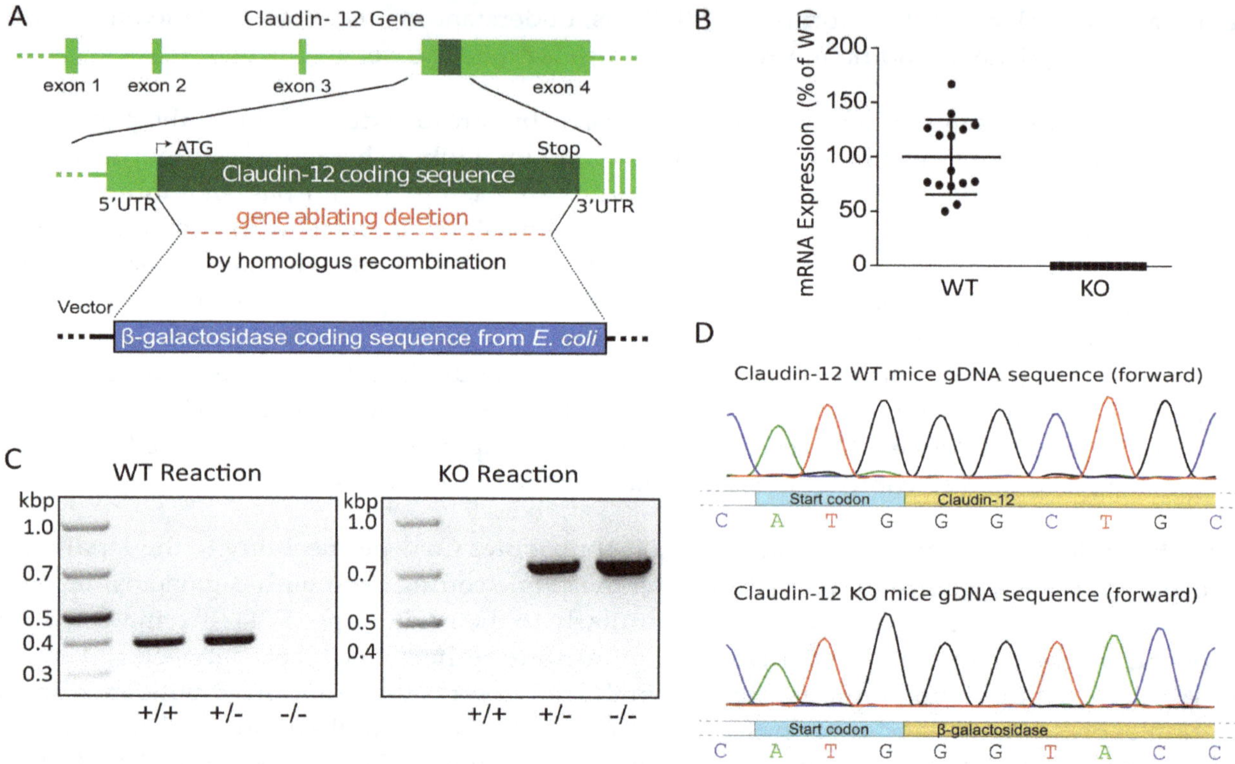

Figure 1. Claudin-12 knockout mouse model. (**A**) Claudin-12 gene deletion targeting strategy. (**B**) Relative claudin-12 mRNA expression normalized to GAPDH, from kidneys of wild type (WT) or claudin-12 knockout mice (KO). (**C**) Genotyping reactions of wild type (+/+), heterozygous (+/−), or claudin-12 knockout (−/−) mice using specific primers for the coding exon of claudin-12 (WT reaction) or the β-galactosidase coding sequence from *E. coli* (KO reaction). (**D**) Sequencing results of the gDNA sequence of claudin-12 WT or KO mice.

2.2. Claudin-12 is Expressed in the Renal Proximal Tubule

In order to localize claudin-12 expression in the kidney, we performed X-gal staining of kidney sections on wild type mice and heterozygous littermates (Figure 2). No staining was evident on the wild type kidney section (Figure 2A). However, we observed prominent staining in the knockout renal cortex, consistent with claudin-12 promoter activity (Figure 2B). Interestingly, the greatest amount of color production was in the juxtamedullary region, with virtually no color produced in the medulla. Higher power images of the cortical region clearly demonstrate β-galactosidase expression concentrated in some, but not all, of the tubules present in the juxtamedullary cortex (Figure 2C). The tubules with greatest β-galactosidase expression were the largest, had an obvious brush border, and were the most abundant in the cortex, consistent with claudin-12 promoter activity in the proximal tubule.

In order to more precisely define claudin-12 renal cortical expression, we performed immunofluorescence localization with tubule segment-specific markers on renal sections from heterozygous mice, after X-gal staining. Aquaporin I (AQP1), a marker of the proximal tubule in the renal cortex [18], colocalized with X-gal production (Figure 3A), confirming predominant localization to the proximal tubule. AQP1 is also present in the thin descending limb in loop of Henle [18], but we limited our assessment to the cortex. For the identification of the thick ascending limb, distal convoluted tubule, and collecting duct, we performed immunostaining with the sodium–potassium–chloride cotransporter II (NKCC2, Figure 3B), the sodium–chloride cotransporter (NCC, Figure 3C), and carbonic anhydrase II (CAII, Figure 3D), respectively. Although CAII is present in the proximal tubule, expression is very low and the greatest expression by far is in the collecting duct (intercalated cells); thus, CAII can be used as a marker of this segment [19]. We observed AQP1 co-staining with X-gal stained tubules. However, none of the other markers demonstrated significant

X-gal co-staining, which is consistent with predominant, if not exclusive, expression of claudin-12 in the renal proximal tubule. Importantly, this method of colocalization cannot exclude low levels of claudin-12 mRNA expression.

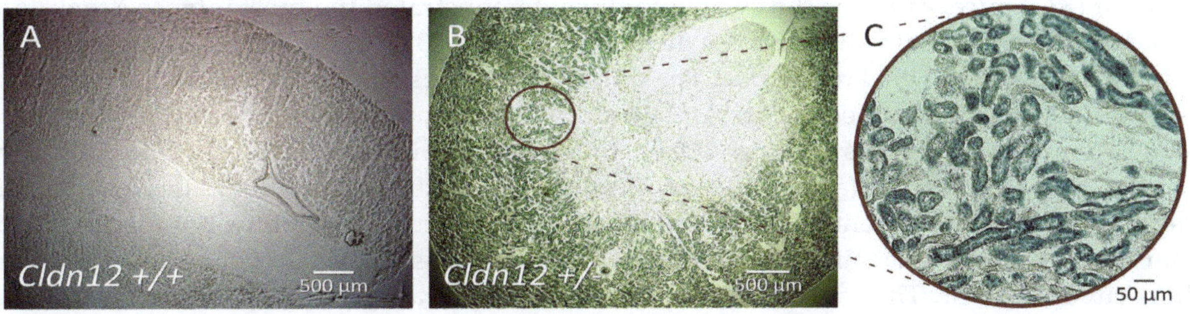

Figure 2. X-gal staining of kidney sections from wild type (**A**) and heterozygous (**B**) mice. β-galactosidase is expressed in the renal cortex in knockout kidney slices and not in wild type ones. Higher power image (**C**).

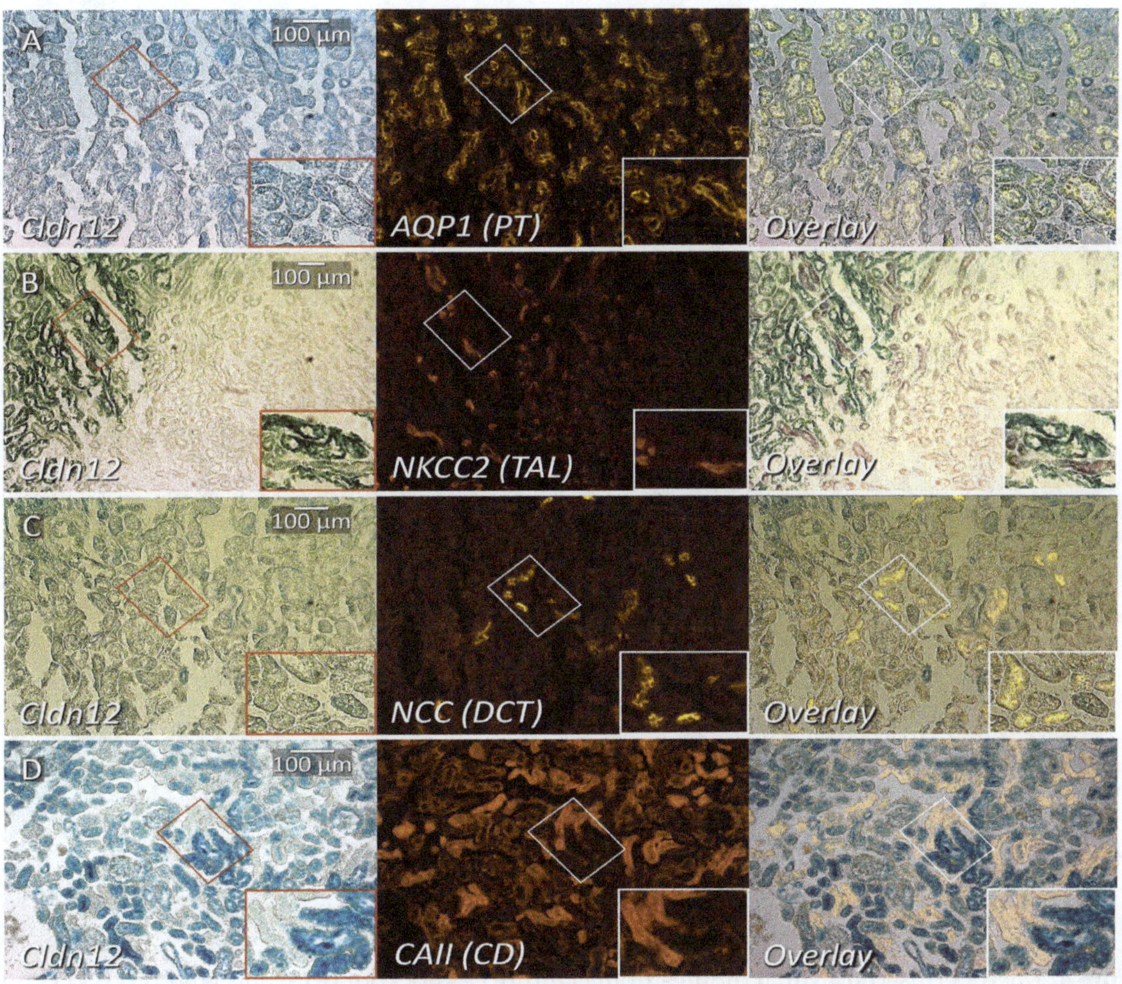

Figure 3. Co-staining of cortical renal tubule markers with X-gal on kidney slices from claudin-12 heterozygous mice. X-gal staining representing claudin-12 promoter expression (Cld12) co-stained with: (**A**) aquaporin-1 (AQP1), (**B**) sodium–potassium–chloride cotransporter 2 (NKCC2), (**C**) sodium–chloride cotransporter (NCC), or (**D**) carbonic anhydrase II (CAII); markers of the proximal tubule (PT), thick ascending limb (TAL), distal collecting tubule (DCT) and collecting duct respectively. The box in the right lower corner is a digitally magnified image from the smaller area depicted with the same shape.

2.3. *Claudin-12 Deletion Decreases Sodium and Calcium Permeability of the Proximal Tubule*

We next turned our attention to the putative role of claudin-12 in the proximal tubule. To this end, we microperfused freshly isolated straight proximal tubules (Figure 4A) from either wild type (WT) or knockout (KO) mice, as this is the segment which demonstrated the most intense X-gal staining. We recorded transepithelial voltage across the tubule (Figure 4B,C) and used a current pulse (13 nA) to calculate the transepithelial resistance. The transepithelial resistance, transepithelial voltage, and short-circuit current determined before and after the addition of ouabain were indistinguishable between WT and KO animals (Figure 4D–F). Ouabain inhibits the Na^+/K^+ATPase, eliminating transepithelial Na^+ transport (as it is the driving force for vectorial Na^+ flux). Importantly, as we were applying a dilution potential across the tubule to measure permeability, not flux, this did not alter our measurement. Ouabain dramatically reduced the transepithelial voltage and consequently the short-circuit current. This is consistent with transcellular transport being substantially reduced, as would be expected. Thus, any subsequent changes in transepithelial voltage generated after the addition of ouabain are primarily due to the paracellular movement of ions induced by the change in the basolateral solution composition. Importantly, the decrease in transcellular transport was not statistically different between WT and KO tubules.

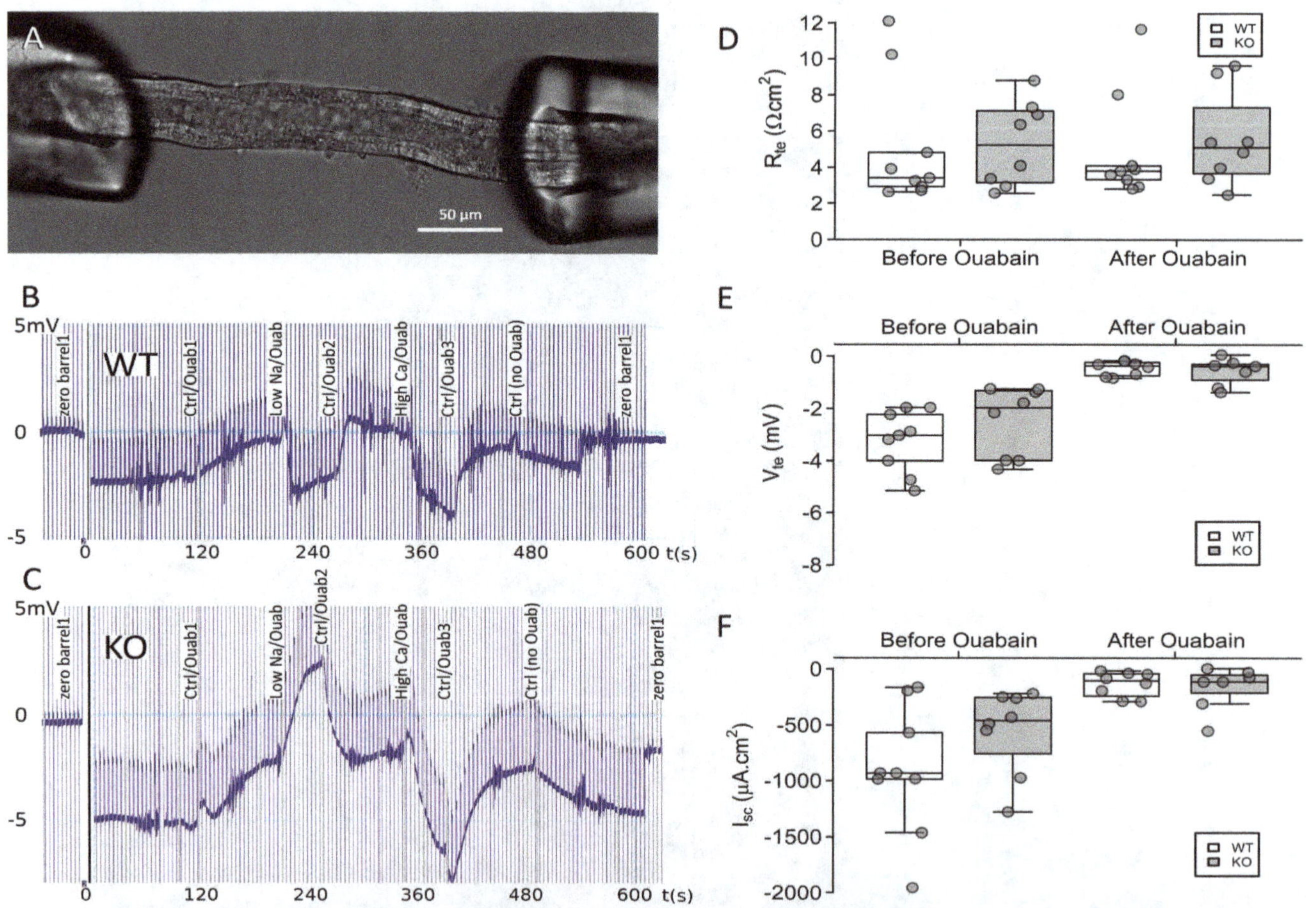

Figure 4. Microperfused proximal tubules from claudin-12 knockout (KO) and wild type (WT) mice. (**A**) Representative image of a proximal tubule perfused ex vivo. (**B**,**C**) Representative recording traces from original experiments in proximal tubules from WT (**B**) and KO (**C**) animals. Basolateral addition of ouabain, and solutions containing a low sodium or high calcium concentration altered the trace by first reducing transcellular transport, and then generating diffusion potentials. (**D**–**F**) Transepithelial resistance (R_{te}), transepithelial voltage (V_{te}), and short-circuit current (I_{sc}) before and after the addition of ouabain.

Next, we imposed a sodium chloride concentration gradient, by adding a solution containing low sodium to the basolateral side, and measured the diffusion potential generated across the tubule. With this result and the resistance, we calculated the permeability ratio of sodium to chloride (Figure 5A, P_{Na}/P_{Cl}). Interestingly we observed an increase in the potential difference (PD) across the tubule in the KO mice, as opposed to a decrease in PD in the WT animals, consistent with altered relative sodium to chloride permeability. We then imposed a calcium to sodium diffusion gradient across the tubule, recorded the diffusion potential generated and used this to calculate the permeability ratio of calcium to sodium (Figure 5A, P_{Ca}/P_{Na}). From these ratios and the resistance, we were able to calculate the absolute permeabilities of the tubule to sodium, chloride, and calcium (Figure 5B). Straight proximal tubules from claudin-12 KO mice displayed a reduced sodium relative to chloride permeability ratio (1.27 ± 0,05 in the WT compared with 0.89 ± 0.17 in the KO), which was the result of decreased sodium permeability ($3.8 \pm 1.3 \times 10^{-4}$ cm/s in WT vs. $2.5 \pm 0.4 \times 10^{-4}$ cm/s in KO), and not increased chloride permeability (Figure 5). Hence, the deletion of claudin-12 changes the selectivity of the straight proximal tubule to relatively more anion permeable, i.e., claudin-12 confers cation selectivity to the straight portion of the renal proximal tubule.

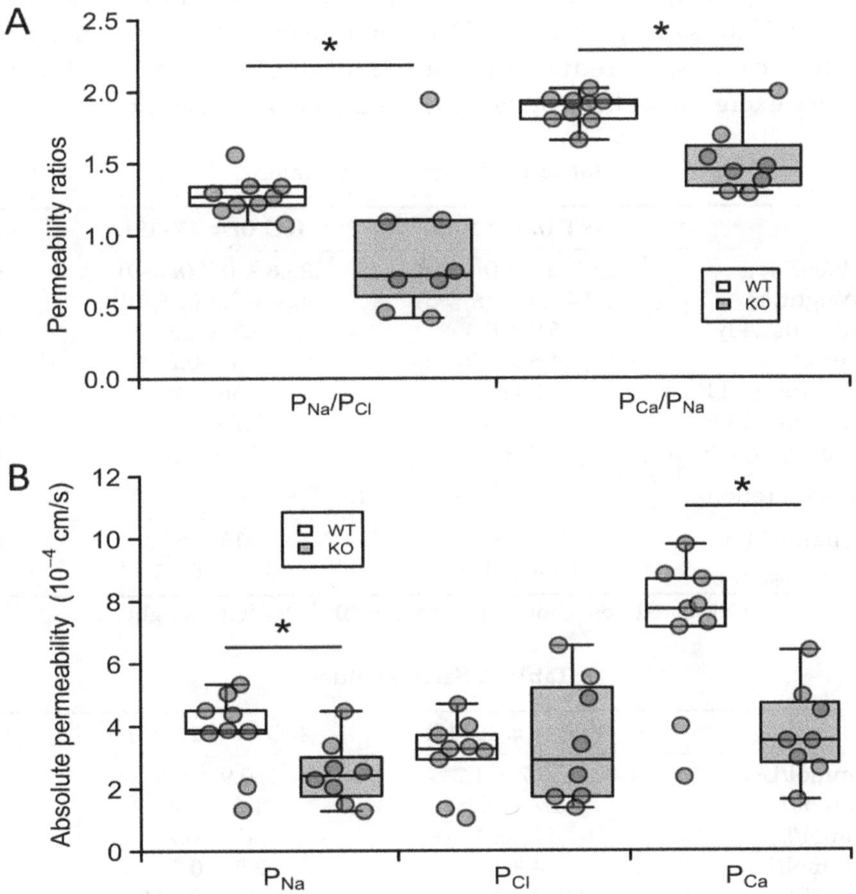

Figure 5. Ion permeabilities in proximal tubules from claudin-12 knockout (KO) and wild type (WT) mice. (**A**) Sodium–chloride and calcium–sodium permeability ratios. (**B**) Absolute permeabilities to sodium, chloride, and calcium. *—$p < 0.05$.

We next examined the calcium permeability characteristics of the proximal tubule. Interestingly, straight proximal tubules from wild type mice displayed a calcium to sodium permeability ratio of about 2 ($P_{Ca}/P_{Na} = 1.87 \pm 0.04$, Figure 5A), inferring that this part of the nephron is approximately twice as permeable to calcium as to sodium. Further, the deletion of claudin-12 resulted in a greater reduction in permeability to calcium than to sodium ($P_{Ca}/P_{Na} = 1.51 \pm 0.08$, Figure 5A), which was also reflected in decreased absolute calcium permeability ($7.1 \pm 0.8 \times 10^{-4}$ cm/s in the WT compared with $3.8 \pm 0.5 \times 10^{-4}$ cm/s in the KO) (Figure 5B, P_{Ca}). These results indicate that the straight portion of the proximal tubule has selective calcium permeability and that claudin-12 confers some, but not all, of the selective and total permeability to calcium.

2.4. Claudin-12 Null Mice Do not Have Hypercalciuria

We hypothesized that reduced calcium permeability in the proximal tubule of claudin-12 knockout mice would result in increased urinary calcium excretion and/or hormonal compensation. We therefore housed mice in metabolic cages with water and chow provided ad libitum. The claudin-12 knockout mice weighed the same as their wild type littermates and no differences were observed in water or chow ingested, urine volume, or fecal weight (Table 1). Further, analysis of serum at the end of the metabolic cage experiments failed to identify significant differences in plasma electrolytes, including Ca^{2+}, creatinine, glucose, and blood urea nitrogen (BUN) (Table 2). Importantly, we did not find differences in the circulating levels of parathyroid hormone (PTH), FGF23, nor the active form of vitamin D, 1,25 dihydroxy vitamin D_3/calcitriol (Table 2). Finally, we examined urinary and fecal excretion. We did not detect a significant difference in the urinary excretion of all electrolytes examined, including Ca^{2+} and Na^+ (Table 3). We also did not observe an alteration in urinary excretion of creatinine, consistent with the same glomerular filtration rate in both genotypes (Table 3). In summary, claudin-12 knockout mice, despite reduced proximal tubular Na^+ and Ca^{2+} permeability, do not display altered urinary excretion of these ions, or hormonal compensation.

Table 1. Metabolic cage data.

	WT (n = 12–13)	KO (n = 17–19)	p-Value
Female Weight, g	26.8 ± 1.3 (n = 10)	26.6 ± 0.7 (n = 9)	0.918
Male Weight, g	29.4 ± 0.9 (n = 3)	30.9 ± 0.8 (n = 10)	0.337
H_2O drunk, mL/24 h	6.0 ± 0.4	6.3 ± 0.3	0.426
Chow eaten, g	4.8 ± 0.2	5.3 ± 0.2	0.068
$Ca^{2+}_{ingested}$ [1], μmol/24 h	1138 ± 46	1266 ± 46	0.068
Urine volume, mL/24 h	1.6 ± 0.2	1.7 ± 0.2	0.603
Fecal excretion [2], g/24 h	5.8 ± 0.3	6.4 ± 0.3	0.162
Fecal calcium excretion			
Ca^{2+}_{feces}, μmol/24 h	1065 ± 53	1116 ± 57	0.536
$Ca^{2+}_{feces}/Ca^{2+}_{ingested}$	0.94 ± 0.04	0.88 ± 0.03	0.177

[1] Calculated from chow eaten per animal. [2] Dry feces weight.

Table 2. Serum values.

	WT (n = 12–13)	KO (n = 17–19)	p-Value
Na^+, mmol/L	150.7 ± 1.2	150.9 ± 0.6	0.820
K^+, mmol/L	4.7 ± 0.2	4.6 ± 0.2	0.784
Cl^-, mmol/L [1]	119 ± 1	117 ± 0.7	0.180
Ca^{2+}, mmol/L	2.4 ± 0.2	2.5 ± 0.1	0.369
Cr, μmol/L	0.030 ± 0.004	0.036 ± 0.004	0.315
Glucose, mmol/L	8.8 ± 0.5	9.3 ± 0.4	0.389
BUN, mmol/L [1]	29.2 ± 1.5	29.3 ± 1.5	0.954
PTH, pg/mL	421 ± 85	332 ± 52	0.370
FGF23, pg/mL	273 ± 16	306 ± 14	0.173
Vitamin D3, pg/mL	81 ± 15	80 ± 11	0.971

[1] WT n = 6, age (WT&KO) between 3 and 6 months.

Wild type and knockout mice were then fed a normal (0.6% wt/wt), low (0.01% wt/wt) or high (2.0% wt/wt) calcium diet, to evaluate if altering calcium intake would alter urinary calcium excretion (Table 4). The diets increased or reduced urinary calcium excretion as expected, but no clear differences between genotypes were observed for either FE_{Ca} or plasma Ca^{2+} levels between genotypes on the different diets.

Table 3. Urinary ion excretion.

	WT (n = 12)	KO (n = 13)	p-Value
Na^+/Creatinine	31 ± 8	39 ± 9	0.495
Cl^-/Creatinine	140 ± 45	136 ± 33	0.951
K^+/Creatinine	99 ± 25	93 ± 21	0.855
Ca^{2+}/Creatinine	0.24 ± 0.05	0.22 ± 0.04	0.819
PO_4^{3-}/Creatinine	5.4 ± 0.8	5.0 ± 0.7	0.688
Creatinine, μmol/24 h	12 ± 2	16 ± 3	0.377

Table 4. Urine ion excretion normalized to the creatinine of WT and KO mice on different calcium-containing diets.

	Normal			High Ca			Low Ca		
	WT (n = 12)	KO (n = 8)	p-Value	WT (n = 14)	KO (n = 15)	p-Value	WT (n = 15)	KO (n = 11)	p-Value
Urine Ca^{2+}/Cr	0.88 ± 0.08	0.79 ± 0.07	0.456	2.65 ± 0.48	3.10 ± 0.52	0.555	0.62 ± 0.03	0.60 ± 0.01	0.614
FE Ca	0.63 ± 0.15	0.37 ± 0.21	0.323	1.88 ± 0.46	1.67 ± 0.31	0.703	0.55 ± 0.14	0.31 ± 0.07	0.178

2.5. Renal Compensation in Claudin-12 KO Mice

Having failed to identify urinary wasting of Ca^{2+} or Na^+ or apparent hormonal compensation, we turned our attention to possible intrarenal compensatory mechanisms. Firstly, we examined the cortical protein expression of Claudin-10 and -2, which are also known to be expressed in the proximal tubule (Figure 6A–D). We found that both were significantly reduced in expression, in the KO mice.

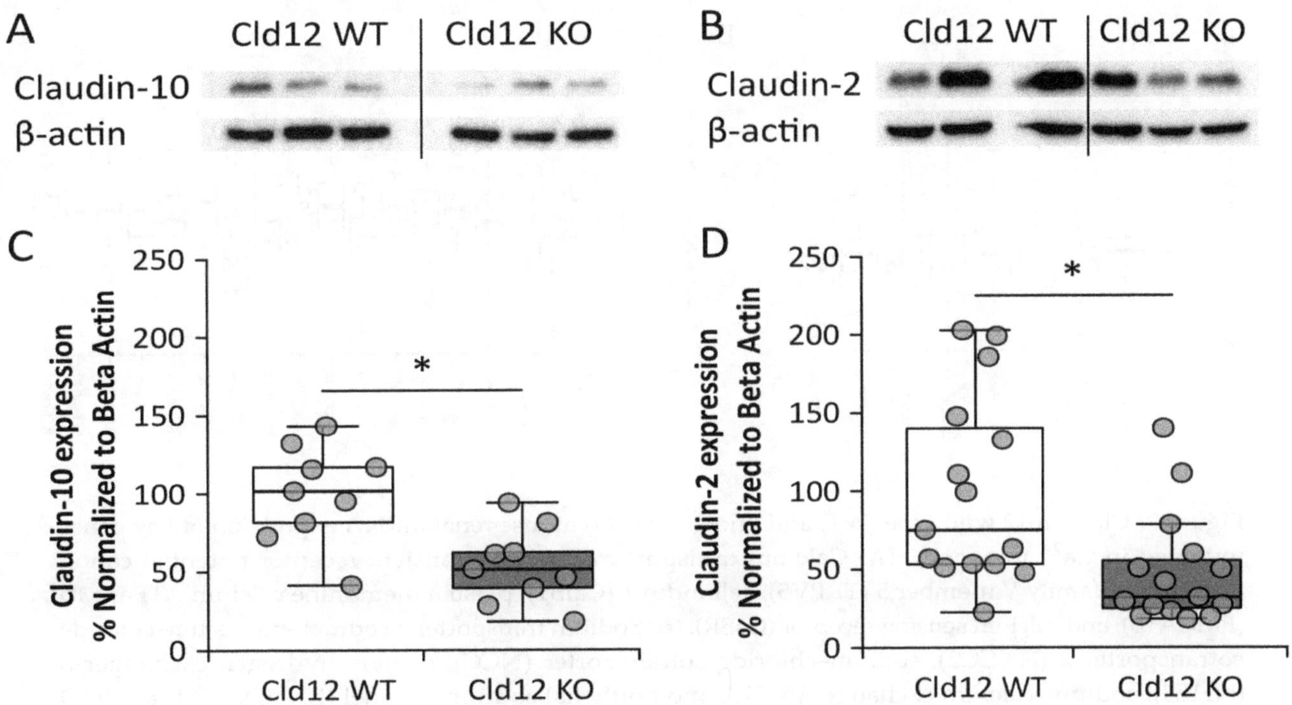

Figure 6. Claudin-12 wild type (WT) and knockout (KO) mouse renal cortical protein expression. (**A**,**C**) Claudin-10 and (**B**,**D**) Claudin-2. * $p < 0.05$.

We also isolated RNA from the whole kidney of wild type and claudin-12 KO mice and performed a quantitative real-time PCR. We did not find a significant difference in the expression of genes mediating transcellular Ca^{2+} reabsorption from the distal convoluted tubule except for PMCA1b, which was slightly reduced (Figure 7A). We also failed to identify significant differences in the expression of apical sodium transport proteins along the nephron (Figure 7B). Further, consistent with the absence of a significant difference in circulating calcitriol levels, we did not find differences in the expression of renal enzymes regulating the amount of this hormone in plasma, the vitamin D receptor, the calcium sensing receptor (CaSR), nor in klotho (Figure 7C). Finally, we assessed the expression of renal claudins (Figure 7D) and found that claudin-1, -8, and -14 had a significantly reduced expression in the knockout animals. Claudin-14 reduction is particularly interesting as it is only expressed in the cortical TAL [20], the nephron segment responsible for the reabsorption of around 25% of filtered calcium [1]. When present, claudin-14 attenuates calcium reabsorption from the TAL via the paracellular pathway [15]. Reduced expression in the knockout mice may therefore reflect increased TAL calcium reabsorption, in compensation for decreased proximal tubule calcium reabsorption.

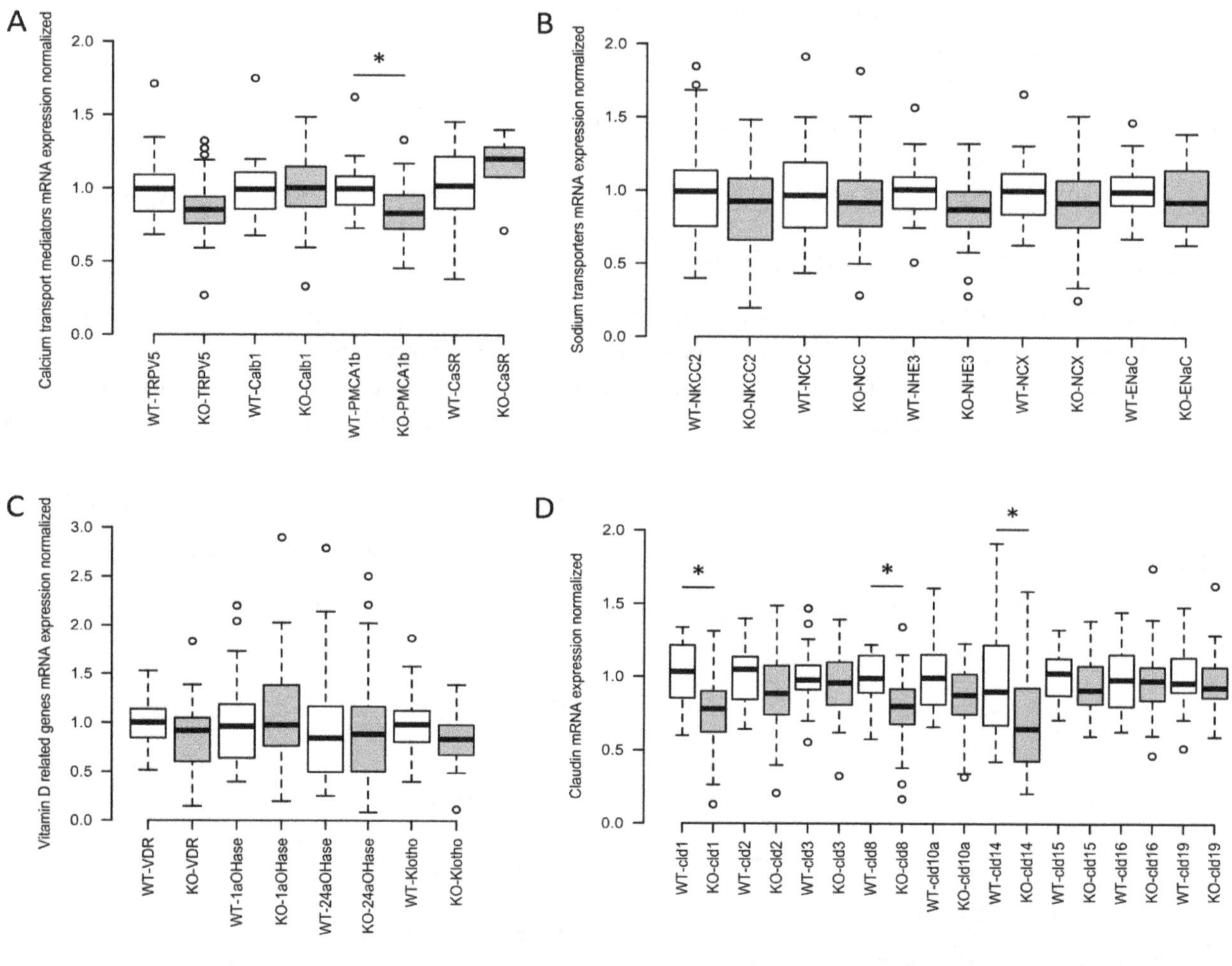

Figure 7. Claudin-12 wild type (WT) and knockout (KO) mouse renal mRNA expression of key genes involved in Ca^{2+} transport. (**A**) Calcium transport mediators: transient receptor potential cation channel subfamily V member 5 (TRPV5), calbindin 1 (Calb1), plasma membrane calcium ATPase 1b (PMCA1b), and calcium sensing receptor (CaSR). (**B**) Sodium transporters: sodium–potassium–chloride cotransporter 2 (NKCC2), sodium–chloride cotransporter (NCC), sodium-hydrogen exchanger 3 (NHE3), sodium–calcium exchanger (NCX), and epithelial sodium channel (ENaC). (**C**) Vitamin D related genes: vitamin D receptor (VDR), 25-hydroxyvitamin D 1 alpha hydroxylase (1aOHase), 24-hydroxylase (24aOHase), and klotho. (**D**) Renal claudins involved in cation transport normalized to 18S as house-keeping gene: claudins 1, 2, 3, 8, 10a, 14, 15, 16, and 19. $n = 26$ and 35, respectively. *—significant with Benjamini–Hochberg critical value for false discovery rate of 0.05. ° represents values that are above or below 1.5 times the interquartile range.

3. Discussion

Evidence supports a role for claudin-12 in contributing paracellular Ca^{2+} permeability in the intestine [11]. Claudin-12 mRNA is expressed in the kidney, and a proximal tubule cell culture model [10,13,21]. The proximal tubule is the location of the greatest amount of Ca^{2+} reabsorption along the nephron and this occurs via the paracellular pathway [1,3–5]. To date the tight junction proteins contributing calcium permeability to this segment have been incompletely defined. We therefore hypothesized that claudin-12 contributes paracellular permeability to the proximal tubule and generated a claudin-12 KO mouse, by replacing the only coding exon of the claudin-12 gene with β-galactosidase to assess this possibility. This model enabled us to localize *Cldn12* to the renal proximal tubule, and by perfusing tubules ex vivo from wild type mice and claudin-12 KO animals, we confirmed that *Cldn12* does indeed contribute Ca^{2+} and Na^{+} permeability to this segment. However, the deletion of *Cldn12* does not induce hypercalciuria or an increase in calciotropic hormone levels.

Direct measurement of Ca^{2+} permeability of the straight portion of the proximal tubule of wild type and claudin-12 KO mice revealed relative selectivity for Ca^{2+} over Na^{+}, which was attenuated in the absence of claudin-12. This strongly supports *Cldn12* contributing to the Ca^{2+} permeability of the proximal tubule. It is surprising then that claudin-12 null mice did not display hypercalciuria in response to a failure to reabsorb Ca^{2+} from the proximal straight tubule. Further, there appeared to be no systemic compensation for reduced paracellular Ca^{2+} permeability in the proximal nephron, as calciotropic hormone levels were also unchanged. What then may explain our observations? Potentially altered permeability of the proximal tubule is compensated by an increased driving force for Ca^{2+} across this segment. This could occur via generating a more lumen positive potential difference across the segment, or increasing water reabsorption from the proximal tubule, which would in turn either enhance the Ca^{2+} concentration gradient or convective driving force for paracellular Ca^{2+} [22]. We did not observe a significant difference in baseline transepithelial voltage (Vte) across tubules perfused ex vivo between wild type and knockout mice (Vte WT = -3.2 ± 0.4 mV, $n = 9$ and KO = -2.5 ± 0.5 mV, $n = 8$, $p = 0.253$), making the former explanation unlikely. Further, we did not observe altered expression of the major Na^{+} transporter in the proximal tubule, sodium hydrogen exchanger isoform 3(NHE3), suggesting that this is not the explanation, although altered NHE3 activity can occur in the absence of altered expression [23–25]. Instead, we observed decreased *Cldn14* expression, consistent with increased Ca^{2+} reabsorption from the TAL in compensation. Claudin-14 interacts in the cortical TAL (cTAL) tight junction with the claudin-16/19 complex, reducing permeability to calcium, and thus calcium absorption from this segment [15,26]. This suggests that increased Ca^{2+} reabsorption from the TAL is compensating for reduced proximal tubule Ca^{2+} reabsorption.

How might *Cldn14* expression be altered in the absence of changes in circulating Ca^{2+} or PTH levels? The expression of Cldn14 is increased by the activation of the calcium sensing receptor (CaSR) in the basolateral membrane of the TAL [15]. Plasma Ca^{2+} levels were not different between wild type and claudin-12 KO mice, however, we were unable to assess the local concentration of Ca^{2+} in the peritubular space of the cortical TAL (where *Cldn14* is expressed [20]). Interestingly, the cTAL runs linearly into the juxtamedullary cortex, and Figure 3 reveals that the cTAL directly abuts tubules with significant *Cldn12* promoter activity. Consistent with this observation, microdissection of straight segments of proximal tubules frequently demonstrated the close adherence of a cTAL (Figure 8). Perhaps, then, reduced reabsorption of Ca^{2+} from the straight portion of the proximal tubule, leads to a lower concentration of Ca^{2+} in the juxtamedullary interstitium, and thus reduced CaSR activation? This would result in increased cTAL Ca^{2+} permeability and increased reabsorption from this segment, in compensation. The proposed proximal tubule–cTAL crosstalk would explain our lack of alteration in urinary Ca^{2+} excretion and the lack of hormonal compensation. However, this hypothesis requires further experimental testing to support it.

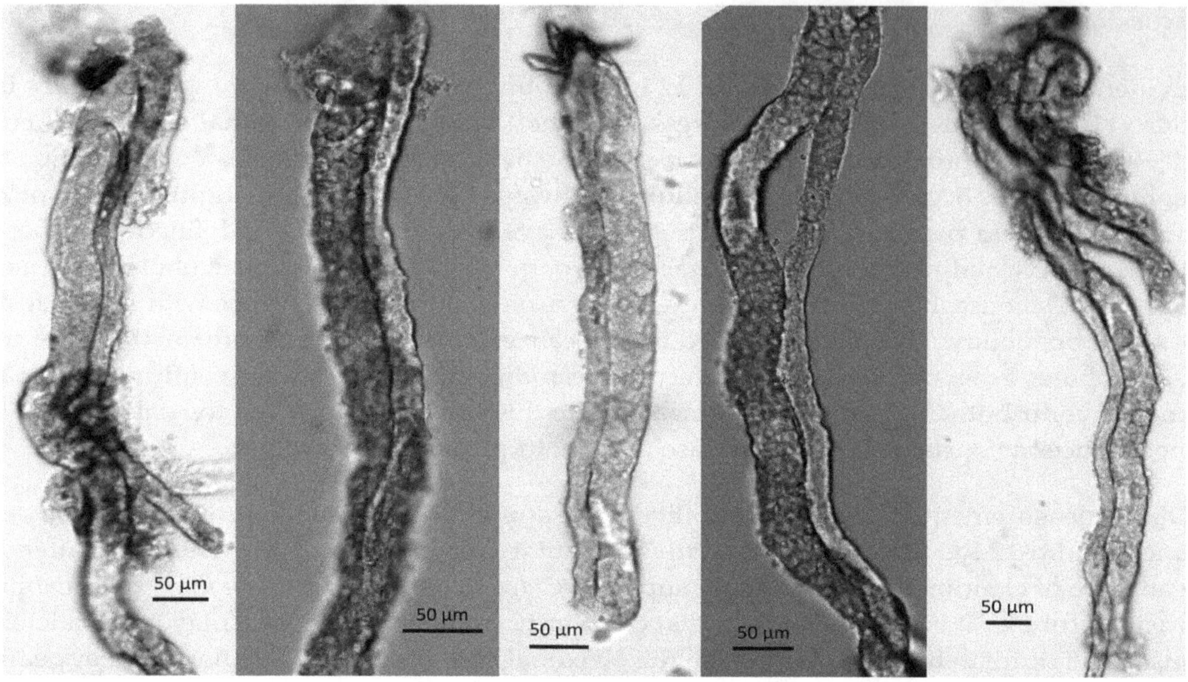

Figure 8. Examples of cTAL (thinner tubule) attached to the straight portion of the pars recta of the proximal tubule (thicker tubule).

Claudin-2 has also been implicated in paracellular intestinal Ca^{2+} absorption. In addition, this protein has been clearly localized to the proximal tubule, where it contributes paracellular Na^{+} permeability [9,11,27]. Proximal tubule paracellular Na^{+} permeability contributes a significant amount of net Na^{+} reabsorption from this segment. Using the paracellular pathway and the concentration gradient/convection forces induced by the considerable water flux across this segment increases the efficiency of this process, thereby reducing the energy required [28]. Interestingly, claudin-2 KO mice display hypercalciuria [9]. It is therefore tempting to speculate that claudin-2 also contributes proximal tubule Ca^{2+} permeability as we observe for claudin-12. Further specific studies will be required to support this. Why claudin-2 KO mice demonstrate hypercalciuria and claudin-12 null mice do not is unclear, but is perhaps due to a greater abundance of claudin-2 in the proximal tubule.

PTH is released from the parathyroid gland in response to low plasma calcium levels. It acts directly on the proximal tubule to inhibit transcellular sodium reabsorption and likely paracellular $Ca2^{+}$ reabsorption [22]. It also acts on the distal convoluted tubule/connecting tubule to increase Ca^{2+} reabsorption [29]. FGF23 is another phosphocalciotropic hormone that acts on the proximal tubule to inhibit sodium phosphate cotransport, and on the distal convoluted tubule/connecting tubule to augment Ca^{2+} absorption [30,31]. Although we did not detect a significant difference in the circulating amounts of these hormones between wild type and claudin-12 KO mice, there were large variations in plasma levels. It is thus possible that alterations in these hormones could be compensating for the reduced paracellular Ca^{2+} permeability in the proximal tubule.

The lack of a specific claudin-12 antibody has limited our ability to confirm protein localization. This is despite trying several commercially available antibodies and generating two of our own. It should be emphasized that we are not the only group to experience this difficulty [12,17]. Regardless, we have shown clearly the absence of claudin-12 DNA in the knockout mice and the inability to detect messenger RNA in the kidney of knockout mice. Moreover, the claudin-12 KO mice that we generated do not have alterations in the endogenous promoter, and we replaced the coding sequence with a reporter β-galactosidase, which clearly demonstrates an expression in the renal cortex, as seen in Figure 2. We are therefore confident that *Cldn12* is expressed in the renal proximal tubule where it confers Ca^{2+} permeability. We should emphasize that we cannot exclude lower levels of expression in other nephron segments.

In conclusion, using a mutant mouse engineered to have β-galactosidase replace the coding exon of *Cldn12*, we were able to localize claudin-12 expression to the renal proximal tubule. Microperfusion studies on this model revealed reduced Na^{+} and Ca^{2+} permeability consistent with claudin-12 conferring paracellular Ca^{2+} permeability to the proximal tubule. Surprisingly the knockout mice did not have increased urinary Ca^{2+} excretion, or alterations in calciotropic hormone levels. We propose that this is due to a proximal tubule–TAL crosstalk leading to reduced *Cldn14* expression and consequently increased Ca^{2+} reabsorption from the TAL, in compensation for decreased proximal tubule Ca^{2+} reabsorption.

4. Materials and Methods

4.1. Generation of Cld12 KO Mice

A claudin-12 null strain was generated through the UC Davies Knock Out Mouse Project (KOMP). The gene encoding claudin-12 is located on chromosome 5, and contains four exons with the coding sequence encoded by exon four (NCBI Gene: 64945). A targeting vector (Velocigene cassette ZEN-Ub1, KOMP Repository category number 13208L1) was used to replace exon four with the neomycin resistance gene and the lacZ reporter gene by homologous recombination in VGB6 embryonic stem (ES) cells derived from the mouse strain C57BL/6NTac. The mouse strain used for this research project, KOMP ES cell line $Cldn12^{tm1(KOMP)Vlcg}$, RRID:MMRRC_053773-UCD, was obtained from the Mutant Mouse Resource and Research Center (MMRRC) at the University of California at Davis, an NIH-funded strain repository, and was donated to the MMRRC by The KOMP Repository, UC Davis Mouse Biology Program. This created a 760-bp deletion between positions 5507663–5508422 of Chromosome 5 resulting in heterozygotes (HET) ES cells (Figure 1A). Subsequently, the recombinant ES cells were used to generate HET mice. Littermates were used as controls. Intercrossing of heterozygotes yielded claudin-12 deficient mice (KO). Genotypes were confirmed by RT-PCR, and gene sequencing (Figure 1B–D).

4.2. Metabolic Cage Studies

Metabolic Cage studies were performed as previously reported [2,15]. Three month-old wild type (WT), or claudin-12 deficient (KO) mice were placed in metabolic cages for three days with free access to normal rodent diet chow (LabDiet® 5001, Fort Worth, TX, USA) and water. Weight, chow eaten, and water consumed were measured daily. Urine and feces were collected every 24 h for analysis. Only samples from day three were used for analysis. Then, animals were anesthetized using pentobarbital sodium, and blood was collected for analysis of serum creatinine via high-performance liquid chromatography. We also performed a blood gas at that time. Kidneys were harvested and snap frozen in liquid nitrogen and stored at −80 °C. Metabolic cage studies on wild type and claudin-12 KO mice fed different calcium containing diets (low = 0.01% TD. 95027, normal = 0.6% TD.97191, or high = 2% TD.00374, all from Harlan Laboratories, Madison, WI, USA) were performed as previously described [15]. All experimental procedures were approved on 3 October 2013 by the Animal Care and Use Committee for Health Sciences at the University of Alberta (protocol number 00000213).

4.3. Measurement of Urinary, Fecal, and Plasma Electrolytes

Serum electrolytes, blood urea nitrogen, and glucose were measured with a handheld blood gas analyzer (Vet Scan i-STAT1 Analyzer, Abaxis, Union City, CA, USA) as previously described [15]. Feces from the last 24-h collection were dried for 48 h in an incubator (Labline, Imperial III Incubator, Mumbai, India) at 55 °C, and then homogenized with a mortar and pestle. A sample of 0.7 g per mouse was then taken and solubilized in nitric acid for elemental analysis. Nitric acid (70%) was added to the samples in two sequential steps: the first 0.3 mL were added when the sample was at 65 °C, then 0.7

mL were added once fumes subsided after about 30 min. The mix was then heated to 85 °C and boiled until the fumes stopped (approximately 1 h later). Subsequently, 1 mL of 30% H_2O_2 was added, again in two steps: first, 0.2 mL were added while the sample was still at 85 °C. Once boiling stopped, the sample was allowed to cool for 10 min before adding the remaining 0.8 mL. After another 10 min at room temperature, the samples were heated again to 85 °C until the reaction stopped (approximately 1 h later). The samples were adjusted to 5 mL with ddH_2O and used for analysis. The calcium content was measured using the QuantiChrom Ca^{2+} Assay Kit (Catalog number: DICA-500) from BioAssay Systems (Hayward, CA, USA).

Urine electrolyte content was measured by ion chromatography. All experiments were performed with a Dionex Aquion Ion Chromatography (IC) System (category number 22176-60004, Thermo Fisher Scientific Inc., Mississauga, ON, Canada) equipped with an autosampler. The cation and anion eluents were prepared as per the manufacturer's instructions. The anion eluent consisted of a 4.5 mM Na_2CO_3/1.5 mM $NaHCO_3$ in ddH_2O, and the cation eluent consisted of 20 mM Methanesulfonic acid in ddH2O. All reagents and samples were filtered with a 0.2 µm syringe filter (category number 03-391-1B, Thermo Fisher Scientific Inc., Mississauga, ON, Canada), and stored in high density polyethylene containers that had been thoroughly cleaned with deionized water to avoid any traces of ions. Calibration standard curves were produced using Dionex five anion standard (category number 037157, Thermo Fisher Scientific Inc., Mississauga, ON, Canada) and Dionex six cation-I standard (category number 040187, Thermo Fisher Scientific Inc., Mississauga, ON, Canada). Thermo Scientific Chromeleon 7 Chromatography Data System (CDS) software was used for automation and data handling. High-performance liquid chromatography (HPLC) was used to detect creatinine concentration in urine and blood samples. All experiments were performed with a Dionex UltiMate 3000 HPLC System (Thermo Scientific, ISO-3100SD PUMP category number 5040.0011; TCC-3000SD Column Thermostat, category number 5730.0010, and VWD-3100 DETECTOR, category number 5074.0005, Thermo Fisher Scientific Inc., Mississauga, ON, Canada). UV detection of samples occurred at 216 nm and the pump flow rate was set at 0.2 mL/min accordingly. The HPLC eluent was prepared as per the manufacturer's instructions. The eluent consisted of acetonitrile (category number A-0626-17, Thermo Scientific)/Trifluoroacetic acid category number T-3258-PB05, Thermo Scientific, Mississauga, ON, Canada) in ddH_2O. All reagents and samples were filtered using 0.2 µM syringe filters (Thermo Scientific, category number 03-391-1B, Mississauga, ON, Canada) and stored in glass containers that had been thoroughly cleaned with deionized water to avoid sample adsorption. Calibration standard curves were produced using creatinine from Acros Organics, NJ, USA, category number AC228940250. Thermo Scientific Chromeleon 7 Chromatography Data System (CDS) software was used for automation and data handling.

4.4. Measurement of Hormone Levels

PTH was measured by ELISA according the manufacturer's instructions (mouse PTH 1-84 ELISA Kit, Immutopics Inc., category number 60-2305, San Diego, CA, USA) and Vitamin D levels were determined by a radioimmunoassay (1,25-Dihydroxy Vitamin D RIA kit, Immunodiagnostic systems, category number AA-54F1, Immunodiagnostic Systems Inc, Gaithersburg, MD, USA), according to the manufacturer's instructions as previously reported [2,15]. FGF23 was quantified by ELISA according to the manufacturer's instructions (FGF-23 ELISA Kit, Kainos, category number CY-4000, Tokyo, Japan).

4.5. X-Gal Staining and Immunofluorescence Microscopy on Renal Sections

We performed X-gal staining on 8-µm renal cryosections fixed with periodate–lysine–paraformaldehyde and prepared as previously reported [2]. Sections were first washed three times with fresh phosphate buffered saline, then rinsed briefly with distilled water. Next sections were incubated for approximately 3 h in X-gal solution, which consisted of a 1:40 dilution of pure X-gal (Sigma-Aldrich, Oakville ON, Canada) dissolved in 4% DMSO with a dilution buffer consisting of potassium ferricyanide crystalline 5 mM, potassium ferricyanide trihydrate 5 mM, $MgCl_2$ 2 mM, and PBS 100 mM, to allow the development of a blue color. Wild type littermate sections were taken as a negative control. Then, sections were rinsed with PBS twice for 5 min before mounting with DAKO (Carpinteria, CA, USA) or undergoing immunostaining, essentially as previously reported [19,32]. In brief, antigen retrieval was performed with sodium citrate, and then washed three times with TN buffer (0.1 M Tris/HCl and 0.15 M NaCl, pH 7.6). Then the section was incubated with 0.3% H_2O_2 in TN buffer for 30 min. Sections were then washed three times with TN buffer before blocking for 1 h with TN buffer containing 0.5% Blocking Reagent (Perkin Elmer, MA, USA) and 0.05% Tween 20. Next, primary antibodies were applied in TN buffer (1:1500 anti-NKCC2, Developmental Studies Hybridoma Bank University of Iowa, USA, 1:500 anti- CAII (Santa Cruz Biotechnology Inc, Dallas, TX, USA, 1:500 anti-AQP1, Santa Cruz Biotechnology Inc., Dallas, TX USA or 1:500 Anti-NCC, Stress Marq Biosciences Inc. Victoria, BC, Canada), overnight all diluted in TN buffer. The slides were then washed three times with TN buffer and then a Cy3 conjugated secondary antibody of either donkey anti-mouse (NKCC2, CAII), anti-rabbit (NCC), or anti-goat (AQP1) was applied in the TN buffer containing 0.05% Tween-20 (TNT) for an hour. Finally, the slides were washed three times with TNT buffer before being mounted and visualized with an Olympus BX51 microscope equipped with a X-cite Series 120Q light source (Lumen Dynamics Inc. Mississauga, ON, Canada) and an Infinity 3, 1.4 Megapixel Cooled, color Camera (Lumenera, Ottawa, ON, Canada).

4.6. Proximal Tubular Perfusion and Electrophysiology

Proximal tubules were isolated and microperfused essentially as previously described [33]. To this end, kidneys were collected from three month-old claudin-12 WT and KO mice, swiftly decapsulated and sliced transversely. Cortical and juxtamedullary straight proximal tubules where mechanically dissected from the slices at 4 °C on a Leica M165C dissecting microscope and transferred to an Axi Observer Zeiss microscope for microperfusion. A Vestavia Scientific, LLC (Atlanta, GA, USA) microperfusion system of concentric micropipettes was used to hold and microperfuse the tubules as described previously [34]. The glass pipettes where built on a Luigs & Neumann GmbH microforge (Ratingen, Germany). We used for the external holding pipettes sodalime glass capillaries (Hilgenberg, Germany, category number 1409679), and elongated double-barreled glass capillary (Theta-Bo-glasscapillaries, Hilgenberg, Germany, category number 1402401) for the double-barreled perfusion pipettes. A silver wire of 0.2-mm diameter was inserted in one of the barrels of the perfusion pipette to be later connected to a pulse generator. We injected currents of 13 nA for one second every other second via a 150-MΩ resistor into the perfused tubules. The generated voltage deflections where used to later calculate the transepithelial resistance using the cable equation. The perfusion system was connected to a pressure pump that was designed at the Institute of Physiology of the University of Kiel, Germany, to keep a constant luminal flow through the perfused tubules. The bath and tubule lumen (via one of the barrels of the perfusion pipette) where connected via agar bridges (3M KCl, 3% agar) to silver–chloride electrodes, Ag/AgCl (RC-3, World Precision Instruments, Sarasota, FL, USA), which were themselves connected to ground, an amplifier, and to an impedance converter. The signal was digitalized with a PowerLab 8/SP and recorded with LabChart data acquisition and analysis software (LabChart 4.2, ADInstruments–North America, Colorado Springs, CO, USA). This enabled the recording of the transepithelial voltage and the transepithelial voltage deflections throughout the course the experiments. The length and diameter of the tubules were measured on the digital images taken using Zen 2.6 Blue Edition Carl Zeiss Microscopy modular image-processing and analysis software (Carl Zeiss Canada, Toronto, ON, Canada). Images were captured with a Hamamatsu ORCA-Flash4.0 V3 digital camera, model C13440-20CU (Hamamatsu Corporation - Americas, Bridgewater, NJ, USA).

In order to measure calcium permeability, tubules were first perfused with modified Ringer's control solution (132 mM NaCl, 0.4 mM KH_2PO_4, 1.4 mM K_2HPO_4, 1.2 mM $MgCl_2 \cdot 6H_2O$, 5 mM Na-Acetate $\cdot$ $3H_2O$, 2.5 mM $NaHCO_3$, 1.3 mM $CaCl_2$, 5 mM L-alanine, and 5 mM glucose) in a running bath of the same control solution. Subsequently, the bath solution was exchanged to the same control solution containing ouabain (10 μM). Once voltage recordings stabilized, we changed the basolateral solution to first induce a NaCl dilution potential, and then a Na^+-Ca^{2+} bi-ionic diffusion potential (all the solutions contained 10 μM ouabain) (Figure 4B,C). The 70 mM low sodium solution (65 mM NaCl, 0.4 mM KH_2PO_4, 1.4 mM $K_2HPO_4 \cdot 3H_2O$, 1.2 mM $MgCl_2 \cdot 6H_2O$, 5 mM Na-Acetate $\cdot$ $3H_2O$, 2.5 mM $NaHCO_3$, 1.3 mM $CaCl_2$, 5 mM L-alanine, 5 mM Glucose, and 112 mannitol) and the 72.5 mM high calcium containing solution (3.6 mM KCl, 72.5 mM $CaCl_2 \cdot 2H_2O$, 5 mM glucose, and 3mM HEPES) were designed using the control solution as the base. The sodium–chloride permeability ratios and the calcium–sodium permeability ratios were then calculated using the Goldmann–Hodgin–Katz flux equation. The absolute permeability to sodium was calculated using the simplified Kimizuka–Koketsu equation, and the absolute permeabilities to chloride and calcium obtained from this value and the permeability ratios [35,36].

4.7. *Quantitative Real-Time PCR and Immunoblotting*

Quantitative real-time PCR and immunoblotting were performed essentially as previously described [2,15,16,21]. Total mRNA was extracted from $\frac{1}{2}$ of a snap frozen kidney from three month-old mice using the TRIzol Reagent (Invitrogen, Carlsbad, CA, USA) according to the manufacturer's instructions. After treatment with DNAse (ThermoScientific, Vilnius, LT, USA, category number EN052), cDNA was reverse transcribed from the isolated total mRNA. Primers and probes were obtained from Applied Biosystems Inc. (Foster City, CA, USA). A quantity of 5 μL (125 ng of cDNA) was used as a template to determine the gene expression by qPCR. TaqMan universal qPCR master mix (Applied Biosystems Inc, Foster City, CA, USA), primer, probe, and RNAse-free water were mixed together, then added to cDNA that had been placed in a 384-well plate (Applied Biosystems Inc, Foster City, CA, USA). Internal control mRNA levels of 18S ribosomal RNA were measured. Expression levels were quantified with an ABI Prism 7900 HT Sequence Detection System (Applied Biosystems Inc, Foster City, CA, USA), and 18S was used for normalization of RNA as none of the experimental perturbations resulted in a significant change in its expression. Immunoblotting was carried out using an anti-claudin-10 (Invitrogen Inc., category number 38-8400, a subsidy of Thermo Fischer Inc., Mississauga, ON, Canada) or anti-claudin-2 (Thermo Fischer Inc., category number 32-5600, Mississauga, ON, Canada) and appropriate secondary antibodies, and then quantified as previously described [2].

4.8. *Statistical Analysis*

The differences between group mean were assessed by Student's t test with a Bonferroni correction for multiple comparisons applied when needed. The Shapiro–Wilk test was used to evaluate the data for normal distribution when pertinent. The results are presented as mean ± SEM (n = number of animals throughout) and considered significant at $p < 0.05$. Renal mRNA expression was analyzed using the Benjamini–Hochberg procedure, with a critical value for false discovery rate of 0.05 [37].

Author Contributions: Conceptualization and design, A.P., W.P., D.O., M.U., M.R.B., M.F., H.D., E.C. and R.T.A.; experimental data acquisition, A.P., W.P., D.O., M.U. and M.F.; data analysis and interpretation, A.P., W.P., D.O., M.U., M.R.B., M.F., E.C. and R.T.A.; figure preparation, A.P., W.P. and M.U.; manuscript drafting, A.P. and R.T.A.; editing and revision of the manuscript, A.P., W.P., D.O., M.U., M.R.B., M.F., H.D., E.C. and R.T.A.; funding acquisition, R.T.A. All authors have read and agreed to the published version of the manuscript.

Abbreviations

PT	Renal proximal tubule
KO	Null (knock-out) mice
WT	Wild-type mice
P_X	Paracellular permeability to ion X
(c)TAL	(cortical) Thick ascending limb of the Henle's loop
PTH	Parathyroid hormone
FGF23	Fibroblast growth factor 23
ES	Embryonic stem cells

HET	Heterozygotes
AQP1	Aquaporin I
NKCC2	Sodium–potassium–chloride cotransporter II
NCC	Sodium–chloride cotransporter
CAII	Carbonic anhydrase II
BUN	Blood urea nitrogen
PMCA1b	Plasma membrane Ca^{2+} ATPase, isoform 1b
NHE3	Sodium–hydrogen exchanger type 3
CaSR	Calcium sensing receptor

References

1. Suki, W.N. Calcium transport in the nephron. *Am. J. Physiol.* **1979**, *237*, F1–F6. [CrossRef] [PubMed]
2. Pan, W.; Borovac, J.; Spicer, Z.; Hoenderop, J.G.; Bindels, R.J.; Shull, G.E.; Doschak, M.R.; Cordat, E.; Alexander, R.T. The epithelial sodium/proton exchanger, NHE3, is necessary for renal and intestinal calcium (re)absorption. *Am. J. Physiol.* **2012**, *302*, F943–F956. [CrossRef] [PubMed]
3. Brunette, M.; Aras, M. A microinjection study of nephron permeability to calcium and magnesium. *Am. J. Physiol.* **1971**, *221*, 1442–1448. [CrossRef] [PubMed]
4. Frick, A.; Rumrich, G.; Ullrich, K.J.; Lassiter, W.E. Microperfusion study of calcium transport in the proximal tubule of the rat kidney. *Pflug. Arch. Gesamte Physiol. Menschen Tiere* **1965**, *286*, 109–117. [CrossRef] [PubMed]
5. Murayama, Y.; Morel, F.; Le Grimellec, C. Phosphate, calcium and magnesium transfers in proximal tubules and loops of Henle, as measured by single nephron microperfusion experiments in the rat. *Pflug. Arch.* **1972**, *333*, 1–16. [CrossRef]
6. Edwards, A.; Bonny, O. A model of calcium transport and regulation in the proximal tubule. *Am. J. Physiol.* **2018**, *315*, F942–F953. [CrossRef]
7. Worcester, E.M.; Coe, F.L.; Evan, A.P.; Bergsland, K.J.; Parks, J.H.; Willis, L.R.; Clark, D.L.; Gillen, D.L. Evidence for increased postprandial distal nephron calcium delivery in hypercalciuric stone-forming patients. *Am. J. Physiol.* **2008**, *295*, F1286–F1294. [CrossRef]
8. Yu, A.S. Claudins and the kidney. *J. Am. Soc. Nephrol.* **2015**, *26*, 11–19. [CrossRef]
9. Muto, S.; Hata, M.; Taniguchi, J.; Tsuruoka, S.; Moriwaki, K.; Saitou, M.; Furuse, K.; Sasaki, H.; Fujimura, A.; Imai, M.; et al. Claudin-2-deficient mice are defective in the leaky and cation-selective paracellular permeability properties of renal proximal tubules. *Proc. Natl. Acad. Sci. USA* **2010**, *107*, 8011–8016. [CrossRef] [PubMed]
10. Beggs, M.R.; Appel, I.; Svenningsen, P.; Skjodt, K.; Alexander, R.T.; Dimke, H. Expression of transcellular and paracellular calcium and magnesium transport proteins in renal and intestinal epithelia during lactation. *Am. J. Physiol.* **2017**, *313*, F629–F640. [CrossRef]
11. Fujita, H.; Sugimoto, K.; Inatomi, S.; Maeda, T.; Osanai, M.; Uchiyama, Y.; Yamamoto, Y.; Wada, T.; Kojima, T.; Yokozaki, H.; et al. Tight junction proteins claudin-2 and -12 are critical for vitamin D-dependent Ca^{2+} absorption between enterocytes. *Mol. Biol. Cell* **2008**, *19*, 1912–1921. [CrossRef]
12. Kiuchi-Saishin, Y.; Gotoh, S.; Furuse, M.; Takasuga, A.; Tano, Y.; Tsukita, S. Differential expression patterns of claudins, tight junction membrane proteins, in mouse nephron segments. *J. Am. Soc. Nephrol.* **2002**, *13*, 875–886. [PubMed]
13. Borovac, J.; Barker, R.S.; Rievaj, J.; Rasmussen, A.; Pan, W.; Wevrick, R.; Alexander, R.T. Claudin-4 forms a paracellular barrier, revealing the interdependence of claudin expression in the loose epithelial cell culture model opossum kidney cells. *Am. J. Physiol. Cell Physiol.* **2012**, *303*, C1278–C1291. [CrossRef] [PubMed]
14. Abuazza, G.; Becker, A.; Williams, S.S.; Chakravarty, S.; Truong, H.T.; Lin, F.; Baum, M. Claudins 6, 9 and 13 are developmentally expressed renal tight junction proteins. *Am. J. Physiol. Renal Physiol.* **2006**, *291*, F1132–F1141. [CrossRef] [PubMed]
15. Dimke, H.; Desai, P.; Borovac, J.; Lau, A.; Pan, W.; Alexander, R.T. Activation of the Ca^{2+}-sensing receptor increases renal claudin-14 expression and urinary Ca^{2+} excretion. *Am. J. Physiol. Renal Physiol.* **2013**, *304*, F761–F769. [CrossRef]
16. Krishnan, D.; Pan, W.; Beggs, M.R.; Trepiccione, F.; Chambrey, R.; Eladari, D.; Cordat, E.; Dimke, H.; Alexander, R.T. Deficiency of Carbonic Anhydrase II Results in a Urinary Concentrating Defect. *Front. Physiol.* **2017**, *8*, 1108. [CrossRef]

17. Castro Dias, M.; Coisne, C.; Baden, P.; Enzmann, G.; Garrett, L.; Becker, L.; Holter, S.M.; German Mouse Clinic, C.; Hrabe de Angelis, M.; Deutsch, U.; et al. Claudin-12 is not required for blood-brain barrier tight junction function. *Fluids Barriers CNS* **2019**, *16*, 30. [CrossRef]
18. Sabolic, I.; Valenti, G.; Verbavatz, J.M.; Van Hoek, A.N.; Verkman, A.S.; Ausiello, D.A.; Brown, D. Localization of the CHIP28 water channel in rat kidney. *Am. J. Physiol.* **1992**, *263*, C1225–C1233. [CrossRef]
19. Vilas, G.; Krishnan, D.; Loganathan, S.K.; Malhotra, D.; Liu, L.; Beggs, M.R.; Gena, P.; Calamita, G.; Jung, M.; Zimmermann, R.; et al. Increased water flux induced by an aquaporin-1/carbonic anhydrase II interaction. *Mol. Biol. Cell* **2015**, *26*, 1106–1118. [CrossRef]
20. Gong, Y.; Renigunta, V.; Himmerkus, N.; Zhang, J.; Renigunta, A.; Bleich, M.; Hou, J. Claudin-14 regulates renal Ca^{++} transport in response to CaSR signalling via a novel microRNA pathway. *EMBO J.* **2012**, *31*, 1999–2012. [CrossRef] [PubMed]
21. Rievaj, J.; Pan, W.; Cordat, E.; Alexander, R.T. The Na^+/H^+ exchanger isoform 3 is required for active paracellular and transcellular Ca^{2+} transport across murine cecum. *Am. J. Physiol. Gastrointest. Liver Physiol.* **2013**, *305*, G303–G313. [CrossRef] [PubMed]
22. Alexander, R.T.; Dimke, H.; Cordat, E. Proximal tubular NHEs: Sodium, protons and calcium? *Am. J. Physiol. Renal Physiol.* **2013**, *305*, F229–F236. [CrossRef] [PubMed]
23. Schultheis, P.J.; Clarke, L.L.; Meneton, P.; Miller, M.L.; Soleimani, M.; Gawenis, L.R.; Riddle, T.M.; Duffy, J.J.; Doetschman, T.; Wang, T.; et al. Renal and intestinal absorptive defects in mice lacking the NHE3 Na^+/H^+ exchanger. *Nat. Genet.* **1998**, *19*, 282–285. [CrossRef] [PubMed]
24. Leong, P.K.; Devillez, A.; Sandberg, M.B.; Yang, L.E.; Yip, D.K.; Klein, J.B.; McDonough, A.A. Effects of ACE inhibition on proximal tubule sodium transport. *Am. J. Physiol. Renal Physiol.* **2006**, *290*, F854–F863. [CrossRef]
25. Riquier-Brison, A.D.; Leong, P.K.; Pihakaski-Maunsbach, K.; McDonough, A.A. Angiotensin II stimulates trafficking of NHE3, NaPi2, and associated proteins into the proximal tubule microvilli. *Am. J. Physiol. Renal Physiol.* **2010**, *298*, F177–F186. [CrossRef]
26. Plain, A.; Wulfmeyer, V.C.; Milatz, S.; Klietz, A.; Hou, J.; Bleich, M.; Himmerkus, N. Corticomedullary difference in the effects of dietary Ca^{2+} on tight junction properties in thick ascending limbs of Henle's loop. *Pflug. Arch.* **2016**, *468*, 293–303. [CrossRef]
27. Enck, A.H.; Berger, U.V.; Yu, A.S. Claudin-2 is selectively expressed in proximal nephron in mouse kidney. *Am. J. Physiol. Renal Physiol.* **2001**, *281*, F966–F974. [CrossRef]
28. Pei, L.; Solis, G.; Nguyen, M.T.; Kamat, N.; Magenheimer, L.; Zhuo, M.; Li, J.; Curry, J.; McDonough, A.A.; Fields, T.A.; et al. Paracellular epithelial sodium transport maximizes energy efficiency in the kidney. *J. Clin. Investig.* **2016**, *126*, 2509–2518. [CrossRef]
29. Greger, R.; Lang, F.; Oberleithner, H. Distal site of calcium reabsorption in the rat nephron. *Pflug. Arch.* **1978**, *374*, 153–157. [CrossRef]
30. Gattineni, J.; Bates, C.; Twombley, K.; Dwarakanath, V.; Robinson, M.L.; Goetz, R.; Mohammadi, M.; Baum, M. FGF23 decreases renal NaPi-2a and NaPi-2c expression and induces hypophosphatemia in vivo predominantly via FGF receptor 1. *Am. J. Physiol. Renal Physiol.* **2009**, *297*, F282–F291. [CrossRef]
31. Olauson, H.; Lindberg, K.; Amin, R.; Jia, T.; Wernerson, A.; Andersson, G.; Larsson, T.E. Targeted deletion of Klotho in kidney distal tubule disrupts mineral metabolism. *J. Am. Soc. Nephrol.* **2012**, *23*, 1641–1651. [CrossRef]
32. Ure, M.E.; Heydari, E.; Pan, W.; Ramesh, A.; Rehman, S.; Morgan, C.; Pinsk, M.; Erickson, R.; Herrmann, J.M.; Dimke, H.; et al. A variant in a cis-regulatory element enhances claudin-14 expression and is associated with pediatric-onset hypercalciuria and kidney stones. *Hum. Mutat.* **2017**, *38*, 649–657. [CrossRef] [PubMed]
33. Wiebe, S.A.; Plain, A.; Pan, W.; O'Neill, D.; Braam, B.; Alexander, R.T. NHE8 attenuates Ca^{2+} influx into NRK cells and the proximal tubule epithelium. *Am. J. Physiol. Renal Physiol.* **2019**, *317*, F240–F253. [CrossRef] [PubMed]
34. Greger, R. Cation selectivity of the isolated perfused cortical thick ascending limb of Henle's loop of rabbit kidney. *Pflug. Arch.* **1981**, *390*, 30–37. [CrossRef] [PubMed]
35. Gunzel, D.; Stuiver, M.; Kausalya, P.J.; Haisch, L.; Krug, S.M.; Rosenthal, R.; Meij, I.C.; Hunziker, W.; Fromm, M.; Muller, D. Claudin-10 exists in six alternatively spliced isoforms that exhibit distinct localization and function. *J. Cell Sci.* **2009**, *122*, 1507–1517. [CrossRef] [PubMed]

36. Kimizuka, H.; Koketsu, K. Ion transport through cell membrane. *J. Theor. Biol.* **1964**, *6*, 290–305. [CrossRef]
37. Thissen, D.; Steinberg, L.; Kuang, D. Quick and easy implementation of the Benjamini-Hochberg procedure for controlling the false positive rate in multiple comparisons. *J. Educ. Behav. Stat.* **2002**, *27*, 77–83. [CrossRef]

5

Enteropathogenic *Escherichia coli* (EPEC) Recruitment of PAR Polarity Protein Atypical PKCζ to Pedestals and Cell–Cell Contacts Precedes Disruption of Tight Junctions in Intestinal Epithelial Cells

Rocio Tapia [1,†], Sarah E. Kralicek [1,†] and Gail A. Hecht [1,2,3,*]

1 Department of Medicine, Division of Gastroenterology and Nutrition, Loyola University Chicago, Maywood, IL 60153, USA; rtpastrana@luc.edu (R.T.); sknopf@luc.edu (S.E.K.)
2 Department of Microbiology and Immunology, Loyola University Chicago, Maywood, IL 60153, USA
3 Edward Hines Jr. VA Hospital, Hines, IL 60141, USA
* Correspondence: ghecht@lumc.edu
† These authors contributed equally to this work.

Abstract: Enteropathogenic *Escherichia coli* (EPEC) uses a type three secretion system to inject effector proteins into host intestinal epithelial cells, causing diarrhea. EPEC induces the formation of pedestals underlying attached bacteria, disrupts tight junction (TJ) structure and function, and alters apico-basal polarity by redistributing the polarity proteins Crb3 and Pals1, although the mechanisms are unknown. Here we investigate the temporal relationship of PAR polarity complex and TJ disruption following EPEC infection. EPEC recruits active aPKCζ, a PAR polarity protein, to actin within pedestals and at the plasma membrane prior to disrupting TJ. The EPEC effector EspF binds the endocytic protein sorting nexin 9 (SNX9). This interaction impacts actin pedestal organization, recruitment of active aPKCζ to actin at cell–cell borders, endocytosis of JAM-A S285 and occludin, and TJ barrier function. Collectively, data presented herein support the hypothesis that EPEC-induced perturbation of TJ is a downstream effect of disruption of the PAR complex and that EspF binding to SNX9 contributes to this phenotype. aPKCζ phosphorylates polarity and TJ proteins and participates in actin dynamics. Therefore, the early recruitment of aPKCζ to EPEC pedestals and increased interaction with actin at the membrane may destabilize polarity complexes ultimately resulting in perturbation of TJ.

Keywords: enteropathogenic *E. coli* (EPEC); tight junctions (TJ); polarity; atypical aPKCζ; transepithelial electrical resistance (TER); sorting nexin 9 (SNX9); EspF

---

1. Introduction

Enteropathogenic *Escherichia coli* (EPEC) delivers bacterial effector proteins into host intestinal epithelial cells (IECs) through a type III secretion system (TTSS), inducing actin pedestal formation, attaching and effacing lesions, and physiological changes in IECs that contribute to diarrhea [1]. EPEC alters the architecture and barrier function of tight junctions (TJ) [2,3] although the mechanisms are not well understood.

TJ are localized at the most apical region of the lateral membrane and constitute a paracellular diffusion barrier modulating the flow of ions and solutes. These structures consist of integral membrane proteins (claudin family, occludin, tricellulin, MarvelD3, and JAM-A) that interact with adhesion molecules of adjacent cells and with intracellular domains that associate with cytoplasmic adaptor proteins (MAGUK family, cingulin, paracingulin, MAGI-1-3, and MUPP-1) [4,5]. TJ also constitute a fence contributing to the maintenance of apico-basal polarity by restricting the intermixing of apical and lateral plasma membrane components. Three main protein complexes control epithelial polarity,

Crumbs (Crb3/Pals1/Patj), PAR (Par3/Par6/aPKCζ/Cdc42), and Scribble (Scrib/Lgl/Dlg). Apico-basal polarity contributes to cell morphology, directional vesicle transportation, ion and solute transport, and specific localization of proteins and lipids to different membrane domains [6,7].

The interdependence between apico-basal polarity complexes and TJ is well established. Alterations in Crb3 expression or reduced expression of Patj/Pals1 impair apical polarity and TJ development [8–12]. Inhibition of aPKCζ activity, impaired phosphorylation of Par3, as well as deletion of the aPKCζ binding domain of Par6, delays TJ assembly [13–16]. aPKC activity also maintains TJ integrity and membrane localization of occludin and ZO-1 [17]. Downregulation of Scrib or Dlg compromises TJ establishment [18–20]. In contrast, increased expression of Scrib in MCF10A cells promotes the formation of functional TJ [21]. These data demonstrate that polarity complexes are crucial to TJ assembly, maintenance, and function.

EPEC effectors perturb TJ structure and function and alter apico-basal polarity of IECs. EspF perturbs barrier function in vivo and in vitro by redistributing TJ proteins from the cell–cell contacts, decreasing transepithelial electrical resistance (TER), and increasing paracellular permeability [2,22–24]. Map increases permeability to charged and non-charged molecules, indicating a failure in gate function [24,25]. NleA mislocalizes occludin and ZO-1 from the cell–cell contacts leading to barrier dysfunction [26]. EspG also contributes to leaky barrier, perturbs microtubule networks, and induces cytoplasmic accumulation of occludin and delays TJ recovery [27–29].

EPEC infection causes progressive redistribution of the basolateral proteins, β1-integrin and Na^+/K^+ ATPase, to the apical compartment and the mislocalization of TJ proteins, occludin, claudin-1, and ZO-1 from cell–cell contacts to the lateral membrane and cytoplasm [22,23,26,30–32], suggesting that cell polarity is altered. We recently reported that EPEC drives Crb3 and Pals1 away from the apical membrane, and cell–cell contacts into the cytoplasm of IECs and EspF is crucial for this phenotype [32]. EspF is a multifunctional molecule that interacts with several host proteins including actin, profilin, Arp2, N-WASP, SNX9, Abcf2, cytokeratin 18, 14-3-3, WIPF1, SNX18, and SNX33 [33–38]. EspF interacts with the SH3 domain of sorting nexin 9 (SNX9) through its RxAPxxP motif [33,35]. The interaction of EspF with SNX9 promotes the formation of elongated plasma membrane tubules, as well as the internalization of EPEC into IECs [39]. EspF/SNX9 complex is required for impairment of both cell polarity and altered TJ structure and function [32,33,40].

Despite extensive investigation into the mechanisms by which EPEC effectors directly perturb TJ, no such evidence has been reported. In view of the interdependence of polarity and TJ complexes, we hypothesized that EPEC-induced disruption of intestinal epithelial cell TJ structure and function stems from the initial targeting of polarity complexes. This study examines the effect of EPEC on the PAR complex with particular focus on aPKCζ, which phosphorylates several targets crucial for the establishment and maintenance of apico–basal polarity and TJ function. The data presented herein support the notion that EPEC-induced perturbation of TJ is a downstream consequence of EspF-induced disruption of the PAR polarity complex, particularly the recruitment of aPKCζ to actin-rich pedestals, and its increased co-localization with actin at the membrane.

2. Results

2.1. *EPEC Disrupts PAR Polarity Complexes In Vivo and In Vitro*

EPEC alters the localization of Crb complex proteins resulting in perturbed cell polarity [32]. Here, we investigate the effect of EPEC on PAR polarity components. EPEC does not change the localization of Par3 in murine colonocytes as compared with uninfected (UI) tissues (Figure 1A). In contrast, colonocytes of EPEC-infected mice show that Par6 is redistributed from the apical membrane to the cytoplasm (Figure 1A). Similarly, phosphorylated aPKCζ–T410, aPKCζ–T560, and total aPKCζ in the apical membrane of colonocytes of EPEC-infected mice appear diffuse and shift from the apical to the lateral membrane and into the cytoplasm compared to its restricted localization to the apical membrane in UI cells (Figure 1A). To begin to explore the mechanisms involved in PAR complex perturbation caused by EPEC, in vitro models were used. SKCO-15 monolayers were infected for 1-2 h and the localization and protein levels of PAR complex proteins were examined. In accordance with in vivo data, Par3 localization in cultured IECs was unchanged by EPEC infection (Figure 1B). In contrast, EPEC displaced Par6 from cell–cell contacts into the cytoplasm (Figure 1B). Interestingly, aPKCζ displayed a ringlet pattern around attached bacteria (Figure 1B, zoom). Expression levels of PAR complex proteins are not altered by EPEC infection (Figure 1C). These data demonstrate that EPEC alters the distribution of the PAR complex proteins, Par6 and aPKCζ without changing their expression levels.

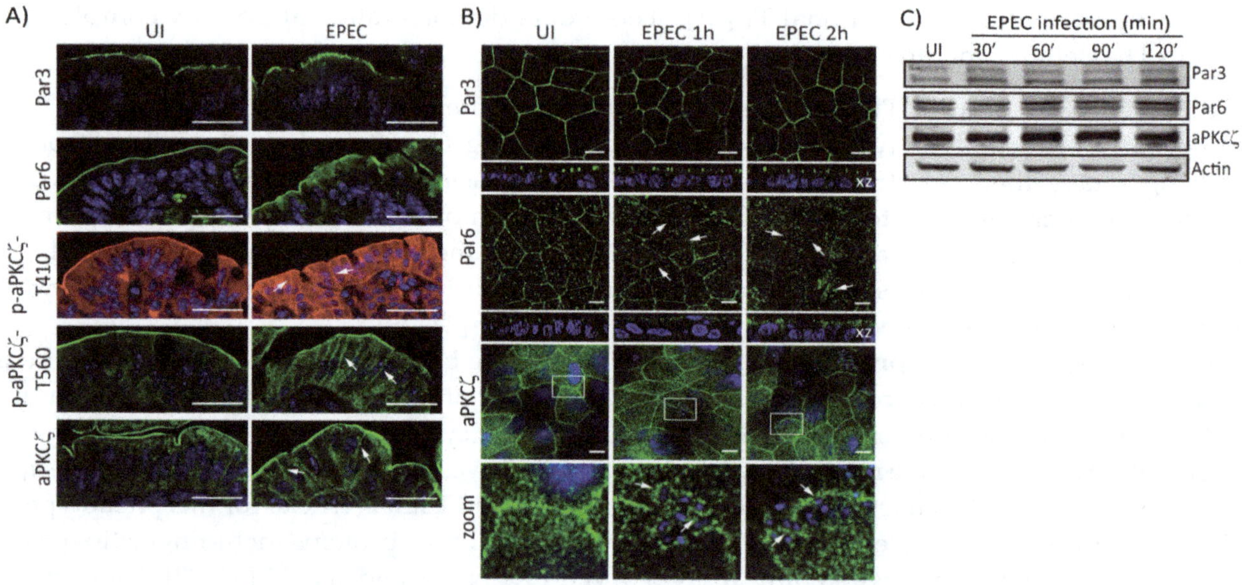

Figure 1. EPEC alters PAR complex localization. (**A**) Immunofluorescence microscopy of murine colonic epithelial tissue 3 days post-infection with EPEC or not (uninfected—UI). Par3 localization is unchanged. Par6 accumulates within the cytoplasm. Phosphorylated aPKCζ–T410, aPKCζ–T560, and total aPKCζ mislocalize from the apical membrane and accumulate within the cytoplasm and lateral membrane (arrows). Scale bar: 40 μm. (**B** and **C**) SKCO-15 monolayers were infected, or not (UI), with EPEC at indicated times, then the localization and protein levels of PAR complex proteins were assessed by immunofluorescence and western blot, respectively. (**B**) Par3 localization is unaltered by EPEC infection, and Par6 redistributes from cell–cell contacts to the cytoplasm (arrows). aPKCζ localizes under attached bacteria (arrows, inset zoom). Hoechst was used to stain host and bacteria nuclei (blue). Scale bar: 10 μm. (**C**) EPEC does not change the expression levels of PAR complex proteins; β-actin was used as a loading control.

2.2. *EspF, Via Its SNX9-Binding Domain, Impacts the Structural Organization of aPKCζ and F-actin at Pedestals*

EPEC induces the translocation of aPKCζ from the cytoplasm to the plasma membrane [41,42]. aPKCζ localization under attached EPEC in SKCO-15 cells suggests it is recruited to actin-rich pedestals. Indeed, aPKCζ co-localizes with filamentous actin (F-actin) around attached bacteria as early as 30 min and becomes more evident at 60 min post-infection (Figure 2A). The pattern around attached bacteria is highly organized in which actin surrounds bacteria, then a ring of aPKCζ and F-actin co-localization, and finally aPKCζ alone as seen in Figure 2A zoom, and depicted in the schematic in Figure 2B. Confocal microscopy confirms pedestal organization at 60 min post-infection consisting of attached bacteria atop actin, an interface of actin and aPKCζ co-localization, and then a region of aPKCζ alone (Figure 2A z-stack and Figure 2B). This structural organization is maintained at 2 h post-infection (Figure 2C zoom and z-stack, Figure 2B,D). Interestingly, aPKCζ is also seen within large filopodia induced by attached EPEC co-localizing with actin (Figure 2E).

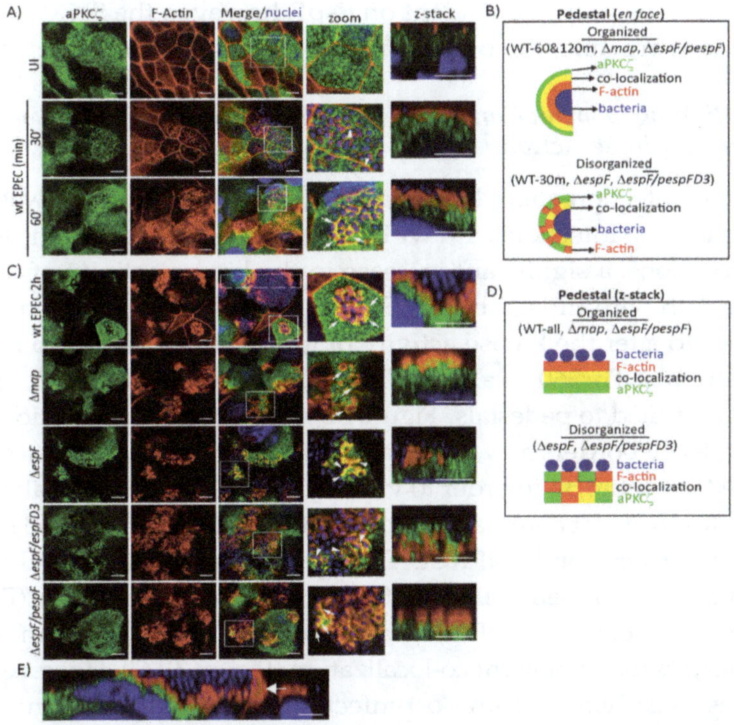

Figure 2. EspF and its SNX9-binding domain are crucial for the structural organization of F-actin and aPKCζ within EPEC pedestals. SKCO-15 monolayers were infected with wt EPEC, Δ*map*, Δ*espF*, Δ*espF/pespF*D3, or Δ*espF/pespF* to assess F-actin and aPKCζ co-localization. (**A**) F-actin and aPKCζ are recruited to and co-localize around attached bacteria at 30 and 60 min post-infection with wt EPEC. Zoom and z-stack images at 60 min display EPEC pedestals with attached bacteria surrounded by F-actin, an interface of co-localized F-actin and aPKCζ, then aPKCζ alone. (**B**) Schematic of F-actin (red), aPKCζ (green), and co-localizing interface (yellow) organization around bacteria after infection with wt EPEC and mutant strains. (**C**) aPKCζ recruitment to pedestals is maintained at 2 h post-infection with wt EPEC. Zoom and z-stacks reveal the robust organization of F-actin and aPKCζ co-localization. Deletion of *map* (Δ*map*) does not alter aPKCζ recruitment to pedestals. Infection with Δ*espF* and Δ*espF/pespF*D3 causes disorganization of co-localized F-actin and aPKCζ within pedestals. Complementation of *espF* (Δ*espF/pespF*) restores the organized phenotype. Squares correspond to zoom and z-stack areas in all except wt EPEC, in which a large rectangle corresponds to z-stack image. Arrows and arrowheads indicate organized and disorganized pedestals, respectively. (**D**) Schematic representation of pedestals showing the organized and disorganized localization of F-actin (red), aPKCζ (green), and co-localizing interface (yellow) after infection with wt EPEC and mutant strains. (**E**) aPKCζ also co-localizes with F-actin in filopodia (arrow) 2 h post-infection with wt EPEC. Scale bars: 10 μm (en face); 5 μm (z-stack).

We previously demonstrated that the effectors Map and EspF, through its interaction with SNX9, disrupt the Crb polarity complex [32]. Therefore, we investigated the contribution of Map and EspF to the recruitment of aPKCζ to pedestals. aPKCζ recruitment to EPEC pedestals is not diminished by deletion of *map* (Δ*map*) or *espF* (Δ*espF*), or by mutation of the SNX9-binding domain of EspF (Δ*espF/pespF*D3) (Figure 2C, green channel). However, the localization of F-actin and aPKCζ within pedestals is disorganized when infected with Δ*espF* and Δ*espF/pespF*D3, but not Δ*map*, compared to infection with wild-type (wt) EPEC, seen both en face and within z-stack images (Figure 2B–D). Complementation of *espF* (Δ*espF/pespF*) restores the organization of actin and aPKCζ within pedestals (Figure 2C, zoom and z-stack, Figure 2B,D).

Interestingly, infection of T84 monolayers with Δ*espF/pespF*D3 produces aPKCζ aggregates under attached bacteria in actin pedestals similar to wt EPEC and Δ*espF/pespF*, while only infection with Δ*espF* induces less co-localization of aPKCζ and actin (Supplemental Figure S1A,B). These data indicate that aPKCζ recruitment to pedestals is not dependent on EspF, however, the SNX9-binding domain of EspF is required for the structural organization of actin and aPKCζ within pedestals in a cell-specific manner.

2.3. *EspF and Its SNX9-Binding Domain Contribute to the Recruitment of Phosphorylated aPKCζ–T560 to Pedestals without Altering Kinase Activity*

To determine if EPEC infection alters kinase activity, total PKC activity was measured in T84 and SKCO-15 monolayers following infection with wt EPEC or EspF mutant strains for 1–2 h. In agreement with our previous observations, a significant increase in the kinase activity of total PKC in T84 cells is observed following infection with EPEC and EspF mutant strains (Supplementary Figure S1C) [41,42]. Interestingly, EPEC failed to alter the kinase activity of total PKC in SKCO-15 monolayers (Figure 3A). aPKCζ autophosphorylation at Thr560 (p-aPKCζ–T560) activates kinase activity [43,44]. We questioned whether active aPKCζ is recruited to pedestals. Figure 3B shows the co-localization of p-aPKCζ-T560 with F-actin in EPEC at the cell–cell borders and pedestals. Heat maps were generated by merging the green and red channels depicted in Figure 3B in order to visually quantitate the co-localization of p-aPKCζ–T560 with F-actin; white indicates the most abundant co-localization and blue-black the least to none (Figure 3C). To more specifically quantify F-actin and p-aPKCζ–T560 co-localization, immunofluorescence intensity was measured in regions of interest within cell–cell contacts, the cytoplasm, or pedestals (Figure 3D). In uninfected cells, there is no significant association of F-actin and p-aPKCζ–T560 at cell membranes or within the cytoplasm (Figure 3D), despite their apparent co-localization (Figure 3B,C). However, co-localization within the membrane intensifies as early as 15 min post-infection with wt EPEC (Figure 3B,C) and significant correlation between F-actin and p-aPKCζ–T560 is present at 30 min post-infection (Figure 3D). Interestingly, at both 15 and 30 min post-infection with wt EPEC, there is significant association between p-aPKCζ–T560 and F-actin within pedestals (Figure 3B–D) and confocal microscopy confirms this co-localization (Figure 3E). Interestingly, both Δ*espF*and Δ*espF/pespF*D3 diminish p-aPKCζ–T560 and F-actin co-localization at both the plasma membrane and within pedestals 30 min post-infection (Figure 3B–E). Complementation of Δ*espF*with wt EspF (Δ*espF/pespF*) restores the interaction of F-actin and p-aPKCζ–T560 at cell–cell borders and pedestals (Figure 3B–E). Together, these results suggest that although PKC activity is unchanged by EPEC infection in SKCO-15 cells, active aPKCζ localization within the cell is altered and is dependent on EspF and its SNX9-binding domain for the recruitment of p-aPKCζ–T560 to actin at cell–cell borders and EPEC-induced pedestals.

2.4. *SNX9-Binding Domain of EspF Is Essential to Disrupt TJ Structure and Function in SKCO-15 Monolayers*

EPEC effectors contribute to loss of intestinal epithelial TJ structure and function, and EspF is largely responsible for this phenotype [2]. To determine the individual contribution of EspF to barrier disruption, TER was measured in SKCO-15 monolayers harboring pRetroX–Tight–Pur–EspF-HA with and without inducible expression. EspF-HA expression levels were determined by western blot of control and Tet–On cells incubated with increasing amounts of doxycycline (dox) (Figure 4A). The basal expression of EspF–HA is observed in the absence of dox compared to control cells and increases at 500 ng/mL of dox (Figure 4A). Control and EspF–HA cells were grown in Transwells with and without dox (500 ng/mL) for three days, then TER was determined. Interestingly, uninduced (-dox) cells transfected with EspF–HA show a significant reduction in TER values as compared to control monolayers (647 ± 44 ohms.cm^2 vs. 1545 ± 22 ohms.cm^2) (Figure 4B), likely due to basal expression of EspF (Figure 4A) as previously reported [32]. Induced expression of EspF–HA (+dox) significantly decreases TER compared to uninduced (-dox) (308 ± 86 ohms.cm^2 vs. 647 ± 44 ohms.cm^2) (Figure 4B) indicating that EspF alone significantly reduces TER in SKCO-15 monolayers.

To assess in detail the EspF motifs involved in perturbation of TJ barrier function, SKCO-15 monolayers were infected with wt EPEC, Δ*espF*, Δ*espF/pespF*, or Δ*espF* complemented to express specific site-directed EspF mutations (Δ*espF/pespF*D3, Δ*espF/pSer47A*, and Δ*espF/pSer47/50A*). In addition to binding SNX9, EspF binds to 14-3-3 family proteins [36]. EPEC strains expressing EspF harboring 14-3-3ζ binding motif mutations (*ΔespF/pSer47A* and *ΔespF/pSer47/50A)* were used as controls because of their inability to perturb the Crb polarity complex [32]. Infection with *ΔespF/pSer47A* and *ΔespF/pSer47/50A* perturbed barrier function to a similar degree as wt EPEC (–41.0 ± 6.0% and –40.0 ± 5.0% vs. –38.0 ± 7.0%, change from baseline, respectively) (Figure 4C). In contrast, *ΔespF* and *ΔespF/pespF*D3 mutants induced minimal change in TER at 2 h compared to wt EPEC (–9.0 ± 7.0% and –7.0 ± 6.0% vs. –38.0 ± 7.0%, change from baseline, respectively) (Figure 4C). As reported previously in T84 cells [33], *ΔespF/pespF*D3 decreased TER to the same extent as wt EPEC (–35.7 ± 5.8% vs. –32.3 ± 5.1%, change from baseline) (Supplemental Figure S1D). Infection with *ΔespF* attenuates the decrease in resistance (–18.3 ± 5.1%, change from baseline) and complementation of *espF* (*ΔespF/pespF*) restores this phenotype (–38.6 ± 7.8%, change from baseline) (Supplemental Figure S1D), indicating the cell-specific responses to the EspF SNX9-binding domain mutant.

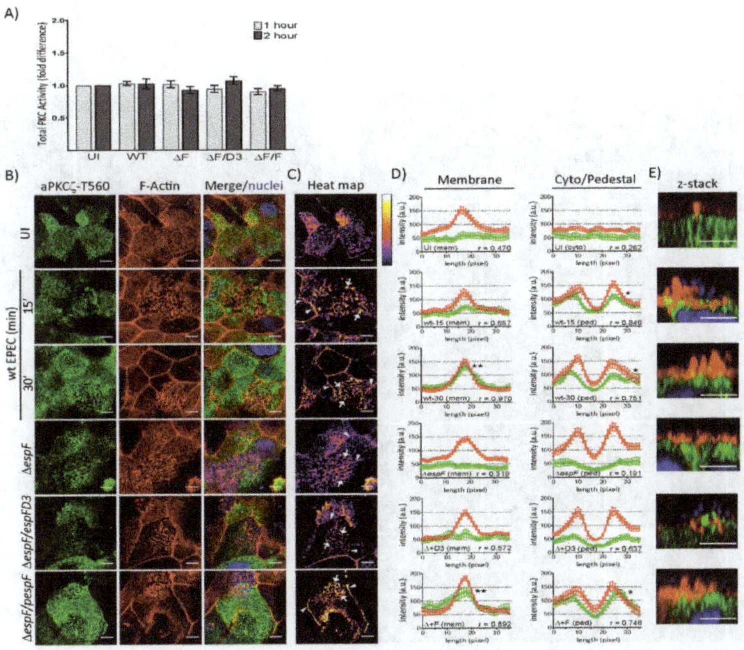

Figure 3. EspF and its SNX9-binding domain contribute to the co-localization of phosphorylated aPKCζ–T560 and F-actin at the plasma membrane and within pedestals. SKCO-15 monolayers plated on Transwells were infected with wt EPEC or EspF mutant strains, then PKC kinase activity, and F-actin and p-aPKCζ-T560 co-localization were assessed. (**A**) Fold change in PKC activity does not change at 1 or 2 h post-infection with EPEC (WT), *ΔespF* (ΔF), *ΔespF* complemented with mutant *espF* (ΔF/D3) or wt *espF* (ΔF/F) compared to UI monolayers. (**B–E**) p-aPKCζ–T560 is recruited to and co-localizes with F-actin at the cell–cell contacts and within pedestals 15 and 30 min post-infection with wt EPEC. p-aPKCζ–T560 recruitment is reduced following infection with *ΔespF* or *ΔespF/pespF*D3. Complementation of *espF* (*ΔespF/pespF*) restores F-actin and p-aPKCζ–T560 co-localization levels similar to wt. (**C**) Heat map generated from merged images in B indicating the intensity of F-actin and p-aPKCζ–T560 co-localization at the plasma membrane (arrowheads) and pedestals (arrows). (**D**) Immunofluorescence (IF) quantification of images represented in B. IF intensity in arbitrary units (a.u.) plotted against length (pixel) of p-aPKCζ–T560 (green) and F-actin (red) at the membrane and within the cell. Graphs represent data from >15 regions of interest from three biological replicates. Average peak intensity for p-aPKCζ–T560 reveals significantly higher levels at the membrane with wt and *ΔespF/pespF* at 30 min post-infection and significantly higher levels within pedestals with wt-15, wt-30, and *ΔespF/pespF* 30 min post-infection compared to uninfected controls; one-way ANOVA, * $p < 0.01$ and ** $p < 0.001$. r = Pearson correlation coefficient indicating an association between F-actin and p-aPKCζ–T560 localization at the membrane and within pedestals in an EspF and SNX9-binding domain dependent manner. (**E**) Z-stack images reveal F-actin and p-aPKCζ–T560 co-localization within pedestals. Scale bars: 10 μm (en face); 5 μm (z-stack).

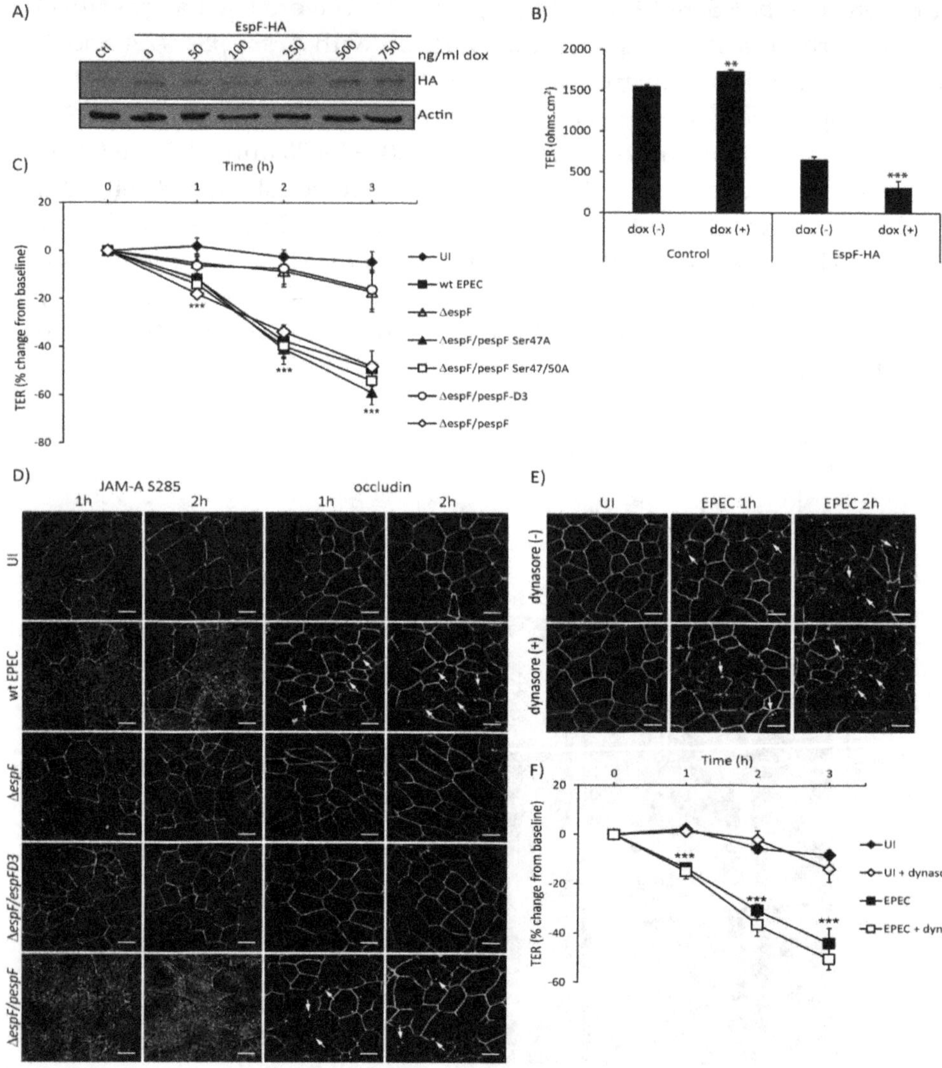

Figure 4. The SNX9 binding domain of EspF is crucial for disruption of TJ structure and function. (**A** and **B**) To generate Tet–On cells, EspF–HA was cloned into the doxycycline-inducible p–RetroX–Tight–Pur vector and transfected or not (control – Ctl) into wild-type SKCO-15 cells. Tet–On SKCO-15 cells were plated in absence of doxycycline (dox). (**A**) EspF–HA expression was induced with increasing concentrations of dox for three days. Cells were processed for western blot and detection of HA was performed. (**B**) Tet–On SKCO-15 cells were cultured on Transwells for 1 week (-dox) then expression of pRetroX–Tight–Pur–EspF-HA was induced (+dox) for three days. Basal expression of EspF–HA induces a significant reduction in TER (-dox) as compared to controls. Induction of EspF–HA (+dox) significantly reduces TER even further. (**C** and **D**) SKCO-15 monolayers were infected with wt EPEC, Δ*espF*, or Δ*espF*complemented with wt *espF* or site-directed *espF* mutants (Δ*espF/pespF*, Δ*espF/pespF*D3, Δ*espF/pSer47A*, or Δ*espF/pSer47/50A*). TER and localization of JAM-A S285 and occludin were determined. (**C**) Infection with Δ*espF/pespF*, Δ*espF/pSer47A* and Δ*espF/pSer47/50A*mutant strains reduce TER to levels similar to those induced by wt EPEC infection. Infection with Δ*espF*or Δ*espF/pespF*D3 significantly blunt the decrease in TER compared to wt EPEC. ***$p < 0.001$. (**D**) Infection with wt EPEC and Δ*espF/pespF*redistributes JAM-A S285 and occludin from intercellular junctions to the cytoplasm. Infection with Δ*espF*or Δ*espF/pespFD3*does not alter the localization of JAM-A S285 and occludin. Scale bar, 10 μm. (**E** and **F**) SKCO-15 monolayers were treated with or without dynasore 1 h prior to EPEC infection, then occludin localization and TER were assessed. (**E**) Blocking dynamin-dependent endocytosis with dynasore fails to prevent occludin internalization induced by wt EPEC infection. Arrows indicate the endocytosis of occludin triggered by EPEC infection. Scale bar, 10 μm. (**F**) Dynasore does not prevent the drop in TER induced by EPEC. TER reported as percent change from baseline. *$p < 0.01$, ***$p < 0.001$.

We next questioned whether EspF and the SNX9-binding domain play a role in the maintenance of TJ structure and function. We evaluated the distribution of phosphorylated JAM-A at Ser285 and occludin, which are both involved in the assembly and maintenance of TJ. Interestingly, Δ*espF*/*pespF*D3 ablates the redistribution of JAM-A S285 and occludin from the membrane to the cytoplasm caused by wt EPEC in a similar manner to those cells infected with Δ*espF* (Figure 4D). We previously demonstrated that the EspF/SNX9 interaction is crucial for the endocytosis of Crb3 in a clathrin-dependent manner [32]. We therefore investigated whether the redistribution of occludin caused by EPEC is mediated by a similar endocytic pathway. SKCO-15 monolayers were incubated, or not, with dynasore, a dynamin inhibitor that plays an important role in clathrin-dependent endocytosis, and infected, or not, with EPEC for 1–2 h. Occludin localization and TER measurements were followed. Treatment with dynasore did not prevent the endocytosis of occludin (Figure 4E) or the decrease in TER caused by EPEC at 1–2 h post-infection (–15.0 ± 3.0% vs. –14.0 ± 2.0%; –36.0 ± 5.0% vs. –30.0 ± 2.0%, with and without dynasore at 1 and 2 h post-infection, respectively) (Figure 4F). These results indicate that occludin endocytosis and barrier dysfunction mediated by EPEC occurs via a dynamin-independent pathway.

2.5. *Temporal Disruption of TJ Proteins Mediated by EPEC Infection*

The interplay between cell polarity and TJ establishment and function is well defined. Active aPKCζ targets TJ proteins and participates in the assembly and maintenance of barrier function. EPEC impairs both cell polarity and intestinal epithelial TJ barrier function. To determine if the temporal sequence of TJ disruption following EPEC infection correlates with aPKCζ redistribution, SKCO-15 monolayers were infected with EPEC and the localization of TJ proteins and TER were analyzed over time. EPEC does not alter the localization of total JAM-A from the cell–cell contacts (Figure 5A). In contrast, JAM-A S285 is internalized by 30 min and TER significantly decreases 45 min post-infection (Figure 5A,C). Phosphorylation of JAM-A at tyrosine 280 (JAM-A Y280) is related to loss of barrier function [45]. Interestingly, we found that JAM-A Y280 is detectable only after 1–2 h EPEC infection, corresponding to leaky TJ (Figure 5A). Occludin disruption was seen at 1 h post-infection (Figure 5B) corresponding with a more significant drop in TER (–14.0 ± 8.0%, change from baseline) (Figure 5C). ZO-1 was the last TJ protein altered by EPEC moving into the cytoplasm at 2 h post-infection (Figure 5B) and corresponding to a profound drop in TER (–29.0 ± 5.0%, change from baseline) (Figure 5C).

To further delineate the temporal relationship of EPEC-induced recruitment of p-aPKCζ-T560 to pedestals and the redistribution of occludin from the cell–cell borders, localization of F-actin, p-aPKCζ-T560, and occludin, was assessed from 5–120 min post-infection. EPEC induces the recruitment of F-actin almost immediately upon bacterial attachment (5–15 min) (Figure 6A). Recruitment of p-aPKCζ–T560 to pedestals occurs as early as 5–15 min, and its co-localization with F-actin is seen by 15–30 min post-infection (Figure 6A). Consistent with our previous results, occludin remains at the cell borders at early time points after infection being displaced from the membrane at 60–120 min post-infection (Figure 6A). To examine if inhibition of aPKCζ activity affects its recruitment to EPEC pedestals and subsequent impairment in barrier function, SKCO-15 monolayers were incubated or not (-) with aPKCζ pseudosubstrate (PS) prior to EPEC infection, then localization of p-aPKCζ–T410 and TER were determined. Monolayers treated with PS alone show the redistribution of p-aPKCζ–T410 from the apical membrane and cell–cell contacts to the lateral membrane domain and cytoplasm, and cell death is apparent (Figure 6B). Interestingly, PS treatment does not alter the recruitment of p-aPKCζ–T410 to EPEC pedestals (Figure 6B). In agreement with previous reports [17], inhibition of aPKCζ with PS disrupts barrier function (–28.0% ± 4.0 and –73.0% ± 3.0 change from baseline, PS 5 μM and PS 10 μM, respectively) (Figure 6C). EPEC infection does not alter the drop in TER caused by PS (Figure 6C). Interestingly, monolayers incubated with PS alone (5μM) begin to recover TER (Figure 6C). In contrast, EPEC infection (2 h) in the presence of 5 μM PS causes a progressive drop in TER (–46.0% ± 4.0 change from baseline) similar to infection with EPEC alone (–45.0% ± 2.0 change from baseline) (Figure 6C), suggesting that the displacement of p-aPKCζ–T410 away from TJ results in a perturbation of barrier function. Together these results demonstrate the progressive dismantling of TJ by

EPEC and the corresponding impact on barrier function, events that occur after the recruitment of active aPKCζ to actin pedestals.

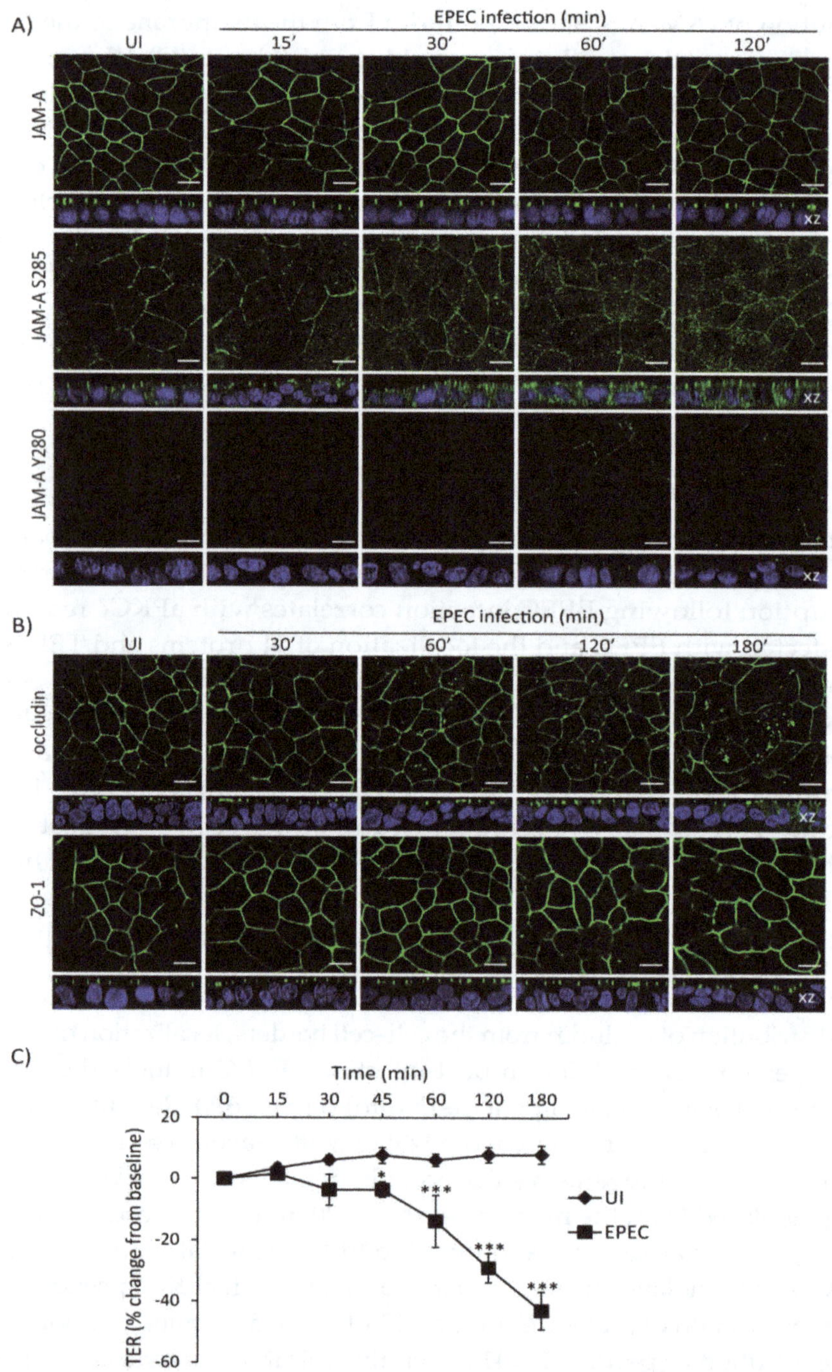

Figure 5. Temporal redistribution of TJ proteins and barrier dysfunction caused by EPEC. (**A–C**) SKCO-15 cells were plated on Transwells and infected or not (UI) with EPEC. Total JAM-A, JAM-A S285, JAM-A Y280, occludin, and ZO-1 localization and TER were determined. (**A**) EPEC does not alter the distribution of total JAM-A. In contrast, JAM-A S285 is displaced from the cell–cell contacts to the cytoplasm at 30 min post-infection. Tyrosine phosphorylation of JAM-A Y280 is apparent at 60–120 min post-infection. Scale bars, 10 μm. (**B**) EPEC induces the endocytosis of occludin and ZO-1 at 1 and 2 h post-infection, respectively. Scale bars, 10 μm. (**C**) TER drops significantly as early as 45 min post-infection and progressively decreases over time as more TJ proteins are displaced. TER reported as percent change from baseline. * $p < 0.01$, *** $p < 0.001$.

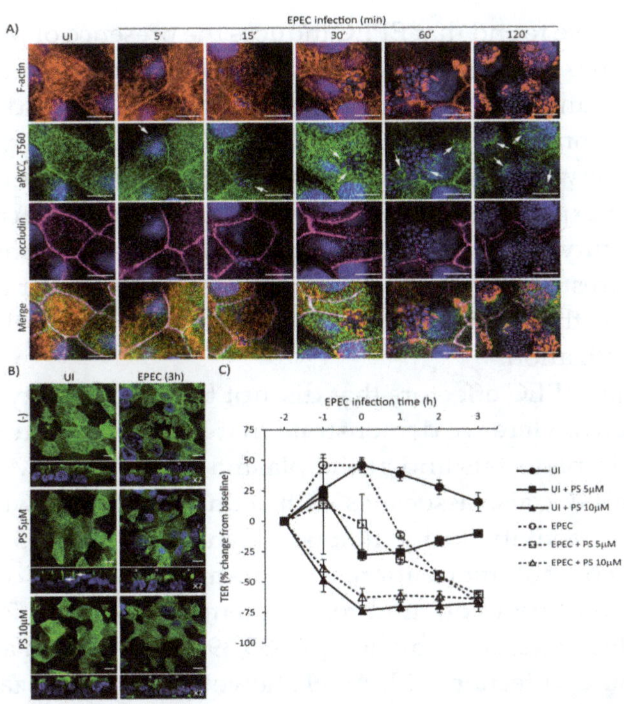

Figure 6. EPEC induces the recruitment of p-aPKCζ–T560 to actin pedestals prior to redistribution of occludin from cell borders. (**A**) SKCO-15 cells were plated on Transwells and infected or not (UI) with EPEC for 5–120 min. Immunofluorescence microscopy for F-actin, p-aPKCζ–T560, and occludin was performed. p-aPKCζ–T560 is recruited to pedestals co-localizing with actin consistently as early as 15 min and increases as infection progresses. Occludin localization is not displaced from cell borders until 1–2 h post-infection. Arrows indicate the presence of p-aPKCζ–T560 within EPEC pedestals. Scale bars, 10 μm. (**B** and **C**) SKCO-15 cells were plated on Transwells and pre-incubated or not with aPKCζ pseudosubstrate (PS) for 1 h prior to EPEC infection. Immunofluorescence of p-aPKCζ–T410 was performed and TER was measured. (**B**) Treatment with PS redistributes p-aPKCζ–T410 from the apical membrane and intercellular junctions to the lateral membrane (arrows) and cytoplasm without altering recruitment to EPEC pedestals (stars). Scale bars, 10 μm. (**C**) Inhibition of aPKCζ results in rapid disruption of barrier function in both UI and EPEC-infected monolayers.

3. Discussion

The present study provides evidence that EPEC perturbs PAR polarity complex integrity and induces the recruitment of aPKCζ to actin pedestals almost immediately following EPEC attachment. The SNX9-binding domain of EspF is important for the recruitment and organization of p-aPKCζ–T560 organization within EPEC pedestals, and the endocytosis of TJ proteins. We speculate that the very early recruitment of active aPKCζ to actin within pedestals and cell–cell contacts triggers downstream signalling events that ultimately disrupt intestinal epithelial TJ structure and function.

In view of the well-established interdependence between polarity and TJ complexes, we questioned if the apico-basal polarity defects caused by EPEC might be upstream of events that lead to TJ barrier perturbation. We provide evidence that EPEC perturbs PAR polarity complex integrity, mislocalizes active aPKCζ to the lateral membrane in vivo, and induces aPKCζ recruitment to pedestals and co-localization with actin at cell–cell borders almost immediately following EPEC attachment in vitro. These events precede the progressive dismantling of TJ proteins from the cell–cell contacts and barrier dysfunction. This is the first report correlating the temporal disruption of the PAR polarity complex with TJ disassembly by EPEC. Phosphorylated JAM-A S285, which is involved in TJ assembly [46], is displaced from cell–cell contacts at early times following EPEC infection. Subsequently, occludin then ZO-1 are displaced from TJ temporally correlating with the progressive loss of TER. Furthermore, pedestal formation and recruitment of active aPKCζ to pedestals occurs prior to occludin mislocalization. Interestingly, we found that EPEC induces the presence of phosphorylated JAM-A Y280 at intercellular contacts at times that correspond to barrier loss. Inflammatory stimuli increase JAM-A Y280 phosphorylation by balancing activity of the Src kinase, YES-1, and the phosphatase PTPN13, ultimately leading to disruption of barrier function [45]. EPEC induces a pro-inflammatory response through several signaling pathways resulting in the activation of NF-κB, ERK1/2, p38, JNK and PKCζ, and the upregulation of IL-8 expression, which contribute to intestinal barrier dysfunction [42,47–49]. EPEC also regulates the activity of multiple kinases, including the Src family, that contribute to actin polymerization [50,51]. Interestingly, Src family kinases contribute to Tir phosphorylation and actin pedestal formation [52,53]. Further analysis is required to understand how these pathways are involved in aPKCζ activity and relocalization.

EspF and Map are major EPEC effectors that disrupt the Crb polarity complex and TJ, however, the mechanisms are not known. Here we demonstrate that EspF likely through binding SNX9 recruits active aPKCζ to actin within pedestals and at the plasma membrane. Map does not have a role in aPKCζ recruitment, but as aPKCζ also associates with actin in filopodia, we cannot discard the notion that Map may regulate aPKCζ activity during filopodia dynamics [33,54,55]. The EspF/SNX9/N-WASP complex participates in F-actin polymerization, membrane remodeling during EPEC pathogenesis, impairment of TJ structure and function, and recruitment of ZO-1 and ZO-2 to pedestals, but this complex does not affect EPEC pedestal formation [33–35,39,40]. This is in agreement with our data in which clear pedestals form after infection with Δ*espF*, however, the organization of total aPKCζ within pedestals and the recruitment of active aPKCζ to actin within pedestals and at cell–cell borders are severely altered following infection with Δ*espF* and Δ*espF/pespFD3*. Interestingly, EspF of rabbit EPEC (REPEC) and EPEC is involved in pedestal maturation [34,56], suggesting that EspF coordination of aPKCζ and actin within pedestals and at the membrane influences downstream signaling events that lead to TJ disruption.

EspF interacts with 14-3-3ζ [36], a protein that binds Par3 to regulate cell polarity [57]. This interaction is not involved in TJ barrier disruption by EPEC as determined by infection with Δ*espF/pSer47A* and Δ*espF/pSer47/50A* and is consistent with our finding that Par3 localization remains unchanged after EPEC infection. EspF also forms a complex with SNX9 and N-WASP. This binding is required to disrupt Crb3, ZO-1, and E-cadherin from cell–cell contacts increasing their cytoplasmic accumulation, thus leading to impaired cell polarity and barrier function [32,40]. We found that EspF is essential for the displacement of JAM-A S285 and occludin from cell–cell contacts, as well as disruption of barrier function in SKCO-15 cells, as occurs in Caco-2 and T84 cells [33,40]. While the EspF–SNX9 interaction is important for barrier disruption in SKCO-15 and Caco-2, it does not play a role in T84 cells, as infection with the EspF-D3 mutant failed to protect against a loss of TER and the redistribution of occludin in polarized T84 cells [33]. Furthermore, examination of pedestals in EPEC-infected T84 cells reveals that aPKCζ and actin co-localization is not impacted by the EspF-D3 mutant. PKC activity post-EPEC-infection also differs between T84 and SKCO-15 cells [41,42]. Although all derived from colonic cancer cell lines, Caco-2 when grown for extended periods differentiate into small intestine-like cells, whereas SKCO-15 and T84 cells are more colonic in nature. Several studies have focused on the biochemical and structural differences between these colonic cell lines [58–60]. In addition, SNX9 expression levels differ between colon cancer cell lines, as well as having other varying redundant sorting nexin proteins [61–63]. These data highlight the cell-specific aPKCζ responses and different mechanisms that contribute to TJ barrier dysfunction and lends support to the notion that mislocalization of aPKCζ activity contributes to TJ disruption in SKCO-15 cells.

EPEC modulates clathrin-mediated endocytosis, thus it has been suggested that EPEC-induced TJ disassembly may occur via this mechanism. For instance, EspF recruits clathrin, AP2, early (Rab5a and EEA1) and recycling (Rab4a, Rab11a, Rab11b, FIP2, Myo5b) endocytic proteins to sites of infection [38]. The EspF binding partner SNX9 is recruited to clathrin-coated pits and associates with N-WASP, dynamin, Arp2/3 and other associated proteins to promote endocytosis of plasma membrane receptors [64–69]. In addition, EPEC mediates Crb3 endocytosis in a dynamin-dependent manner [32].

However, data presented herein indicate that EPEC-induced endocytosis of occludin and disruption of barrier function occur via a dynamin-independent mechanism indicating that an alternative endocytic pathway is responsible and supports aPKCζ involvement in TJ disassembly.

Interestingly, aPKCζ has direct and indirect roles in the formation and maintenance of polarity and TJ structure and function. aPKCζ phosphorylates the TJ proteins occludin, JAM-A, claudin-4, and ZO-1 to establish and maintain TJ structure and barrier function [17,46,70,71]. Silencing of aPKCζ or inhibition of its kinase activity with PS are associated with dephosphorylation of occludin and ZO-1, delayed assembly and perturbed maintenance of barrier function [17]. We found that PS treatment mislocalizes p-aPKCζ–T410 to the lateral membrane and cytoplasm but does not affect its recruitment to EPEC pedestals. In addition, when aPKCζ is absent from the lateral membrane, whether due to high concentrations of PS or recruitment to EPEC pedestals, barrier function is severely altered. aPKCζ also phosphorylates polarity proteins including Crb3, Par3, Lgl, and Par1b, and is dependent on interactions with Par6 and Cdc42 for its activity at the apical domain [13,72–75]. aPKC activation during endothelial morphogenesis is determined by the adaptor protein Nck [76]. After Nck is recruited to actin pedestals [77,78], it could serve as a hub for aPKCζ pedestal activity. Therefore, one could speculate that the early recruitment of aPKCζ to pedestals redirects its kinase activity away from TJ and polarity complexes, thus disrupting apico-basal polarity and barrier function.

Besides phosphorylation of polarity and TJ proteins, aPKCζ plays a role in actin dynamics. In *Drosophila*, Cdc42 and active aPKC cause the interaction of DSH3PX1/Dock/Wasp/Arp2/3, homologues of human SNX9/Nck/N-WASP/Arp2/3. This complex is involved in actin rearrangement and adherens junction stability [79–81]. Interestingly, we found that the EspF–SNX9 interaction was essential not only for the highly organized structure of aPKCζ and actin within pedestals but also for increased aPKCζ and actin co-localization at the membrane. aPKC activates ezrin, a plasma membrane protein that links and stabilizes the actin cytoskeleton to the membrane facilitating microvilli formation and endocytosis [82–84]. Ezrin and SNX9 are enriched on curved membranes; ezrin may interact with SNX9 via its curvature-sensing I-BAR domain facilitating ezrin to tether and close membranes during membrane tubule formation [62,85]. EspF/SNX9 binding is involved in F-actin polymerization and promotes the formation of tubular vesicles in EPEC-infected cells [33,35,39]. We previously determined that ezrin recruitment to the cytoskeleton is dependent on EspF, and ezrin activation contributes to disruption of TJ barrier function [86]. EspF, through its interaction with several host proteins including actin, profilin, Arp2, N-WASP, Abcf2, cytokeratin 18, WIPF1, SNX9/18/33, regulates cytoskeletal dynamics [33,34,36–38]. Together, these data indicate that through EspF interactions with actin-binding proteins and its structured localization of aPKCζ with actin at the membrane, aPKCζ activity likely participates in actin dynamics controlling endocytic pathways and TJ maintenance following EPEC infection.

The recruitment of cytoskeletal, adaptor, and signaling proteins into pedestals, as well as rearrangement of the actin cytoskeleton in host cells during EPEC infection are crucial steps in EPEC pathogenesis [77,87–90]. Together the results reported herein suggest a mechanism by which EPEC perturbs intestinal epithelial TJ (Figure 7). EPEC recruits active aPKCζ to interact with actin within pedestals and at the membrane of cell–cell contacts immediately following bacterial attachment. The recruitment of active aPKCζ away from polarity complexes and into highly organized actin pedestals via an EspF–SNX9 dependent process may be the initial upstream event that triggers downstream pathways that alter actin dynamics and disrupt intestinal epithelial TJ structure and function thus contributing to EPEC pathogenesis.

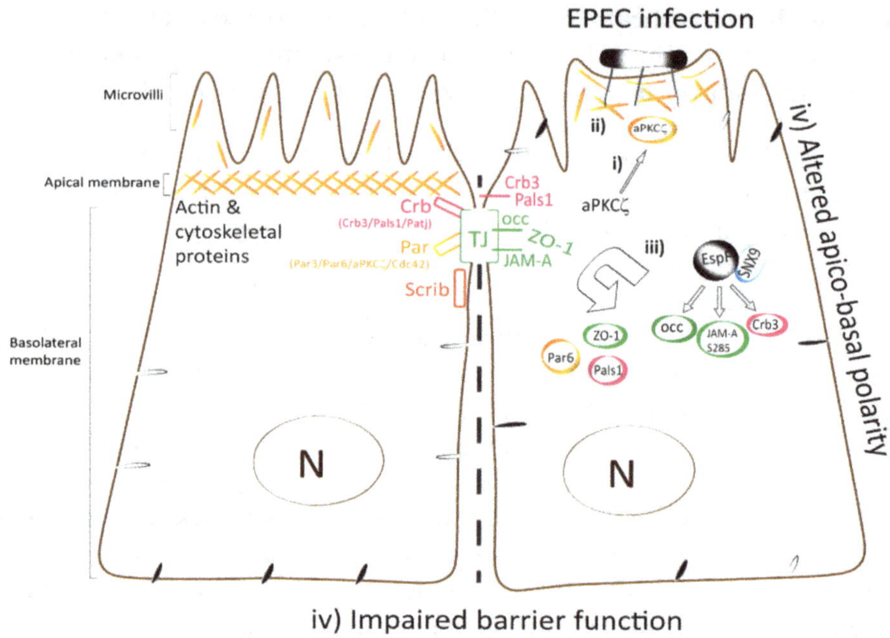

Figure 7. Model displaying the temporal mislocalization of aPKCζ leading to the dismantling of polarity and TJ complexes. Polarized epithelial cells consist of the apical membrane facing the lumen and the basolateral domain in contact with the underlying basement membrane. TJ, localized at the most apical area of the lateral membrane, contribute to the maintenance of apico-basal polarity. Polarity complexes (Crb, PAR, and Scrib) localize along the lateral membrane and modulate TJ assembly and function. (**i**) EPEC infection induces the recruitment of active aPKCζ to actin pedestals and at plasma membrane almost immediately upon bacterial attachment. (**ii**) Actin and aPKCζ are highly organized within EPEC pedestals via a mechanism that is dependent on the EspF/SNX9 binding domain. (**iii**) TJ (JAM-A S285, occludin, and ZO-1) and polarity (Par6, Crb3, and Pals1) proteins are displaced from cell–cell contacts and internalized into the cytoplasm. (**iv**) Apico-basal polarity, evidenced by redistribution of basolateral proteins to the apical membrane [32], and TJ are perturbed. These events are largely dependent on EspF–SNX9 binding domain, although other effectors may contribute.

4. Materials and Methods

4.1. Tissue Culture

SKCO-15 cell line derived from human adenocarcinoma of the colon displays typical intestinal epithelial cell features with adhesion complexes and microvilli [59,60]. SKCO-15 cultures were grown in Dulbecco's Modified Eagle Medium (DMEM) complemented with 10% fetal bovine serum with antibiotics as previously described [32]. T84 cells were grown in low glucose DMEM/Ham's F12 (Gibco) medium complemented with 10% newborn calf serum (Gibco, Life Technologies, Carlsbad, CA, USA) and antibiotics as previously reported [91]. Cells were grown at 37 °C in 5% CO2, once monolayers were confluent (7–10 days), media was replaced with bacterial growth medium 24 h prior to infection.

4.2. Antibodies and Reagents

Par3 (07-330, EMD Millipore), Par6 (ab49776 and ab6022, Abcam, Cambridge, MA, USA), aPKCζ (sc-17781, Santa Cruz Biotechnology, Dallas, TX, USA), p-aPKCζ–T560 (ab62372, Abcam, Cambridge, MA, USA), p-aPKCζ–T410 (sc-12894R, Santa Cruz Biotechnology, Dallas, TX, USA), actin (A2066, Sigma-Aldrich, St. Louis, MO, USA), F-actin BODIPY 558/568 Phalloidin (B3475, Invitrogen, Life Technologies Carlsbad, CA, USA), occludin (33-1500, Invitrogen, Life Technologies, Carlsbad, CA, USA), JAM-A S285 (sc-17430, Santa Cruz Biotechnology, Dallas, TX, USA), JAM-A Y280 (600-401-GN5, Rockland Immunochemicals Inc, Limerick, PA), JAM-A (361700, Invitrogen, Life Technologies, Carlsbad, CA, USA), and ZO-1 (61-7300, Invitrogen, Life Technologies,

Carlsbad, CA, USA). Secondary antibodies used for immunofluorescence were Alexa Fluor (Life Technologies, Carlsbad, CA, USA). Dynasore (Sigma-Aldrich, St. Louis, MO, USA) was used at 80 μM. aPKCζ pseudosubstrate (539624, Millipore-Sigma, Burlington, MA, USA) was used at 5 and 10 μM.

4.3. *Bacterial Culture*

The following EPEC strains were used: wt EPEC 0127:H6 E2348/69, Δ*espF* [92], Δ*espF*/p*espF*, Δ*map* [93], Δ*espF*/p*espF*D3 [33], Δ*espF/pSer47A* and Δ*espF/pSer47/50A* [Hecht unpublished]. Bacterial cultures were grown overnight in Luria-Bertani broth with appropriate selective antibiotics. For infections, bacterial cultures were processed as previously reported [32,91]. Monolayers plated on Transwells (Costar #3740), or six-well plates were infected with bacterial strains at a multiplicity of infection of 50. Infected cells were incubated at 37 °C in 5% CO2 for indicated times.

4.4. *Murine Infection*

Male mice C57BL/6J from 8 to 10 weeks old were used (Jackson Laboratory, Bar Harbor, ME, USA), and housed in a specific pathogen-free facility at Loyola University Chicago (LUC) Medical Campus. LUC Animal Care and Use Committee approved all animal protocols. Animals were infected with EPEC by oral gavage and sacrificed on day three post-infection [94]. Intestinal tissues were processed for immunofluorescence as reported [91].

4.5. *Immunofluorescence*

Cells were plated on coverslips or Transwells and fixed with cold methanol at −20 °C for 10 min or with 4% PFA for 15–30 min and permeabilized with 0.1% Triton X-100 (5–15 min). Samples were blocked and incubated with primary antibodies overnight at 4 °C with Invitrogen blocking solution, washed with PBS, then processed for immunofluorescence. Paraffin embedded intestinal sections of infected mice were processed as previously described [91].

4.6. *Imaging*

Slides were analyzed using a confocal Leica TCS SPE DMI 4000B (LAS X software, Leica, Wetzlar, Germany) microscope. Z-stack images were acquired in 0.33 μm sections and processed using the 3-D volume setting of LasX software. Images were processed using Adobe Photoshop CC 2018 and FIJI-ImageJ-64 software. Heat maps were generated from red and green channels using ImageJ software (Analyze> Colocalization> colocalization threshold > show colocalized pixel map), then channels were split and co-localization channel assigned LUT > Fire and auto-contrasted. Quantitation of p-PKCζ–T560 and F-actin were accomplished with ImageJ. Green and red channels were analyzed separately. A 35-pixel length line was drawn and centered over cell–cell contacts or pedestals, and the line duplicated in each channel (Edit > selection > restore selection). Pixel intensity was measured over the length of the line (Analyze > plot profile), recorded for at least 15 regions of interest from membranes and pedestals of three biological replicates and statistical analysis was performed as described below.

4.7. *Western Blot Analysis*

SKCO-15 monolayers were rinsed twice with cold PBS, protein extraction was achieved with RIPA buffer containing protease inhibitors as previously reported [32]. Cell lysates were processed by electrophoresis (SDS-PAGE), transferred to Immobilon membrane (IPFL00010, Millipore, Burlington, MA, USA). Membranes were incubated with primary antibodies overnight at 4 °C and LI-COR secondary antibodies for 1 h at room temperature. Immunoblotting was performed using LI-COR Odyssey system (LI-COR Bioscience, Lincoln, NE, USA)

4.8. Generation of Tet-On System

SKCO-15 cells were co-transfected with p-RetroX–Tet–On Advanced and pRetroX–Tight–Pur–EspF–HA [32]. To generate stable cell lines, transfected SKCO-15 cells were selected with G418 and puromycin for 2–3 weeks. Expression of the transgene pRetroX–Tight–Pur–EspF–HA was induced in presence of doxycycline (500 ng/mL) for three days [32].

4.9. Measurement of Transepithelial Electrical Resistance

Wild type and Tet–On SKCO-15 cells (3×10^5 cells) were plated in triplicate on Transwell filters in DMEM. TER was measured using cellZscope (nanoAnalytics, Munster, Germany) for 5–7 days. Cells were infected with wt EPEC or mutant strains and TER measurements were taken at indicated intervals. Expression of pRetroX–Tight–Pur–EspF–HA was induced with doxycycline (+dox) and measurements were followed for 3–6 days. TER values were normalized to UI monolayers.

4.10. PKC activity Assays

SKCO-15 and T84 cells were grown in 6-well plates and infected for the indicated time points. PKC Kinase Activity Assay Kit (ab139437, Abcam, Cambridge, MA) was used according to the manufacturer's instructions.

4.11. Statistical Analysis

All experiments were performed in triplicate. TER results are the mean ± SEM of three independent experiments performed in triplicate. *P* values were calculated by ANOVA Tukey's Multiple Comparison Test and Pearson correlation co-efficient using GraphPad Prism v7.

Author Contributions: Conceptualization, R.T., S.E.K. and G.A.H.; methodology, R.T., S.E.K. and G.A.H.; validation, R.T., S.E.K.; formal analysis, R.T., S.E.K.; investigation, R.T., S.E.K.; resources, G.A.H.; data curation, R.T., S.E.K.; writing—original draft preparation, review and editing, R.T., S.E.K. and G.A.H.; visualization, R.T., S.E.K. and G.A.H.; supervision, G.A.H.; project administration, G.A.H.; funding acquisition, G.A.H. All authors have read and agreed to the published version of the manuscript.

Abbreviations

aPKCζ	Atypical protein kinase C zeta
EPEC	Enteropathogenic *Escherichia coli*
F-actin	Filamentous actin
JAM-A S285	Junctional adhesion molecule-A phosphorylated at Serine 285
JAM-A Y280	Junctional adhesion molecule-A phosphorylated at Tyrosine 280
p-aPKCζ-T410	aPKCζ phosphorylated at Threonine 410
p-aPKCζ-T560	aPKCζ phosphorylated at Threonine 560
PS	aPKCζ pseudosubstrate inhibitor
SNX9	Sorting nexin 9
TJ	Tight junctions
TER	Transepithelial electrical resistance
ZO-1	Zonula Occludens -1

References

1. Pearson, J.S.; Giogha, C.; Wong Fok Lung, T.; Hartland, E.L. The genetics of enteropathogenic escherichia coli virulence. *Annu. Rev. Genet.* **2016**, *50*, 493–513. [CrossRef]
2. McNamara, B.P.; Koutsouris, A.; O'Connell, C.B.; Nougayrede, J.P.; Donnenberg, M.S.; Hecht, G. Translocated EspF protein from enteropathogenic escherichia coli disrupts host intestinal barrier function. *J. Clin. Invest.* **2001**, *107*, 621–629. [CrossRef]
3. Muza-Moons, M.M.; Schneeberger, E.E.; Hecht, G.A. Enteropathogenic escherichia coli infection leads to appearance of aberrant tight junctions strands in the lateral membrane of intestinal epithelial cells. *Cell. Microbiol.* **2004**, *6*, 783–793. [CrossRef]
4. Tsukita, S.; Tanaka, H.; Tamura, A. The claudins: From Tight junctions to biological systems. *Trends Biochem. Sci.* **2019**, *44*, 141–152. [CrossRef]
5. Van Itallie, C.M.; Anderson, J.M. Architecture of tight junctions and principles of molecular composition. *Semin. Cell Dev. Biol.* **2014**, *36*, 157–165. [CrossRef]
6. Wen, W.; Zhang, M. Protein complex assemblies in epithelial cell polarity and asymmetric cell division. *J. Mol. Biol.* **2018**, *430*, 3504–3520. [CrossRef]
7. Rodriguez-Boulan, E.; Macara, I.G. Organization and execution of the epithelial polarity programme. *Nat. Rev. Mol. Cell Biol.* **2014**, *15*, 225–242. [CrossRef]
8. Hurd, T.W.; Gao, L.; Roh, M.H.; Macara, I.G.; Margolis, B. Direct interaction of two polarity complexes implicated in epithelial tight junction assembly. *Nat. Cell Biol.* **2003**, *5*, 137–142. [CrossRef] [PubMed]
9. Roh, M.H.; Fan, S.; Liu, C.J.; Margolis, B. The Crumbs3-Pals1 complex participates in the establishment of polarity in mammalian epithelial cells. *J. Cell. Sci.* **2003**, *116*, 2895–2906. [CrossRef] [PubMed]
10. Fogg, V.C.; Liu, C.J.; Margolis, B. Multiple regions of Crumbs3 are required for tight junction formation in MCF10A Cells. *J. Cell. Sci.* **2005**, *118*, 2859–2869. [CrossRef] [PubMed]
11. Straight, S.W.; Shin, K.; Fogg, V.C.; Fan, S.; Liu, C.J.; Roh, M.; Margolis, B. Loss of PALS1 expression leads to tight junction and polarity defects. *Mol. Biol. Cell* **2004**, *15*, 1981–1990. [CrossRef] [PubMed]
12. Tilston-Lunel, A.M.; Haley, K.E.; Schlecht, N.F.; Wang, Y.; Chatterton, A.L.; Moleirinho, S.; Watson, A.; Hundal, H.S.; Prystowsky, M.B.; Gunn-Moore, F.J.; et al. Crumbs 3b promotes tight junctions in an Ezrin-dependent manner in mammalian cells. *J. Mol. Cell. Biol.* **2016**, *8*, 439–455. [CrossRef] [PubMed]
13. Nagai-Tamai, Y.; Mizuno, K.; Hirose, T.; Suzuki, A.; Ohno, S. Regulated protein-protein interaction between aPKC and PAR-3 plays an essential role in the polarization of epithelial cells. *Genes Cells* **2002**, *7*, 1161–1171. [CrossRef] [PubMed]
14. Suzuki, A.; Yamanaka, T.; Hirose, T.; Manabe, N.; Mizuno, K.; Shimizu, M.; Akimoto, K.; Izumi, Y.; Ohnishi, T.; Ohno, S. Atypical protein kinase c is involved in the evolutionarily conserved par protein complex and plays a critical role in establishing epithelia-specific junctional structures. *J. Cell Biol.* **2001**, *152*, 1183–1196. [CrossRef] [PubMed]
15. Chen, X.; Macara, I.G. Par-3 controls tight junction assembly through the Rac exchange factor Tiam1. *Nat. Cell Biol.* **2005**, *7*, 262–269. [CrossRef] [PubMed]
16. Yamanaka, T.; Horikoshi, Y.; Suzuki, A.; Sugiyama, Y.; Kitamura, K.; Maniwa, R.; Nagai, Y.; Yamashita, A.; Hirose, T.; Ishikawa, H.; et al. PAR-6 regulates aPKC activity in a novel way and mediates cell-cell contact-induced formation of the epithelial junctional complex. *Genes Cells* **2001**, *6*, 721–731. [CrossRef]
17. Jain, S.; Suzuki, T.; Seth, A.; Samak, G.; Rao, R. Protein kinase czeta phosphorylates occludin and promotes assembly of epithelial tight junctions. *Biochem. J.* **2011**, *437*, 289–299. [CrossRef]
18. Qin, Y.; Capaldo, C.; Gumbiner, B.M.; Macara, I.G. The mammalian scribble polarity protein regulates epithelial cell adhesion and migration through E-cadherin. *J. Cell Biol.* **2005**, *171*, 1061–1071. [CrossRef]
19. Stucke, V.M.; Timmerman, E.; Vandekerckhove, J.; Gevaert, K.; Hall, A. The MAGUK protein MPP7 binds to the polarity protein hDlg1 and facilitates epithelial tight junction formation. *Mol. Biol. Cell* **2007**, *18*, 1744–1755. [CrossRef]
20. Ivanov, A.I.; Young, C.; Den Beste, K.; Capaldo, C.T.; Humbert, P.O.; Brennwald, P.; Parkos, C.A.; Nusrat, A. Tumor suppressor scribble regulates assembly of tight junctions in the intestinal epithelium. *Am. J. Pathol.* **2010**, *176*, 134–145. [CrossRef]
21. Elsum, I.A.; Martin, C.; Humbert, P.O. Scribble regulates an EMT polarity pathway through modulation of MAPK-ERK signaling to mediate junction formation. *J. Cell. Sci.* **2013**, *126*, 3990–3999. [CrossRef] [PubMed]

22. Shifflett, D.E.; Clayburgh, D.R.; Koutsouris, A.; Turner, J.R.; Hecht, G.A. Enteropathogenic E. Coli disrupts tight junction barrier function and structure in Vivo. *Lab. Invest.* **2005**, *85*, 1308–1324. [CrossRef] [PubMed]
23. Guttman, J.A.; Li, Y.; Wickham, M.E.; Deng, W.; Vogl, A.W.; Finlay, B.B. Attaching and effacing pathogen-induced tight junction disruption in Vivo. *Cell. Microbiol.* **2006**, *8*, 634–645. [CrossRef] [PubMed]
24. Dean, P.; Kenny, B. Intestinal barrier dysfunction by Enteropathogenic Escherichia Coli is mediated by two effector molecules and a bacterial surface protein. *Mol. Microbiol.* **2004**, *54*, 665–675. [CrossRef]
25. Singh, A.P.; Aijaz, S. Generation of a MDCK cell line with constitutive expression of the Enteropathogenic E. Coli effector protein map as an in Vitro model of pathogenesis. *Bioengineered* **2015**, *6*, 335–341. [CrossRef]
26. Thanabalasuriar, A.; Koutsouris, A.; Weflen, A.; Mimee, M.; Hecht, G.; Gruenheid, S. The bacterial virulence factor NleA is required for the disruption of intestinal tight junctions by Enteropathogenic Escherichia Coli. *Cell. Microbiol.* **2010**, *12*, 31–41. [CrossRef]
27. Tomson, F.L.; Viswanathan, V.K.; Kanack, K.J.; Kanteti, R.P.; Straub, K.V.; Menet, M.; Kaper, J.B.; Hecht, G. Enteropathogenic Escherichia Coli EspG disrupts microtubules and in conjunction with Orf3 enhances perturbation of the tight junction barrier. *Mol. Microbiol.* **2005**, *56*, 447–464. [CrossRef]
28. Matsuzawa, T.; Kuwae, A.; Abe, A. Enteropathogenic Escherichia Coli Type III effectors EspG and EspG2 alter epithelial paracellular permeability. *Infect. Immun.* **2005**, *73*, 6283–6289. [CrossRef]
29. Glotfelty, L.G.; Zahs, A.; Hodges, K.; Shan, K.; Alto, N.M.; Hecht, G.A. Enteropathogenic E. Coli effectors EspG1/G2 disrupt microtubules, contribute to tight junction perturbation and inhibit restoration. *Cell. Microbiol.* **2014**, *16*, 1767–1783. [CrossRef]
30. Thanabalasuriar, A.; Kim, J.; Gruenheid, S. The inhibition of COPII trafficking is important for intestinal epithelial tight junction disruption during Enteropathogenic Escherichia Coli and Citrobacter Rodentium infection. *Microbes Infect.* **2013**, *15*, 738–744. [CrossRef]
31. Muza-Moons, M.M.; Koutsouris, A.; Hecht, G. Disruption of cell polarity by Enteropathogenic Escherichia Coli enables basolateral membrane proteins to migrate apically and to potentiate physiological consequences. *Infect. Immun.* **2003**, *71*, 7069–7078. [CrossRef] [PubMed]
32. Tapia, R.; Kralicek, S.E.; Hecht, G.A. EPEC Effector EspF Promotes Crumbs3 Endocytosis and Disrupts Epithelial Cell Polarity. *Cell. Microbiol.* **2017**, *19*, e12757. [CrossRef] [PubMed]
33. Alto, N.M.; Weflen, A.W.; Rardin, M.J.; Yarar, D.; Lazar, C.S.; Tonikian, R.; Koller, A.; Taylor, S.S.; Boone, C.; Sidhu, S.S.; et al. The type III effector EspF coordinates membrane trafficking by the spatiotemporal activation of two eukaryotic signaling pathways. *J. Cell Biol.* **2007**, *178*, 1265–1278. [CrossRef] [PubMed]
34. Peralta-Ramirez, J.; Hernandez, J.M.; Manning-Cela, R.; Luna-Munoz, J.; Garcia-Tovar, C.; Nougayrede, J.P.; Oswald, E.; Navarro-Garcia, F. EspF interacts with nucleation-promoting factors to recruit junctional proteins into pedestals for pedestal maturation and disruption of paracellular permeability. *Infect. Immun.* **2008**, *76*, 3854–3868. [CrossRef] [PubMed]
35. Marches, O.; Batchelor, M.; Shaw, R.K.; Patel, A.; Cummings, N.; Nagai, T.; Sasakawa, C.; Carlsson, S.R.; Lundmark, R.; Cougoule, C.; et al. EspF of Enteropathogenic Escherichia Coli binds sorting nexin 9. *J. Bacteriol.* **2006**, *188*, 3110–3115. [CrossRef]
36. Viswanathan, V.K.; Lukic, S.; Koutsouris, A.; Miao, R.; Muza, M.M.; Hecht, G. Cytokeratin 18 interacts with the Enteropathogenic Escherichia Coli Secreted Protein F (EspF) and is redistributed after infection. *Cell. Microbiol.* **2004**, *6*, 987–997. [CrossRef]
37. Nougayrede, J.P.; Foster, G.H.; Donnenberg, M.S. Enteropathogenic Escherichia Coli effector EspF Interacts with host protein Abcf2. *Cell. Microbiol.* **2007**, *9*, 680–693. [CrossRef]
38. Kassa, E.G.; Zlotkin-Rivkin, E.; Friedman, G.; Ramachandran, R.P.; Melamed-Book, N.; Weiss, A.M.; Belenky, M.; Reichmann, D.; Breuer, W.; Pal, R.R.; et al. Enteropathogenic Escherichia Coli remodels host endosomes to promote endocytic turnover and breakdown of surface polarity. *PLoS Pathog.* **2019**, *15*, e1007851. [CrossRef]
39. Weflen, A.W.; Alto, N.M.; Viswanathan, V.K.; Hecht, G.E. Coli secreted protein f promotes EPEC invasion of intestinal epithelial cells via an SNX9-Dependent mechanism. *Cell. Microbiol.* **2010**, *12*, 919–929. [CrossRef]
40. Garber, J.J.; Mallick, E.M.; Scanlon, K.M.; Turner, J.R.; Donnenberg, M.S.; Leong, J.M.; Snapper, S.B. Attaching-and-Effacing pathogens exploit junction regulatory activities of N-WASP and SNX9 to disrupt the intestinal barrier. *Cell. Mol. Gastroenterol. Hepatol.* **2017**, *5*, 273–288. [CrossRef]

41. Tomson, F.L.; Koutsouris, A.; Viswanathan, V.K.; Turner, J.R.; Savkovic, S.D.; Hecht, G. Differing roles of protein kinase C-Zeta in disruption of tight junction barrier by Enteropathogenic and Enterohemorrhagic Escherichia Coli. *Gastroenterology* **2004**, *127*, 859–869. [CrossRef] [PubMed]
42. Savkovic, S.D.; Koutsouris, A.; Hecht, G. PKC zeta participates in activation of inflammatory response induced by Enteropathogenic E. Coli. *Am. J. Physiol. Cell. Physiol.* **2003**, *285*, C512–C521. [CrossRef] [PubMed]
43. Keranen, L.M.; Dutil, E.M.; Newton, A.C. Protein Kinase C is regulated in Vivo by Three functionally distinct phosphorylations. *Curr. Biol.* **1995**, *5*, 1394–1403. [CrossRef]
44. Tsutakawa, S.E.; Medzihradszky, K.F.; Flint, A.J.; Burlingame, A.L.; Koshland, D.E., Jr. Determination of in Vivo phosphorylation sites in Protein Kinase C. *J. Biol. Chem.* **1995**, *270*, 26807–26812. [CrossRef] [PubMed]
45. Fan, S.; Weight, C.M.; Luissint, A.C.; Hilgarth, R.S.; Brazil, J.C.; Ettel, M.; Nusrat, A.; Parkos, C.A. Role of JAM-A tyrosine phosphorylation in epithelial barrier dysfunction during intestinal inflammation. *Mol. Biol. Cell* **2019**, *30*, 566–578. [CrossRef] [PubMed]
46. Iden, S.; Misselwitz, S.; Peddibhotla, S.S.; Tuncay, H.; Rehder, D.; Gerke, V.; Robenek, H.; Suzuki, A.; Ebnet, K. aPKC Phosphorylates JAM-A at Ser285 to promote cell contact maturation and tight junction formation. *J. Cell Biol.* **2012**, *196*, 623–639. [CrossRef]
47. Savkovic, S.D.; Koutsouris, A.; Hecht, G. Attachment of a noninvasive enteric pathogen, Enteropathogenic Escherichia Coli, to cultured human intestinal epithelial monolayers induces transmigration of neutrophils. *Infect. Immun.* **1996**, *64*, 4480–4487. [CrossRef]
48. Savkovic, S.D.; Koutsouris, A.; Hecht, G. Activation of NF-kappaB in intestinal epithelial cells by Enteropathogenic Escherichia Coli. *Am. J. Physiol.* **1997**, *273*, C1160–C1167. [CrossRef]
49. Czerucka, D.; Dahan, S.; Mograbi, B.; Rossi, B.; Rampal, P. Implication of mitogen-activated protein kinases in T84 cell responses to Enteropathogenic Escherichia Coli infection. *Infect. Immun.* **2001**, *69*, 1298–1305. [CrossRef]
50. Young, J.C.; Clements, A.; Lang, A.E.; Garnett, J.A.; Munera, D.; Arbeloa, A.; Pearson, J.; Hartland, E.L.; Matthews, S.J.; Mousnier, A.; et al. The Escherichia Coli Effector EspJ Blocks Src Kinase activity Via Amidation and ADP Ribosylation. *Nat. Commun.* **2014**, *5*, 5887. [CrossRef]
51. Pollard, D.J.; Berger, C.N.; So, E.C.; Yu, L.; Hadavizadeh, K.; Jennings, P.; Tate, E.W.; Choudhary, J.S.; Frankel, G. Broad-spectrum regulation of nonreceptor tyrosine kinases by the bacterial ADP-Ribosyltransferase EspJ. *MBio* **2018**, *9*. [CrossRef] [PubMed]
52. Phillips, N.; Hayward, R.D.; Koronakis, V. Phosphorylation of the Enteropathogenic E. Coli receptor by the Src-Family Kinase C-Fyn triggers actin pedestal formation. *Nat. Cell Biol.* **2004**, *6*, 618–625. [CrossRef] [PubMed]
53. Hayward, R.D.; Hume, P.J.; Humphreys, D.; Phillips, N.; Smith, K.; Koronakis, V. Clustering transfers the translocated escherichia coli receptor into lipid rafts to stimulate reversible activation of C-Fyn. *Cell. Microbiol.* **2009**, *11*, 433–441. [CrossRef] [PubMed]
54. Huang, Z.; Sutton, S.E.; Wallenfang, A.J.; Orchard, R.C.; Wu, X.; Feng, Y.; Chai, J.; Alto, N.M. Structural insights into Host GTPase isoform selection by a Family of bacterial GEF mimics. *Nat. Struct. Mol. Biol.* **2009**, *16*, 853–860. [CrossRef] [PubMed]
55. Berger, C.N.; Crepin, V.F.; Jepson, M.A.; Arbeloa, A.; Frankel, G. The Mechanisms used by Enteropathogenic Escherichia Coli to control filopodia dynamics. *Cell. Microbiol.* **2009**, *11*, 309–322. [CrossRef] [PubMed]
56. Ugalde-Silva, P.; Navarro-Garcia, F. Coordinated transient interaction of ZO-1 and Afadin is required for pedestal maturation induced by EspF from Enteropathogenic Escherichia Coli. *Microbiologyopen* **2019**, *8*, e931. [CrossRef] [PubMed]
57. Hurd, T.W.; Fan, S.; Liu, C.J.; Kweon, H.K.; Hakansson, K.; Margolis, B. Phosphorylation-dependent binding of 14-3-3 to the polarity protein Par3 regulates cell polarity in Mammalian Epithelia. *Curr. Biol.* **2003**, *13*, 2082–2090. [CrossRef]
58. Devriese, S.; Van den Bossche, L.; Van Welden, S.; Holvoet, T.; Pinheiro, I.; Hindryckx, P.; De Vos, M.; Laukens, D. T84 monolayers are superior to Caco-2 as a model system of Colonocytes. Histochem. *Cell Biol.* **2017**, *148*, 85–93.
59. Le Bivic, A.; Real, F.X.; Rodriguez-Boulan, E. Vectorial targeting of apical and basolateral plasma membrane proteins in a human adenocarcinoma epithelial cell line. *Proc. Natl. Acad. Sci. USA* **1989**, *86*, 9313–9317. [CrossRef]

60. Yoo, B.K.; Yanda, M.K.; No, Y.R.; Yun, C.C. Human intestinal epithelial cell line SK-CO15 is a new model system to study Na(+)/H(+) exchanger 3. *Am. J. Physiol. Gastrointest. Liver Physiol.* **2012**, *303*, G180–G188. [CrossRef]
61. Zhang, J.; Zhang, X.; Guo, Y.; Xu, L.; Pei, D. Sorting nexin 33 induces mammalian cell micronucleated phenotype and actin polymerization by interacting with Wiskott-Aldrich syndrome protein. *J. Biol. Chem.* **2009**, *284*, 21659–21669. [CrossRef] [PubMed]
62. Park, J.; Kim, Y.; Lee, S.; Park, J.J.; Park, Z.Y.; Sun, W.; Kim, H.; Chang, S. SNX18 shares a redundant role with SNX9 and modulates endocytic trafficking at the plasma membrane. *J. Cell. Sci.* **2010**, *123*, 1742–1750. [CrossRef] [PubMed]
63. Haberg, K.; Lundmark, R.; Carlsson, S.R. SNX18 is an SNX9 paralog that acts as a membrane Tubulator in AP-1-positive endosomal trafficking. *J. Cell. Sci.* **2008**, *121*, 1495–1505. [CrossRef] [PubMed]
64. Hua, Y.; Yan, K.; Wan, C. Clever cooperation: Interactions between EspF and host proteins. *Front. Microbiol.* **2018**, *9*, 2831. [CrossRef]
65. Badour, K.; McGavin, M.K.; Zhang, J.; Freeman, S.; Vieira, C.; Filipp, D.; Julius, M.; Mills, G.B.; Siminovitch, K.A. Interaction of the Wiskott-Aldrich syndrome protein with sorting Nexin 9 is required for CD28 endocytosis and cosignaling in T cells. *Proc. Natl. Acad. Sci. USA* **2007**, *104*, 1593–1598. [CrossRef]
66. Merrifield, C.J.; Qualmann, B.; Kessels, M.M.; Almers, W. Neural Wiskott Aldrich Syndrome Protein (N-WASP) and the Arp2/3 complex are recruited to sites of clathrin-mediated endocytosis in cultured fibroblasts. *Eur. J. Cell Biol.* **2004**, *83*, 13–18. [CrossRef]
67. Benesch, S.; Polo, S.; Lai, F.P.; Anderson, K.I.; Stradal, T.E.; Wehland, J.; Rottner, K. N-WASP deficiency impairs EGF internalization and Actin assembly at clathrin-coated pits. *J. Cell. Sci.* **2005**, *118*, 3103–3115. [CrossRef]
68. Yarar, D.; Waterman-Storer, C.M.; Schmid, S.L. SNX9 couples actin assembly to phosphoinositide signals and is required for membrane remodeling during endocytosis. *Dev. Cell.* **2007**, *13*, 43–56. [CrossRef]
69. Lundmark, R.; Carlsson, S.R. SNX9—A prelude to vesicle release. *J. Cell. Sci.* **2009**, *122*, 5–11. [CrossRef]
70. Andreeva, A.Y.; Krause, E.; Muller, E.C.; Blasig, I.E.; Utepbergenov, D.I. Protein kinase C regulates the phosphorylation and cellular localization of occludin. *J. Biol. Chem.* **2001**, *276*, 38480–38486. [CrossRef]
71. Aono, S.; Hirai, Y. Phosphorylation of claudin-4 is required for tight junction formation in a human keratinocyte cell line. *Exp. Cell Res.* **2008**, *314*, 3326–3339. [CrossRef] [PubMed]
72. Sotillos, S.; Diaz-Meco, M.T.; Caminero, E.; Moscat, J.; Campuzano, S. DaPKC-dependent phosphorylation of crumbs is required for epithelial cell polarity in drosophila. *J. Cell Biol.* **2004**, *166*, 549–557. [CrossRef] [PubMed]
73. Hurov, J.B.; Watkins, J.L.; Piwnica-Worms, H. Atypical PKC phosphorylates PAR-1 kinases to regulate localization and activity. *Curr. Biol.* **2004**, *14*, 736–741. [CrossRef] [PubMed]
74. Plant, P.J.; Fawcett, J.P.; Lin, D.C.; Holdorf, A.D.; Binns, K.; Kulkarni, S.; Pawson, T. A Polarity complex of mPar-6 and atypical PKC binds, phosphorylates and regulates mammalian Lgl. *Nat. Cell Biol.* **2003**, *5*, 301–308. [CrossRef]
75. Tobias, I.S.; Newton, A.C. Protein scaffolds control localized protein kinase Czeta activity. *J. Biol. Chem.* **2016**, *291*, 13809–13822. [CrossRef]
76. Chaki, S.P.; Barhoumi, R.; Rivera, G.M. Actin remodeling by Nck regulates endothelial lumen formation. *Mol. Biol. Cell* **2015**, *26*, 3047–3060. [CrossRef]
77. Gruenheid, S.; DeVinney, R.; Bladt, F.; Goosney, D.; Gelkop, S.; Gish, G.D.; Pawson, T.; Finlay, B.B. Enteropathogenic E. Coli Tir binds Nck to initiate actin pedestal formation in host cells. *Nat. Cell Biol.* **2001**, *3*, 856–859. [CrossRef]
78. Nieto-Pelegrin, E.; Kenny, B.; Martinez-Quiles, N. Nck adaptors, besides promoting N-WASP mediated actin-nucleation activity at pedestals, influence the cellular levels of Enteropathogenic Escherichia Coli Tir effector. *Cell Adhes Migr.* **2014**, *8*, 404–417. [CrossRef]
79. Worby, C.A.; Simonson-Leff, N.; Clemens, J.C.; Huddler, D., Jr.; Muda, M.; Dixon, J.E. Drosophila Ack targets its substrate, the sorting nexin DSH3PX1, to a protein complex involved in axonal guidance. *J. Biol. Chem.* **2002**, *277*, 9422–9428. [CrossRef]
80. Worby, C.A.; Simonson-Leff, N.; Clemens, J.C.; Kruger, R.P.; Muda, M.; Dixon, J.E. The sorting nexin, DSH3PX1, connects the axonal guidance receptor, dscam, to the actin cytoskeleton. *J. Biol. Chem.* **2001**, *276*, 41782–41789. [CrossRef]

81. Georgiou, M.; Marinari, E.; Burden, J.; Baum, B. Cdc42, Par6, and aPKC regulate Arp2/3-mediated endocytosis to control local adherens junction stability. *Curr. Biol.* **2008**, *18*, 1631–1638. [CrossRef] [PubMed]
82. Wald, F.A.; Oriolo, A.S.; Mashukova, A.; Fregien, N.L.; Langshaw, A.H.; Salas, P.J. Atypical protein kinase C (Iota) activates Ezrin in the apical domain of intestinal epithelial cells. *J. Cell. Sci.* **2008**, *121*, 644–654. [CrossRef] [PubMed]
83. Rouven Bruckner, B.; Pietuch, A.; Nehls, S.; Rother, J.; Janshoff, A. Ezrin is a major regulator of membrane tension in epithelial cells. *Sci. Rep.* **2015**, *5*, 14700. [CrossRef] [PubMed]
84. Crawley, S.W.; Mooseker, M.S.; Tyska, M.J. Shaping the intestinal brush border. *J. Cell Biol.* **2014**, *207*, 441–451. [CrossRef] [PubMed]
85. Tsai, F.C.; Bertin, A.; Bousquet, H.; Manzi, J.; Senju, Y.; Tsai, M.C.; Picas, L.; Miserey-Lenkei, S.; Lappalainen, P.; Lemichez, E.; et al. Ezrin enrichment on curved membranes requires a specific conformation or interaction with a curvature-sensitive partner. *Elife* **2018**, *7*, e37262. [CrossRef] [PubMed]
86. Simonovic, I.; Arpin, M.; Koutsouris, A.; Falk-Krzesinski, H.J.; Hecht, G. Enteropathogenic Escherichia Coli activates ezrin, which participates in disruption of tight junction barrier function. *Infect. Immun.* **2001**, *69*, 5679–5688. [CrossRef]
87. Law, H.T.; Chua, M.; Moon, K.M.; Foster, L.J.; Guttman, J.A. Mass spectrometry-based proteomics identification of Enteropathogenic Escherichia Coli pedestal constituents. *J. Proteome Res.* **2015**, *14*, 2520–2527. [CrossRef]
88. Campellone, K.G.; Giese, A.; Tipper, D.J.; Leong, J.M. A tyrosine-phosphorylated 12-amino-acid sequence of Enteropathogenic Escherichia Coli Tir Binds the host adaptor protein Nck and is required for Nck localization to actin pedestals. *Mol. Microbiol.* **2002**, *43*, 1227–1241. [CrossRef]
89. Kalman, D.; Weiner, O.D.; Goosney, D.L.; Sedat, J.W.; Finlay, B.B.; Abo, A.; Bishop, J.M. Enteropathogenic E. Coli acts through WASP and Arp2/3 complex to form actin pedestals. *Nat. Cell Biol.* **1999**, *1*, 389–391. [CrossRef]
90. Lommel, S.; Benesch, S.; Rottner, K.; Franz, T.; Wehland, J.; Kuhn, R. Actin pedestal formation by Enteropathogenic Escherichia Coli and intracellular motility of Shigella Flexneri are abolished in N-WASP-defective cells. *EMBO Rep.* **2001**, *2*, 850–857. [CrossRef]
91. Kralicek, S.E.; Nguyen, M.; Rhee, K.J.; Tapia, R.; Hecht, G. EPEC NleH1 is significantly more effective in reversing colitis and reducing mortality than NleH2 via differential effects on host signaling pathways. *Lab. Invest.* **2018**, *98*, 477–488. [CrossRef] [PubMed]
92. McNamara, B.P.; Donnenberg, M.S. A novel proline-rich protein, EspF, is secreted from Enteropathogenic Escherichia Coli via the Type III export pathway. *FEMS Microbiol. Lett.* **1998**, *166*, 71–78. [CrossRef] [PubMed]
93. Kenny, B.; Jepson, M. Targeting of an Enteropathogenic Escherichia Coli (EPEC) effector protein to host mitochondria. *Cell. Microbiol.* **2000**, *2*, 579–590. [CrossRef] [PubMed]
94. Rhee, K.J.; Cheng, H.; Harris, A.; Morin, C.; Kaper, J.B.; Hecht, G. Determination of spatial and temporal colonization of Enteropathogenic E. Coli and Enterohemorrhagic E. Coli in mice using bioluminescent in Vivo imaging. *Gut Microbes* **2011**, *2*, 34–41. [CrossRef]

6

Regional Differences in Tight Junction Protein Expression in the Blood–DRG Barrier and their Alterations after Nerve Traumatic Injury in Rats

Thomas J. Lux [1], Xiawei Hu [1], Adel Ben-Kraiem [1], Robert Blum [2], Jeremy Tsung-Chieh Chen [1,*,†] and Heike L. Rittner [1,*,†]

1 Department of Anaesthesiology, University Hospital of Wuerzburg, 97074 Wuerzburg, Germany; thomas.lux@stud-mail.uni-wuerzburg.de (T.J.L.); xiawei_hu@163.com (X.H.); Ben_A@ukw.de (A.B.-K.)
2 Institute of Clinical Neurobiology, University Hospital of Wuerzburg, 94074 Wuerzburg, Germany; Blum_R@ukw.de
* Correspondence: chen_j1@ukw.de (J.T.-C.C.); rittner_h@ukw.de (H.L.R.).
† These authors contributed equally to this work.

Abstract: The nervous system is shielded by special barriers. Nerve injury results in blood–nerve barrier breakdown with downregulation of certain tight junction proteins accompanying the painful neuropathic phenotype. The dorsal root ganglion (DRG) consists of a neuron-rich region (NRR, somata of somatosensory and nociceptive neurons) and a fibre-rich region (FRR), and their putative epi-/perineurium (EPN). Here, we analysed blood–DRG barrier (BDB) properties in these physiologically distinct regions in Wistar rats after chronic constriction injury (CCI). *Cldn5*, *Cldn12*, and *Tjp1* (rats) mRNA were downregulated 1 week after traumatic nerve injury. Claudin-1 immunoreactivity (IR) found in the EPN, claudin-19-IR in the FRR, and ZO-1-IR in FRR-EPN were unaltered after CCI. However, laser-assisted, vessel specific qPCR, and IR studies confirmed a significant loss of claudin-5 in the NRR. The NRR was three-times more permeable compared to the FRR for high and low molecular weight markers. NRR permeability was not further increased 1-week after CCI, but significantly more CD68^{+} macrophages had migrated into the NRR. In summary, NRR and FRR are different in naïve rats. Short-term traumatic nerve injury leaves the already highly permeable BDB in the NRR unaltered for small and large molecules. Claudin-5 is downregulated in the NRR. This could facilitate macrophage invasion, and thereby neuronal sensitisation and hyperalgesia. Targeting the stabilisation of claudin-5 in microvessels and the BDB barrier could be a future approach for neuropathic pain therapy.

Keywords: tight junction; claudin-5; neuropathic pain; nerve injury; dorsal root ganglion

1. Introduction

Neuropathic pain is defined as a "pain caused by a lesion or disease of the somatosensory nervous system". The etiologies range from metabolic conditions like diabetes mellitus, autoimmune disorders, infectious diseases and chemotherapy-induced neuropathies to traumatic nerve injury and postoperative conditions. Overall, 7–10% of the general population are affected, and their quality of life is significantly lowered [1]. The heterogeneity of etiologies leading to neuropathic pain indicates that a multitude of not yet understood factors participate in the pathogenesis of neuropathic pain [2]. A causative therapy is currently often impossible [3]. While conventional analgesics, e.g., cyclooxygenase inhibitors such as ibuprofen, have little effect on neuropathic pain, symptomatic treatments include antidepressants, antiepileptics, and opioids, but the efficacy of these drugs is moderate [2]. Attempts to develop pathophysiology-oriented drugs have not delivered the desired results at this point. A better understanding of the mechanisms behind neuropathic phenotypes, therefore, is necessary to find new diagnostic and therapeutic approaches.

Several distinct animal models are used for research in neuropathy. Models are separated into peripheral nerve injury, central pain, drug-induced neuropathy, and disease-induced approaches [4]. Since the introduction of the chronic constriction injury (CCI) model for traumatic neuropathy in 1988 [1], it has been commonly used to understand the pathophysiology of mononeuropathies. Animals are subjected to four ligations of the sciatic nerve, which cause focal ischemia, intraneural edema, and Wallerian degeneration, and subsequently, regeneration. This results in a behavioral phenotype with signs of spontaneous pain and thermal and mechanical hypersensitivity [2]. Hypersensitivity reaches its maximum after 7 d [5], but nociceptive thresholds return to normal within the next 3–6 weeks depending on the noxious stimulus [6]. The local (inflammatory) reaction caused by cell damage at the ligature site does not explain the phenotype after CCI sufficiently. Changes in the dorsal root ganglion (DRG), such as infiltration of macrophages (ED1 for CD68 in rats) and CD8+ lymphocytes have been reported [7]. Furthermore, sprouting of sympathetic axons into the DRG after CCI has been observed [8]. These findings are thought to participate in the pathogenesis of traumatic mononeuropathies. Only about 30% of the neurons of the L4/5 DRGs are afferences from the sciatic nerve. Since other areas of the L4/5 dermatome show few to no changes after CCI, a combination of neuronal and environmental changes must be elementary. Electrophysiological and histopathological studies identified reduced nerve conduction velocity and axonal damage, as correlates of A-fiber damage and sensitization of C-fibers, as probable causes for this phenotype.

Even slight environmental changes disrupt physiological function of neural tissues. Therefore, the entry points to the nervous system are protected. Highly specialized barriers guard myelinated fibers (myelin barrier, MB), the peripheral nerve (blood–nerve barrier), the DRG (blood–DRG barrier, BDB), the spinal cord (blood–spinal-cord barrier), and the brain (blood–brain barrier). The names originate from the protected region, and each of them can be subdivided into the different tissue interfaces, e.g., the endothelial vessels or the perineurium, each fulfilling different physiological demands. Key features for this sealing function are tight junction proteins. The specific tight junction protein composition determines the barriers characteristics. While many of these characteristics are similar in the blood–nerve barrier and blood–brain barrier, distinct features, such as absence of several neurotransmitter transporters and lack of astrocytes in the periphery, qualify the blood–nerve barrier as a specialized, unique structure [9].

After axonal injury or acute demyelination, extravasation of blood-borne molecules such as albumin and intraneural edema as correlates of blood–nerve barrier leakage can be observed [10]. Studies of the blood–nerve barrier in nerve injury models have revealed downregulation of *Cldn1*, *Cldn5*, *Ocln*, *Tjp1* (ZO-1) and *Jam3* (JAMC) mRNA in the sciatic nerve [10–14] as well as *Tjp1* (ZO-1), *Cldn5* and *Ocln* in the spinal cord of rats [15,16]. Similar studies observed reductions of *Cldn1* and *Tjp1* (ZO-1) mRNA and immunoreactivity (IR) in the sciatic nerves of mice [13]. Endothelial cells of the blood–nerve barrier are disrupted with increased permeability as soon as 6 h after CCI surgery, while the neuropathic phenotype develops over days [9]. So, barrier disruption occurs early after nerve injury even before hypersensitivity.

Surprisingly, the BDB has barely been studied before. Regions in the DRG can be divided: Somata of primary sensory neurons reside in DRGs (neuron rich region (NRR)) in contrast to fiber rich regions (FRR). It is known that the BDB is considerably more permeable and contains a higher density of capillaries [17–19]. Claudin-5 IR was detected in the NRR, while claudin-1 and occludin were found in the FRR [19]. However, no quantitative data of either protein or mRNA in all areas, including epi-/perineurium (EPN), are currently available in naïve animals or after neuropathy. Furthermore, whole tissue analysis can be insensitive to small changes of specific barriers and novel techniques, making selective analysis necessary.

In this study, we wanted to fully characterize the BDB and its alteration in neuropathy. To this end, we defined four regions in the DRG: the neuron-rich and the fiber-rich regions (NRR,

FRR), and their putative epi-/perineurial regions (NRR-EPN, FRR-EPN). We used these regions and region-selective techniques to analyze typical tight junction proteins known from the blood nerve and myelin barrier in control rats and in neuropathy and evaluated functional properties of the BDB. Tight junction proteins, including claudin-1, were detected in the nerve perineurium, claudin-5 was found in endoneurial vessels, ZO-1 was ubiquitously detected in the nerve, and claudin-12 and 19 were present in Schwann cells.

2. Results

2.1. Claudin-1, Claudin-19, and ZO-1 Immunoreactivity Is Tissue Specific in Rat DRGs

To characterize the BDB and its molecular structure, we quantified the immunoreactivity (IR) in the DRG, considering its different regions. After separating the DRGs distinct areas, claudin-1, claudin-5, claudin-12, claudin-19, and ZO-1, IR was semi-quantified and compared between the NRR, FRR, and the putative EPN of DRGs after CCI.

While claudin-1 IR was up to five times higher in the EPN as in the inner regions of the DRG (Figure 1b,f), claudin-19 IR was highest in the FRR (Figure 1d,f) and ZO-1 IR was highest in the FRR-EPN (Figure 1e,f). Claudin-5 and claudin-12 expressions were not region specific. The mean intensity of claudin-5 IR was low, but areas with strong signals, mostly associated with claudin-1 signals, were observed (Figure 3). In brightfield images, these areas resembled vessels. This was also seen for ZO-1-IR, which was expressed in the structures resembling capillaries and in the EPN. Claudin-12 was not only found in putative Schwann cell structures, but neurons as well. In contrast, claudin-19-IR was detected in typical paranodal structures of Schwann cells.

2.2. Nocifensive Responsiveness after Nerve Injury

We next wanted to test whether neuropathy results in alterations of the BDB. To demonstrate the validity of the neuropathy CCI model, mechanical nociceptive thresholds were recorded for both rear limbs before and one week after CCI via the von-Frey test.

While there was no change in the sensitivity in the contralateral paw, the limb with nerve injury exhibited a decrease of the mechanical withdrawal threshold to 34% of the previous baseline value (Figure 2), indicating mechanical hypersensitivity in neuropathy.

2.3. Downregulation of Tight Junction Protein mRNA Expression after CCI in Rodents

For evaluation of tight junction protein expression patterns during CCI, we analyzed the mRNA levels of *Cldn1, Cldn5, Cldn12, Cldn19,* and *Tjp1* (ZO-1) in rats' whole DRGs. The expression levels of *Cldn5, Cldn12,* and *Tjp1* (ZO-1) mRNA were reduced by 28%, 45% and 26%, respectively after CCI compared to sham (Figure 3).

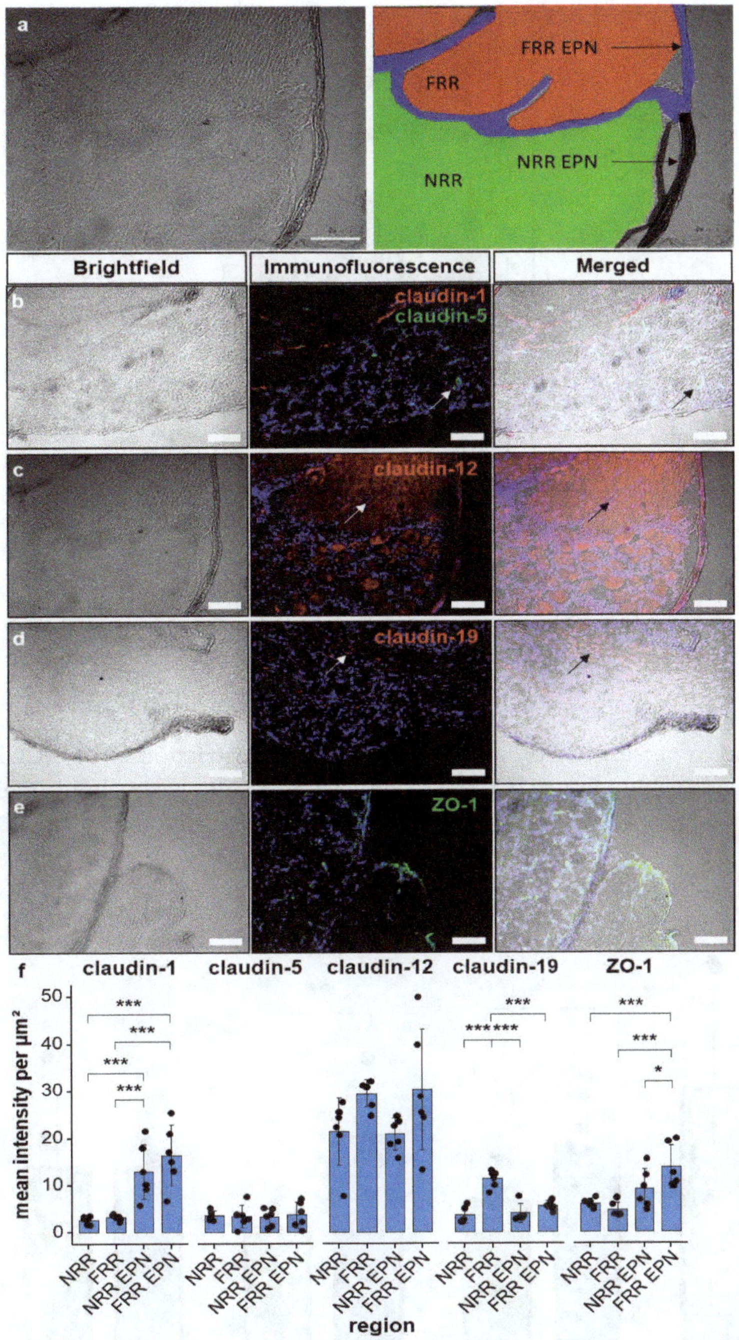

Figure 1. Claudin-1 immunoreactivity (IR) and ZO-1 IR are preferentially found in epi-/perineurium in rats' dorsal root ganglia (DRGs), while claudin-19 IR is most abundant in the fiber-rich region. Classification into neuron rich region (NRR, green), NRR epi-/perineurium (EPN, black), fiber rich region (FRR, red), and FRR-EPN (blue) is shown in (**a**). Control DRG sections (CL) from Wistar rats were immunostained. IRs of claudin-1, claudin-5, and ZO-1 were quantified and compared between NRR and FRR, together with their putative EPN. Representative stainings for claudin-1 (red), claudin-5 (green) (**b**), claudin-12 (**c**), claudin-19 (**d**), and ZO-1 (**e**) are shown. Arrows point to structures identified as vessels (**b**), Schwann cells (**c**), and paranodes of Schwann cells (**d**). Quantification of the signal intensity in the specified areas (**f**). (Scale bars = 100 µm; $n = 5$ or 6; claudin-1: NRR versus NRR-EPN, NRR versus FRR-EPN, FRR versus NRR-EPN, and FRR versus FRR-EPN: $p < 0.0001$; ZO-1: FRR-EPN versus NRR: $p = 0.0003$; FRR-EPN versus NRR-EPN: $p = 0.0138$; FRR-EPN versus NRR: $p = 0.00018$. No normal distribution: ZO-1 FRR. No variance homogeneity: claudin-1. Two-way ANOVA, Tukey HSD. * $p < 0.05$, *** $p < 0.001$).

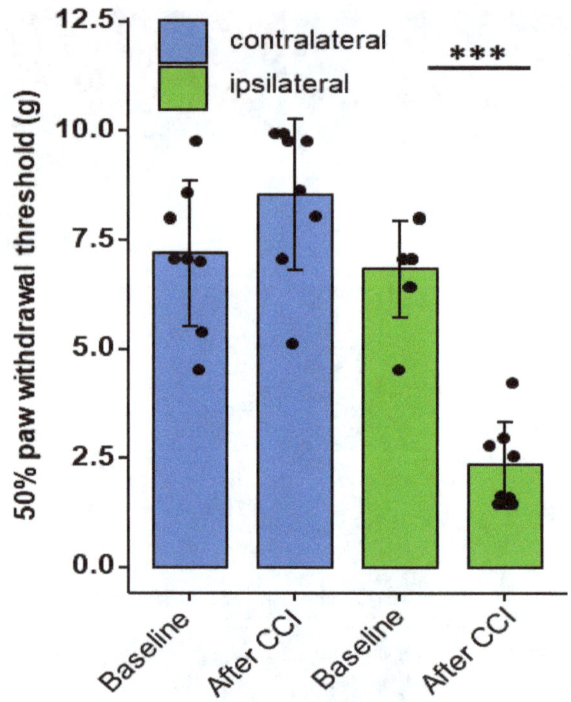

Figure 2. Mechanical hypersensitivity in rats after chronic constriction injury (CCI). Male Wistar rats underwent chronic constriction injury (CCI). Mechanical nociceptive thresholds were evaluated before and one week after CCI (n = 8, Von-Frey test; repeated measures ANOVA, *** $p < 0.001$).

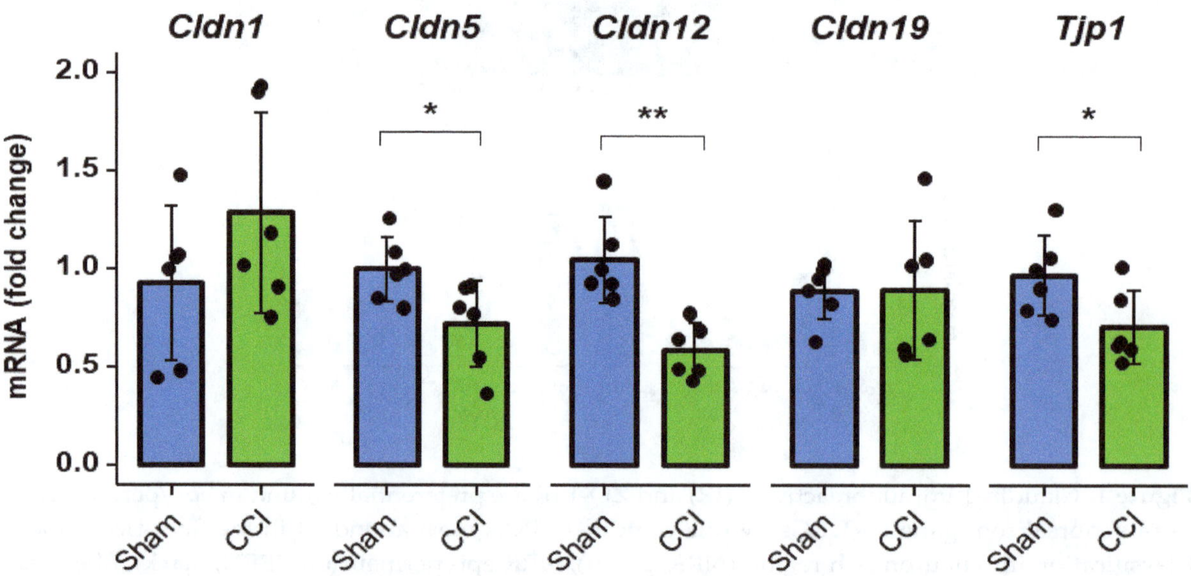

Figure 3. Tight junction protein mRNA expression in rat DRGs is downregulated one week after CCI. Wistar rats were treated with CCI. Relative mRNA expressions of *Cldn1*, *Cldn5*, *Cldn12*, *Cldn19*, and *Tjp1* (ZO-1) in comparison to sham operated animals were analyzed with qPCR after one week. (n = 6; *Cldn5*: $p = 0.0338$; *Cldn12* $p = 0.0021$, *Tjp1* (ZO-1) $p = 0.0419$. No variance homogeneity: *Cldn19* rats. Welch two-sample t-test, Wilcoxon rank sum test: * $p < 0.05$, ** $p < 0.01$).

2.4. Claudin-5 Expression in Vessels Is Reduced Only in the NRR after CCI

Comparison of the tight junction protein IR between naive and CCI rats in the distinct regions did not reveal significant differences (Figures 4 and 5). Only a tendency was observed for claudin-12 in the FRR and claudin-19 for the FRR-EPN ($p = 0.057$).

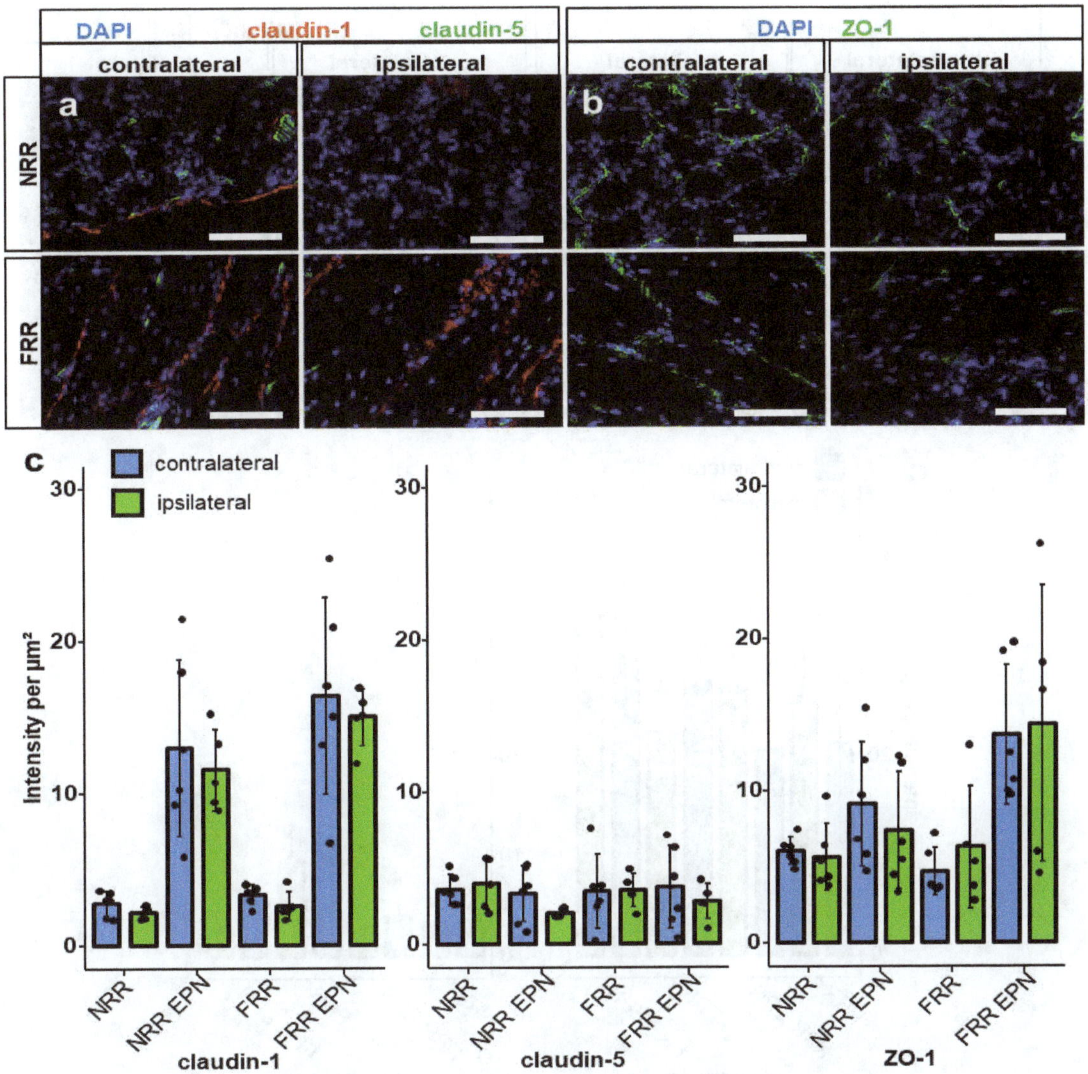

Figure 4. Claudin-1 and ZO-1 IR are unchanged after CCI. DRG sections from Wistar rats with 7 d CCI and naïve controls were immunostained. IRs of claudin-1, claudin-5, and ZO-1 were quantified and compared between IL and CL DRGs after CCI in the neuron rich region (NRR) and fiber rich region (FRR), and their putative EPN are presented as light intensity per μm^2. Representative sections of claudin-1 (red) and claudin-5 (green) (**a**), and ZO-1 (green) (**b**) are shown. Quantification of the signal intensity in the specified areas: significant differences between regions were not indicated, as they are already analyzed in Figure 1 (**c**). (Scale bars = 100 μm. $n = 5$ or 6; claudin-1: NRR versus NRR-EPN, NRR versus FRR-EPN, FRR versus NRR-EPN, and FRR versus FRR-EPN: $p < 0.0001$; ZO-1: FRR-EPN versus NRR: $p = 0.0003$; FRR-EPN versus NRR-EPN: $p = 0.0138$; FRR-EPN versus NRR: $p = 0.00018$. No normal distribution: ZO-1 CL FRR. No variance homogeneity: claudin-1 CL. Two-way ANOVA, Tukey HSD).

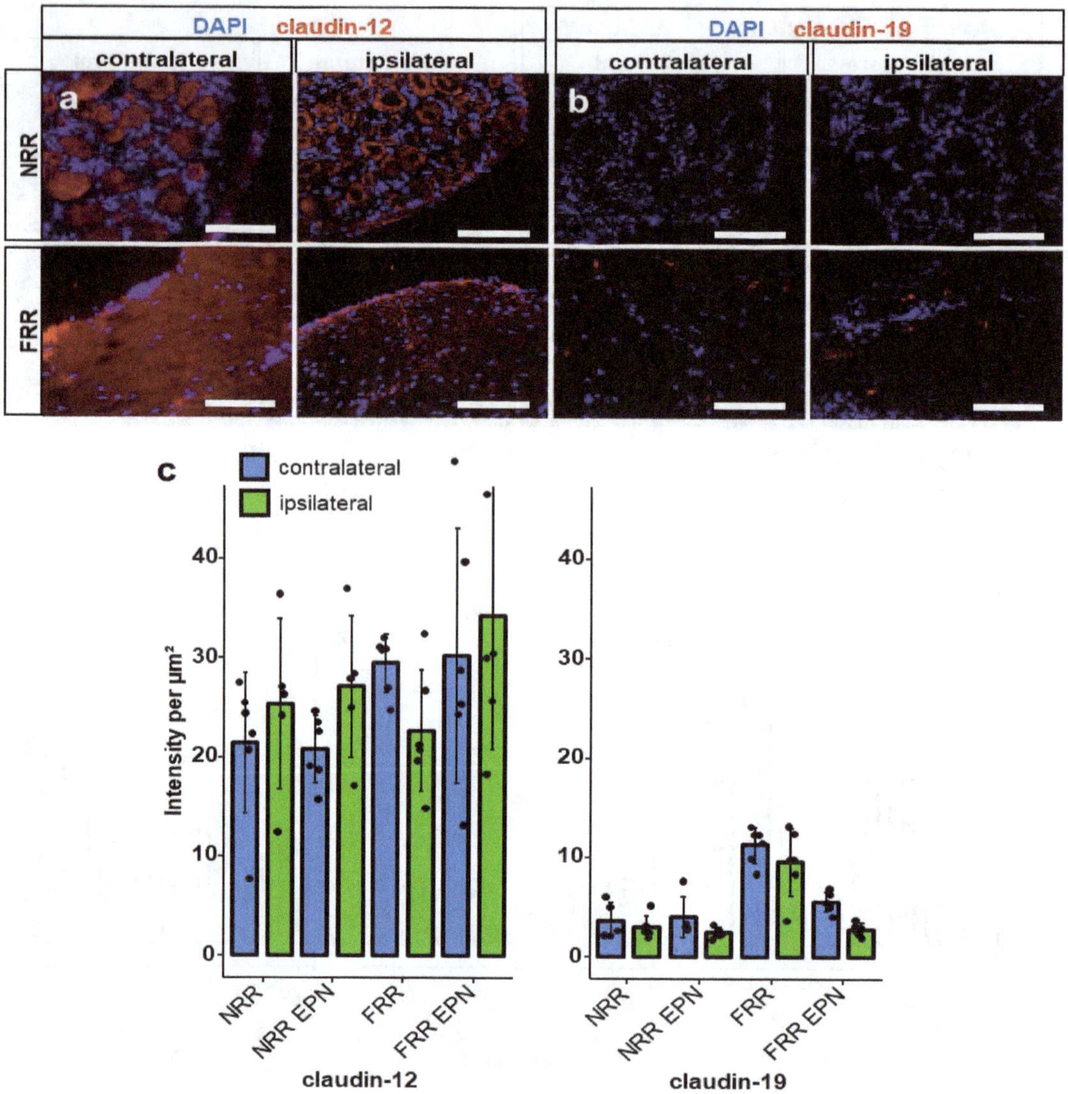

Figure 5. Claudin-19 and claudin-12-IR are unaltered after CCI. After CCI, DRG sections of Wistar rats were immunostained. IRs of claudin-12 and claudin-19 were quantified as light intensity per μm^2. and compared between IL and CL DRGs after CCI in the NRR and FRR, and their putative EPNs. Representative sections of claudin-12 (red) (**a**) and claudin-19 (red) (**b**) are shown. Quantification of the signal intensity in the specified areas: significant differences between regions are not indicated, as they were already analyzed in Figure 1 (**c**). (Scale bars = 100 µm. $n = 5$ or 6; claudin-19: FRR versus NRR, FRR versus NRR-EPN, and FRR versus FRR-EPN: $p < 0.0001$. No normal distribution: claudin-19 CL NRR-EPN. Two-way ANOVA, Tukey HSD).

Since the method is optimized for homogenously expressed proteins and lacks sensitivity for proteins with clustered expression, e.g., in vessels, we analyzed claudin-5 using a different approach. We co-stained the tissue with von Willebrand Factor (vWF) to analyze capillaries in the NRR and FRR (Figure 6). Selective analysis of vessels revealed higher vWF IR in vessels of the FRR compared with the NRR, but no change after CCI. Analysis of claudin-5 within vessels could also show higher IR in the FRR than in the NRR area. Most importantly, we observed a reduction of claudin-5 signal in the vessels of the NRR after CCI.

To verify these results on a mRNA level, we dissected capillaries from the FRR and the NRR by laser dissection in sham and CCI rats 1 d to reflect the mRNA alterations preceding protein expression. No changes were observed for *Vwf* and *Cldn5* mRNA in the FRR, but *Cldn5* was significantly downregulated in the NRR.

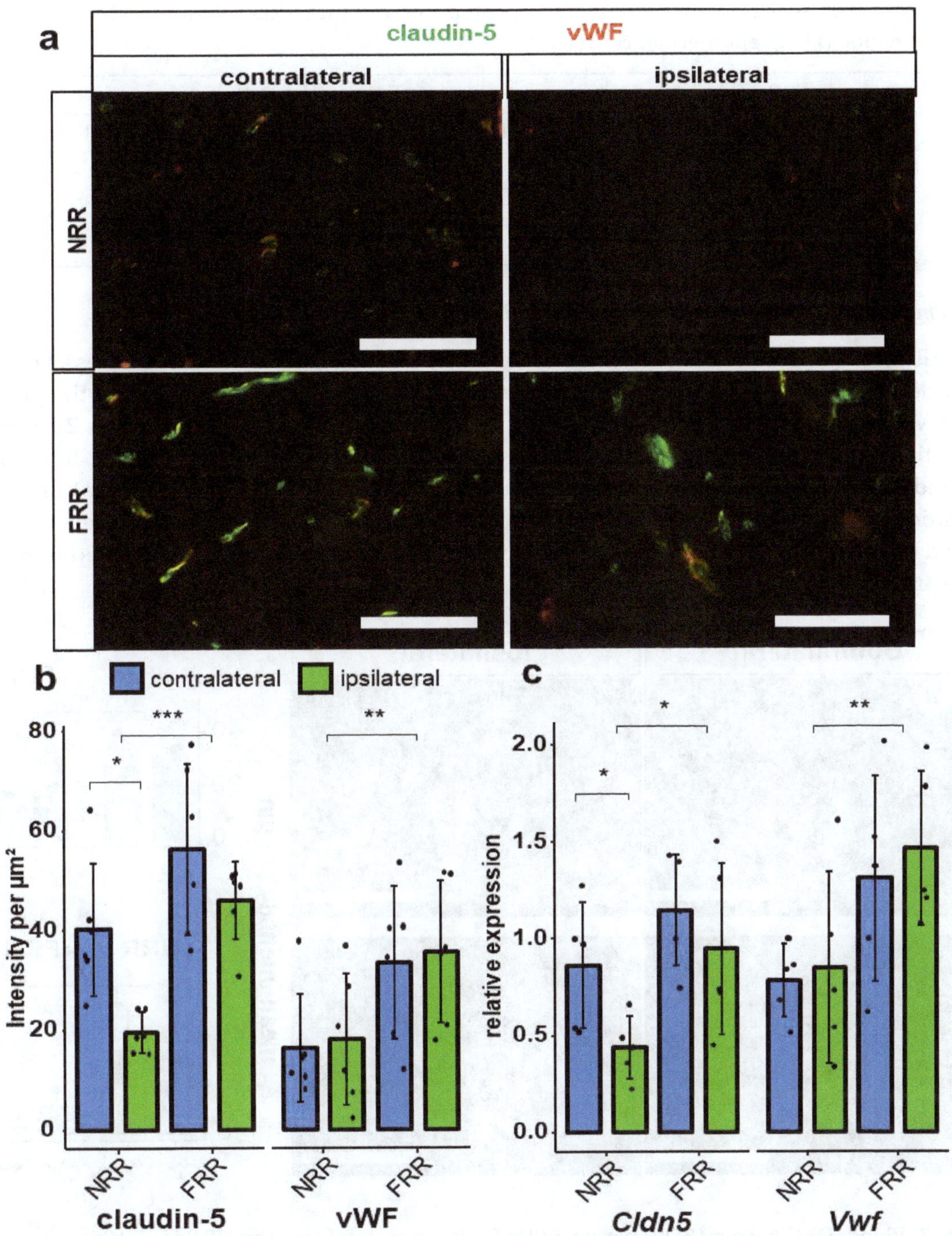

Figure 6. Claudin-5 IR within vessels and *Cldn5* mRNA in the NRR are reduced in rat DRGs. DRG sections from Wistar rats after CCI were immunostained. Quantification of claudin-5 and van Willebrand factor (vWF) IRs and comparison between IL and CL DRGs after CCI in the NRR and FRR as light intensity per μm^2. Tissue samples from the NRR and vessels of the FRR were obtained using laser microdissection (LMD). Claudin-5 and vWF mRNA were quantified as relative expression (to GAPDH) via RT-qPCR. Representative sections of co-stainings with claudin-5 (green) and vWF (red) (**a**). Quantification of the claudin-5 and vWF IR (**b**) $n = 6$; claudin-5: FRR versus NRR, $p = 0.000261$; NRR: CL versus CCI, $p = 0.02985$; vWF: NRR versus FRR, $p = 0.00515$. No normal distribution: claudin-5 FRR CCI, vWF NRR CL. No variance homogeneity: claudin-5 FRR. Two-way ANOVA, Tukey HSD.) And quantification of mRNA expression (**c**) $n = 5$; *Cldn5*: NRR versus FRR, $p = 0.0136$; CL versus IL: $p = 0.0448$; *Vwf*: NRR versus FRR, $p = 0.00838$; two-way ANOVA, Tukey HSD) in the specified areas. (Scale bars = 100 µm. * $p < 0.05$, ** $p < 0.01$, *** $p < 0.001$.) All results are summarized in Table 1.

Table 1. Regional expression of tight junction proteins within rat DRGs. Neuron-rich region (NRR), fiber-rich region (FRR), epi-/perineurium (EPN).

Protein	Expression Pattern
Claudin-1	NRR-EPN and FRR-EPN > NRR and FRR
Claudin-5 (in vessels)	FRR > NRR
ZO-1	FRR-EPN > NRR, FRR, NRR-EPN
Claudin-19	FRR > NRR, NRR-EPN, FRR-EPN

2.5. Permeability of the DRG and Migration of Macrophages

To finally evaluate functional properties of the BDB, we established an assay for vessel permeability with FITC-dextran (70 kDa). After intravenous application, the dye accumulated in the perineurium and within vessels, diffusing a few micrometers into the tissue. Hence, the NRR was 2.35 times more permeable than the FRR; however, no difference after CCI was observed (Figure 7). In the next step, we analyzed the permeability for small molecules using Hoechst reagent (562 Da) after systemic administration in vivo. Hoechst reagent ubiquitously diffused into the tissue and then stained nuclei. We detected significantly higher (3.51 times) permeability in the NRR region compared to the FRR but no change after CCI.

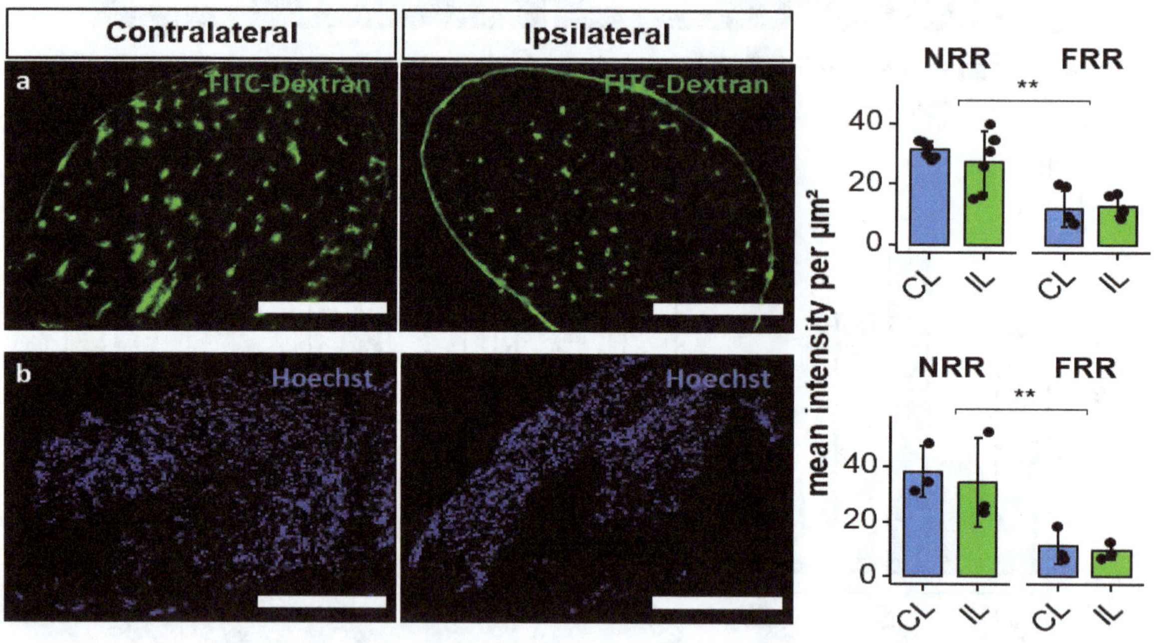

Figure 7. Blood–DRG barrier (BDB) permeability for large molecules in the NRR is higher compared to FRR of DRGs, but has no change after CCI. Wistar rats after 7 d CCI were perfused with FITC-dextran ((**a**), 70 kDa) and Hoechst reagent ((**b**), 562 Da). NRR and FRR of ipsilateral (IL) and contralateral (CL) DRG sections were analyzed. Representative images of the IL show comparable IR. Quantification of FITC-dextran ($n = 6$, NRR versus FRR: $p < 0.001$; two-way ANOVA, Tukey HSD) and Hoechst ($n = 3$, NRR versus FRR $p = 0.002$; two-way ANOVA, Tukey HSD) Immunofluorescence as intensity per μm^2. (Scale bars = 100 µm; ** $p < 0.01$.)

$CD68^+$ cells and CD68 immunoreactivity was quantified in the NRR in naive animals DRGs, and contralateral and ipsilateral DRGs after CCI to assess macrophage migration (Figure 8). $CD68^+$ cells per μm^2 were significantly increased in the IL DRGs after CCI compared to CL, while there was only a tendency ($p = 0.065$) compared to naive DRGs. The anti-CD68 immunoreactivity per μm^2 was significantly higher in IL DRGs after CCI, compared to naive DRGs.

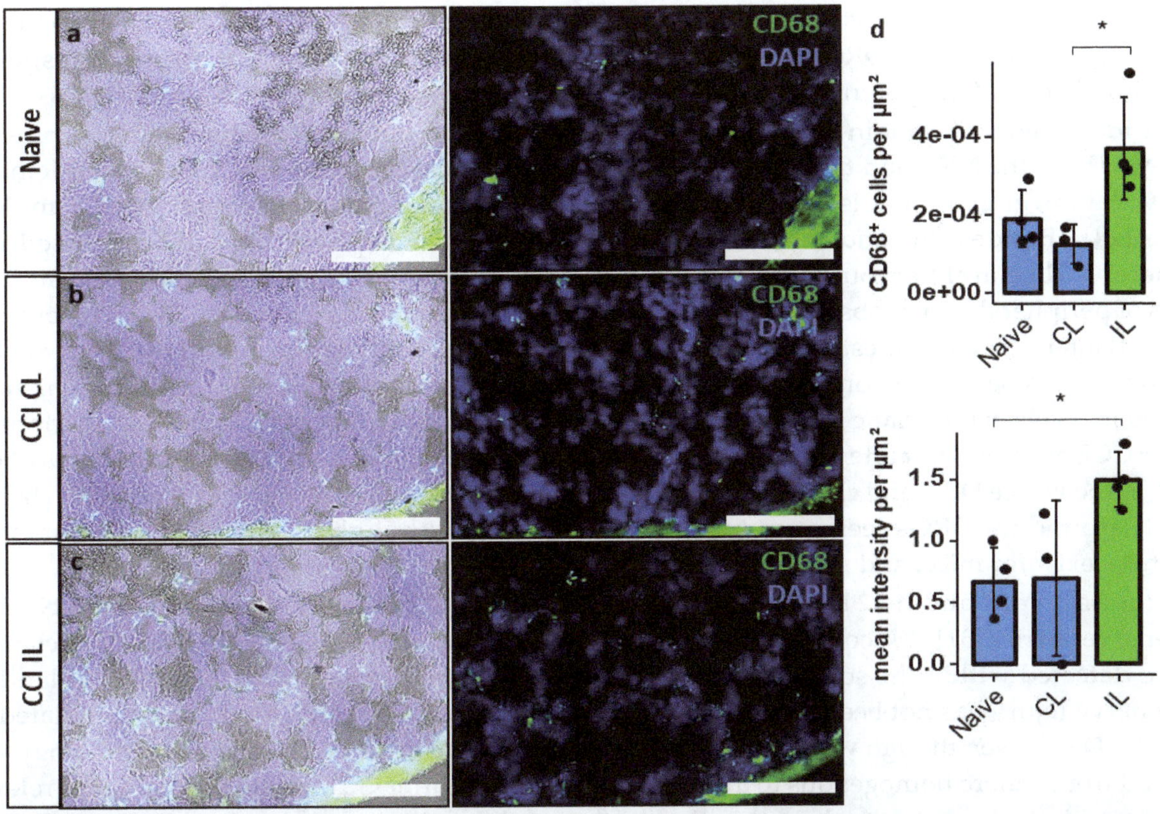

Figure 8. Increased macrophage invasion in the NRR after CCI. DRGs of naive (**a**) and contralateral (CL) (**b**) and ipsilateral (IL) (**c**) DRGs of Wistar rats 7d after CCI were harvested and stained using DAPI and anti-CD68 antibodies (left: immunofluorescence with brightfield; right: only immunofluorescence). In the NRR, CD68+ cells were counted manually, and signal intensity was measured. Both were quantified per μm^2 (**d**). (n = 3 or 4; CD68+ cells per μm^2: CL versus IL, $p = 0.024$; mean intensity per μm^2: Naive versus IL, $p = 0.038$. Two-way ANOVA, Tukey HSD; scale bars = 100 μm; * $p < 0.05$).

3. Discussion

In our study we further characterized the BDB and examined whether traumatic neuropathy not only results in blood–nerve barrier and blood spinal cord barrier breakdown but also affects the BDB. We revealed higher ZO-1 expression in the FRR EPN compared to the NRR and observed regional expression differences for claudin-1 (highest expression in NRR ENP and FRR EPN), claudin-5 (higher expression in the FRR compared to the NRR), and claudin-19 (highest expression in the FRR), as indicated in Table 1. After CCI, claudin-5 IR and *Cldn5* mRNAs were selectively reduced in the vessels of the NRR. The NRR containing somatosensory neurons was 2–3 times more permeable to low and high molecular weight dyes compared to the nerve fiber area with no further increase in permeability after CCI. However more CD68+ macrophages migrated into the NRR.

While the blood–nerve barrier is focus of current literature, the BDB is, so far, almost neglected. As an imperative link between the peripheral nerve and the spinal cord, open questions should be discussed more thoroughly, as they could reveal important knowledge for understanding the peripheral sensory system. The BDB is very similar to the blood–nerve barrier regarding morphology and tight junction protein composition, but its NRR barrier is leaky in its physiological state compared to the blood–nerve barrier. In accordance with our data, intravenously injected Hoechst dye or FITC-dextran diffusion out of capillaries can already be observed in DRGs of naive rats, while dyes only accumulate in peripheral nerves after blood–nerve barrier breakdown [18,20]. A clinically relevant consequence of these characteristics is the accumulation of drugs in the DRG, which can be favorable in case of opioids or undesirable in case of chemotherapeutic drugs causing neuropathies. This hints towards a concept of the DRG being more than a simple relay-station, but being an important signal modulator sensitive to endogenous and exogenous stimuli [21]. In addition to molecules, cells can migrate into the DRG. Macrophages reside in the DRG and further migrate into the DRG to shape an immune response in neuropathy [22,23]. So, opening of the BDB could facilitate these pathophysiological processes.

We identified that claudin-1, a major sealing tight junction protein, is predominantly expressed in the EPN of both the NRR and the FRR, as they show the highest claudin-1 IRs. *Cldn1* KO is lethal due to dehydration, as claudin-1 is key component of the epidermal barrier in the stratum granulosum in mice [24]. Besides this and other vital functions, claudin-1 is physiologically expressed in the peripheral and central nervous system as part of the blood–nerve barrier and blood–brain barrier. Studies of peripheral nerves observed high expression of *Cldn1* in the perineurium [13,25,26], but also in Schmidt-Lantermann incisures, paranodal loops of myelinating Schwann cells, and the mesaxon [8,27]. Previous studies performed on rats [12,28] and mice [13] also described a reduction of *Cldn1* mRNA and protein levels in the sciatic nerve after nerve injury. Downregulation of *Cldn1* mRNA starts around 3 h after CCI, and is lowest after 7 d [12]. In the DRG, we determined no significant differences of *Clnd1* mRNA in the whole DRG and claudin-1-IR in the subregions between naïve and CCI groups. Therefore, the EPN around rat DRGs seems not to be affected by CCI, although it is possible that intraspecies differences exist for mice, and more importantly, humans, in this and other barrier proteins.

Expression of claudin-12 by glial cells has been observed in the blood–brain barrier [29,30] and blood–nerve barrier [31]. We confirmed *Cldn12* mRNA expression in the DRG. High levels of claudn-12 IR were detected within the somata, and in or around the fibers of neurons. The role of claudin-12 during nerve injury has not been examined yet. In our study, *Cldn12* mRNA was downregulated after CCI in the DRG. Even though we could not observe lower claudin-12 IR, the signal morphology seems disturbed from a more homogenous to a clustered distribution after CCI. While the biological relevance of the signal in the soma is unclear, the distribution and the change of claudin-12 after CCI should certainly be investigated further.

Claudin-19 is expressed in myelinating glia cells in the PNS, but not the CNS [32]. Tight junction formation in myelinating cells is disrupted without claudin-19 [33]. *Cldn19* KO mice are fertile and vital, but show the phenotype of a peripheral neuropathy, causing mainly motor defects, but in 50%, nerve conduction deficits as well [32]. In our study, claudin-19 IR was clustered and highest in the FRR, and in the space between the neurons' somata in the NRR. These clusters could very well resemble the paranodal regions, like those already described in the sciatic nerve [31]. In agreement previous findings of claudin-19 expression in vessels and the perineurium, there were low signals in the EPNs of both regions [32], although we have previously seen this in the nerve [34]. Neither *Cldn19* mRNA levels nor claudin-19 IR were altered in the DRG proximal to the CCI injury.

ZO-1, an intracellular tight junction associated protein, forms complexes with claudin-1 and is required for correct arrangement of tight junctions [33]. *Tjp1*-KO is lethal, and disruption has been linked to neurological disorders [35]. ZO-1 is mostly expressed in the perineurium [31], vessels, and myelinated Schwann cells in the sciatic nerves of humans [31] and mice [36]. In our study, ZO-1 IR signal was detected predominantly in the EPN, like claudin-1's IR, and in clustered regions in all areas. The ZO-1 clusters colocalized with claudin-5, and are, therefore, most likely vessels. *Tjp1* was downregulated in whole DRGs. In our IF study, we detected the highest levels of ZO-1 IR expression in the FRR-EPN but no alterations after CCI. These findings match with the existing data from studies in the mouse sciatic nerve [36].

Considered the most important tight junction protein of the brain barrier, *Cldn5* mRNA expression is >100-fold higher than other tight junction protein expression in endothelial cells [37]. While *Cldn5* KO mice are macroscopically vital, increased brain barrier permeation by molecules of <800 Da and death 10 d after birth were observed [38]. In the blood–nerve barrier, *Cldn5* is expressed in the endoneurial vessels, and the myelinating Schwann cells of humans [31] and rats [39]. In costainings, claudin-5 IR colocalized with vWF, indicating vessel-specific location. The density of claudin-5 IR within vessels was higher in the FRR than in the NRR. This was accompanied by lower permeability for FITC-dextran and Hoechst dye in the FRR. Therefore, we assume lower claudin-5 expression in vessels as a key-factor for leakiness of the NRR's BDB under physiological conditions. A downregulation of *Cldn5* mRNA in the blood–nerve barrier has consistently been reported after nerve injury in rats [12,28]. In whole DRGs, we observed a *Cldn5* mRNA downregulation. Regional quantification of mean IR within vessels and regional *qPCR* also revealed claudin-5 downregulation after CCI, but only in the NRR. A lower *Cldn5* expression could significantly impact the function of the BDB, making it even more permeable for potentially harmful substances and immune cell invasion. Indeed, we confirmed increased invasion of macrophages in our model in the ipsilateral DRG after CCI, as shown previously in rats [22] and mice [23]. To evaluate possible BDB breakdown regarding permeability for small and large molecules, we performed a perfusion assay with FITC-dextran and Hoechst reagent as representative molecules with high and low molecular weights. We could detect higher permeation for both reagents in the NRR compared to the FRR but no change 1-week after CCI. Even though claudin-5 downregulation in brain vessels results in barrier breakdown accompanied by permeation of small molecules, there are several explanations as why we could not detect this in the DRG: Firstly, the permeability of the NRR in CL DRGs is high, so additional claudin-5 might have little additional impact. Secondly, our permeability assays were established in experimental setups where the baseline showed little to no signal, and therefore, were optimized to detect differences in "low-signal" scenarios. Therefore, high saturations probably caused lower sensitivity. Thirdly, possible compensatory mechanisms might prevent a further increase of BDB permeability, such that there simply is no change. Further studies investigating the BDB after, e.g., long-term CCI or impact of claudin-5 on immune cell migration could clarify these questions.

The structural peculiarities of the DRG are not very well understood: DRG neurons have no afferent synapses and the precise function of the pseudo-unipolar feature is still not known. The somata reside at the end of the T-stem; nevertheless, afferent spikes are connected to the cell bodies and many DRG neurons exhibit specialized membrane characteristics, necessary for the initiation of action potentials. The role of these membranes in signal transmission or generation is not known. Why is the DRG or at least the NRR not as well protected as the nerve? What are the implications? Indeed, not only capillaries but also satellite cells surrounding neurons are permeable. Marker molecules in the extracellular space can access the soma membrane of DRG neurons by diffusion independent of neuropathy [40]. It is, therefore, possible that DRGs carry out some as-yet unidentified chemosensory functions associated with the body's internal milieu. Interference with this delicate homeostasis might lead to increased sensitivity in neuropathy [41]. Interestingly, selectively silencing ectopic activity in the DRG with low dose local anesthetics in neuropathy reduces hypersensitivity [42]. Hypersensitivity could have been triggered by proalgesic molecules bypassing the BDB in neuropathy.

Both methods used, qPCR and IHC, provide specific insights into biological systems within their limitations: The great sensitivity of qPCR can prove even slight changes, but provides spatial resolution only if laser dissection is performed in addition. Furthermore, qPCR data need to be correlated with protein data to generate viable evidence. Antibody-based protein labelling on the other hand, provides an insight into expression and distribution in situ, if a sensitive and specific antibody is used. In our study mRNA data did not fully match our IHC results. While mRNA levels do not necessarily correlate with protein expression, methodological aspects also must be considered, especially for *Cldn5* expression. Whole DRG and LSM qPCR showed a significant difference between *Cldn5* mRNA levels after CCI. Our first method of semi quantification lacks sensitivity regarding proteins, which are not homogenously distributed within one specified region. Therefore, we opted to analyze claudin-5 IR in the vessel area in addition. In summary, tight junctions in the DRG can only be studied sufficiently if different tissues are analyzed separately.

What are the clinical implications? One obvious problem in pain research is the lack of suitable biomarkers and objective measures of pain—especially possible malfunctions of the pain pathway in the primary afferent neuron in the pain pathway [43]. Thus, imaging of the DRG could fill this gap in certain painful conditions. Indeed, alterations of DRG volume have been noted in chemotherapy-induced neuropathy, and two genetic diseases, Fabry disease and neurofibromatosis [44–48]. Alterations in the permeability at the blood tissue interface have been observed in long-standing Fabry disease in the NRR, indicating a concomitant dysfunctional DRG perfusion [47,48]. In the case of Fabry disease, this observation could to be the consequence of glycolipid accumulation. Unfortunately, pain intensity was not examined in the study. Also, it is not known so far, whether other types of neuropathic pain, including nerve injury or metabolic diseases, also have alterations in the BDB, and specifically, an increased leakage, as observed in the nerve [49]. Nevertheless, magnetic resonance imaging and BDB permeability could be very early in vivo markers for involvement of the peripheral nervous system. Secondly, if BDB breakdown is clinically relevant and contributes to pain generation then stabilization of the BDB (e.g., the wnt [34] and hedgehog pathway [12], as shown for the blood–nerve barrier) could also improve pain relief. Claudin-5 is regulated by steroids [50]. So, one might speculate that the epidural steroids used in the clinic could also improve BDB barrier sealing. Thirdly, the leaky BDB—already under baseline conditions—could facilitate targeting of the ectopic firing from the DRG for pain treatment, e.g., via dilute local anesthetics [42,51].

4. Materials and Methods

4.1. Animals

Twenty 8–12 weeks old male rats Wistar rats (Janvier) were kept in cages of six in a 12 h light cycle with water and food ad libitum. All animal protocols were approved by the local authorities (Regierung Unterfranken, RUF55.2.2-2532-2-612-16, 18 April 2018) and were in accordance with the ARRIVE guidelines.

4.2. Chronic Constriction Injury

Rats were randomized to the surgery and sham control groups. Surgery of the animals was performed under deep isoflurane anesthesia (1.8 vol%, fiO_2). Adequate anesthesia was assumed when paw withdrawal was absent. After skin incision and exposition of the sciatic nerve by blunt preparation, four loose silk ligatures (Perma Silk 6.0, Ethicon Inc., Somerville, NJ, USA) were used in rats with approximately 1 mm spacing in between [10,52]. After loosely tightening the ligatures, the skin was stitched (Prolene 5.0, Ethicon Inc., Somerville, NJ, USA). For sham operations, the same procedure was performed without applying the silk ligatures. All procedures were performed unilaterally, and the following experiments used the IL DRGs. As controls we used sham surgery (qPCR) or CL DRGs.

4.3. Behavioural Tests

Mechanical nociceptive thresholds were tested with the von-Frey test [53]. Rats were positioned on a wire mesh. Filaments were applied in ascending order. The initially used hair value was 1 g. In general, the filaments were applied to the plantar surface of the ipsilateral and contralateral hind paw and were held for 1–3 s until the filaments were bent to an angle of 45°. Each limb was tested three times. We determined the withdrawal threshold of the hind-limbs to a mechanical stimulus by using 50% paw withdrawal threshold (PWT) method.

4.4. Reverse Transcription qPCR (RT-qPCR)

Quantification of mRNA of whole DRGs was performed using qPCR. RNA was first extracted from the samples using TRIzol™ Reagent (Invitrogen, Carlsbad, CA, USA); then, reverse transcribed to cDNA using the High capacity cDNA Reverse Transcriptase Kit (Applied Biosystems, Foster City, CA, USA) according to the manufacturer's instructions. For RT-qPCR, the PowerUpTM SYBR Green Master Mix was used following the manufacturer's protocol with primers for *Cldn1*, *Cldn5*, *Cldn12*, *Cldn19*, and *Tjp1* (ZO-1) (Table 2).

Table 2. List of primers used in qRT-PCR analysis.

Gene	Species	Forward Primers	Reverse Primers
Gapdh	rat	5′-AGTCTACTGGCGTCTTCAC-3′	5′-TCATATTTCTCGTGGTTCAC-3′
Cldn1	rat	5′-GGGACAACATCGTGACTGCT-3′	5′-CCACTAATGTCGCCAGACCTG-3′
Cldn5	rat	5′-AAATTCTGGGTCTGGTGCTG-3′	5′-GCCGGTCAAGGTAACAAAGA-3′
Cldn12	rat	5′-AACTGGCCAAGTGTCTGGTC-3′	5′-AGACCCCCTGAGCTAGCAAT-3′
Cldn19	rat	5′-TGCTGAAGGACCCATCTG-3′	5′-TGTGCTTGCTGTGAGAACTG-3′
ZO-1 (Tjp1)	rat	5′-CACGATGCTCAGAGACGAAGG-3′	5′-TTCTACATATGGAAGTTGGGGATC-3′

Total RNA from capillaries and neuron rich areas from rats DRGs was extracted using the RNeasy® Micro Kit (Qiagen, Venlo, The Netherlands). Total RNA (1 μg) was transcribed to cDNA using the high-capacity cDNA kit (Applied Biosystems, Foster City, CA, USA) following the manufacturer's instructions. *Gapdh* and *18S* were used as reference genes for quantification. Analysis via qPCR was performed with the following primers with the Taqman method: *Cldn5* (Rn01753146_s1, Thermo Scientific, Waltham, MA, USA) and *Vwf* (Rn01492158_m1, Thermo Scientific, Waltham, MA, USA). qPCR analysis was carried out using the StepOnePlus Real-Time PCR System (Applied Biosystems, Foster City, CA, USA) with the following program: 95 °C for 20 s followed by 45 cycles at 95 °C for 1 s and 60 °C for 20 s. Samples were analyzed as triplicates. Relative quantification of mRNA was calculated using the $2^{-\Delta\Delta CT}$ method, in which CT represents the threshold cycle value.

4.5. *Immunofluorescence and Microscopy*

After euthanizing the rats, the DRGs were harvested, embedded in Tissue Tek O.C.T. Compound (Sakura Finetek Europe B.V., AV Alphen aan den Rijn, The Netherlands), and snap frozen in liquid nitrogen. The samples were stored at −20 °C until further processing. Cryosections of 10 μm thickness were cut at −20 °C in a cryostat (Leica Biosystems CM3050 S Research Cryostat, Leica Biosystems Nussloch GmbH, Nussloch, Germany) and the slides were stored at −20 °C.

For fixation, the slides were immersed in 4% paraformaldehyde (Sigma Aldrich, St. Louis, MO, USA) in PBS (Sigma Aldrich, St. Louis, MO, USA). The fixed samples were blocked with 10% donkey serum in 0.3% Triton X-100 (Sigma-Aldrich, St. Louis, MO, USA) in PBS for 1 h at room temperature. After blocking, slides were incubated with the putative primary antibody in 10% donkey serum in PBS for 16 h at 4 °C: rabbit claudin-1 antibody (1:100, Invitrogen, Waltham, MA, USA, number 51-9000), mouse claudin-5-Alexa488 conjugate antibody (1:200, Invitrogen, number 352588), claudin-12 (1:100, IBL, number 18801), rabbit claudin-19 (1:100, gift from Hou, St. Louis, USA [54]), mouse ZO-1-Alexa488 conjugate antibody (1:200, Invitrogen, number 339188), rabbit vWF antibody (1:100, Dako, Santa Clara, CA, USA, A0082), and mouse anti rat CD68 antibody (1:100; Bio-rad, Hercules, CA, USA, MCA341R). Claudin-1 and claudin-5 or vWF and claudin-5 antibodies were applied together as a co-stain. Primary antibody incubation was followed by thorough washing in PBS and incubation with secondary antibodies: for claudin-1, claudin-12, and claudin-19, Alexa fluor 555 donkey anti-rabbit antibodies (1:1000, Life Technologies, Invitrogen, MolecularProbes Inc., Eugene OR, USA, A31572) were used. The anti-claudin-5 and -ZO-1 antibodies were already conjugated with Alexa fluor 488. Before mounting, the samples were washed, and Hoechst 33342 solution was applied for five minutes at room temperature for counterstaining the nuclei (1 μg/mL in PBS, Thermo Scientific, Waltham, MA, USA). The sections were mounted with Vectashield Antifade Mounting Medium (Vector Laboratories, Burlingame, CA, USA).

Imaging of the fluorophore labeled sections was performed within one session with the same settings for each antibody (Biorevo BZ-9000-E, Keyence, Osaka, Japan). All images were saved as RGB 8-bit Tagged Image File Format (TIFF) files for further analysis.

4.6. *Laser Microdissection*

DRGs from Wistar rats were embedded in Tissue-Tek O.C.T compound (Sakura Finetek Europe B.V., AV Alphen aan den Rijn, The Netherlands) and stored at −80 °C. Cryosections of 20 μm were collected on Arcturus® polyethylene naphtalate membrane slides (Applied Biosystems, Foster City, CA, United States) using a cryostat (Leica Biosystems CM3050 S Research Cryostat, Leica Biosystems Nussloch GmbH, Nussloch, Germany). The slides were treated with Rnase AWAY® spray (Sigma Aldrich, St. Louis, MO, USA) before section collection. Before laser microdissection (LMD), sections were stained with toluidine blue. The slides were examined under a microscope coupled with a 355 nm laser (Leica®). Using the Leica LMD V7.6 software. capillaries were delimited and cut in the fiber rich area. Then, the entire neuron rich area was selected and sectioned as demonstrated in Figure 9. Samples were collected by gravity in 0.2 mL PCR SoftTubes (Biozym Scientific GmbH, Hessisch Oldendorf, Niedersachsen, Germany) and stored at −80 °C until reverse transcription qPCR.

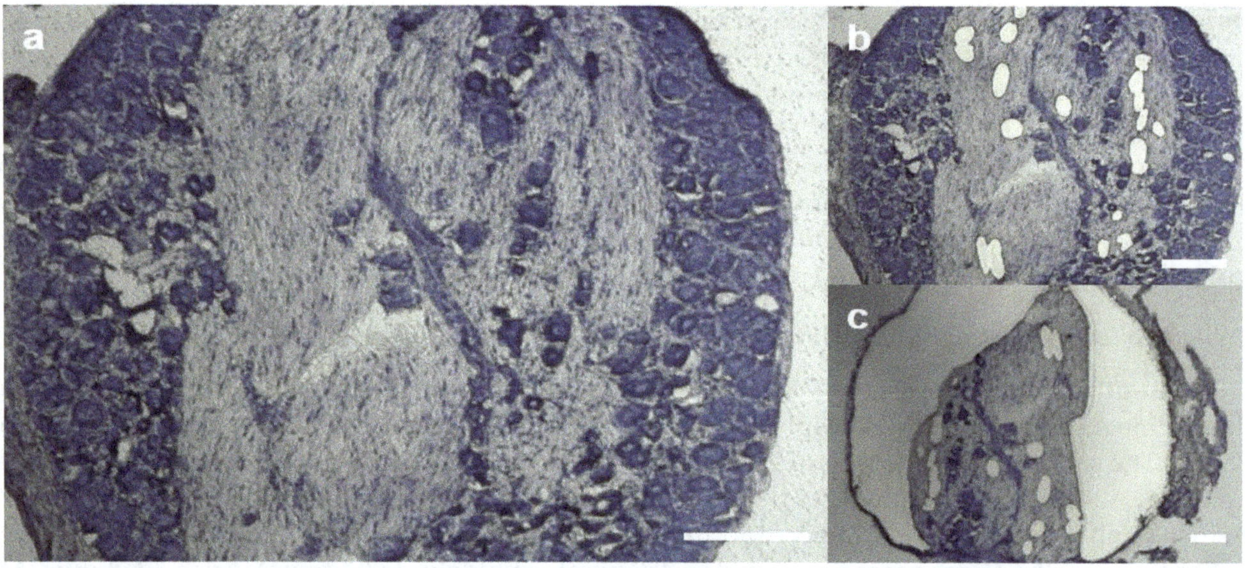

Figure 9. Stepwise sample acquisition of specific rat DRG regions via laser microdissection. Area specific samples of sham operated and rats 1 d after CCI were prepared after toluidine staining. Images of representative sections before dissection (**a**), and after acquisition of vessels in the FRR (**b**) and removal of the NRR (**c**) are shown. Scale bars measure 200 µm.

4.7. Permeability of DRG Capillaries

Anaesthetized rats were laid down in a supine position on a pad. The 5th intercostal space was opened and enlarged by a retractor to open the thorax. The pericardium was stripped, exposing the heart anterior wall, and 12 mL of FITC dextran (70 kDa; 10 mg/mL; Sigma-Aldrich, St. Loui, MO, USA, FD70) or Hoechst 33342 (562 Da; 10 mg/mL; Sigma-Aldrich, St. Louis, MO, USA, 14533) solution was injected into the left ventricle using a syringe. The dye-perfused rats were sacrificed by decapitation after 1 min. DRGs were dissected and embedded in Tissue-Tek. Frozen samples were cut into 10 µm-thick sections on a cryostat at −20 °C. Without any fixation, microscope glass slides containing tissue sections were mounted and were imaged by fluorescence microscopy. The permeability of the DRG was determined by the measuring the fluorescence of FITC dextran in the NRR and FRR.

4.8. Image Analysis

All acquired images were analyzed with Fiji/ImageJ (version 1.52e, Open Source). Brightfield images of rat DRGs were used to distinguish the fiber rich region (FRR) and the neuron rich region (NRR), and their putative epi-/perineuria (FRR-EPN, NRR-EPN), as demonstrated in Figure 1a; and they were saved as regions of interest (ROIs). For vessel specific analysis, ROIs were created using vWF counterstaining and brightfield images.

4.9. Statistical Analysis

RStudio (version 1.1447, Open Source) was used for statistical analysis and plot generation. Datasets were tested for normal distribution and variance homogeneity using the Shapiro–Wilk normality test and Levene's test of equality of variances. Datapoints were distributed normally with homogenous variances if not stated otherwise. Statistical significance of paw withdrawal threshold (Figure 2) was tested via repeated measures ANOVA. mRNA data was tested using Welch's two sample t-test or Wilcoxon rank sum test if criteria for the t-test were not met. Two-factor ANOVA followed by Tukey's test was used for analysis of immunoreactivity studies. All plots show means ± standard deviations. Significance was assumed as $p < 0.05$ (* $p < 0.05$; ** $p < 0.01$; *** $p < 0.001$) [55].

5. Conclusions

Neurons and non-neuronal cells in the DRG regulate the perceptual quality of pain. We found reduced claudin-5 protein and mRNA expression in the NRR of the DRG following peripheral nerve injury by CCI. Measuring dye diffusion, we observed that the permeability in the NRR is higher compared to FRR in the DRG in naïve rats. Despite of reduced claudin-5, permeability to small and large molecules remained unchanged. Nerve injury-induced macrophage accumulation in the NRR was increased, indicating a possible role of tight junction proteins in cell migration.

Further studies of the BDB will decipher its function and role in pain. Whether BDB permeability, e.g., in MR imaging, will be a suitable biomarker for painful affection of the PNS, has to be studied in the future in preclinical models and patient cohorts.

Author Contributions: Conceptualization, T.J.L., J.T.-C.C. and H.L.R.; Data curation, T.J.L. and J.T.-C.C.; Formal analysis, T.J.L. and J.T.-C.C.; Funding acquisition, R.B. and H.L.R.; Investigation, T.J.L., X.H., A.B.-K. and R.B.; Methodology, T.J.L., A.B.-K., R.B. and J.T.-C.C.; Project administration, H.L.R.; Resources, R.B. and H.L.R.; Software, T.J.L.; Supervision, R.B. and H.L.R.; Validation, T.J.L., X.H., J.T.-C.C. and H.L.R.; Visualization, T.J.L.; Writing—original draft, T.J.L.; Writing—review & editing, T.J.L., R.B., J.T.-C.C. and H.L.R. All authors have read and agreed to the published version of the manuscript.

Acknowledgments: We thank Nicola Drosczinsky for acquiring rat DRG PCR data and Isabel U.C. Otto for technical assistance with the immunohistochemistry. We are also acknowledging Solange Sauer for helping with the LSM experiments.

Abbreviations

BDB	blood–DRG barrier
CCI	chronic constriction injury
cldn1	claudin-1
cldn5	claudin-5
cldn12	claudin-12
cldn19	claudin-19
DRG	dorsal root ganglion
EPN	epi-/perineurium
FRR	fiber-rich region
IR	immunoreactivity
LMD	laser microdissection
MB	myelin barrier
NRR	neuron-rich region
vWF	von Willebrandt factor

References

1. Bennett, G.J.; Xie, Y.K. A peripheral mononeuropathy in rat that produces disorders of pain sensation like those seen in man. *Pain* **1988**, *33*, 87–107. [CrossRef]
2. Colloca, L.; Ludman, T.; Bouhassira, D.; Baron, R.; Dickenson, A.H.; Yarnitsky, D.; Freeman, R.; Truini, A.; Attal, N.; Finnerup, N.B.; et al. Neuropathic pain. *Nat. Rev. Dis. Primers* **2017**, *3*, 17002. [CrossRef] [PubMed]
3. Finnerup, N.B.; Attal, N.; Haroutounian, S.; McNicol, E.; Baron, R.; Dworkin, R.H.; Gilron, I.; Haanpää, M.; Hansson, P.; Jensen, T.S.; et al. Pharmacotherapy for neuropathic pain in adults: A systematic review and meta-analysis. *Lancet Neurol.* **2015**, *14*, 162–173. [CrossRef]
4. Joshi, S.K.; Honore, P. Animal models of pain for drug discovery. *Expert Opin. Drug Discov.* **2006**, *1*, 323–334. [CrossRef] [PubMed]
5. De Vry, J.; Kuhl, E.; Franken-Kunkel, P.; Eckel, G. Pharmacological characterization of the chronic constriction injury model of neuropathic pain. *Eur. J. Pharmacol.* **2004**, *491*, 137–148. [CrossRef] [PubMed]
6. Dowdall, T.; Robinson, I.; Meert, T. Comparison of five different rat models of peripheral nerve injury. *Pharmacol. Biochem. Behav.* **2005**, *80*, 93–108. [CrossRef] [PubMed]
7. Hu, P.; Bembrick, A.L.; Keay, K.A.; McLachlan, E.M. Immune cell involvement in dorsal root ganglia and spinal cord after chronic constriction or transection of the rat sciatic nerve. *Brain Behav. Immun.* **2007**, *21*, 599–616. [CrossRef]
8. Ramer, M.S.; Bisby, M.A. Rapid sprouting of sympathetic axons in dorsal root ganglia of rats with a chronic constriction injury. *Pain* **1997**, *70*, 237–244. [CrossRef]
9. Reinhold, A.; Rittner, H. Barrier function in the peripheral and central nervous system—A review. *Pflügers Arch. Eur. J. Physiol.* **2017**, *469*, 123–134. [CrossRef]
10. Reinhold, A.-K.; Yang, S.; Chen, J.T.-C.; Hu, L.; Sauer, R.-S.; Krug, S.M.; Mambretti, E.M.; Fromm, M.; Brack, A.; Rittner, H.L. Tissue plasminogen activator and neuropathy open the blood-nerve barrier with upregulation of microRNA-155-5p in male rats. *Biochim. Et Biophys. Acta (BBA) Mol. Basis Dis.* **2019**, *1865*, 1160–1169. [CrossRef]
11. Moreau, N.; Dieb, W.; Mauborgne, A.; Bourgoin, S.; Villanueva, L.; Pohl, M.; Boucher, Y. Hedgehog Pathway-Mediated Vascular Alterations Following Trigeminal Nerve Injury. *J. Dent. Res.* **2017**, *96*, 450–457. [CrossRef] [PubMed]
12. Moreau, N.; Mauborgne, A.; Bourgoin, S.; Couraud, P.O.; Romero, I.A.; Weksler, B.B.; Villanueva, L.; Pohl, M.; Boucher, Y. Early alterations of Hedgehog signaling pathway in vascular endothelial cells after peripheral nerve injury elicit blood-nerve barrier disruption, nerve inflammation, and neuropathic pain development. *Pain* **2016**, *157*, 827–839. [CrossRef] [PubMed]
13. Reinhold, A.K.; Schwabe, J.; Lux, T.J.; Salvador, E.; Rittner, H.L. Quantitative and Microstructural Changes of the Blood-Nerve Barrier in Peripheral Neuropathy. *Front. Neurosci.* **2018**, *12*, 12. [CrossRef] [PubMed]
14. Lim, T.K.; Shi, X.Q.; Martin, H.C.; Huang, H.; Luheshi, G.; Rivest, S.; Zhang, J. Blood-nerve barrier dysfunction contributes to the generation of neuropathic pain and allows targeting of injured nerves for pain relief. *Pain* **2014**, *155*, 954–967. [CrossRef]
15. Sauer, R.S.; Kirchner, J.; Yang, S.; Hu, L.; Leinders, M.; Sommer, C.; Brack, A.; Rittner, H.L. Blood-spinal cord barrier breakdown and pericyte deficiency in peripheral neuropathy. *Ann. N. Y. Acad. Sci.* **2017**, *1405*, 71–88. [CrossRef]
16. Echeverry, S.; Shi, X.Q.; Rivest, S.; Zhang, J. Peripheral nerve injury alters blood-spinal cord barrier functional and molecular integrity through a selective inflammatory pathway. *J. Neurosci.* **2011**, *31*, 10819–10828. [CrossRef]
17. Jimenez-Andrade, J.M.; Herrera, M.B.; Ghilardi, J.R.; Vardanyan, M.; Melemedjian, O.K.; Mantyh, P.W. Vascularization of the Dorsal Root Ganglia and Peripheral Nerve of the Mouse: Implications for Chemical-Induced Peripheral Sensory Neuropathies. *Mol. Pain* **2008**, *4*, 10. [CrossRef]
18. Abram, S.E.; Yi, J.; Fuchs, A.; Hogan, Q.H. Permeability of injured and intact peripheral nerves and dorsal root ganglia. *Anesthesiology* **2006**, *105*, 146–153. [CrossRef]
19. Hirakawa, H.; Okajima, S.; Nagaoka, T.; Kubo, T.; Takamatsu, T.; Oyamada, M. Regional differences in blood-nerve barrier function and tight-junction protein expression within the rat dorsal root ganglion. *Neuroreport* **2004**, *15*, 405–408. [CrossRef]

20. Liu, Q.; Wang, X.; Yi, S. Pathophysiological Changes of Physical Barriers of Peripheral Nerves After Injury. *Front. Neurosci.* **2018**, *12*. [CrossRef]

21. Bendszus, M.; Stoll, G. Technology Insight: Visualizing peripheral nerve injury using MRI. *Nat. Clin. Pract. Neurol.* **2005**, *1*, 45–53. [CrossRef] [PubMed]

22. Dubovy, P.; Tuckova, L.; Jancalek, R.; Svizenska, I.; Klusakova, I. Increased invasion of ED-1 positive macrophages in both ipsi- and contralateral dorsal root ganglia following unilateral nerve injuries. *Neurosci. Lett.* **2007**, *427*, 88–93. [CrossRef] [PubMed]

23. Simeoli, R.; Montague, K.; Jones, H.R.; Castaldi, L.; Chambers, D.; Kelleher, J.H.; Vacca, V.; Pitcher, T.; Grist, J.; Al-Ahdal, H.; et al. Exosomal cargo including microRNA regulates sensory neuron to macrophage communication after nerve trauma. *Nat. Commun.* **2017**, *8*, 1778. [CrossRef] [PubMed]

24. Furuse, M.; Hata, M.; Furuse, K.; Yoshida, Y.; Haratake, A.; Sugitani, Y.; Noda, T.; Kubo, A.; Tsukita, S. Claudin-based tight junctions are crucial for the mammalian epidermal barrier: A lesson from claudin-1-deficient mice. *J. Cell Biol.* **2002**, *156*, 1099–1111. [CrossRef] [PubMed]

25. Pummi, K.P.; Heape, A.M.; Grenman, R.A.; Peltonen, J.T.; Peltonen, S.A. Tight junction proteins ZO-1, occludin, and claudins in developing and adult human perineurium. *J. Histochem. Cytochem.* **2004**, *52*, 1037–1046. [CrossRef] [PubMed]

26. Hackel, D.; Krug, S.M.; Sauer, R.S.; Mousa, S.A.; Bocker, A.; Pflucke, D.; Wrede, E.J.; Kistner, K.; Hoffmann, T.; Niedermirtl, B.; et al. Transient opening of the perineurial barrier for analgesic drug delivery. *Proc. Natl. Acad. Sci. USA* **2012**, *109*, E2018–E2027. [CrossRef]

27. Parmantier, E.; Lynn, B.; Lawson, D.; Turmaine, M.; Namini, S.S.; Chakrabarti, L.; McMahon, A.P.; Jessen, K.R.; Mirsky, R. Schwann cell-derived Desert hedgehog controls the development of peripheral nerve sheaths. *Neuron* **1999**, *23*, 713–724. [CrossRef]

28. Hirakawa, H.; Okajima, S.; Nagaoka, T.; Takamatsu, T.; Oyamada, M. Loss and recovery of the blood-nerve barrier in the rat sciatic nerve after crush injury are associated with expression of intercellular junctional proteins. *Exp. Cell Res.* **2003**, *284*, 196–210. [CrossRef]

29. Haseloff, R.F.; Dithmer, S.; Winkler, L.; Wolburg, H.; Blasig, I.E. Transmembrane proteins of the tight junctions at the blood-brain barrier: Structural and functional aspects. *Semin Cell Dev. Biol.* **2015**, *38*, 16–25. [CrossRef]

30. Obermeier, B.; Daneman, R.; Ransohoff, R.M. Development, maintenance and disruption of the blood-brain barrier. *Nat. Med.* **2013**, *19*, 1584–1596. [CrossRef]

31. Alanne, M.H.; Pummi, K.; Heape, A.M.; Grenman, R.; Peltonen, J.; Peltonen, S. Tight junction proteins in human Schwann cell autotypic junctions. *J. Histochem. Cytochem.* **2009**, *57*, 523–529. [CrossRef] [PubMed]

32. Miyamoto, T.; Morita, K.; Takemoto, D.; Takeuchi, K.; Kitano, Y.; Miyakawa, T.; Nakayama, K.; Okamura, Y.; Sasaki, H.; Miyachi, Y.; et al. Tight junctions in Schwann cells of peripheral myelinated axons: A lesson from claudin-19-deficient mice. *J. Cell Biol.* **2005**, *169*, 527–538. [CrossRef] [PubMed]

33. Günzel, D.; Yu, A.S.L. Claudins and the Modulation of Tight Junction Permeability. *Physiol. Rev.* **2013**, *93*, 525. [CrossRef] [PubMed]

34. Sauer, R.S.; Krug, S.M.; Hackel, D.; Staat, C.; Konasin, N.; Yang, S.; Niedermirtl, B.; Bosten, J.; Gunther, R.; Dabrowski, S.; et al. Safety, efficacy, and molecular mechanism of claudin-1-specific peptides to enhance blood-nerve-barrier permeability. *J. Control Release* **2014**, *185*, 88–98. [CrossRef] [PubMed]

35. Katsuno, T.; Umeda, K.; Matsui, T.; Hata, M.; Tamura, A.; Itoh, M.; Takeuchi, K.; Fujimori, T.; Nabeshima, Y.; Noda, T.; et al. Deficiency of zonula occludens-1 causes embryonic lethal phenotype associated with defected yolk sac angiogenesis and apoptosis of embryonic cells. *Mol. Biol. Cell* **2008**, *19*, 2465–2475. [CrossRef] [PubMed]

36. Morita, K.; Sasaki, H.; Furuse, M.; Tsukita, S. Endothelial claudin: Claudin-5/TMVCF constitutes tight junction strands in endothelial cells. *J. Cell Biol.* **1999**, *147*, 185–194. [CrossRef]

37. Ohtsuki, S.; Sato, S.; Yamaguchi, H.; Kamoi, M.; Asashima, T.; Terasaki, T. Exogenous expression of claudin-5 induces barrier properties in cultured rat brain capillary endothelial cells. *J. Cell. Physiol.* **2007**, *210*, 81–86. [CrossRef]

38. Nitta, T.; Hata, M.; Gotoh, S.; Seo, Y.; Sasaki, H.; Hashimoto, N.; Furuse, M.; Tsukita, S. Size-selective loosening of the blood-brain barrier in claudin-5-deficient mice. *J. Cell Biol.* **2003**, *161*, 653–660. [CrossRef]

39. Poliak, S.; Matlis, S.; Ullmer, C.; Scherer, S.S.; Peles, E. Distinct claudins and associated PDZ proteins form different autotypic tight junctions in myelinating Schwann cells. *J. Cell Biol.* **2002**, *159*, 361–372. [CrossRef]

40. Shinder, V.; Devor, M. Structural basis of neuron-to-neuron cross-excitation in dorsal root ganglia. *J. Neurocytol.* **1994**, *23*, 515–531. [CrossRef]
41. Devor, M. Unexplained peculiarities of the dorsal root ganglion. *Pain* **1999**, *82* (Suppl. 6), S27–S35. [CrossRef]
42. Yatziv, S.L.; Devor, M. Suppression of neuropathic pain by selective silencing of DRG ectopia using non-blocking concentrations of lidocaine. *Pain* **2019**. [CrossRef] [PubMed]
43. Tracey, I.; Woolf, C.J.; Andrews, N.A. Composite Pain Biomarker Signatures for Objective Assessment and Effective Treatment. *Neuron* **2019**, *101*, 783–800. [CrossRef] [PubMed]
44. Godel, T.; Pham, M.; Heiland, S.; Bendszus, M.; Baumer, P. Human dorsal-root-ganglion perfusion measured in-vivo by MRI. *Neuroimage* **2016**, *141*, 81–87. [CrossRef] [PubMed]
45. Apostolidis, L.; Schwarz, D.; Xia, A.; Weiler, M.; Heckel, A.; Godel, T.; Heiland, S.; Schlemmer, H.P.; Jager, D.; Bendszus, M.; et al. Dorsal root ganglia hypertrophy as in vivo correlate of oxaliplatin-induced polyneuropathy. *PLoS ONE* **2017**, *12*, e0183845. [CrossRef]
46. Godel, T.; Mautner, V.F.; Farschtschi, S.; Pham, M.; Schwarz, D.; Kronlage, M.; Gugel, I.; Heiland, S.; Bendszus, M.; Baumer, P. Dorsal root ganglia volume differentiates schwannomatosis and neurofibromatosis 2. *Ann. Neurol.* **2018**, *83*, 854–857. [CrossRef]
47. Godel, T.; Baumer, P.; Pham, M.; Kohn, A.; Muschol, N.; Kronlage, M.; Kollmer, J.; Heiland, S.; Bendszus, M.; Mautner, V.F. Human dorsal root ganglion in vivo morphometry and perfusion in Fabry painful neuropathy. *Neurology* **2017**, *89*, 1274–1282. [CrossRef]
48. Godel, T.; Kohn, A.; Muschol, N.; Kronlage, M.; Schwarz, D.; Kollmer, J.; Heiland, S.; Bendszus, M.; Mautner, V.F.; Baumer, P. Dorsal root ganglia in vivo morphometry and perfusion in female patients with Fabry disease. *J. Neurol.* **2018**, *265*, 2723–2729. [CrossRef]
49. Baumer, P.; Reimann, M.; Decker, C.; Radbruch, A.; Bendszus, M.; Heiland, S.; Pham, M. Peripheral nerve perfusion by dynamic contrast-enhanced magnetic resonance imaging: Demonstration of feasibility. *Investig. Radiol.* **2014**, *49*, 518–523. [CrossRef]
50. Salvador, E.; Shityakov, S.; Förster, C. Glucocorticoids and endothelial cell barrier function. *Cell Tissue Res.* **2014**, *355*, 597–605. [CrossRef]
51. Deer, T.R.; Pope, J.E.; Lamer, T.J.; Grider, J.S.; Provenzano, D.; Lubenow, T.R.; FitzGerald, J.J.; Hunter, C.; Falowski, S.; Sayed, D.; et al. The Neuromodulation Appropriateness Consensus Committee on Best Practices for Dorsal Root Ganglion Stimulation. *Neuromodulation* **2019**, *22*, 1–35. [CrossRef] [PubMed]
52. Sauer, R.-S.; Rittner, H.L.; Roewer, N.; Sohajda, T.; Shityakov, S.; Brack, A.; Broscheit, J.-A. A Novel Approach for the Control of Inflammatory Pain: Prostaglandin E2 Complexation by Randomly Methylated β-Cyclodextrins. *Anesth. Analg.* **2017**, *124*, 675–685. [CrossRef] [PubMed]
53. Chaplan, S.R.; Bach, F.W.; Pogrel, J.W.; Chung, J.M.; Yaksh, T.L. Quantitative assessment of tactile allodynia in the rat paw. *J. Neurosci. Methods* **1994**, *53*, 55–63. [CrossRef]
54. Hou, J.; Renigunta, A.; Gomes, A.S.; Hou, M.; Paul, D.L.; Waldegger, S.; Goodenough, D.A. Claudin-16 and claudin-19 interaction is required for their assembly into tight junctions and for renal reabsorption of magnesium. *Proc. Natl. Acad. Sci. USA* **2009**, *106*, 15350–15355. [CrossRef]
55. Damoiseaux, J.G.; Döpp, E.A.; Calame, W.; Chao, D.; MacPherson, G.G.; Dijkstra, C.D. Rat macrophage lysosomal membrane antigen recognized by monoclonal antibody ED1. *Immunology* **1994**, *83*, 140–147. [PubMed]

7

Tricellulin Effect on Paracellular Water Transport

Carlos Ayala-Torres, Susanne M. Krug, Jörg D. Schulzke, Rita Rosenthal † and Michael Fromm *,†

Institute of Clinical Physiology/Nutritional Medicine, Medical Department, Division of Gastroenterology, Infectiology and Rheumatology, Charité—Universitätsmedizin Berlin, 12203 Berlin, Germany

* Correspondence: michael.fromm@charite.de

† These authors contributed equally to this work.

Abstract: In epithelia, large amounts of water pass by transcellular and paracellular pathways, driven by the osmotic gradient built up by the movement of solutes. The transcellular pathway has been molecularly characterized by the discovery of aquaporin membrane channels. Unlike this, the existence of a paracellular pathway for water through the tight junctions (TJ) was discussed controversially for many years until two molecular components of paracellular water transport, claudin-2 and claudin-15, were identified. A main protein of the tricellular TJ (tTJ), tricellulin, was shown to be downregulated in ulcerative colitis leading to increased permeability to macromolecules. Whether or not tricellulin also regulates water transport is unknown yet. To this end, an epithelial cell line featuring properties of a tight epithelium, Madin-Darby canine kidney cells clone 7 (MDCK C7), was stably transfected with small hairpin RNA (shRNA) targeting tricellulin, a protein of the tTJ essential for the barrier against passage of solutes up to 10 kDa. Water flux was induced by osmotic gradients using mannitol or 4 and 40 kDa-dextran. Water flux in tricellulin knockdown (KD) cells was higher compared to that of vector controls, indicating a direct role of tricellulin in regulating water permeability in a tight epithelial cell line. We conclude that tricellulin increases water permeability at reduced expression.

Keywords: tricellulin; tricellular tight junction; paracellular water transport; tight epithelium; MDCK C7 cells

---

1. Introduction

For a long time, the tight junction (TJ) was considered as a more or less perfect barrier to impede the paracellular permeation of solutes and water [1]. A few years after the membrane-spanning proteins occludin and the claudin family were identified to form the TJ, the first claudin forming a charge-selective channel was discovered [2] (for review see [3]), explaining how the TJ also can provide a selective pathway for the passive permeation through the paracellular pathway.

Tight junctions are organized in two ways. (i) The bicellular tight junction (bTJ) connects two neighboring cells and (ii) the tricellular tight junction (tTJ) is formed at the site where three cells meet. Ultrastructurally, at the tTJ the bicellular strands extend in the basal direction and form a central tube of assumedly 1 µm in length and 10 nm in diameter [4]. While the bTJs is comprised of the transmembrane proteins occludin and the family of claudins [5], the predominant proteins making up the tTJ are tricellulin and angulins [6].

Tricellulin is a tetraspan protein that plays a critical role in sealing the tTJ against the passage of solutes. If it is fully knocked out, the TJ structure is altered and barrier function breaks down [7]. At low expression rates as found in MDCK II cells, the tTJ opens a diffusional pathway at the tTJ for macromolecules of up to 10 kDa [8]. It would be reasonable if also small solutes could pass then, but in MDCK II cells this was not detectable. This paradox is explained by the fact that the tTJ is too rare to contribute significantly to ion permeability in low-resistance (leaky) epithelia in which the permeability of the bTJ predominates. In the high-resistance (tight) human colon cell line HT-29/B6 also an effect of the tTJ on ion permeability became significant [9]. In intestinal biopsies from patients suffering from ulcerative colitis, tricellulin expression was decreased and the passage of macromolecules was increased [10].

Angulins are unified names for three single-span proteins located at the tTJ, lipolysis-stimulated lipoprotein receptor (LSR, angulin-1), immunoglobulin-like domain-containing receptor 1 (ILDR1, angulin-2), and ILDR2 (angulin-3). Angulin-1 has been shown to recruit tricellulin to the tTJ [11] and the same effect was observed for angulin-2 and -3 [12]. Angulin-1 and -2 are expressed in a complementary way with angulin-3. In addition, angulin-1 and -2 exert a barrier-maintaining effect, while angulin-3 has a weaker effect in this regard [12].

Whether or not the TJ at all is permeable to water was highly controversial for a long time. The transcellular pathway for water via aquaporin channels was established more than 20 years ago [13] and their presence has been described in a wide range of tissues [14]. Another transcellular pathway for water is provided by the Na^+-glucose cotransporter SGLT1 (and several other carriers) which contributes, especially in small intestinal epithelia, significantly to water transport [15].

Regarding the paracellular pathway through the bTJ, Rosenthal et al. have developed an elaborated though difficult to handle technique to measure water transport on monolayers mounted in Ussing chambers [2]. First it was demonstrated that claudin-2, so far identified as a paracellular cation channel [2], also forms a water channel [16]. Cations and water were then shown to pass through the same pore [17]. Recently, a second water channel through the bTJ was identified to be claudin-15, which also was known so far as a cation channel only [18].

With this in mind, three subtypes of channel-forming claudins can be functionally distinguished regarding their permeability to water: (i) proteins forming channels for anions but not for water (claudin-10a and -17) [19,20], (ii) proteins forming channels for cations but not for water (claudin-10b) [16], and (iii) proteins forming channels for cations and also for water (claudin-2 and -15) [16–18]. After it has been demonstrated that the tTJ can provide another paracellular pathway for solutes [8–10,21], the question arises whether or not it also contributes to water passage.

As for the tTJ, Gong et al. provided compelling evidence that water transport is increased in isolated kidney tubules in the absence of angulin-2, but is normally inhibited in presence of angulin-2 [22]. The role of tricellulin has not been elucidated so far. We hypothesize that tricellulin is a direct negative actuator of tTJ water permeability. Therefore, we aimed to characterize the effect of tricellulin expression on water transport in the tight epithelial cell line MDCK C7 and found that the water passage through the tTJ was indeed enhanced under the influence of reduced tricellulin levels.

2. Results

2.1. *Characterization of Tricellulin Knockdown in MDCK C7 Cells*

After transfection of Madin-Darby canine kidney cells clone 7 (MDCK C7) with shRNA targeting tricellulin, puromycin-resistant cell clones were screened for tricellulin knockdown by western blotting. In this study, two knockdown clones and two control clones were investigated (KD 23 and its corresponding control 2, and KD 24 and its corresponding control 9). The two KD clones showed high but different reduction in tricellulin expression: clone KD 23 showed a depression of tricellulin by 30% while clone KD 24 exhibited a drop by 40% (Figure 1a and Table 1).

The transepithelial resistance (TER) was reduced in tricellulin knockdown MDCK C7 cells (Figure 1b, Table 1). KD 24 exhibited a stronger reduction of tricellulin expression and stronger decrease in TER compared to the corresponding control than KD 23. This was due to a huge increase in Na^+ and Cl^- permeability without any preference for anions or cations (Figure 1c, Table 1).

Then we examined the permeability of the four clones to fluorescein isothiocyanate-dextran 4 kDa (FITC–dextran, FD4). As a result, FD4 permeability of KD 23 was not significantly different from its control, but that of clone KD 24 exhibited a strong increase (Figure 1d, Table 1). Although clones KD 23 and KD 24 differed only gradually in reduction of tricellulin expression and TER decrease, the two chosen clones differed in macromolecule permeability.

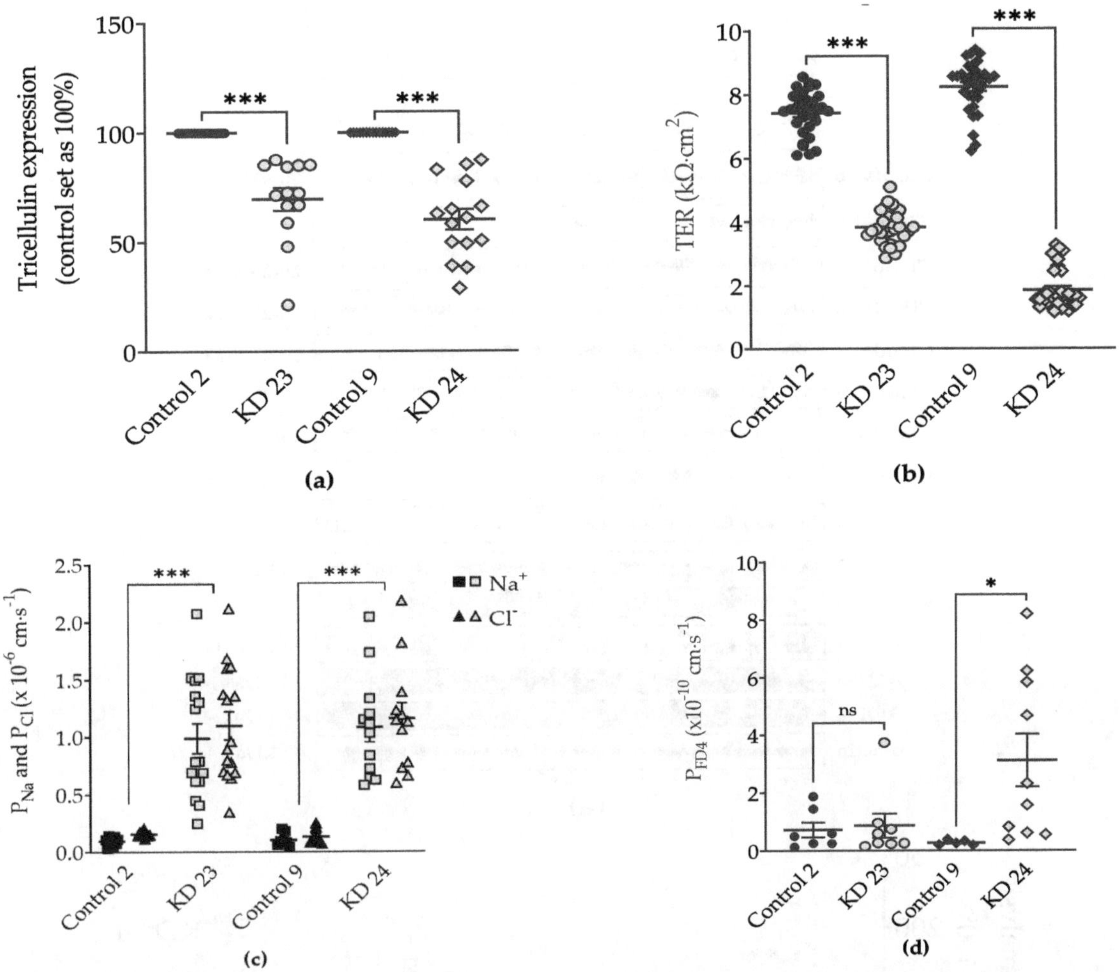

Figure 1. Expression and functional analysis of tricellulin knockdown in MDCK C7 cells. (**a**) Densitometric analysis of tricellulin protein expression levels in stable shTRIC transfectants (KD 23 and KD 24) in comparison to vector-transfected controls (Control 2 and Control 9). shTRIC leads to decreased tricellulin expression ($n = 15$). (**b**) Effect of tricellulin knockdown on transepithelial resistance (TER). Tricellulin knockdown decreases TER in MDCK C7 cells ($n = 24$). (**c**) Effect of tricellulin knockdown on ion permeability. Tricellulin knockdown increases Na^+ and Cl^- permeability to the same extent in MDCK C7 cells ($n = 7$–16). (**d**) Effect of tricellulin knockdown on permeability for 4 kDa-FITC dextran (FD4). Permeability is increased in tricellulin KD 24 cells only ($n = 5$–10). Statistical analysis was performed using Student's t-test between the KD clone and the corresponding control (KD 23 versus control 2, KD 24 versus control 9). (Symbols of (a), (b) and (d): Control 2: ●; KD 23: ○; Control 9: ◆; KD 24: ◇. * $p < 0.05$, *** $p < 0.001$, ns = not significant).

2.2. *Effects of Tricellulin Knockdown on Endogenous Proteins of MDCK C7 Cells*

Knockdown of one TJ protein may cause variation in other proteins, which are involved in transepithelial water transport. Therefore, we examined levels of TJ proteins, such as occludin, several claudins, the three angulins, and of aquaporin (AQP) water channels, AQP1, 3, and 4 (Figure 2a). The densitometric analysis revealed some clonal variability in TJ protein expression between the knockdown clones and the controls. In detail, claudin-1 was increased in the tricellulin KD 23, and occludin and claudin-4 and -8 were decreased in the tricellulin KD 24 (Figure 2b). Clonal variability in protein expression was also observed between both control clones and both KD clones. None of the three angulins were significantly altered. Most importantly, the tricellulin knockdown did not affect the expression of the cell membrane water channels expressed in MDCK C7 cells, AQP1, AQP3, and AQP4 (Figure 2b).

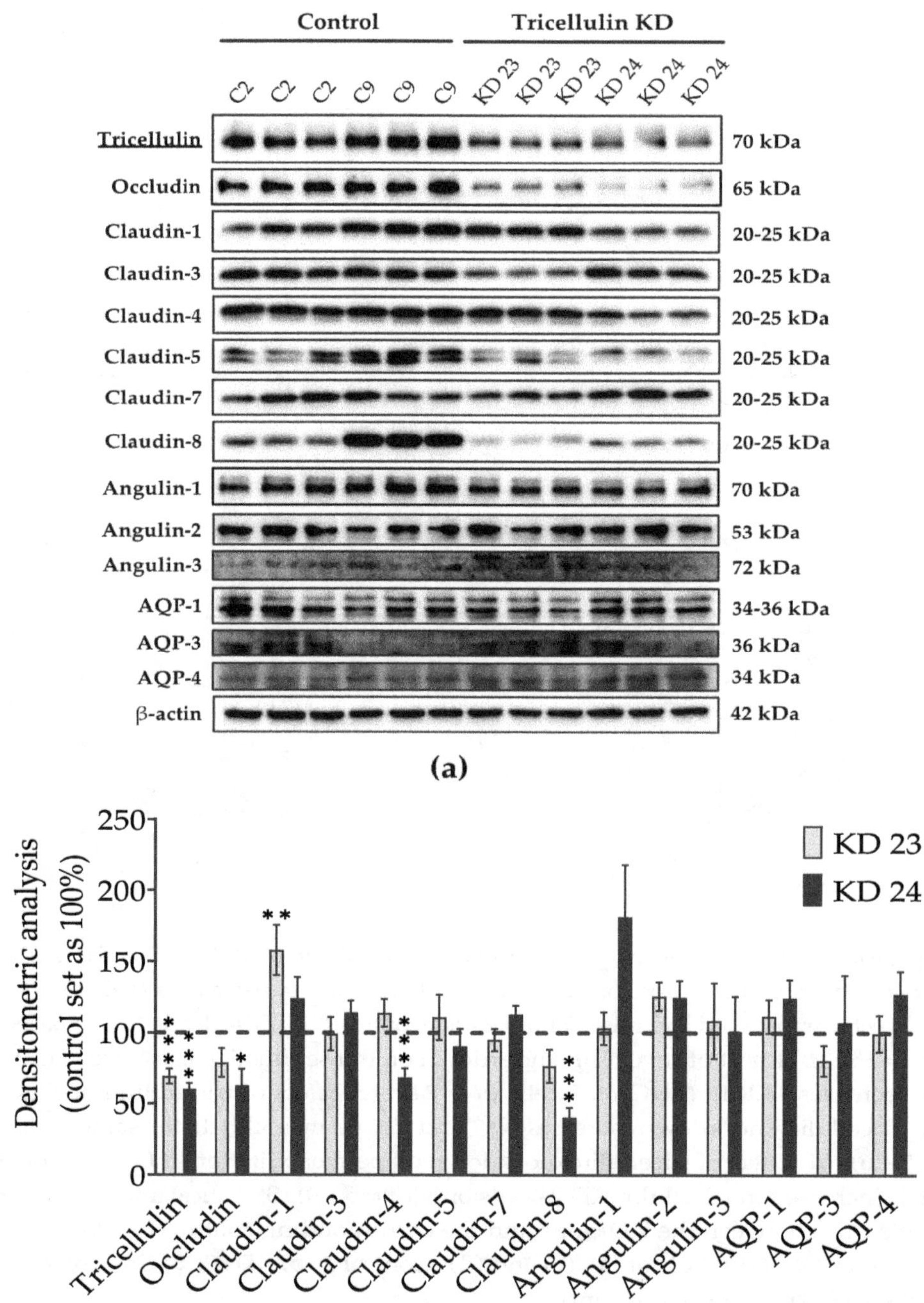

Figure 2. Claudin, occludin, angulin and aquaporin (AQP) expression in tricellulin knockdown MDCK C7 cells. (**a**) Representative western blots. (**b**) Densitometric analysis of protein expression levels in tricellulin knockdown clones KD 23 and KD 24 in comparison to the corresponding vector-transfected controls (n = 9–17, N = 7). β-Actin was used as an internal control for normalization to protein content. Statistical analysis was performed using Student's t-test between the KD clone and the corresponding control (KD 23 versus control 2, KD 24 versus control 9). (* p <0.05, ** p < 0.01, *** p < 0.001).

The proper localization of tricellulin was confirmed for the four clones by immunofluorescence confocal laser-scanning microscopy. Occludin served as a TJ marker (Figure 3).

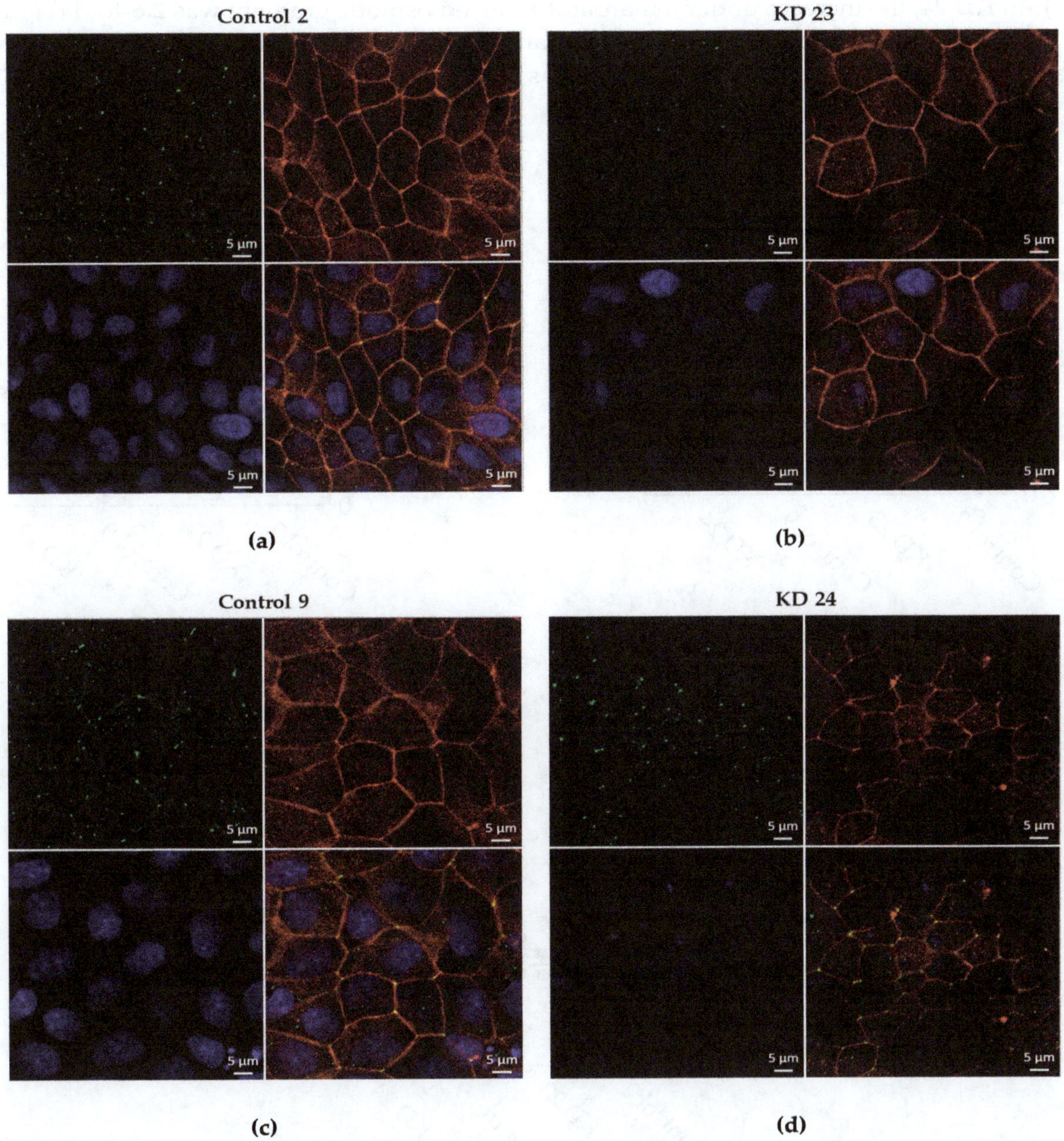

Figure 3. Localization of tricellulin in the four clones used throughout this study. In the KD clones, tricellulin is localized at tricellular tight junction (tTJ) sites. Tricellulin: green; occludin: red; cell nuclei (4′,6-Diamidine-2′-phenylindole dihydrochloride - DAPI): blue. MDCK C7 cells transfected with (**a**,**c**) empty lentiviral vector for shRNA expression (pLKO.1-puro) as a negative control, (**b**) pLKO.1-puro vector containing a sequence for shRNA targeting tricellulin (shRNA 23), and (**d**) pLKO.1-puro vector containing a sequence for shRNA targeting tricellulin (shRNA 24).

2.3. *Effect of Tricellulin Knockdown on Transepithelial Water Transport in MDCK C7 Cells*

To analyze the effect of tricellulin knockdown on water permeability of MDCK C7 cells, water flux was measured after induction with osmotic gradients produced by 100 mM mannitol, by 37 mM 4 kDa-dextran, or by 100 mM 4 kDa-dextran (Figure 4). These concentrations produced measured osmolality gradients of 100 mOsm for 100 mM mannitol and for 37 mM 4 kDa-dextran, and of 900 mOsm for 100 mM 4 kDa-dextran.

Tricellulin KD 23 as well as KD 24 exhibited increased water fluxes compared with their respective vector control clones, however the change obtained by KD 24 was larger under all conditions (Figure 4,

Table 1). In KD 24, the increase under a mannitol-induced osmotic gradient was 2.6-fold (Figure 4a), under a gradient induced by 37 mM 4 kDa-dextran it was 5.2-fold (Figure 4b) and under a gradient induced by 100 mM 4 kDa-dextran the increase was 3.8-fold (Figure 4c), compared to 1.5-fold, 2.1-fold, and 1.4-fold in KD 23, respectively.

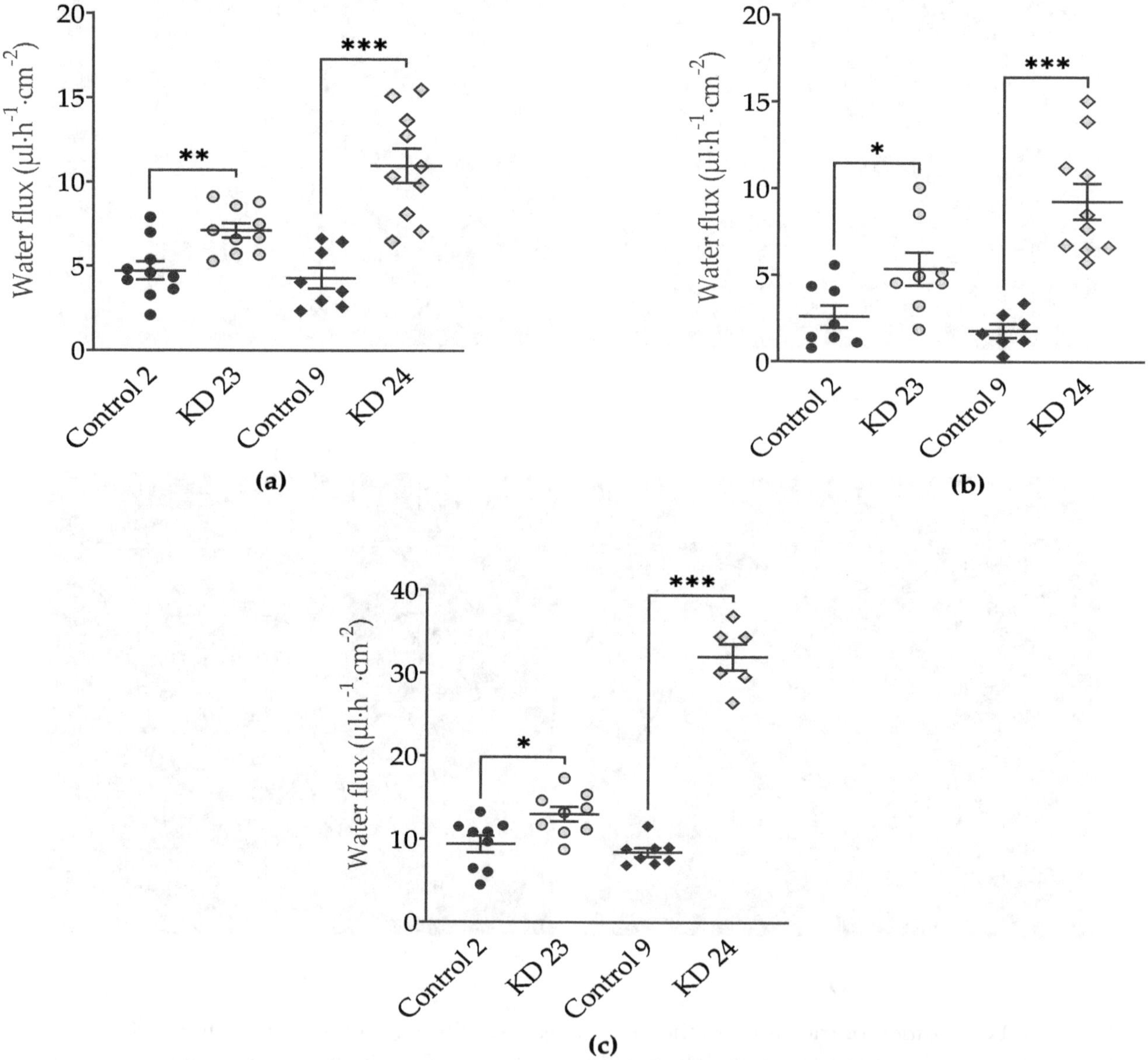

Figure 4. Water flux in tricellulin knockdown MDCK C7 cells stimulated by an osmotic gradient on the apical side of the cell layer. Water flux induced by a gradient of (**a**) 100 mM mannitol (n = 8–10), (**b**) 37 mM 4 kDa-dextran (n = 8–10) and (**c**) 100 mM 4 kDa-dextran (n = 6–9) was increased in both tricellulin knockdown clones, with the stronger increase in clone KD 24 with the higher reduction of tricellulin expression. Statistical analysis was performed using Student's t-test between the KD clone and the corresponding control (KD 23 versus control 2, KD 24 versus control 9). (Control 2: ●; KD 23: ○; Control 9: ◆; KD 24: ◇. * p <0.05, ** p < 0.01, *** p < 0.001).

In a parallel same series of experiments, a 5.5 mM 40 kDa-dextran gradient was applied which produced a measured osmolality gradient of 100 mOsm. In the presence of 100 mOsm 40 kDa-dextran, water flux was lower than in the presence of 100 mOsm mannitol. The 40 kDa-dextran gradient did not significantly change water flux in tricellulin KD 24 compared to control, whereas water flux in KD 23 was increased (Figure 5, Table 1). This finding indicates that the water flux through the tTJ was inhibited in the presence of 40 kDa-dextran in the clone with the strong reduction of tricellulin.

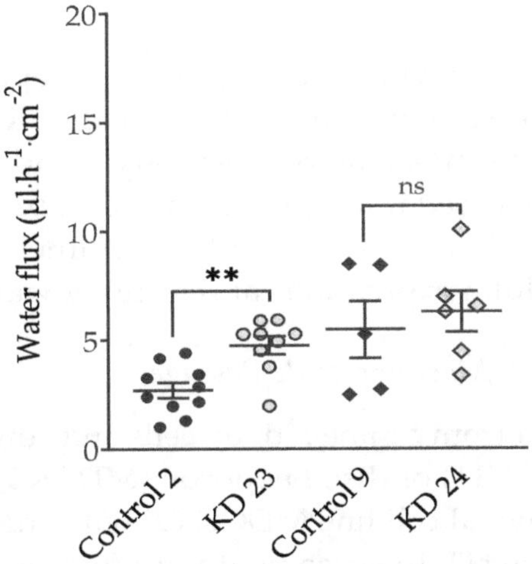

Figure 5. Water flux in tricellulin knockdown MDCK C7 cells stimulated by an osmotic gradient of 5.5 mM 40 kDa-dextran (100 mOsm) on the apical side of the cell layer. No effect could be observed in tricellulin KD 24 cells, whereas in KD 23 with the higher tricellulin expression, the increase was in the same order of magnitude as in the presence of 100 mM (100 mOsm) mannitol. Statistical analysis was performed using Student' *t*-test between the KD clone and the corresponding control (KD 23 versus control 2, KD 24 versus control 9). (Control 2: ●; KD 23: ○; Control 9: ◆; KD 24: ◇. ** $p \leq 0.01$, ns = not significant).

Table 1. Characteristics of MDCK C7 tricellulin knockdown clones and the corresponding controls. Two tricellulin knockdown clones and their corresponding controls were analyzed in this study (Control 2 and KD 23, Control 9 and KD 24). Data of tricellulin expression have been obtained by densitometric analysis of western blots using β-actin for normalization. Data of Na^+/Cl^- permeability and absolute permeabilities for Na^+ and Cl^- (P_{Na}, P_{Cl}) were obtained from dilution potential measurements in the Ussing chamber. Water flux measurements were performed in a modified Ussing chamber with water flux induced by different osmotic gradients. Significances refer to respective controls. *n* number of experiments, * $p \leq 0.05$, ** $p \leq 0.01$, *** $p \leq 0.001$.

		Control 2	KD 23	Control 9	KD 24
Tricellulin Expression (%)		100 (*n* = 15)	69.9 ± 4.8 *** (*n* = 13)	100 (*n* = 15)	60.1 ± 4.6 *** (*n* = 15)
TER (kΩ·cm^2)		7.4 ± 0.2 (*n* = 24)	3.8 ± 0.2 *** (*n* = 24)	8.2 ± 0.2 (*n* = 24)	1.8 ± 0.1 *** (*n* = 24)
P_{FD4} (×10^{-10} cm·s^{-1})		0.72 ± 0.23 (*n* = 7)	0.87 ± 0.39 (*n* = 8)	0.28 ± 0.03 (*n* = 5)	3.11 ± 0.86 * (*n* = 10)
P_{Na} (×10^{-6} cm·s^{-1})		0.10 ± 0.01 (*n* = 9)	0.99 ± 0.13 *** (*n* = 16)	0.11 ± 0.02 (*n* = 7)	1.08 ± 0.13 *** (*n* = 12)
P_{Cl} (×10^{-6} cm·s^{-1})		0.16 ± 0.01 (*n* = 9)	1.10 ± 0.12 *** (*n* = 16)	0.14 ± 0.03 (*n* = 7)	1.16 ± 0.14 *** (*n* = 12)
P_{Na}/P_{Cl}		0.81 ± 0.05 (*n* = 9)	0.94 ± 0.04 (*n* = 16)	0.85 ± 0.07 (*n* = 7)	1.02 ± 0.03 (*n* = 12)
Water flux (µl·h^{-1}·cm^{-2})	100 mM mannitol (100 mOsm)	4.7 ± 0.5 (*n* = 10)	7.1 ± 0.4 ** (*n* = 10)	4.3 ± 0.6 (*n* = 8)	11.0 ± 1.0 *** (*n* = 10)
	37 mM 4 kDa-dextran (100 mOsm)	2.6 ± 0.7 (*n* = 8)	5.4 ± 0.8 * (*n* = 8)	1.8 ± 0.5 (*n* = 7)	9.3 ± 1.0 *** (*n* = 10)
	100 mM 4 kDa-dextran (900 mOsm)	9.4 ± 1.0 (*n* = 9)	13.0 ± 0.8 * (*n* = 9)	8.4 ± 0.5 (*n* = 8)	32.0 ± 1.4 *** (*n* = 6)
	5.5 mM 40 kDa-dextran (100 mOsm)	2.7 ± 0.3 (*n* = 10)	4.7 ± 0.6 ** (*n* = 9)	5.2 ± 1.2 (*n* = 5)	6.3 ± 0.9 (*n* = 6)

3. Discussion

This study shows that tricellulin downregulation in the tight epithelial cell line MDCK C7 resulted in an increased ion permeability, as shown by the reduction of TER and, dependent on the reduction of tricellulin, in an increased macromolecule permeability. Additionally, a decrease in tricellulin expression is associated with an increased transepithelial water flux. Thus, tricellulin is able to regulate osmotically-induced transepithelial water flux in a tight epithelium. This regulation occurs in a reciprocal way, i.e., lowered tricellulin expression causes higher water fluxes.

3.1. Effect of Tricellulin Knockdown on Macromolecule Passage

The MDCK cell line is derived from canine kidney cells and common strains feature properties of either proximal (MDCK II, MDCK C11) or distal nephron (MDCK I, MDCK C7) [23,24].

Transfection of the tight epithelial cell line MDCK C7 with shRNA targeting tricellulin resulted in a reduction of this protein in the tTJ. In agreement with former experiments on cell cultures [10], the reduction in the expression of tricellulin resulted in a lowered transepithelial resistance and the development of a pathway for macromolecules. Complete knockout, however, leads to a disorganization of not only tTJs but also bTJs and complete breakdown of barrier parameters [7]. This is of clinical importance in genetic defects in tricellulin causing hearing loss [25], however, for studies focused selectively on the tTJ, graded knockdown approaches that reduce tricellulin only in part are suitable.

In addition, the permeability to the paracellular marker 4 kDa-dextran was increased by tricellulin knockdown as normally this protein is responsible for blocking the passage of macromolecules with diameters from 1.3 to 4.6 nm (0.9 to 10 kDa) [8]. This is consistent with the present data showing that in tight epithelial cell lines a downregulation of tricellulin strongly regulates the permeability for 4 kDa-macromolecules. It might be surprising that a moderate knockdown of tricellulin by 40% already leads to five times higher permeability for macromolecules. This is explained by the fact that normally the permeability for macromolecules is extremely low (in the order of 10^{-12} cm/s) so that even a slight opening of a pathway for macromolecules causes a significant effect [21]. Furthermore, the reduction by 30% (KD 23) had no significant effect on macromolecule permeability, however, did have an effect on ion permeability, indicating that there is already a significant impairment of the tTJ barrier which however is only affecting small solutes, while 4 kDa-dextran is still not able to pass. This leads to the assumption that the varying range of tricellulin expression may lead to different effects. The limit of macromolecule passage seems to range between a 30% to 40% reduction of tricellulin, while for small solutes the regulation is stricter.

3.2. Effects of Tricellulin Knockdown on Other Proteins of MDCK C7 Cells

Immunoblot analysis with an anti-tricellulin antibody demonstrated a clear reduction of tricellulin expression (Figure 2a). Nevertheless, the generated tricellulin knockdown clones compared with the controls showed clonal variation in other TJ proteins which reached significance for occludin, claudin-1, claudin-4, and claudin-8 (Figure 2a,b). The clonal variation also concerns the two control clones, here the differences are most prominent for claudin-1, claudin-5, and claudin-8. These claudins seem to have no effect on paracellular water transport, since the water transport of the two control clones did not show any difference. In the tricellulin knockdown clones, the expression of occludin, claudin-1, claudin-4, and claudin-8 was changed compared to the corresponding vector control. Considering the functional aspects of the variations it can be said that occludin, which is downregulated only in one of the tricellulin knockdown clones, is assumed to have no effect on TJ barrier function, either genuinely or through compensation of its downregulation by other tight junction proteins [26–28]. All claudins present in MDCK C7 cells are known to behave as barrier formers, acting in concert and compensating their respective up- and down-regulation [29]. Ion and water channel-forming claudins can be disregarded as they are not genuinely present in MDCK C7 cells. A reduced expression of claudin-4 and claudin-8 and an increased expression of claudin-1 was found in one of the knockdown clones, respectively. However, claudin-4 is described to be dispensable for the barrier properties of TJs in wild-type MDCK II cells [30]. Furthermore, claudin-1 and claudin-4 were shown to be dispensable for water barrier formation in human submerged keratinocyte cultures [31]. Therefore, we feel safe to conclude that the observed clonal variation of the claudins in sum is balanced and provides a constant barrier function of the bTJ. Most importantly, the three angulins as well as the aquaporins AQP-1, -3, and -4 did not significantly change. Thus, the changes in ion, macromolecule and water permeability found in the knockdown clones can be assumed to be exclusively due to the reduced tricellulin expression in the tTJ.

That the knockout of a tTJ-related protein, angulin-2, increases water transport was shown in elegant experiments by Gong et al. on isolated mouse kidney tubules [24]. The authors suggest that the barrier function of angulin-2 derives from a trans-interaction between its extracellular immunoglobulin-like domain. All three angulins are known to be able to recruit tricellulin to the tTJ [11]. It is beyond the scope of the present study but will be subject of future investigation, to experimentally clarify whether angulins mediate a direct effect on water permeability independent of tricellulin, or angulins act indirectly via tricellulin regulation.

3.3. *Water Transport as Driven by Different Osmotic Gradients*

In other studies from this lab using the method of generating water flux by an osmotic gradient e.g., by mannitol, this solute was impermeable and could not pass the TJ by itself [16–20]. In the present study, for the first time this was different and more complicated. In the tricellulin KD clones the tTJ central tube is opened, and through this pathway mannitol can diffuse following its concentration gradient. This has two consequences: First, in the course of the experiment, mannitol may be diluted and its gradient is reduced. However, this will be below significance, because the amount of mannitol in the 9 mL bath solution is high compared to the amount diffusing through the tTJ. Second and more disturbing, mannitol would diffuse through the central tube in the opposite direction to the presumed water flux. As long as a molecule like mannitol is small compared to the diameter of the central tube, this may not have a large effect. As this mechanism may alter water transport rates at least slightly, we decided not to calculate water permeability coefficients in this study and present data as fluxes only.

In order to investigate the side effect of mannitol diffusion, we compared data obtained with 100 mM mannitol with those obtained with larger molecules, 37 mM 4 kDa- and 5.5 mM 40 kDa-dextran. For this, dextran concentrations were chosen which all produced an osmolality of 100 mOsm, as used for mannitol. This resulted in lower nominal concentrations due to a known water-sequestering effect of some macromolecules [32].

It turned out that gradients of 100 mM mannitol and 37 mM 4 kDa-dextran produced comparable water fluxes and that an elevated 4 kDa-dextran concentration (100 mM resulting in roughly 900 mOsm) led to higher water fluxes (see Table 1). Since these numbers satisfactorily fit together this indicates the reliability of the methods used.

The most valuable results should have been obtained with an osmotic gradient obtained by 5.5 mM 40-dextran (measured osmolality 100 mOsm). Surprisingly, this was not the case: In the presence of 40 kDa-dextran, water flux was reduced in the control clones as well as in the knockdown clones compared to water fluxes in the presence of mannitol with the same osmolality. An induction of water flux by a gradient with 40 kDa-dextran did not produce significantly changed water fluxes in the KD 24 clone with the stronger tricellulin reduction, whereas KD 23 showed an increased water flux compared to the control clone. As an explanation for this effect, the sizes of the molecule should be related to that of the tTJ central tube. The hydrodynamic radius of 40 kDa-dextran reportedly amounts 6.6 nm [33] while the diameter of the tTJ central tube was assumed to be 10 nm [4]. We suggest that due to its molecular size the 40 kDa-dextran only barely fits through the central tube or even clogs its entrance so that water is hindered to pass. Thus, water flux was reduced in all clones compared to mannitol-induced fluxes. The strongest reduction was observed in the KD 24 clone with the lowest tricellulin expression and possibly the largest diameter of the tTJ central tube. These assumptions are supported by the findings of Krug et al. that the tTJ central tube at low tricellulin levels allows for passage of molecules of 10 kDa but not of 20 kDa [8]. This reasoning finds its equivalent in experiments dealing with the bTJ, where a NaCl gradient was used to drive water flux through the claudin-15 channel. Under this condition, claudin-15-mediated water flux was inhibited by Na^+ diffusion in the opposite direction [18].

3.4. *Effect of Tricellulin Knockdown on Transepithelial Water Transport*

For the full line of experiments, two tricellulin KD clones were selected, which differ in the grade of tricellulin reduction, namely by 30% (clone KD 23) and by 40% (clone KD 24) compared to the respective controls. While both KD clones caused a decrease in TER, only KD 24 produced an increase in permeability to a 4 kDa-macromolecule and, most importantly, an increase of water flux driven by different osmotic gradients as discussed above.

The question arises whether this is an all-or-nothing effect. If yes, one would postulate a threshold of tricellulin depression somewhere between 30% and 40%. In order to have closer look, an additional experiment was performed employing a clone with tricellulin reduced by 35% (clone KD 22). Being aware that this is based on three points only, resulting water flux did appear to correlate with the tricellulin expression, suggesting that the effect of tricellulin KD depression is gradual, starting at the level found for the controls (Figure 6).

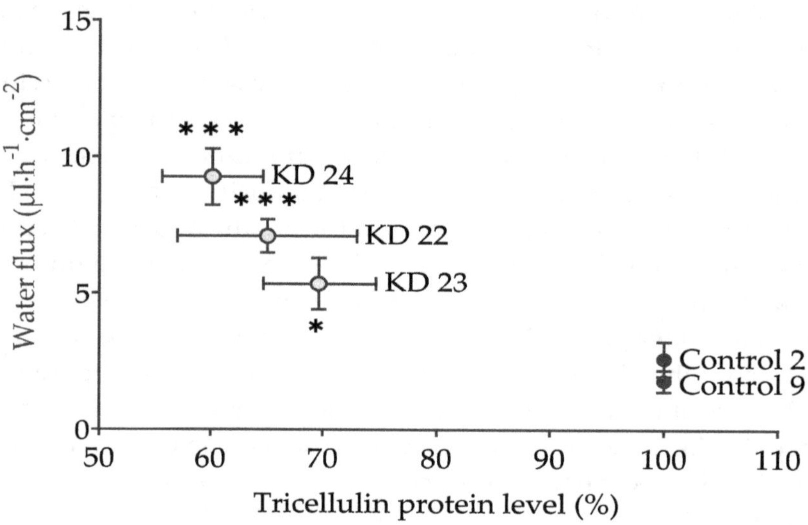

Figure 6. Water flux as a function of the level of tricellulin expression in MDCK C7 cells driven by an osmotic gradient of 37 mM 4 kDa-dextran (100 mOsm). Tricellulin KD 22 (n = 7) is an additional KD clone that was tested only for this purpose (shRNA 22: TRCN0000072635, NM_144724.1-989s1c1, Sigma-Aldrich, Schnelldorf, Germany) and compared with Control 9. Statistical analysis was performed using Student's t-test between the KD clone and the corresponding control (KD 23 versus control 2, KD 24 versus control 9). In case of the KD 22 and KD 24 and their control 9, the Bonferroni-Holm adjustment for multiple testing was applied. (* $p < 0.05$, *** $p < 0.001$).

4. Materials and Methods

4.1. *Cell Culture, Transfection, and TER Measurement*

The kidney cell line MDCK C7 (RRID: CVCL_0423) exhibits a high transepithelial resistance and other basic properties making it an excellent model of a tight epithelium. For stable tricellulin knockdown, MDCK C7 cells were transfected with pLKO.1-puro vector containing a sequence for shRNA targeting tricellulin (shRNA 23: TRCN0000072633, NM_144724.1-2011s1c1; and shRNA 24: TRCN0000072634, NM_144724.1-1097s1c1; Sigma-Aldrich, Schnelldorf, Germany) or pLKO.1-puro empty vector as a negative control (Sigma-Aldrich, Schnelldorf, Germany).

The transfected cells were incubated at 37 °C and 5% CO_2 in sterilization incubators held. For the cultivation of the cells sterile culture vessels made of plastic were used. The cells were cultured in a nutrient medium Earl's salts MEM (minimal essential medium) supplemented with 10% FCS as well as 100 U/mL penicillin/100 μg/mL streptomycin and 1.5 μg/mL of puromycin. Every second to third day the medium was changed. Consumption of nutrients was also visible due to a color change of the medium which contained a detecting pH indicator.

For protein quantification, water flux measurements and electrophysiological studies, cell monolayers were cultured on porous culture plate inserts (Millicell PCF filters, pore size 0.4 μm, effective area 0.6 cm^2, Millipore GmbH, Schwalbach, Germany) for 7–9 days before they were used for experiments.

Transepithelial resistance (TER) was measured at 37 °C using chopstick electrodes (STX2, World Precision Instruments, Friedberg, Germany). Electrodes were reproducibly positioned by a semi-automatic motor-driven device and signals were processed by a low-frequency clamp (both own design). The resistances of the bathing solution and the blank filter support were subtracted from measured values, which finally were converted to $\Omega\ cm^2$.

4.2. *Western Blot Analysis*

Cells grown on culture-plate inserts were scraped and homogenized in total lysis buffer containing 10 mM of Tris, 150 mM of NaCl, 0.5% of Triton X-100, and 0.1% of sodium dodecyl sulphate (SDS), and protease inhibitors (complete™ ethylenediaminetetraacetic acid (EDTA)-free; Roche, Basel, Switzerland). The samples were incubated 2 h at 4 °C (vortexed every 20 min), and then centrifuged at 11,000× *g* during 20 min. The pellet was discarded and the protein concentration in the supernatant was determined by the BCA (bicinchoninic acid) method (reagents were purchased from Pierce (Perbio Science, Bonn, Germany)) and quantified with a plate reader (Tecan Deutschland, Crailsheim, Germany). Aliquots between 10 and 15 μg protein samples were separated by 12% SDS-polyacrylamide gel electrophoresis and then transferred to a PVDF membrane (Perkin Elmer, Rodgau, Germany) for detection of tight-junction-associated marvel proteins (TAMP), claudins, angulins and AQPs. After blocking for 2 h in 1% PVP-40 and 0.05% Tween-20, membranes were incubated overnight with primary antibodies specific for tricellulin (Abfinity, Invitrogen), occludin (Invitrogen), angulin-1, -2, and -3 (Sigma-Aldrich, Germany), claudin-1, -3, -4, -5, -7, and -8 (Invitrogen), and AQP-1, -3, and -4 (Santa Cruz, Heidelberg, Germany). After removing the first antibody and three washing steps, the membranes were incubated 2 h with the second peroxidase-conjugated antibody (anti-mouse or anti-rabbit) in 2% of milk powder prepared in TBST 1X. For detection of the chemiluminescence signal induced by addition of Lumi-LightPLUS western blotting kit (Roche) a Fusion FX7 (Vilber Lourmat, Eberhardzell, Germany) were used. Densitometric analysis was performed with quantification software (Image Studio™ Lite, LI-COR Biosciences, Lincoln, Nebraska USA). Equal protein loading in each lane was verified by comparison with signals for β-actin (Sigma-Aldrich). For western blot analysis, lysates of at least three individual cell cultures were used and one representative experiment is shown.

4.3. *Immunofluorescent Staining*

Immunofluorescence studies were performed on culture-plate inserts. Confluent monolayers were rinsed with phosphate-buffered saline (PBS), fixed with 4% paraformaldehyde for 20 min and permeabilized for 10 min with PBS containing 0.5% (*v/v*) Triton X-100. To block non-specific binding sites, cells were then incubated in PBS containing 1% (*w/v*) BSA and 5% (*v/v*) goat serum (blocking solution; Biochrom) for 60 min. All subsequent washing procedures were performed with this blocking solution. After blocking, cells were incubated 60 min at room temperature with primary antibodies for tricellulin (Abfinity, Invitrogen; 1:250) and occludin (Invitrogen, 1:250), followed by washing steps and incubation during 60 min at room temperature with the respective secondary antibodies (Alexa Fluor 488 goat anti-rabbit and Alexa Fluor 594 goat anti-mouse, each 1:500; Molecular Probes MoBiTec) and 4′,6-diamidino-2-phenylindole (1:1000). Images were obtained with a confocal laser-scanning microscope (LSM 780, Zeiss, Jena, Germany) and processed using ZEN software (Zeiss, Oberkochen, Germany).

4.4. Measurement of 4 kDa-FITC-Dextran Flux

Flux studies were performed in conventional Ussing chambers under voltage clamp conditions. Dextran flux was measured in 5 mL circulating Ringer's containing 37 mM unlabeled 4 kDa-dextran on the apical side. After addition of 100 µL of 37 mM FITC-labeled dialyzed dextran (4 kDa-FITC-dextran; Sigma-Aldrich) to the apical bath, basolateral samples (200 µL) were collected at 0, 20, 40, 60, 80, 100, and 120 min. Tracer fluxes were determined from FITC-dextran samples, which were measured with a fluorometer at 520 nm (Spectramax Gemini, Molecular Devices, Ismaning, Germany). Dextran permeability was calculated from $P = J/\Delta c$ with P = permeability (cm s^{-1}), J = flux (mol h^{-1} cm^{-2}) and c = concentration (mol/L).

4.5. Dilution Potential Measurements

Dilution potential measurements for the determination of ion permeabilities were performed in Ussing chambers modified for cell-culture inserts. Water-jacketed gas lifts kept at 37 °C were filled with 10 mL circulating fluid on each side. The bathing solution contained (in mM) 119 NaCl, 21 $NaHCO_3$, 5.4 KCl, 1.2 $CaCl_2$, 1 $MgSO_4$, 3 HEPES, and 10 D(+)-glucose, and was gassed with 95% O_2 and 5% CO_2 to ensure a pH value of 7.4. All experimental data were corrected for the resistance of the empty filter and the bathing solution. Dilution potentials were measured with modified bathing solution on the apical or basolateral side of the epithelial monolayer. In the modified bathing solution, NaCl was iso-osmotically replaced by mannitol. The ratio of P_{Na} and P_{Cl} and the absolute permeabilities for Na^+ and Cl^- were calculated as described before [19].

4.6. Measurement of Transepithelial Water Transport

Water flux measurements were performed using a modified Ussing chamber with two glass tubes instead of the gas lifts as described before [16–18]. Throughout these experiments, transepithelial resistance (TER, $\Omega \cdot cm^2$), short-circuit current (I_{SC}, $\mu A \cdot cm^{-2}$), and transepithelial voltage (mV) were recorded routinely. Resistances of bathing solution and blank filter support were measured prior to each experiment and subtracted. The stability of TER served as an indicator of cell viability.

Cell filters were mounted in Ussing chambers and perfused with HEPES-buffered solution with the following composition (in mM): 144.8 NaCl, 2.4 Na_2HPO_4, 0.6 NaH_2PO_4, 5.4 KCl, 1.2 $MgCl_2$, 1.2 $CaCl_2$, 10.6 HEPES, and 10 D(+)-glucose. The pH value of the perfusion solution was pH 7.4. A rotary pump ensured constant circulation of the perfusion solution (4.0 $mL \cdot min^{-1}$) and thus a fast fluid exchange in both hemichambers (volume 500 µL) to avoid effects of unstirred layers on water permeability. Water flux was induced by a transepithelial osmotic gradient: (i) 100 mM mannitol, (ii), 37 mM 4 kDa-dextran, (iii) 100 mM 4 kDa-dextran, or (iv) 5.5 mM 40 kDa-dextran. The solution was added in the apical compartment of the Ussing chamber.

The osmolality of the perfusion solutions (mosmol/kg, abbreviated mOsm) was determined using a Vapor Pressure Osmometer (5100B, Wescor, Logan, UT). The osmolality of the HEPES-buffered solution for water flux measurements was 287.4 ± 5.7 mOsm (n = 15). The solutions for measuring water flux ((i) 100 mM mannitol, (ii), 37 mM 4 kDa-dextran, (iii) 100 mM 4 kDa-dextran or (iv) 5.5 mM 40 kDa-dextran) increased the osmolality to 388.2 ± 3.2 mOsm (n = 5), 399.4 ± 2.4 mOsm (n = 10), 877.8 ± 13.7 mOsm (n = 5), and 390.2 ± 2.5 mOsm (n = 5), respectively. Molecular diameters of the dextrans were estimated using the equation: $d_{(Å)} = 0.215 \cdot (MW^{0.587})$.

The fluid level in both glass tubes was monitored by a visual system ColorView XS (Olympus Soft Imaging Solutions GmbH, Munster, Germany), at time 0 min and with intervals of 10 min over a period of 120 min. Transepithelial water flux, given as flux per square centimeter and hour, was calculated after special calibration from the difference between the menisci at the registration times. Fluxes directed from the basolateral to the apical compartment were defined as positive flux.

4.7. Statistical Analysis

Data are expressed as mean values ± SEM (standard error of the mean), indicating n as the number of single measurements, and N as the number of independent experiments, which means independent seeding of cells. Statistical analysis was performed using Student's t-test between the KD clone and the corresponding control (KD 23 versus control 2, KD 24 versus control 9). In case of the KD 22 and KD 24 data of Figure 6, the Bonferroni–Holm adjustment for multiple testing was applied. $p < 0.05$ was considered significant (* $p < 0.05$, ** $p < 0.01$, *** $p < 0.001$).

5. Conclusions

This study analyzed the contribution of the tTJ to water permeability as regulated by its prominent protein tricellulin. Our results demonstrate that tricellulin KD in MDCK C7 cells, which lack the bTJ cation and water channels claudin-2 and -15, results in an increase in transepithelial water transport. This water flux appears to be negatively correlated with the tricellulin expression level. We conclude that in a tight epithelial cell line, tricellulin contributes to regulate water flux through the paracellular pathway.

Author Contributions: Conceptualization, R.R., M.F., S.M.K. and C.A.-T.; Funding Acquisition, M.F., J.D.S.; Methodology, R.R. and M.F.; Investigation, C.A.-T.; Supervision and Validation, R.R., M.F. and S.M.K.; Visualization, C.A.-T.; Writing–Original Draft Preparation, C.A.-T.; Writing–Review and Editing, M.F., R.R., S.M.K. and J.D.S.

Acknowledgments: We would like to thank Britta Jebautzke, In-Fah M. Lee, and Anja Fromm (all three: Institute of Clinical Physiology/Nutritional Medicine, Medical Department, Division of Gastroenterology, Infectiology and Rheumatology, Charité – Universitätsmedizin Berlin, 12203 Berlin, Germany) for their expert technical assistance.

Abbreviations

bTJ	Bicellular tight junction
$d_{(Å)}$	Diameter (Å)
EDTA	Ethylenediaminetetraacetic acid
FD4	4 kDa-FITC-dextran 4 kDa
FITC	fluorescein isothiocyanate-dextran
KD	Knockdown
MDCK C7	Madin-Darby canine kidney cells, clone 7
shRNA	small hairpin RNA
TJ	Tight junction
tTJ	Tricellular tight junction
P	Apparent permeability (cm s^{-1})
PBS	phosphate-buffered saline
SEM	Standard error of the mean
TAMP	tight-junction-associated marvel protein
TER	Transepithelial resistance (Ω cm^2)

References

1. Farquhar, M.G.; Palade, G.E. Junctional complexes in various epithelia. *J. Cell Biol.* **1963**, *17*, 375–412. [CrossRef] [PubMed]
2. Amasheh, S.; Meiri, N.; Gitter, A.H.; Schöneberg, T.; Mankertz, J.; Schulzke, J.D.; Fromm, M. Claudin-2 expression induces cation-selective channels in tight junctions of epithelial cells. *J. Cell Sci.* **2002**, *115*, 4969–4976. [CrossRef] [PubMed]
3. Krug, S.M.; Schulzke, J.D.; Fromm, M. Tight junction, selective permeability, and related diseases. *Semin. Cell Dev. Biol.* **2014**, *36*, 166–176. [CrossRef] [PubMed]

4. Staehelin, L.A. Further observations on the fine structure of freeze-cleaved tight junctions. *J. Cell Sci.* **1973**, *13*, 763–786.
5. Günzel, D.; Fromm, M. Claudins and other tight junction proteins. *Compr. Physiol* **2012**, *2*, 1819–1852. [CrossRef]
6. Higashi, T.; Miller, A.L. Tricellular junctions: How to build junctions at the TRICkiest points of epithelial cells. *Mol. Biol. Cell* **2017**, *28*, 2023–2034. [CrossRef]
7. Ikenouchi, J.; Furuse, M.; Furuse, K.; Sasaki, H.; Tsukita, S.; Tsukita, S. Tricellulin constitutes a novel barrier at tricellular contacts of epithelial cells. *J. Cell Biol.* **2005**, *171*, 939–945. [CrossRef]
8. Krug, S.M.; Amasheh, S.; Richter, J.F.; Milatz, S.; Günzel, D.; Westphal, J.K.; Huber, O.; Schulzke, J.D.; Fromm, M. Tricellulin forms a barrier to macromolecules in tricellular tight junctions without affecting ion permeability. *Mol. Biol. Cell* **2009**, *20*, 3713–3724. [CrossRef]
9. Krug, S.M. Contribution of the tricellular tight junction to paracellular permeability in leaky and tight epithelia. *Ann. N. Y. Acad. Sci.* **2017**, *1397*, 219–230. [CrossRef]
10. Krug, S.M.; Bojarski, C.; Fromm, A.; Lee, I.M.; Dames, P.; Richter, J.F.; Turner, J.R.; Fromm, M.; Schulzke, J.D. Tricellulin is regulated via interleukin-13-receptor alpha2, affects macromolecule uptake, and is decreased in ulcerative colitis. *Mucosal Immunol.* **2018**, *11*, 345–356. [CrossRef]
11. Masuda, S.; Oda, Y.; Sasaki, H.; Ikenouchi, J.; Higashi, T.; Akashi, M.; Nishi, E.; Furuse, M. LSR defines cell corners for tricellular tight junction formation in epithelial cells. *J. Cell Sci.* **2011**, *124*, 548–555. [CrossRef] [PubMed]
12. Higashi, T.; Tokuda, S.; Kitajiri, S.; Masuda, S.; Nakamura, H.; Oda, Y.; Furuse, M. Analysis of the 'angulin' proteins LSR, ILDR1 and ILDR2–tricellulin recruitment, epithelial barrier function and implication in deafness pathogenesis. *J. Cell Sci.* **2013**, *126*, 966–977. [CrossRef] [PubMed]
13. Agre, P.; Preston, G.M.; Smith, B.L.; Jung, J.S.; Raina, S.; Moon, C.; Guggino, W.B.; Nielsen, S. Aquaporin CHIP: The archetypal molecular water channel. *Am. J. Physiol.* **1993**, *265*, F463–F476. [CrossRef] [PubMed]
14. Day, R.E.; Kitchen, P.; Owen, D.S.; Bland, C.; Marshall, L.; Conner, A.C.; Bill, R.M.; Conner, M.T. Human aquaporins: Regulators of transcellular water flow. *Biochim. Biophys. Acta* **2014**, *1840*, 1492–1506. [CrossRef] [PubMed]
15. Zeuthen, T.; Gorraitz, E.; Her, K.; Wright, E.M.; Loo, D.D. Structural and functional significance of water permeation through cotransporters. *Proc. Natl. Acad. Sci. USA* **2016**, *113*, E6887–E6894. [CrossRef] [PubMed]
16. Rosenthal, R.; Milatz, S.; Krug, S.M.; Oelrich, B.; Schulzke, J.D.; Amasheh, S.; Günzel, D.; Fromm, M. Claudin-2, a component of the tight junction, forms a paracellular water channel. *J. Cell Sci.* **2010**, *123*, 1913–1921. [CrossRef]
17. Rosenthal, R.; Günzel, D.; Krug, S.M.; Schulzke, J.D.; Fromm, M.; Yu, A.S. Claudin-2-mediated cation and water transport share a co Günzel (with u-umlaut)mmon pore. *Acta Physiol.* **2017**, *219*, 521–536. [CrossRef]
18. Rosenthal, R.; Günzel, D.; Piontek, J.; Krug, S.M.; Ayala-Torres, C.; Hempel, C.; Theune, D.; Fromm, M. Claudin-15 forms a water channel through the tight junction with distinct function compared to claudin-2. *Acta Physiol.* **2019**, e13334. [CrossRef]
19. Krug, S.M.; Günzel, D.; Conrad, M.P.; Rosenthal, R.; Fromm, A.; Amasheh, S.; Schulzke, J.D.; Fromm, M. Claudin-17 forms tight junction channels with distinct anion selectivity. *Cell. Mol. Life Sci.* **2012**, *69*, 2765–2778. [CrossRef]
20. Rosenthal, R.; Günzel, D.; Theune, D.; Czichos, C.; Schulzke, J.D.; Fromm, M. Water channels and barriers formed by claudins. *Ann. N. Y. Acad. Sci.* **2017**, *1397*, 100–109. [CrossRef]
21. Krug, S.M.; Hayaishi, T.; Iguchi, D.; Watari, A.; Takahashi, A.; Fromm, M.; Nagahama, M.; Takeda, H.; Okada, Y.; Sawasaki, T.; et al. Angubindin-1, a novel paracellular absorption enhancer acting at the tricellular tight junction. *J. Control. Release* **2017**, *260*, 1–11. [CrossRef] [PubMed]
22. Gong, Y.; Himmerkus, N.; Sunq, A.; Milatz, S.; Merkel, C.; Bleich, M.; Hou, J. ILDR1 is important for paracellular water transport and urine concentration mechanism. *Proc. Natl. Acad. Sci. USA* **2017**, *114*, 5271–5276. [CrossRef] [PubMed]
23. Wegmann, M.; Nüsing, R.M. Prostaglandin E2 stimulates sodium reabsorption in MDCK C7 cells, a renal collecting duct principal cell model. *Prostaglandins Leukot Essent Fatty Acids* **2003**, *69*, 315–322. [CrossRef] [PubMed]

24. Zak, J.; Schneider, S.W.; Eue, I.; Ludwig, T.; Oberleithner, H. High-resistance MDCK-C7 monolayers used for measuring invasive potency of tumour cells. *Pflügers Archiv Eur. J. Physiol.* **2000**, *440*, 179–183. [CrossRef] [PubMed]
25. Riazuddin, S.; Ahmed, Z.M.; Fanning, A.S.; Lagziel, A.; Kitajiri, S.; Ramzan, K.; Khan, S.N.; Chattaraj, P.; Friedman, P.L.; Anderson, J.M.; et al. Tricellulin is a tight-junction protein necessary for hearing. *Am. J. Hum. Genet.* **2006**, *79*, 1040–1051. [CrossRef]
26. Saitou, M.; Furuse, M.; Sasaki, H.; Schulzke, J.D.; Fromm, M.; Takano, H.; Noda, T.; Tsukita, S. Complex phenotype of mice lacking occludin, a component of tight junction strands. *Mol. Biol. Cell* **2000**, *11*, 4131–4142. [CrossRef]
27. Schulzke, J.D.; Gitter, A.H.; Mankertz, J.; Spiegel, S.; Seidler, U.; Amasheh, S.; Saitou, M.; Tsukita, S.; Fromm, M. Epithelial transport and barrier function in occludin-deficient mice. *Biochim. Biophys. Acta* **2005**, *1669*, 34–42. [CrossRef]
28. Raleigh, D.R.; Marchiando, A.M.; Zhang, Y.; Shen, L.; Sasaki, H.; Wang, Y.; Long, M.; Turner, J.R. Tight junction-associated MARVEL proteins marveld3, tricellulin, and occludin have distinct but overlapping functions. *Mol. Biol. Cell* **2010**, *21*, 1200–1213. [CrossRef]
29. Günzel, D.; Yu, A.S. Claudins and the modulation of tight junction permeability. *Physiol. Rev.* **2013**, *93*, 525–569. [CrossRef]
30. Weber, C.R.; Tokuda, S.; Hirai, T.; Furuse, M. Claudin-4 knockout by TALEN-mediated gene targeting in MDCK cells: Claudin-4 is dispensable for the permeability properties of tight junctions in wild-type MDCK cells. *PLoS ONE* **2017**, *12*, e0182521. [CrossRef]
31. Kirschner, N.; Rosenthal, R.; Furuse, M.; Moll, I.; Fromm, M.; Brandner, J.M. Contribution of tight junction proteins to ion, macromolecule, and water barrier in keratinocytes. *J. Investig. Dermatol.* **2013**, *133*, 1161–1169. [CrossRef] [PubMed]
32. Rudan-Tasic, D.; Klofutar, C. Osmotic Coefficients of Aqueous Solutions of Some Poly(oxyethylene) Glycols at the Freezing Point of Solution. *Chem. Mon.* **2004**, *135*, 773–784. [CrossRef]
33. Wen, H.; Hao, J.; Li, S.K. Characterization of human sclera barrier properties for transscleral delivery of bevacizumab and ranibizumab. *J. Pharm. Sci.* **2013**, *102*, 892–903. [CrossRef] [PubMed]

8

Mouse Models of Human Claudin-Associated Disorders: Benefits and Limitations

Murat Seker [1], Cármen Fernández-Rodríguez [2], Luis Alfonso Martínez-Cruz [2] and Dominik Müller [1,*]

1 Department of Pediatric Gastroenterology, Nephrology and Metabolism, Charité—Universitätsmedizin Berlin, Charité, 13353 Berlin, Germany; murat.seker@charite.de
2 CIC BioGUNE, Bizkaia Science and Technology Park, 801A, 48160 Derio, Spain; cfernandez@cicbiogune.es (C.F.-R.); amartinez@cicbiogune.es (L.A.M.-C.)
* Correspondence: dominik.mueller@charite.de

Abstract: In higher organisms, epithelia separate compartments in order to guarantee their proper function. Such structures are able to seal but also to allow substances to pass. Within the paracellular pathway, a supramolecular structure, the tight junction transport is largely controlled by the temporospatial regulation of its major protein family called claudins. Besides the fact that the expression of claudins has been identified in different forms of human diseases like cancer, clearly defined mutations in the corresponding claudin genes have been shown to cause distinct human disorders. Such disorders comprise the skin and its adjacent structures, liver, kidney, the inner ear, and the eye. From the phenotype analysis, it has also become clear that different claudins can cause a complex phenotype when expressed in different organs. To gain deeper insights into the physiology and pathophysiology of claudin-associated disorders, several mouse models have been generated. In order to model human disorders in detail, they have been designed either as full knockouts, knock-downs or knock-ins by a variety of techniques. Here, we review human disorders caused by CLDN mutations and their corresponding mouse models that have been generated thus far and assess their usefulness as a model for the corresponding human disorder.

Keywords: tight junction; claudin; mutations; kidney; liver; skin; human; mice; disease

---

1. Introduction

In higher organisms epithelia, endothelia and mesothelia are essential to separate different compartments in order to guarantee their proper function. Such structures can be found ubiquitously, like in skin which separates the body from the surrounding environment, the lung (air/blood) and the intestine (gut lumen/blood). Further, epithelial structures form the blood-brain-barrier [1] and are also essential for liver and kidney function [2]. Epithelia consist of monolayer or multilayer structures [3] but a common prerequisite for their function is polarization, i.e., a clear orientation of apical to basolateral requiring a well-defined maintenance molecular machinery [4]. Moreover, besides providing a tight barrier (e.g., within the bladder), epithelial structures are able to regulate coordinated transport across cellular barriers. Transport across epithelial barriers is provided in general via two systems, the paracellular and the transcellular pathway. Whereas the first provides the organism with a maximum of resorptive capacity by minimal energy expenditure, the latter aims at fine-tuning depending on the actual needs of the organism, regulated also by a subset of various local, regional and global mechanisms [5].

In more detail, the transcellular pathway consists of apical uptake, intracellular buffering, transport and basolateral extrusion mechanisms. This pathway is highly energy dependent and mostly driven by the basolateral Na^+-K^+-ATPase. One of the advantages of such a pathway is its controllability on several of the steps, e.g., by hormones (e.g., 1,25(OH)$_2$D$_3$ or Parathyroid Hormone (PTH)) and therefore also aims at fine-tuning to provide the organism according to its actual needs and also on middle and long term regulation [6]. Especially for these pathways, it has been shown that mutations in human gene cause distinct monogenetic disorders, like Bartter's [7] and Gitelman's Syndrome [8] with a clear phenotype-genotype correlation. In order to study such genetic effects, mouse models have been generated to study successfully the consequences of human mutations [9–13].

Considering the paracellular pathway, enormous progress has been achieved during the last two decades as it became clear that this pathway, although basically driven by a given electrochemical gradient only, is regulated by several structures and mechanisms. In regions where concentration gradients of solutes across the epithelial layer are high, like the proximal tubule of the kidney, paracellular pathways provide the organism with maximal reabsorption by minimal energy expenditure. More distantly, crosstalk takes place before the transcellular pathway provides the major mechanism. A key component of the paracellular pathway is a supramolecular structure, called tight junction (TJ). The TJ consists of several membrane-bound proteins and their intracellular adapter and scaffolding proteins [14]. The major proteins essential for the TJ are claudins (lat. *claudere*, i.e., to seal). The family of claudins is currently included 27 members in eukaryotes. The claudin proteins are membrane-bound and span four times the plasma membrane with an intracellular C- and N-terminal part. Each of the four provides a functional entity, i.e., either to act as pore-forming or as a sealing component (for a detailed review: [15]). Classification of claudins has been made based on different factors such as sequence similarity and functionality. In this review, we used the latter one which was based on sealing or pore-forming capabilities of claudins reviewed by Krause et al. [16].

In 1999, the group of Lifton showed that the disorder *Familial Hypercalciuria, Hypomagnesemia with Nephroclacinosis (FHHNC)* is caused by mutations in the *CLDN16* gene. This observation provided the first evidence that TJ disorders cause human disorders and diseases [17]. Since then, several other TJ disorders have been shown to cause, when mutated, (*CLDN1, CLDN2 CLDN9 CLDN10, CLDN14, CLDN16, CLDN19*), human disorders [18–23]. Furthermore, expression of claudins are influenced by many factors (e.g., smoking, diet changes, alcohol intake) [24–26] and numerous associations of disordered claudin expression and disease have been described. (for detail reviews see [27,28]).

To model such disorders, genetically modified mice have been generated. Initially, embryonic stem cell technology was used to create such models [29] and later, other approaches have been established successfully [30]. Besides global deletions, regional and local variations of murine gene expression have been established. Although not reported yet, based on current developments, it can be expected that CRISPR/Cas9 technology will take place as a standard procedure [31].

However, even though such models provide insights into physiology and pathophysiology and may open new avenues for the development of therapeutical interventions, they face limitations. Here, we provide an overview of current human claudin-associated disorders and their corresponding mouse models, their impact on physiology and pathophysiology.

2. Claudin Mutations Causing Human Disorders

2.1. Claudin 1

CLDN1 consists of four exons encoding a 211 amino acid protein that conveys barrier properties [16]. It has been shown that claudin-1 interacts with claudin-3 and claudin-5, which are also barrier–forming claudins [32,33]. In the skin, *CLDN1* is expressed in the stratum corneum, stratum basale, stratum granulosum and stratum spinosum [34]. Further, it is expressed in the kidney [35], gall bladder [36], human ovarian epithelium [37] and the inner ear [38]. The expression of *CLDN1* is increased in colorectal cancer [39], lung carcinoma [40], cervical cancer [41] and reduced in larynx cancer [42]. Evans et al. showed that *CLDN1* is involved in Hepatitis C Virus entry into intestinal cells, which is presumably depending on the first extracellular loop (ECL, Figure 1) [43].

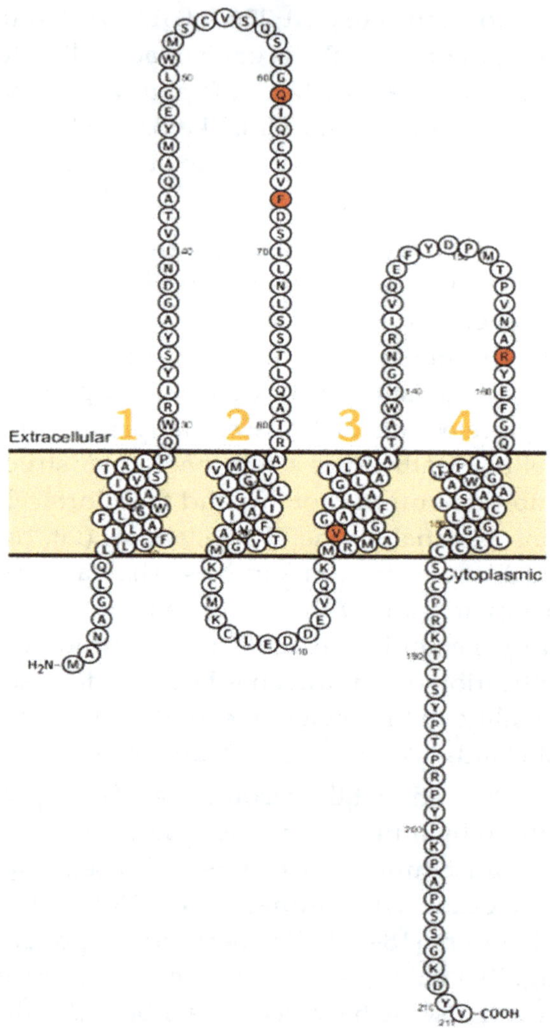

Figure 1. Topology of *CLDN1*. FASTA sequence of *CLDN1* (Uniprot accession no: O95832; plotted by Protter software http://wlab.ethz.ch/protter/). Each amino acid is shown as a single letter code and numbers (orange) indicate transmembrane domains. Mutations shown to be involved in human diseases are shown in red.

In 2002, Baala et al. described a family with autosomal dominant ichthyosis, alopecia, leucocytic vacuoles and sclerosing cholangitis (ILVASC; or neonatal ichthyosis with sclerosing cholangitis (NISCH-Syndrome OMIM 607626) assigned to chromosome 3q27–28 [44]. In 2004 Hadj-Rabia et al. identified in the same kindred a frameshift mutation in *CLDN1* leading to a premature stop codon at position 67 [18] (Figure 1). This finding, that ILVASC/NISCH syndrome is caused by mutations in *CLDN1* has been later confirmed by others [18,44–46]. In 2006, a neonate patient with erythroderma, massive lamellar desquamation and alopecia were reported. The hepatologic aspects were icterus, hyperbilirubinemia and increased biliary acids, liver biopsy showed panlobular cholestasis with acute hepatitis. All reported cases showed that *CLDN1* mutations in humans were not lethal and did not affect fertility.

Furuse and colleagues generated *Cldn1* deficient mice, which exhibited low body weight and died within the first day after birth, possibly attributable to an excessive trans-epithelial water loss (TEWL). *Cldn1* deficient mice displayed also altered lipid composition and defects of the stratum corneum of the skin. On the other hand, *CLDN1* deficient patients also displayed mild wrinkled skin and hyperproliferation of keratinocytes. However, only some patients had liver cell injury [18]. Whether knockout mice were affected by hepatic abnormalities has not been reported, probably due to the perinatal death.

Since the mouse model was limited because of its early lethality, a KD mouse approach was used, that showed reduced expression levels of *Cldn1* and *Cldn1* levels were associated with the severity of the phenotype [47]. *Cldn1* KD mice were born with wrinkled skin similar to KO mice, however, morphological examinations at 8 weeks of age revealed a normal development, which might be explained by the low but still expressed *Cldn1*. The underlying mechanism of TEWL was further investigated by Hirano et al. by using tamoxifen-induced epidermis-specific deletion of *Cldn1* in adult mice. Four days after induction, claudin-1 began depleting from basal layers and was undetectable in granular layers at day eight and tight junction leakage was observed. Neither TEWL nor stratum corneum defects were observed until day 18 suggesting that TJ deterioration is a prerequisite for stratum corneum defects [48].

2.2. Claudin 2

Claudin-2 is one of the two claudins that were initially identified by Furuse et al. in 1998 [49]. In humans, its gene contains two exons encoding a 203 amino acid protein that constitutes cation-selective pores [50]. It is expressed in rat brain [51], proximal tubules of the kidney [52] liver, and epididymis [53].

Its expression is increased in colorectal cancer [54], active ulcerative colitis [55], a severe form of the coeliac disease [56] and inflammatory bowel disease [57] whereas it is downregulated in breast tumors and osteosarcoma [58,59]. It has been shown that miR-16 modulates *CLDN2* expression and causes dysfunctional barrier properties in inflammatory bowel disease [60].

Cldn2 KO mice have been generated by Muto et al. [61]. Mice were morphologically normal at birth. The authors did not observe renal morphological abnormalities under both light and electron microscope. However, further analysis in proximal tubules revealed reduced absorption of Na^+ and higher urinary fractional excretion of Ca^{2+}. The same mice were investigated by a different group, focusing on small intestine showing that *Cldn2* deficient mice have slightly larger intestine and longer intestinal villi. They further demonstrated *Cldn2* dependent Na^+ selective intestinal paracellular permeability. Matsumoto et al. focused on liver and biliary tissues of *Cldn2* deficient mice and similar to the kidney, no obvious morphological abnormalities were observed [62]. The detailed physiological analysis revealed a decreased bile flow in *Cldn2* deficient mice and four weeks under a lithogenic diet, KO mice developed gallstones as a consequence of altered bile composition and flow rate [62]. As *Cldn2* and *Cldn15* play an important role in paracellular ion flow, Wada et al. generated *Cldn2*$^{-/-}$ *Cldn15*$^{-/-}$ double KO (dKO) mice [63]. Deficiency of both genes caused early death by week three, which was attributed to overt hypoglycemia of the dKO animals presumably caused by a disrupted Na^+ flow, which is important for intestinal glucose absorption. Transgenic mice with colon specific overexpression of *Cldn2* displayed an enlarged colon and RNAs of genes involved in cell proliferation were found to be increased. In contrast, for inflammation related RNAs, the opposite was shown, in line with the fact that mice were protected against experimentally induced colitis [64].

Recently a missense mutation of Cldn2 associated with obstructive azoospermia in a four generation spanning family has been identified (Figure 2). Further, it has been shown that different claudins (*CLDN1, CLDN2, CLDN3, CLDN4* and *CLDN7)* are expressed in human epididymal tight junctions [65]. Except for *CLDN2* all these claudins cause, when eliminated, a disruption of the epithelial barrier. Therefore, it has been assumed that malfunction of these claudins might cause infertility in men but has not been proven so far.

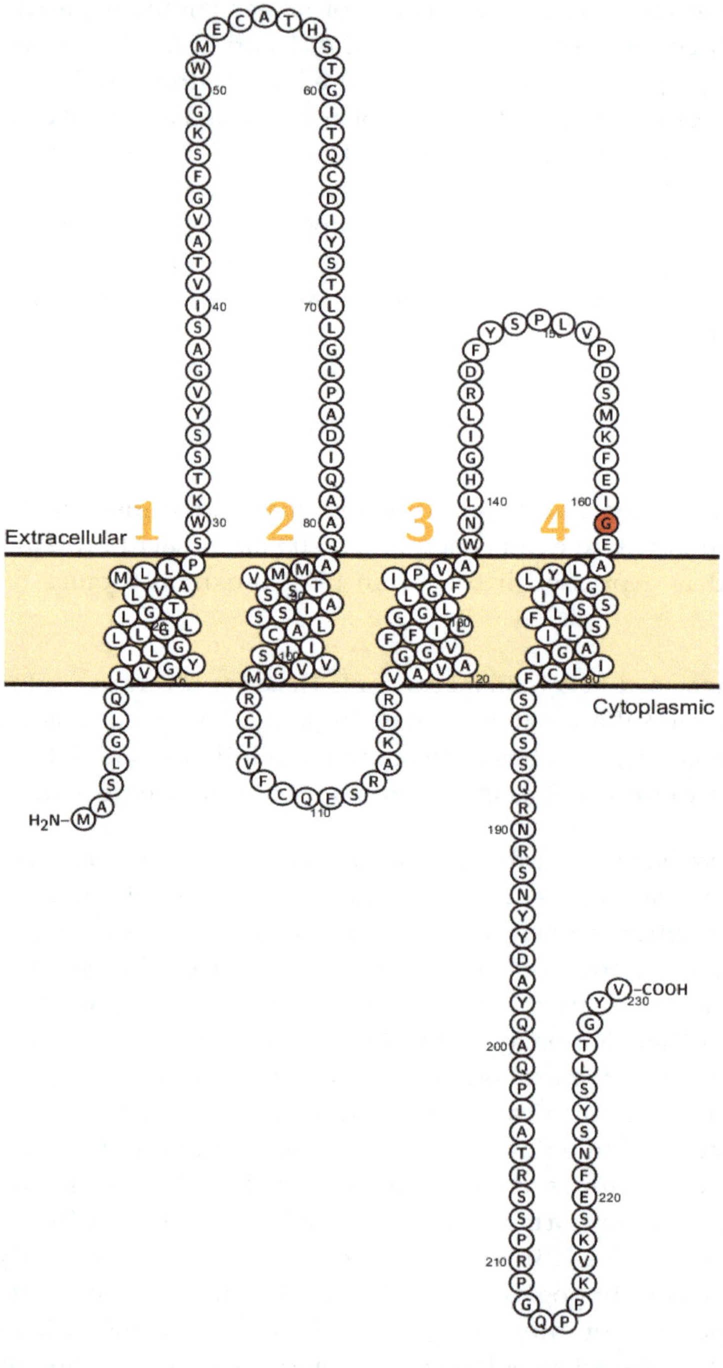

Figure 2. Topology of *CLDN2*. FASTA sequence of *CLDN2* (Uniprot accession no: P57739). Each amino acid is shown as a single letter code and numbers (orange) indicate transmembrane domains. Mutations shown to be involved in human diseases are shown in red.

2.3. Claudin 9

The Claudin-9 gene consists of a single exon encoding a 217 amino acid protein. It constitutes a barrier to K^+ and Na^+. Its expression was found in the inner ear, the liver [66] and the kidney [67].

Zhang et al. showed that claudin-9, together with claudin-6 mediates Hepatitis C Virus (HCV) entry into hepatoma cell lines [66]. Fofana et al. generated monoclonal antibodies against *CLDN9* which inhibit HCVpp (HCV pseudo particles) entry to *CLDN9* expressing cell lines [68].

The expression of *CLDN9* is increased in gastric cancer [69] and promotes cell proliferation and migration of lung metastasis [70]. In cervical carcinoma, RNA expression levels were found to be decreased [71].

Sineni et al. described patients with inherited autosomal recessive hearing loss who had a truncated variant of claudin-9 (p.L29fs). The mutation is located at the beginning of the first ECL (Figure 3). The truncated protein was not detected at the plasma membrane, indicating a dysfunctional TJ and thereby affecting the peri- and endolymphatic ion composition in the inner ear as the cause of hearing impairment [19]. Patients did not display coordination disturbances. This pathophysiological principle has been demonstrated in the past when mutations in BSND have been identified in patients with Bartter syndrome Type 4 [72].

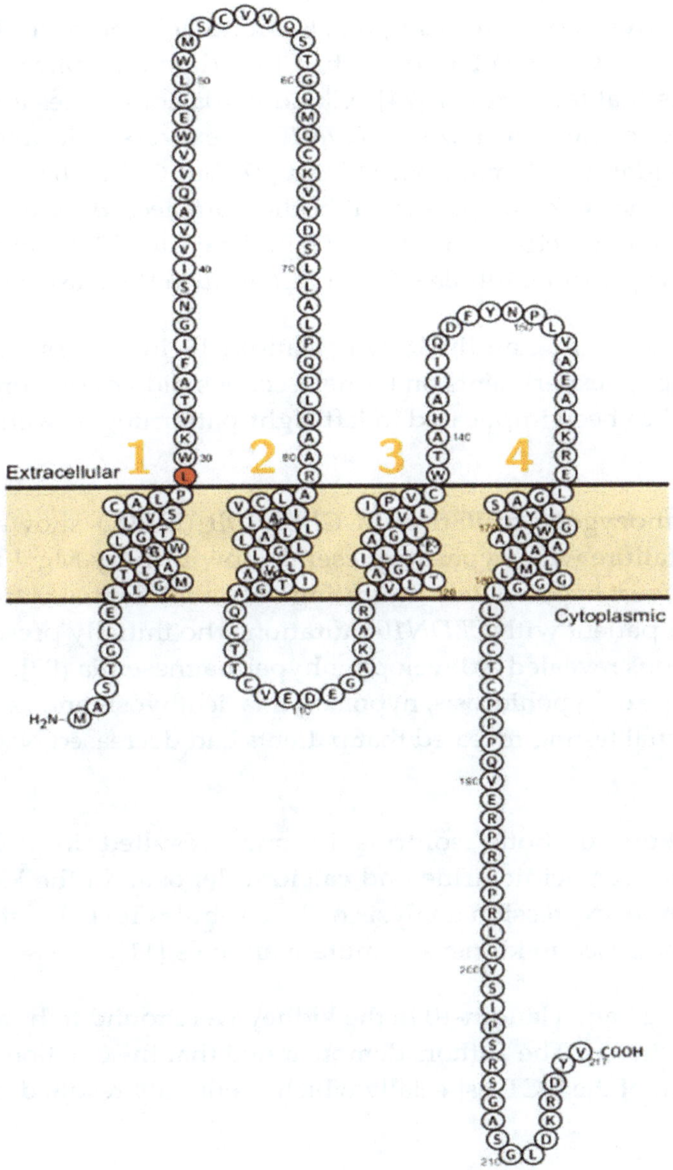

Figure 3. Topology of *CLDN9*. FASTA sequence of *CLDN9* (Uniprot accession no: O95484). Each amino acid is shown as a single letter code and numbers (orange) indicate transmembrane domains. Mutations shown to be involved in human diseases are shown in red.

Mice that carry a missense mutation in *Cldn9* were shown to be deaf. Further analysis showed that Reissner's membrane was morphologically normal when compared to wild type mice and that only the basal but not the apical part of the cochlea was morphologically affected. Inner ear development was normal until postnatal day 14 (P14) then degeneration of the organ of Corti was observed. Since claudin-9 acts as a barrier for Na^+ and K^+, heterologous expression of the mutant in MDCK cells did not affect membrane targeting, indicating that this protein was not functional even when properly targeted and inserted into the TJ. In both, patients and mice hearing loss are age dependent. In mice at P14, outer hair cells (OHC) appeared normal whereas severe degeneration of OHC was observed at P80. In patients, a younger affected sibling had moderate hearing loss when compared to the older sibling. Neither mutant mice nor human patients had balance problems. Although *Cldn9* is expressed in various tissues, the authors did not report any abnormalities of other organs (e.g., kidney) yet.

2.4. Claudin 10

CLDN10 consists of five exons encoding a 228 amino acid protein. Initially, there were two isoforms described [73], later Günzel et al. described four different splice variants mostly localized at TJs except for the ones that lack exon 4 [74]. Claudin-10a constitutes a pore for anions whereas claudin-10b forms a pore for cations [16,73]. *CLDN10b* is expressed in many tissues including the brain, lung [75], salivary gland [76], mammary gland [77] but *CLDN10a* is exclusively expressed in the kidney [73]. There are no isoform-specific antibodies, but according to RNA hybridization data, *Cldn10b* is highly expressed in medulla and *Cldn10a* in the cortex [73] and based on RNA-Seq data, *Cldn10a* is expressed in the proximal tubule (PT) and *Cldn10b* in thick ascending limb (TAL) [11].

Both isoforms interact with claudin-18 and claudin-19 in a yeast two-hybrid analysis [78], whereas immunofluorescence experiments on kidney sections did not demonstrate colocalization [79]. Additionally, claudin-10 has been implicated in left-right-patterning as well as in tumor progression and invasiveness [80].

Patients with a homozygous mutation in *CLDN10B* (N48K) showed anhidrosis, alacrima, dry mouth, and kidney failure with hypermagnesemia, low urinary Mg^{2+} and Ca^{2+} [81] (Figure 4). Patients did not display overt hypokalemia indicating only a mild renal electrolyte wasting although Meyers et al. described a patient with *CLDN10* mutation who initially presented with hypokalemia and follow-up examinations revealed a developing hypermagnesemia [82]. In fact, *CLDN10* patients exhibit a considerable range of hypohidrosis, hypolacrymia, ichthyosis and xerostomia and a decreased amount of saliva. Functional testing revealed that patients had decreased NaCl absorption in the TAL, too [82].

TAL specific knockout of both isoforms in mice resulted in polyuria, polydipsia and hypermagnesemia. Moreover, acidic urine and calcium deposits in the kidney (nephrocalcinosis) were observed. Quantitative expression analysis of TAL tubules revealed that expression of *Cldn10*, *Cldn16* and *Cldn19* was increased in kidneys of mutant animals [11].

Mice lacking claudin-16 and claudin-10 in the kidney were found to have normal Mg^{2+} in serum and absence of nephrocalcinosis. The authors demonstrated that the deletion of *Cldn10* and the loss of *Cldn16* led to an expansion of the DCT, especially which eventually resulted in increased resorption of Mg^{2+} [83].

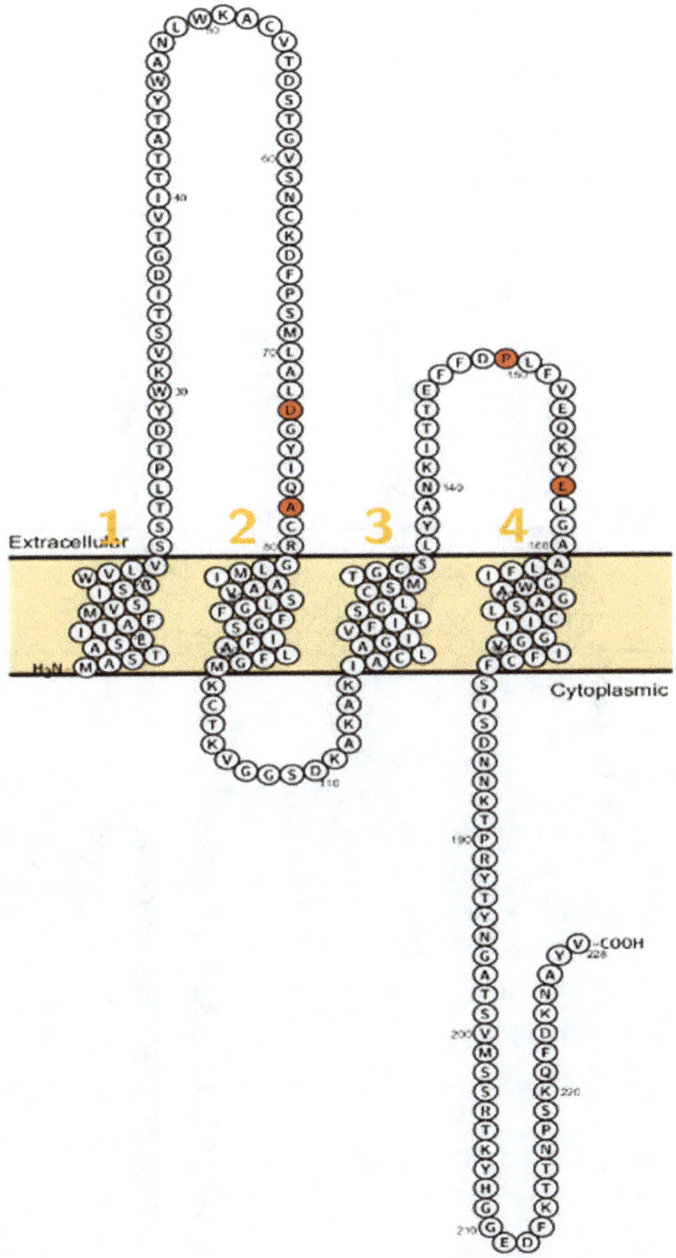

Figure 4. Topology of *CLDN10*. FASTA sequence of *CLDN10* (Uniprot accession no: P78369). Each amino acid is shown as a single letter code and numbers (orange) indicate transmembrane domains. Mutations shown to be involved in human diseases are shown in red.

2.5. Claudin 14

CLDN14 is located on human chromosome 21 and consists of two exons encoding a 239 amino acid protein. It has five protein-coding transcript variants and is classified as a barrier forming claudin [15].

Overexpression of *CLDN14* is associated with gastric and hepatocellular forms of cancer [84].

Cldn14 expression was found in outer and inner ear hair cells and also in the TAL, the DCT [85] and in the liver [12].

Wilcox et al. reported on deaf patients from two large consanguineous families revealing a mutation (V85D) in *CLDN14*. This mutation is predicted to interfere with a phosphorylation site and this residue is conserved between *CLDN3* and *CLDN9* which are expressed in the inner ear too [86]. Immunocytochemistry experiments showed that the mutant failed to localize on the membrane [87]. After this initial study, several other mutations (W56, R81H, G232R) have been reported (Figure 5).

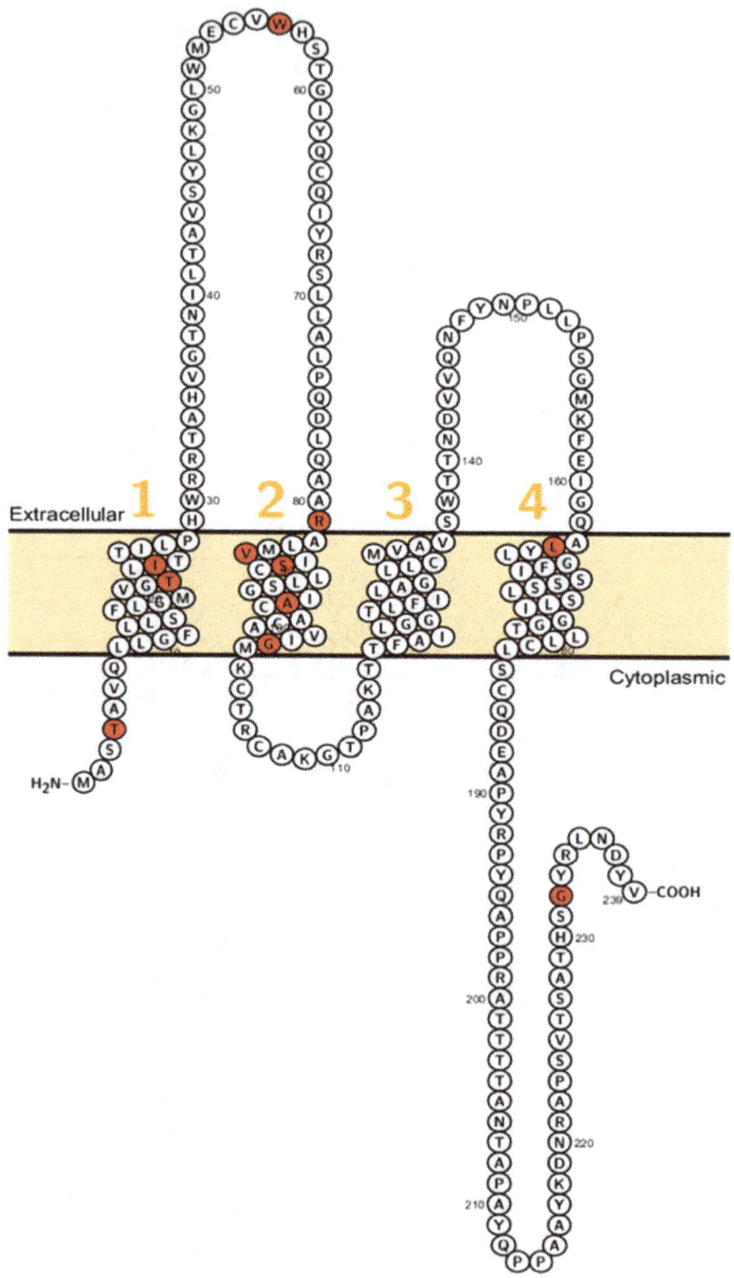

Figure 5. Topology of *CLDN14*. FASTA sequence of *CLDN14* (Uniprot accession no: O95500). Each amino acid is shown as a single letter code and numbers (orange) indicate transmembrane domains. Mutations shown to be involved in human diseases are shown in red.

Examination of *Cldn14* mice that included kidney and liver along with 41 other tissues revealed no morphological differences between knockout and wild type animals. Inner hair cells and outer hair cells (OHC) seemed to develop normally. However, the rapid loss of OHC characterized by disorganization and loss of stereocilia was observed after the first week. Auditory brainstem response (ABR), tests revealed that knockout mice suffer from hearing loss by the third week of age [12].

Although initially no kidney abnormalities were observed neither in mice nor in patients, upon a high calcium diet, *Cldn14* deficient mice developed hypomagnesemia and hypocalciuria. Moreover, it has been shown that two microRNA regulate extracellular Ca^{2+} levels through suppressing *Cldn14* expression under normal diet conditions [85].

Transgenic overexpression of *Cldn14* in TAL of kidney resulted in hypercalciuria and hypermagnesuria [88].

In humans, the hearing loss occurred at the prelingual stage and the severity of hearing loss depends on the mutation. A patient with non-common mutation (A163V) was reported with hearing ability until the age of three, in line with the observation in mice [89].

2.6. Claudin 16

Claudin-16 is a pore forming claudin encoded by *CLDN16* gene [16]. It is expressed in duodenum, jejunum, ileum, colon [90], in tooth germ [91], in salivary glands [92], and also has been found in the endolymphatic duct of the inner ear [93]. The by far highest expression has been found in the thick ascending limb (TAL) of the kidney [94].

Several studies related overexpression of *CLDN16* with different cancer types including ovarian cancer [95] and papillary thyroid carcinoma [96]. Also, claudin-16 might play a role in cell proliferation and differentiation, e.g., breast cancer [97,98].

In the kidney, Claudin-16 is crucial for TJ-specific ion transport in the TAL and handles approximately 25% Ca^{2+} and 70% of Mg^{2+} reabsorption [99]. However, the exact role of claudin-16 remains a matter of debate. Hou and colleagues showed that claudin-16 enhances the permeability of monovalent cations, including Na^{+}, than that of divalent cations, as Mg^{2+} (<50%) [90]. When *Cldn16* was deleted in mice, a cation-to-anion selectivity (P_{Na}/P_{Cl}) but no divalent-to-monovalent cation selectivity (P_{Mg}/P_{Na}) was observed [100]. Other groups have reported a selectivity of Claudin-16 for divalent cations Mg^{2+} and Ca^{2+} [101,102]. Hou et al. have described an interaction with claudin-19 as being necessary for correct TJ integration [78]. Additionally, it has been reported that claudin-14 reduces the cation permeability of claudin-16, but not for the claudin-19 in transfected LLC-PK1 cells. Based on this hypothesis, claudin-14 might act as a negative regulator of divalent cation permeation [85]. Split-ubiquitin yeast 2-hybrid (Y2H) membrane protein interaction assay showed that claudin-14 interacts with the claudin-16 [79]. Other known interactions reported are with syntaxin-8 [102] and with PDZ domain containing RING finger 1 (encoded by *PDZRN*) [103] by electrophysiological experiments [104].

Mutations in human *CLDN16* cause an autosomal recessive disorder called *Familial hypomagnesemia with hypercalcinuria and nephrocalcinosis* (FHHNC) [105] (Figure 6). Other phenotypic features are incomplete distal tubular acidosis, impaired bone homeostasis [94], hypocitraturia and hyperuricemia which can be considered a secondary effect of renal insufficiency. To date, about 66 mutations including missense/nonsense, splicing, small deletion and small indels have been identified. Although still a matter of debate, it is believed that claudin-16 and claudin-19 form a heterodimer/tetramer essential for the divalent cation selectivity of the paracellular channels at the TAL and it has been demonstrated that some mutations disrupt this interaction [106].

Cldn16 deficient mice exhibited hypomagnesia and hypercalciuria and a low urinary pH but did not show nephrocalcinosis which could be explained by the upregulation of several genes (e.g., *Trpm6, Trpv5, Cnnm2*) involved in calcium and magnesium transport or the altered pH. Moreover, mice did not show renal insufficiency [13].

Besides, knockdown mice were generated by RNA interference technology [100] which phenocopied the main features of FHHNC including hypercalcinuria, hypomagnesemia, nephrocalcinosis and urinary Ca^{2+} and Mg^{2+} wasting without showing nephrocalcinosis or renal insufficiency. *Cldn16* KD animals showed increase of 1,25-dihydroxychlecalciferol [100].

For Cldn16 Patients, a genotype-phenotype correlation has been reported [94]. In sharp contrast to humans, although mice recapitulate many features of FHHNC like hypomagnesemia and hypercalciuria, they do not display nephrocalcinosis nor renal insufficiency [100]. Most of the patients with FHHNC nephrocalcinosis develop end stage renal disease with a need of renal transplantation [107]. Recently, amelogenesis imperfecta has been related to the absence of Claudin-16 in the ameloblasts in humans and mice [105] as an additional role of Cldn16 deficiency.

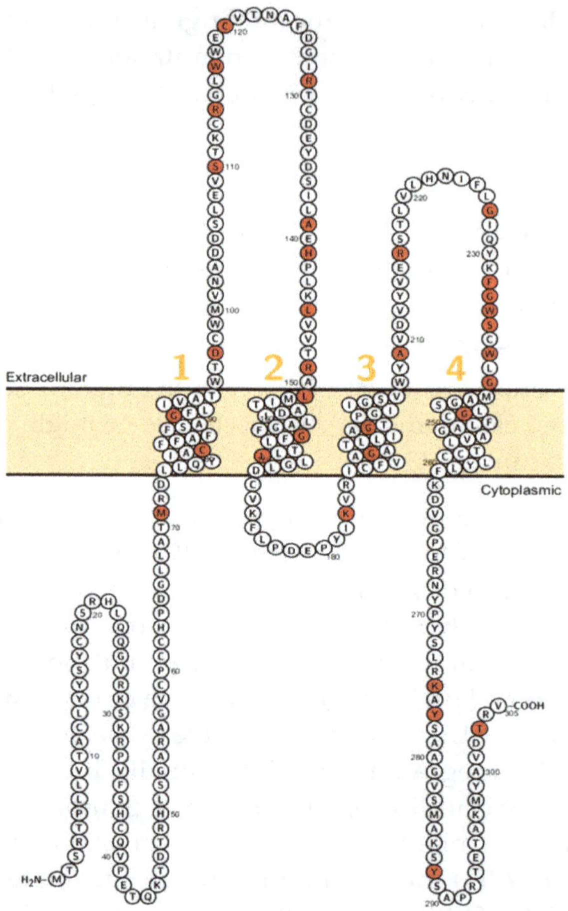

Figure 6. Topology of *CLDN16*. FASTA sequence of *CLDN16* (Uniprot accession no: Q9Y517). Each amino acid is shown as a single letter code and numbers (orange) indicate transmembrane domains. Mutations shown to be involved in human diseases are shown in red.

2.7. Claudin 19

The CLDN-19 gene encodes a 224 amino acid protein. Claudin-19 is a barrier forming claudin and is expressed in the thick ascending limb of the kidney and the retinal pigment epithelium (RPE) and the sheath of myelinated peripheral nerves. Additionally its expression was reported detected in the inner ear [38], stomach [108] and lung [109]. Like claudin-16, claudin-19 plays a major role in the permeability and selectivity of the TJ in TAL.

Claudin-19 interacts with claudin-16, both are involved in the reabsorption of divalent cations in the TAL [79,110]. Gong et al. isolated a stable dimer of claudin-16 with claudin-19 from transfected HEK293 cells and Sf9 cells [85]. The dimerization occurs through the cis-association of the third and the fourth transmembrane domain in both proteins [111]. This hypothesis is complemented by in vivo transgenic animal models, deletion of claudin-16 rendered claudin-19 delocalization from the TJ and vice versa [78]. Selective mutations disrupt the dimerization, triggering a loss of the transport function with FHHNC disease is the consequence [112].

Moreover, claudin-19 and ZO-1 are found by co-immunoprecipitation forming a complex in Madin-Darby canine kidney (MDCK) cells [97]. Meier et al. reported patients presenting by a phenotype of FHHNC disease but with the ocular disease [113,114]. All individuals of these families suffered from severe visual impairment, characterized by macular colobomata, significant myopia, and horizontal nystagmus. The genotype of these patients did not show a mutation in *CLDN16*, but Konrad and colleagues identified three mutations (G20D, Q57E) in *CLDN19*, recapitulating CLDN16 defects but, ocular defects, too. Other reported mutations (L90P and G123R) disrupt the interaction between claudin-16 and claudin-19 [23,112] (Figure 7).

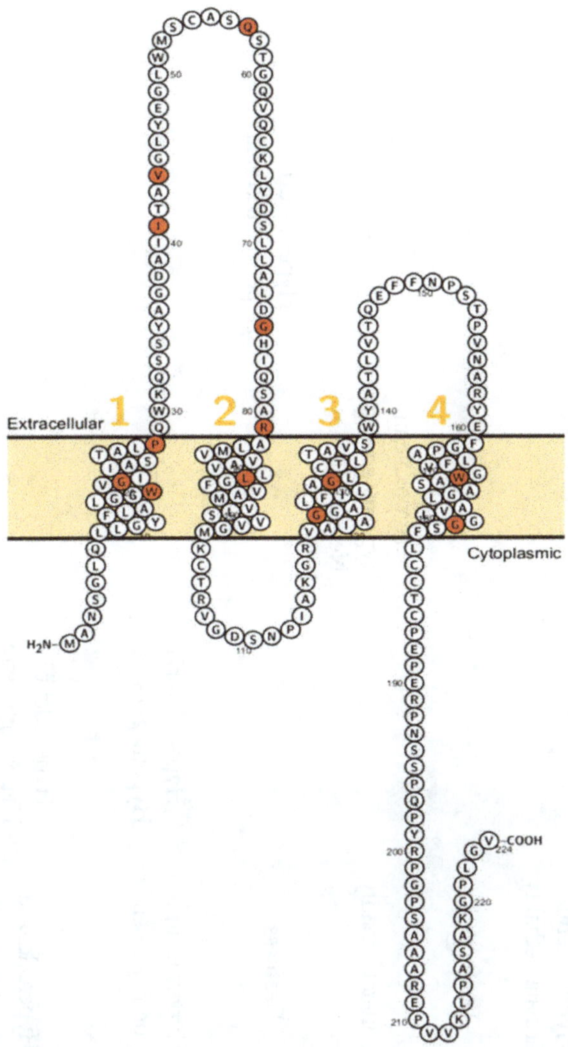

Figure 7. Topology of *CLDN19*. FASTA sequence of *CLDN19* (Uniprot accession no: Q8N6F1). Each amino acid is shown as a single letter code and numbers (orange) indicate transmembrane domains. Mutations shown to be involved in human diseases are shown in red.

Cldn19 KD mice showed a reduction of Mg^{2+} in plasma and excessive losses of Ca^{2+} and Mg^{2+} levels in kidney similar to *Cldn16* KO mice [78]. These mutations are the cause of FHHNC diseases, in which interaction between claudin-16 and-19 is disrupted. The dissociation of these proteins can trigger a loss of the transport function of them and cause FHHNC [112]. On the other hand, knockout mouse of *Cldn19* provokes renal reabsorption deficiency [79].

3. Discussion

Twenty years ago, the seminal work of Lifton's group described mutations in a gene coding for a TJ protein (initially named paracellin-1 and thereafter classified as claudin-16) as being causative for a human disorder (FHHNC) [99]. *CLDN16*, being the first claudin of a continuously expanding group of claudins causing human disorders, has been subject to various investigations on expression, function and also on possibilities of pharmaceutical interventions [115]. To gain further insights, mouse models, either as knockdown or knock out have been established (Table 1). Such models recapitulated some (Hypercalciuria, Hypomagnesemia) but not all (Nephrocalcinosis and renal insufficiency) of the hallmarks of human FHHNC, demonstrating the benefits but also the limitations of such models of TJ disorders. Recently the association of claudin-16 and amelogenesis imperfecta in men and mice has been shown [91].

Table 1. Overview of claudins causing human diseases and corresponding mouse models. The existence (by publication) is indicated by (+) while (−) indicates the absence of a corresponding model.

Human			Mouse			
Claudin	**Human Disorder**	**Hallmarks**	**Knockout**	**Knockdown**	**Conditional Knockout**	**Transgenic Overexpression**
1	NISCH Syndrome	Ichthyosis	Ichthyosis	Ichthyosis	+	+
2	Obstructive azoospermia	Male infertility	($Cldn2^{-/-}$) Larger intestine ($Cldn2^{-/-}$ $Cldn15^{-/-}$) Hypoglycemia Perinatal lethality	–	–	Larger colon
9	Nonsyndromic sensorioneural deafness	Deafness	Deafness	–	–	–
10	HELIX syndrome	Hypokalemia Hypermagnesemia Nephrocalcinosis	Perinatal lethality	–	Hypokalemi Hypermagnesemia Nephrocalcinosis	–
14	Nonsyndromic sensorioneural deafness	Deafness	Deafness	–	–	Hypercalcuria Hypermagnesemia
16	FHHNC	Hypercalciuria Renal insufficiency Hypomagnesemia Nephrocalcinosis	Hypercalciuria Hypomagnesemia	Hypercalciuria Hypomagnesemia	–	–
19	FHHNC + Eye Involvement	Hypercalciuria Renal insufficiency Hypomagnesemia Nephrocalcinosis Eye involvement	Hypercalciuria Hypomagnesemia	Hypercalciuria Hypomagnesemia Eye abnormalities	–	–

As the example of claudin-16 demonstrates, disorders that are caused by mutations in the corresponding claudin genes are not restricted to a given organ but rather should be considered as 'claudinopathies'. This view would not restrict the consequences to tissue-specific local or regional expression but appreciates the function and loss in all of the organs and tissues where a given claudin is expressed. Moreover, in a given organ, multiple different claudins are expressed and can cause, when mutated a similar phenotype albeit having different underlying pathophysiology. For example, kidney stones are a worldwide problem that affects 12% of the world population regardless of gender [116,117]. Besides *CLDN16* and *CLDN19*, in 2009, a genome-wide association study on an Icelandic and Dutch population revealed that a *CLDN14* variant (rs219780[C]) is associated with kidney stones too.

In this perspective, animal models, especially mice with all their limitations, are indispensable to study the consequences of claudin deficiency.

References

1. Daneman, R.; Prat, A. The blood-brain barrier. *Cold Spring Harb. Perspect. Biol.* **2015**, *7*, a020412. [CrossRef]
2. Baud, L. Renal Epithelial Cells: Differentiation and Plasticity. *J. Am. Soc. Nephrol.* **2003**, *14*, S1–S2. [CrossRef]
3. Macara, I.G.; Guyer, R.; Richardson, G.; Huo, Y.; Ahmed, S.M. Epithelial homeostasis. *Curr. Biol.* **2014**, *24*, R815–R825. [CrossRef]
4. Roignot, J.; Peng, X.; Mostov, K. Polarity in Mammalian Epithelial Morphogenesis. *Cold Spring Harb. Perspect. Biol.* **2013**, *5*, a013789. [CrossRef] [PubMed]
5. Karasov, W.H. Integrative physiology of transcellular and paracellular intestinal absorption. *J. Exp. Biol.* **2017**, *220*, 2495–2501. [CrossRef]
6. Sterling, T.M.; Nemere, I. Calcium uptake and membrane trafficking in response to PTH or 25(OH)D3 in polarized intestinal epithelial cells. *Steroids* **2007**, *72*, 151–157. [CrossRef] [PubMed]
7. Cunha, T.D.S.; Heilberg, I.P. Bartter syndrome: Causes, diagnosis, and treatment. *Int. J. Nephrol. Renov. Dis.* **2018**, *11*, 291–301. [CrossRef]
8. Knoers, N.V.; Levtchenko, E.N. Gitelman syndrome. *Orphanet J. Rare Dis.* **2008**, *3*, 22. [CrossRef]
9. Furuse, M.; Hata, M.; Furuse, K.; Yoshida, Y.; Haratake, A.; Sugitani, Y.; Noda, T.; Kubo, A.; Tsukita, S. Claudin-based tight junctions are crucial for the mammalian epidermal barrier: A lesson from claudin-1-deficient mice. *J. Cell Biol.* **2002**, *156*, 1099–1111. [CrossRef]
10. Nakano, Y.; Kim, S.H.; Kim, H.M.; Sanneman, J.D.; Zhang, Y.; Smith, R.J.; Marcus, D.C.; Wangemann, P.; Nessler, R.A.; Banfi, B. A claudin-9-based ion permeability barrier is essential for hearing. *PLoS Genet.* **2009**, *5*, e1000610. [CrossRef]
11. Breiderhoff, T.; Himmerkus, N.; Stuiver, M.; Mutig, K.; Will, C.; Meij, I.C.; Bachmann, S.; Bleich, M.; Willnow, T.E.; Müller, D. Deletion of claudin-10 (Cldn10) in the thick ascending limb impairs paracellular sodium permeability and leads to hypermagnesemia and nephrocalcinosis. *Proc. Natl. Acad. Sci. USA* **2012**, *109*, 14241–14246. [CrossRef] [PubMed]
12. Ben-Yosef, T.; Belyantseva, I.A.; Saunders, T.L.; Hughes, E.D.; Kawamoto, K.; Van Itallie, C.M.; Beyer, L.A.; Halsey, K.; Gardner, D.J.; Wilcox, E.R.; et al. Claudin 14 knockout mice, a model for autosomal recessive deafness DFNB29, are deaf due to cochlear hair cell degeneration. *Hum. Mol. Genet.* **2003**, *12*, 2049–2061. [CrossRef]
13. Will, C.; Breiderhoff, T.; Thumfart, J.; Stuiver, M.; Kopplin, K.; Sommer, K.; Gunzel, D.; Querfeld, U.; Meij, I.C.; Shan, Q.; et al. Targeted deletion of murine Cldn16 identifies extra- and intrarenal compensatory mechanisms of Ca^{2+} and Mg^{2+} wasting. *Am. J. Physiol. Ren. Physiol.* **2010**, *298*, F1152–F1161. [CrossRef] [PubMed]
14. Tsukita, S.; Furuse, M. The structure and function of claudins, cell adhesion molecules at tight junctions. *Ann. N. Y. Acad. Sci.* **2000**, *915*, 129–135. [CrossRef]
15. Günzel, D.; Yu, A.S.L. Claudins and the Modulation of Tight Junction Permeability. *Physiol. Rev.* **2013**, *93*, 525–569. [CrossRef] [PubMed]
16. Krause, G.; Winkler, L.; Mueller, S.L.; Haseloff, R.F.; Piontek, J.; Blasig, I.E. Structure and function of claudins. *Biochim. Biophys. Acta* **2008**, *1778*, 631–645. [CrossRef]

17. Sanjad, S.A.; Hariri, A.; Habbal, Z.M.; Lifton, R.P. A novel PCLN-1 gene mutation in familial hypomagnesemia with hypercalciuria and atypical phenotype. *Pediatr. Nephrol.* **2007**, *22*, 503–508. [CrossRef]
18. Hadj-Rabia, S.; Baala, L.; Vabres, P.; Hamel-Teillac, D.; Jacquemin, E.; Fabre, M.; Lyonnet, S.; De Prost, Y.; Munnich, A.; Hadchouel, M.; et al. Claudin-1 gene mutations in neonatal sclerosing cholangitis associated with ichthyosis: A tight junction disease. *Gastroenterology* **2004**, *127*, 1386–1390. [CrossRef]
19. Sineni, C.J.; Yildirim-Baylan, M.; Guo, S.; Camarena, V.; Wang, G.; Tokgoz-Yilmaz, S.; Duman, D.; Bademci, G.; Tekin, M. A truncating CLDN9 variant is associated with autosomal recessive nonsyndromic hearing loss. *Hum. Genet.* **2019**, *138*, 1071–1075. [CrossRef]
20. Hadj-Rabia, S.; Brideau, G.; Al-Sarraj, Y.; Maroun, R.C.; Figueres, M.-L.; Leclerc-Mercier, S.; Olinger, E.; Baron, S.; Chaussain, C.; Nochy, D.; et al. Multiplex epithelium dysfunction due to CLDN10 mutation: The HELIX syndrome. *Genet. Med.* **2017**, *20*, 190–201. [CrossRef]
21. Wilcox, E.R.; Burton, Q.L.; Naz, S.; Riazuddin, S.; Smith, T.N.; Ploplis, B.; Belyantseva, I.; Ben-Yosef, T.; Liburd, N.A.; Morell, R.J.; et al. Mutations in the Gene Encoding Tight Junction Claudin-14 Cause Autosomal Recessive Deafness DFNB29. *Cell* **2001**, *104*, 165–172. [CrossRef]
22. Hampson, G.; A Konrad, M.; Scoble, J. Familial hypomagnesaemia with hypercalciuria and nephrocalcinosis (FHHNC): Compound heterozygous mutation in the claudin 16 (CLDN16) gene. *BMC Nephrol.* **2008**, *9*, 12. [CrossRef] [PubMed]
23. Konrad, M.; Schaller, A.; Seelow, D.; Pandey, A.V.; Waldegger, S.; Lesslauer, A.; Vitzthum, H.; Suzuki, Y.; Luk, J.M.; Becker, C.; et al. Mutations in the tight-junction gene claudin 19 (CLDN19) are associated with renal magnesium wasting, renal failure, and severe ocular involvement. *Am. J. Hum. Genet.* **2006**, *79*, 949–957. [CrossRef] [PubMed]
24. De Santis, S.; Cavalcanti, E.; Mastronardi, M.; Jirillo, E.; Chieppa, M. Nutritional Keys for Intestinal Barrier Modulation. *Front. Immunol.* **2015**, *6*, 612. [CrossRef] [PubMed]
25. Merikallio, H.; Kaarteenaho, R.; Pääkkö, P.; Lehtonen, S.; Hirvikoski, P.; Mäkitaro, R.; Harju, T.; Soini, Y. Impact of smoking on the expression of claudins in lung carcinoma. *Eur. J. Cancer* **2011**, *47*, 620–630. [CrossRef] [PubMed]
26. Ramalingam, A.; Wang, X.; Gabello, M.; Valenzano, M.C.; Soler, A.P.; Ko, A.; Morin, P.J.; Mullin, J.M. Dietary methionine restriction improves colon tight junction barrier function and alters claudin expression pattern. *Am. J. Physiol. Physiol.* **2010**, *299*, C1028–C1035. [CrossRef]
27. Soini, Y. Claudins in lung diseases. *Respir. Res.* **2011**, *12*, 70. [CrossRef]
28. Barmeyer, C.; Schulzke, J.D.; Fromm, M. Claudin-related intestinal diseases. *Semin. Cell Dev. Biol.* **2015**, *42*, 30–38. [CrossRef]
29. Limaye, A.; Hall, B.; Kulkarni, A.B. Manipulation of mouse embryonic stem cells for knockout mouse production. *Curr. Protoc. Cell Biol.* **2009**, *44*, 19.13.1–19.13.24.
30. Kooij, G.; Kopplin, K.; Blasig, R.; Stuiver, M.; Koning, N.; Goverse, G.; van der Pol, S.M.; van Het Hof, B.; Gollasch, M.; Drexhage, J.A.; et al. Disturbed function of the blood-cerebrospinal fluid barrier aggravates neuro-inflammation. *Acta Neuropathol.* **2014**, *128*, 267–277. [CrossRef]
31. Adli, M. The CRISPR tool kit for genome editing and beyond. *Nat. Commun.* **2018**, *9*, 1911. [CrossRef] [PubMed]
32. Daugherty, B.L.; Ward, C.; Smith, T.; Ritzenthaler, J.D.; Koval, M. Regulation of Heterotypic Claudin Compatibility. *J. Biol. Chem.* **2007**, *282*, 30005–30013. [CrossRef] [PubMed]
33. Piontek, J.; Fritzsche, S.; Cording, J.; Richter, S.; Hartwig, J.; Walter, M.; Yu, D.; Turner, J.R.; Gehring, C.; Rahn, H.-P.; et al. Elucidating the principles of the molecular organization of heteropolymeric tight junction strands. *Cell. Mol. Life Sci.* **2011**, *68*, 3903–3918. [CrossRef] [PubMed]
34. Zorko, M.S.; Veraniä, P.; Leskovec, N.K.; Pavloviä, M.D.; Pavlović, M.D.; Veranič, P. Expression of tight-junction proteins in the inflamed and clinically uninvolved skin in patients with venous leg ulcers. *Clin. Exp. Dermatol.* **2009**, *34*, e949–e952. [CrossRef] [PubMed]
35. Reyes, J.L.; Lamas, M.; Martin, D.; Namorado, M.D.C.; Islas, S.; Luna, J.; Tauc, M.; Gonzalez-Mariscal, L.; Gonz, L. The renal segmental distribution of claudins changes with development. *Kidney Int.* **2002**, *62*, 476–487. [CrossRef]
36. Laurila, J.J.; Karttunen, T.; Koivukangas, V.; Laurila, P.A.; Syrjala, H.; Saarnio, J.; Soini, Y.; Ala-Kokko, T.I. Tight junction proteins in gallbladder epithelium: Different expression in acute acalculous and calculous cholecystitis. *J. Histochem. Cytochem.* **2007** *55*, 567–573. [CrossRef]

37. Zhu, Y.; Brannstrom, M.; Janson, P.O.; Sundfeldt, K. Differences in expression patterns of the tight junction proteins, claudin 1, 3, 4 and 5, in human ovarian surface epithelium as compared to epithelia in inclusion cysts and epithelial ovarian tumours. *Int. J. Cancer* **2006**, *118*, 1884–1891. [CrossRef]
38. Kitajiri, S.-I.; Furuse, M.; Morita, K.; Saishin-Kiuchi, Y.; Kido, H.; Ito, J.; Tsukita, S. Expression patterns of claudins, tight junction adhesion molecules, in the inner ear. *Hear. Res.* **2004**, *187*, 25–34. [CrossRef]
39. Mori, M.; Nakagawa, S.; Miyoshi, N.; Ishii, H.; Mimori, K.; Tanaka, F.; Sekimoto, M.; Doki, Y. Expression of CLDN1 in colorectal cancer: A novel marker for prognosis. *Int. J. Oncol.* **2011**, *39*, 791–796. [CrossRef]
40. Zhang, Z.; Wang, A.; Sun, B.; Zhan, Z.; Chen, K.; Wang, C. Expression of CLDN1 and CLDN10 in lung adenocarcinoma in situ and invasive lepidic predominant adenocarcinoma. *J. Cardiothorac. Surg.* **2013**, *8*, 95. [CrossRef]
41. Zhang, W.-N.; Li, W.; Wang, X.-L.; Hu, Z.; Zhu, D.; Ding, W.-C.; Liu, D.; Li, K.-Z.; Ma, D.; Wang, H. CLDN1 expression in cervical cancer cells is related to tumor invasion and metastasis. *Oncotarget* **2016**, *7*, 87449–87461. [CrossRef] [PubMed]
42. Zhou, S.; Piao, X.; Wang, C.; Wang, R.; Song, Z. Identification of claudin1, 3, 7 and 8 as prognostic markers in human laryngeal carcinoma. *Mol. Med. Rep.* **2019**, *20*, 393–400. [PubMed]
43. Evans, M.J.; Von Hahn, T.; Tscherne, D.M.; Syder, A.J.; Panis, M.; Wölk, B.; Hatziioannou, T.; McKeating, J.A.; Bieniasz, P.D.; Rice, C.M. Claudin-1 is a hepatitis C virus co-receptor required for a late step in entry. *Nature* **2007**, *446*, 801–805. [CrossRef]
44. Baala, L.; Hadj-Rabia, S.; Hamel-Teillac, D.; Hadchouel, M.; Prost, C.; Leal, S.M.; Jacquemin, E.; Sefiani, A.; De Prost, Y.; Courtois, G.; et al. Homozygosity mapping of a locus for a novel syndromic ichthyosis to chromosome 3q27-q28. *J. Investig. Dermatol.* **2002**, *119*, 70–76. [CrossRef] [PubMed]
45. Kirchmeier, P.; Sayar, E.; Hotz, A.; Hausser, I.; Islek, A.; Yilmaz, A.; Artan, R.; Fischer, J. Novel mutation in the CLDN1 gene in a Turkish family with neonatal ichthyosis sclerosing cholangitis (NISCH) syndrome. *Br. J. Dermatol.* **2014**, *170*, 976–978. [CrossRef] [PubMed]
46. Vreeburg, M.; Wagner, A.; Van Geel, M.; Nagtzaam, I.F.; Peeters, V.P.M.; Steijlen, P.M.; Van Steensel, M.A.M. Novel CLDN1 mutation in ichthyosis-hypotrichosis-sclerosing cholangitis syndrome without signs of liver disease. *Br. J. Dermatol.* **2018**, *178*, e202–e203.
47. Tokumasu, R.; Yamaga, K.; Yamazaki, Y.; Murota, H.; Suzuki, K.; Tamura, A.; Bando, K.; Furuta, Y.; Katayama, I.; Tsukita, S. Dose-dependent role of claudin-1 in vivo in orchestrating features of atopic dermatitis. *Proc. Natl. Acad. Sci. USA* **2016**, *113*, E4061–E4068. [CrossRef]
48. Hirano, T.; Yokouchi, M.; Atsugi, T.; Amagai, M.; Kubo, A. Epidermis-specific ablation of claudin-1 in adult mice demonstrates the essential role of a tight junction barrier in skin homeostasis. *J. Dermatol. Sci.* **2016**, *84*, e37. [CrossRef]
49. Furuse, M.; Fujita, K.; Hiiragi, T.; Fujimoto, K.; Tsukita, S. Claudin-1 and -2: Novel integral membrane proteins localizing at tight junctions with no sequence similarity to occludin. *J. Cell Biol.* **1998**, *141*, 1539–1550. [CrossRef]
50. Amasheh, S.; Meiri, N.; Gitter, A.H.; Schöneberg, T.; Mankertz, J.; Schulzke, J.D.; Fromm, M. Claudin-2 expression induces cation-selective channels in tight junctions of epithelial cells. *J. Cell Sci.* **2002**, *115*, 4969–4976. [CrossRef]
51. Lamas, M.; González-Mariscal, L.; Gutiérrez, R. Presence of claudins mRNA in the brain. Selective modulation of expression by kindling epilepsy. *Mol. Brain Res.* **2002**, *104*, 250–254. [CrossRef]
52. Enck, A.H.; Berger, U.V.; Yu, A.S.L. Claudin-2 is selectively expressed in proximal nephron in mouse kidney. *Am. J. Physiol. Physiol.* **2001**, *281*, F966–F974. [CrossRef]
53. Guan, X.; Inai, T.; Shibata, Y. Segment-specific expression of tight junction proteins, claudin-2 and -10, in the rat epididymal epithelium. *Arch. Histol. Cytol.* **2005**, *68*, 213–225. [CrossRef] [PubMed]
54. Kinugasa, T.; Huo, Q.; Higashi, D.; Shibaguchi, H.; Kuroki, M.; Tanaka, T.; Futami, K.; Yamashita, Y.; Hachimine, K.; Maekawa, S.; et al. Selective up-regulation of claudin-1 and claudin-2 in colorectal cancer. *Anticancer Res.* **2007**, *27*, 3729–3734. [CrossRef]
55. Oshima, T.; Miwa, H.; Joh, T. Changes in the expression of claudins in active ulcerative colitis. *J. Gastroenterol. Hepatol.* **2008**, *23*, S146–S150. [CrossRef]
56. Szakál, D.N.; Győrffy, H.; Arató, A.; Cseh, Á.; Molnár, K.; Papp, M.; Dezsofi, A.; Veres, G. Mucosal expression of claudins 2, 3 and 4 in proximal and distal part of duodenum in children with coeliac disease. *Virchows Archiv.* **2010**, *456*, 245–250. [CrossRef]

57. Weber, C.R.; Nalle, S.C.; Tretiakova, M.; Rubin, D.T.; Turner, J.R. Claudin-1 and claudin-2 expression is elevated in inflammatory bowel disease and may contribute to early neoplastic transformation. *Lab. Investig.* **2008**, *88*, 1110–1120. [CrossRef]

58. Kim, T.H.; Huh, J.H.; Lee, S.; Kang, H.; I Kim, G.; An, H.J. Down-regulation of claudin-2 in breast carcinomas is associated with advanced disease. *Histopathology* **2008**, *53*, 48–55. [CrossRef]

59. Zhang, X.; Wang, H.; Li, Q.; Li, T. CLDN2 inhibits the metastasis of osteosarcoma cells via down-regulating the afadin/ERK signaling pathway. *Cancer Cell Int.* **2018**, *18*, 160. [CrossRef]

60. Martínez, C.; Rodiño-Janeiro, B.K.; Lobo, B.; Stanifer, M.L.; Klaus, B.; Granzow, M.; González-Castro, A.M.; Salvo-Romero, E.; Alonso-Cotoner, C.; Pigrau, M.; et al. miR-16 and miR-125b are involved in barrier function dysregulation through the modulation of claudin-2 and cingulin expression in the jejunum in IBS with diarrhoea. *Gut* **2017**, *66*, 1537–1538. [CrossRef]

61. Muto, S.; Hata, M.; Taniguchi, J.; Tsuruoka, S.; Moriwaki, K.; Saitou, M.; Furuse, K.; Sasaki, H.; Fujimura, A.; Imai, M.; et al. Claudin-2–deficient mice are defective in the leaky and cation-selective paracellular permeability properties of renal proximal tubules. *Proc. Natl. Acad. Sci. USA* **2010**, *107*, 8011–8016. [CrossRef] [PubMed]

62. Matsumoto, K.; Imasato, M.; Yamazaki, Y.; Tanaka, H.; Watanabe, M.; Eguchi, H.; Nagano, H.; Hikita, H.; Tatsumi, T.; Takehara, T.; et al. Claudin 2 Deficiency Reduces Bile Flow and Increases Susceptibility to Cholesterol Gallstone Disease in Mice. *Gastroenterology* **2014**, *147*, 1134–1145.e10. [CrossRef] [PubMed]

63. Wada, M.; Tamura, A.; Takahashi, N.; Tsukita, S. Loss of claudins 2 and 15 from mice causes defects in paracellular Na+ flow and nutrient transport in gut and leads to death from malnutrition. *Gastroenterology* **2013**, *144*, 369–380. [CrossRef] [PubMed]

64. Ahmad, R.; Chaturvedi, R.; Olivares-Villagómez, D.; Habib, T.; Asim, M.; Shivesh, P.; Polk, D.B.; Wilson, K.T.; Washington, M.K.; Van Kaer, L.; et al. Targeted colonic claudin-2 expression renders resistance to epithelial injury, induces immune suppression, and protects from colitis. *Mucosal Immunol.* **2014**, *7*, 1340–1353. [CrossRef] [PubMed]

65. Dubé, É.; Dufresne, J.; Chan, P.T.; Hermo, L.; Cyr, D.G. Assessing the Role of Claudins in Maintaining the Integrity of Epididymal Tight Junctions Using Novel Human Epididymal Cell Lines1. *Biol. Reprod.* **2010**, *82*, 1119–1128. [CrossRef]

66. Zheng, A.; Yuan, F.; Li, Y.; Zhu, F.; Hou, P.; Li, J.; Song, X.; Ding, M.; Deng, H. Claudin-6 and Claudin-9 Function as Additional Coreceptors for Hepatitis C Virus. *J. Virol.* **2007**, *81*, 12465–12471. [CrossRef]

67. Abuazza, G.; Becker, A.; Williams, S.S.; Chakravarty, S.; Truong, H.-T.; Lin, F.; Baum, M. Claudins 6, 9, and 13 are developmentally expressed renal tight junction proteins. *Am. J. Physiol. Physiol.* **2006**, *291*, F1132–F1141. [CrossRef]

68. Fofana, I.; Zona, L.; Thumann, C.; Heydmann, L.; Durand, S.C.; Lupberger, J.; Blum, H.E.; Pessaux, P.; Gondeau, C.; Reynolds, G.M.; et al. Functional Analysis of Claudin-6 and Claudin-9 as Entry Factors for Hepatitis C Virus Infection of Human Hepatocytes by Using Monoclonal Antibodies. *J. Virol.* **2013**, *87*, 10405–10410. [CrossRef]

69. Zavala-Zendejas, V.E.; Torres-Martinez, A.C.; Salas-Morales, B.; Fortoul, T.I.; Montano, L.F.; Rendon-Huerta, E.P. Claudin-6, 7, or 9 overexpression in the human gastric adenocarcinoma cell line AGS increases its invasiveness, migration, and proliferation rate. *Cancer Investig.* **2011**, *29*, 1–11. [CrossRef]

70. Sharma, R.K.; Chheda, Z.S.; Das Purkayastha, B.P.; Gomez-Gutierrez, J.G.; Jala, V.R.; Haribabu, B. A spontaneous metastasis model reveals the significance of claudin-9 overexpression in lung cancer metastasis. *Clin. Exp. Metastasis* **2016**, *33*, 263–275. [CrossRef]

71. Zhu, J.; Wang, R.; Cao, H.; Zhang, H.; Xu, S.; Wang, A.; Liu, B.; Wang, Y.; Wang, R. Expression of claudin-5, -7, -8 and -9 in cervical carcinoma tissues and adjacent non-neoplastic tissues. *Int. J. Clin. Exp. Pathol.* **2015**, *8*, 9479–9486. [PubMed]

72. Birkenhäger, R.; Otto, E.; Schürmann, M.J.; Vollmer, M.; Ruf, E.-M.; Maier-Lutz, I.; Beekmann, F.; Fekete, A.; Omran, H.; Feldmann, D.; et al. Mutation of BSND causes Bartter syndrome with sensorineural deafness and kidney failure. *Nat. Genet.* **2001**, *29*, 310–314. [CrossRef] [PubMed]

73. Van Itallie, C.M.; Rogan, S.; Yu, A.; Vidal, L.S.; Holmes, J.; Anderson, J.M. Two splice variants of claudin-10 in the kidney create paracellular pores with different ion selectivities. *Am. J. Physiol. Physiol.* **2006**, *291*, F1288–F1299. [CrossRef] [PubMed]

74. Günzel, D.; Stuiver, M.; Kausalya, P.J.; Haisch, L.; Krug, S.; Rosenthal, R.; Meij, I.C.; Hunziker, W.; Fromm, M.;

Muller, D. Claudin-10 exists in six alternatively spliced isoforms that exhibit distinct localization and function. *J. Cell Sci.* **2009**, *122*, 1507–1517. [CrossRef]

75. Zemke, A.C.; Snyder, J.C.; Brockway, B.L.; Drake, J.A.; Reynolds, S.D.; Kaminski, N.; Stripp, B.R. Molecular staging of epithelial maturation using secretory cell-specific genes as markers. *Am. J. Respir. Cell Mol. Biol.* **2009**, *40*, 340–348. [CrossRef]
76. Hashizume, A.; Ueno, T.; Furuse, M.; Tsukita, S.; Nakanishi, Y.; Hieda, Y. Expression patterns of claudin family of tight junction membrane proteins in developing mouse submandibular gland. *Dev. Dyn.* **2004**, *231*, 425–431. [CrossRef]
77. Jakab, C.; Halász, J.; Szasz, A.M.; Batmunkh, E.; Kiss, A.; Schaff, Z.; Rusvai, M.; Galfi, P.; Kulka, J. Expression and localisation of claudin-1,-2,-3,-4,-5,-7 and-10 proteins in the normal canine mammary gland. *Acta Vet. Hung.* **2008**, *56*, 341–352. [CrossRef]
78. Hou, J.; Renigunta, A.; Gomes, A.S.; Hou, M.; Paul, D.L.; Waldegger, S.; Goodenough, D.A. Claudin-16 and claudin-19 interaction is required for their assembly into tight junctions and for renal reabsorption of magnesium. *Proc. Natl. Acad. Sci. USA* **2009**, *106*, 15350–15355. [CrossRef]
79. Milatz, S.; Himmerkus, N.; Wulfmeyer, V.C.; Drewell, H.; Mutig, K.; Hou, J.; Breiderhoff, T.; Muller, D.; Fromm, M.; Bleich, M.; et al. Mosaic expression of claudins in thick ascending limbs of Henle results in spatial separation of paracellular Na^+ and Mg^{2+} transport. *Proc. Natl. Acad. Sci. USA* **2017**, *114*, e219–e227. [CrossRef]
80. Schumann, S.; Buck, V.U.; Classen-Linke, I.; Wennemuth, G.; Grümmer, R. Claudin-3, claudin-7, and claudin-10 show different distribution patterns during decidualization and trophoblast invasion in mouse and human. *Histochem. Cell Biol.* **2015**, *144*, 571–585. [CrossRef]
81. Klar, J.; Piontek, J.; Milatz, S.; Tariq, M.; Jameel, M.; Breiderhoff, T.; Schuster, J.; Fatima, A.; Asif, M.; Sher, M.; et al. Altered paracellular cation permeability due to a rare CLDN10B variant causes anhidrosis and kidney damage. *PLoS Genet.* **2017**, *13*, e1006897. [CrossRef] [PubMed]
82. Meyers, N.; Nelson-Williams, C.; Malaga-Dieguez, L.; Kaufmann, H.; Loring, E.; Knight, J.; Lifton, R.P.; Trachtman, H. Hypokalemia Associated with a Claudin 10 Mutation: A Case Report. *Am. J. Kidney Dis.* **2019**, *73*, 425–428. [CrossRef] [PubMed]
83. Breiderhoff, T.; Himmerkus, N.; Drewell, H.; Plain, A.; Gunzel, D.; Mutig, K.; Willnow, T.E.; Muller, D.; Bleich, M. Deletion of claudin-10 rescues claudin-16-deficient mice from hypomagnesemia and hypercalciuria. *Kidney Int.* **2018**, *93*, 580–588. [CrossRef] [PubMed]
84. Gao, M.; Li, W.; Wang, H.; Wang, G. The distinct expression patterns of claudin-10, -14, -17 and E-cadherin between adjacent non-neoplastic tissues and gastric cancer tissues. *Diagn. Pathol.* **2013**, *8*, 205. [CrossRef] [PubMed]
85. Gong, Y.; Renigunta, V.; Himmerkus, N.; Zhang, J.; Renigunta, A.; Bleich, M.; Hou, J. Claudin-14 regulates renal Ca^{++} transport in response to CaSR signalling via a novel microRNA pathway. *EMBO J.* **2012**, *31*, 1999–2012. [CrossRef] [PubMed]
86. Lee, K.; Ansar, M.; Andrade, P.B.; Khan, B.; Santos-Cortez, R.L.P.; Ahmad, W.; Leal, S.M. Novel CLDN14 mutations in Pakistani families with autosomal recessive non-syndromic hearing loss. *Am. J. Med Genet. Part A* **2012**, *158*, 315–321. [CrossRef]
87. Wattenhofer, M.; Reymond, A.; Falciola, V.; Charollais, A.; Caille, D.; Borel, C.; Lyle, R.; Estivill, X.; Petersen, M.B.; Meda, P.; et al. Different mechanisms preclude mutant CLDN14 proteins from forming tight junctions in vitro. *Hum. Mutat.* **2005**, *25*, 543–549. [CrossRef]
88. Gong, Y.; Hou, J. Claudin-14 underlies Ca(+)(+)-sensing receptor-mediated Ca(+)(+) metabolism via NFAT-microRNA-based mechanisms. *J. Am. Soc. Nephrol.* **2014**, *25*, 745–760. [CrossRef]
89. Pater, J.A.; Benteau, T.; Griffin, A.; Penney, C.; Stanton, S.G.; Predham, S.; Kielley, B.; Squires, J.; Zhou, J.; Li, Q.; et al. A common variant in CLDN14 causes precipitous, prelingual sensorineural hearing loss in multiple families due to founder effect. *Hum. Genet.* **2017**, *136*, 107–118. [CrossRef]
90. Hou, J. Claudins and mineral metabolism. *Curr. Opin. Nephrol. Hypertens.* **2016**, *25*, 308–313. [CrossRef]
91. Bardet, C.; Courson, F.; Wu, Y.; Khaddam, M.; Salmon, B.; Ribes, S.; Thumfart, J.; Yamaguti, P.M.; Rochefort, G.Y.; Figueres, M.L.; et al. Claudin-16 Deficiency Impairs Tight Junction Function in Ameloblasts, Leading to Abnormal Enamel Formation. *J. Bone Miner. Res.* **2016**, *31*, 498–513. [CrossRef] [PubMed]
92. Kriegs, J.O.; Homann, V.; Kinne-Saffran, E.; Kinne, R.K.H. Identification and subcellular localization of paracellin-1 (claudin-16) in human salivary glands. *Histochem. Cell Biol.* **2007**, *128*, 45–53. [CrossRef]

93. Runggaldier, D.; Pradas, L.G.; Neckel, P.H.; Mack, A.F.; Hirt, B.; Gleiser, C. Claudin expression in the rat endolymphatic duct and sac—First insights into regulation of the paracellular barrier by vasopressin. *Sci. Rep.* **2017**, *7*, 45482. [CrossRef] [PubMed]
94. Weber, S.; Schlingmann, K.P.; Peters, M.; Nejsum, L.N.; Nielsen, S.; Engel, H.; Grzeschik, K.H.; Seyberth, H.W.; Gröne, H.J.; Nüsing, R.; et al. Primary gene structure and expression studies of rodent paracellin-1. *J. Am. Soc. Nephrol.* **2001**, *12*, 2664–2672. [PubMed]
95. Rangel, L.B.; Sherman-Baust, C.A.; Wernyj, R.P.; Schwartz, D.R.; Cho, K.R.; Morin, P.J. Characterization of novel human ovarian cancer-specific transcripts (HOSTs) identified by serial analysis of gene expression. *Oncogene* **2003**, *22*, 7225–7232. [CrossRef] [PubMed]
96. Fluge, Ø.; Bruland, O.; Akslen, L.A.; Lillehaug, J.R.; Varhaug, J.E. Gene Expression in Poorly Differentiated Papillary Thyroid Carcinomas. *Thyroid* **2006**, *16*, 161–175.
97. Lee, N.P.; Tong, M.K.; Leung, P.P.; Chan, V.W.; Leung, S.; Tam, P.-C.; Chan, K.-W.; Lee, K.-F.; Yeung, W.S.; Luk, J.M. Kidney claudin-19: Localization in distal tubules and collecting ducts and dysregulation in polycystic renal disease. *FEBS Lett.* **2006**, *580*, 923–931. [CrossRef]
98. Martin, T.A.; Harrison, G.M.; Watkins, G.; Jiang, W.G. Claudin-16 reduces the aggressive behavior of human breast cancer cells. *J. Cell. Biochem.* **2008**, *105*, 41–52. [CrossRef] [PubMed]
99. Simon, D.B.; Lu, Y.; Choate, K.A.; Velazquez, H.; Al-Sabban, E.; Praga, M.; Casari, G.; Bettinelli, A.; Colussi, G.; Rodriguez-Soriano, J.; et al. Paracellin-1, a renal tight junction protein required for paracellular Mg^{2+} resorption. *Science* **1999**, *285*, 103–106. [CrossRef]
100. Hou, J.; Shan, Q.; Wang, T.; Gomes, A.S.; Yan, Q.; Paul, D.L.; Bleich, M.; Goodenough, D.A. Transgenic RNAi Depletion of Claudin-16 and the Renal Handling of Magnesium. *J. Biol. Chem.* **2007**, *282*, 17114–17122. [CrossRef]
101. Kausalya, P.J.; Amasheh, S.; Gunzel, D.; Wurps, H.; Muller, D.; Fromm, M.; Hunziker, W. Disease-associated mutations affect intracellular traffic and paracellular Mg^{2+} transport function of Claudin-16. *J. Clin. Investig.* **2006**, *116*, 878–891. [CrossRef] [PubMed]
102. Ikari, A.; Hirai, N.; Shiroma, M.; Harada, H.; Sakai, H.; Hayashi, H.; Suzuki, Y.; Degawa, M.; Takagi, K. Association of Paracellin-1 with ZO-1 Augments the Reabsorption of Divalent Cations in Renal Epithelial Cells. *J. Biol. Chem.* **2004**, *279*, 54826–54832. [CrossRef] [PubMed]
103. Marunaka, K.; Furukawa, C.; Fujii, N.; Kimura, T.; Furuta, T.; Matsunaga, T.; Endo, S.; Hasegawa, H.; Anzai, N.; Yamazaki, Y.; et al. The RING finger- and PDZ domain-containing protein PDZRN3 controls localization of the Mg(2+) regulator claudin-16 in renal tube epithelial cells. *J. Biol. Chem.* **2017**, *292*, 13034–13044. [CrossRef]
104. Hou, J.; Renigunta, V.; Nie, M.; Sunq, A.; Himmerkus, N.; Quintanova, C.; Bleich, M.; Renigunta, A.; Wolf, M.T.F. Phosphorylated claudin-16 interacts with Trpv5 and regulates transcellular calcium transport in the kidney. *Proc. Natl. Acad. Sci. USA* **2019**, *116*, 19176–19186. [CrossRef] [PubMed]
105. Gimenez-Mascarell, P.; Schirrmacher, C.E.; Martínez-Cruz, L.A.; Muller, D. Novel Aspects of Renal Magnesium Homeostasis. *Front. Pediatr.* **2018**, *6*, 77. [CrossRef]
106. Yamaguti, P.M.; Dos Santos, P.A.C.; Leal, B.S.; Santana, V.B.B.D.M.; Mazzeu, J.F.; Acevedo, A.C.; Neves, F.D.A.R. Identification of the first large deletion in the CLDN16 gene in a patient with FHHNC and late-onset of chronic kidney disease: Case report. *BMC Nephrol.* **2015**, *16*, 92. [CrossRef]
107. Konrad, M.; Hou, J.; Weber, S.; Dötsch, J.; Kari, J.A.; Seeman, T.; Kuwertz-Bröking, E.; Peco-Antic, A.; Tasic, V.; Dittrich, K.; et al. CLDN16 genotype predicts renal decline in familial hypomagnesemia with hypercalciuria and nephrocalcinosis. *J. Am. Soc. Nephrol.* **2008**, *19*, 171–181. [CrossRef]
108. Niimi, T.; Nagashima, K.; Ward, J.M.; Minoo, P.; Zimonjic, D.B.; Popescu, N.C.; Kimura, S. claudin-18, a Novel Downstream Target Gene for the T/EBP/NKX2.1 Homeodomain Transcription Factor, Encodes Lung- and Stomach-Specific Isoforms through Alternative Splicing. *Mol. Cell. Biol.* **2001**, *21*, 7380–7390. [CrossRef]
109. Wolburg, H.; Wolburg-Buchholz, K.; Liebner, S.; Engelhardt, B. Claudin-1, claudin-2 and claudin-11 are present in tight junctions of choroid plexus epithelium of the mouse. *Neurosci. Lett.* **2001**, *307*, 77–80. [CrossRef]
110. Fromm, M.; Piontek, J.; Rosenthal, R.; Gunzel, D.; Krug, S.M. Tight junctions of the proximal tubule and their channel proteins. *Pflüg. Arch.* **2017**, *469*, 877–887. [CrossRef]
111. Gong, Y.; Renigunta, V.; Zhou, Y.; Sunq, A.; Wang, J.; Yang, J.; Renigunta, A.; Baker, L.A.; Hou, J. Biochemical

and biophysical analyses of tight junction permeability made of claudin-16 and claudin-19 dimerization. *Mol. Biol. Cell* **2015**, *26*, 4333–4346. [CrossRef] [PubMed]
112. Hou, J.; Renigunta, A.; Konrad, M.; Gomes, A.S.; Schneeberger, E.E.; Paul, D.L.; Waldegger, S.; Goodenough, D.A. Claudin-16 and claudin-19 interact and form a cation-selective tight junction complex. *J. Clin. Investig.* **2008**, *118*, 619–628. [CrossRef] [PubMed]
113. Meier, W.; Blumberg, A.; Imahorn, W.; De Luca, F.; Wildberger, H.; Oetliker, O. Idiopathic hypercalciuria with bilateral macular colobomata: A new variant of oculo-renal syndrome. *Helvetica Paediatr. Acta* **1979**, *34*, 257–269.
114. Rodriguez-Soriano, J.; Vallo, A. Pathophysiology of the renal acidification defect present in the syndrome of familial hypomagnesaemia-hypercalciuria. *Pediatr. Nephrol.* **1994**, *8*, 431–435. [CrossRef] [PubMed]
115. Takahashi, A.; Kondoh, M.; Suzuki, H.; Yagi, K. Claudin as a target for drug development. *Curr. Med. Chem.* **2011**, *18*, 1861–1865. [CrossRef] [PubMed]
116. Pawar, A.S.; Thongprayoon, C.; Cheungpasitporn, W.; Sakhuja, A.; Mao, M.A.; Erickson, S.B. Incidence and characteristics of kidney stones in patients with horseshoe kidney: A systematic review and meta-analysis. *Urol. Ann.* **2018**, *10*, 87–93.
117. Alelign, T.; Petros, B. Kidney Stone Disease: An Update on Current Concepts. *Adv. Urol.* **2018**, *2018*, 3068365. [CrossRef]

9

Antagonistic Effects of IL-4 on IL-17A-Mediated Enhancement of Epidermal Tight Junction Function

Matthew G. Brewer [1], Takeshi Yoshida [1], Fiona I. Kuo [1], Sade Fridy [1], Lisa A. Beck [1] and Anna De Benedetto [1,2,*]

[1] Department of Dermatology, University of Rochester Medical Center, Rochester, NY 14642, USA
[2] Department of Dermatology, College of Medicine University of Florida, Gainesville, FL 32610, USA
* Correspondence: adebenedetto@ufl.edu

Abstract: Atopic dermatitis (AD) is the most common chronic and relapsing inflammatory skin disease. AD is typically characterized by skewed T helper (Th) 2 inflammation, yet other inflammatory profiles (Th1, Th17, Th22) have been observed in human patients. How cytokines from these different Th subsets impact barrier function in this disease is not well understood. As such, we investigated the impact of the canonical Th17 cytokine, IL-17A, on barrier function and protein composition in primary human keratinocytes and human skin explants. These studies demonstrated that IL-17A enhanced tight junction formation and function in both systems, with a dependence on STAT3 signaling. Importantly, the Th2 cytokine, IL-4 inhibited the barrier-enhancing effect of IL-17A treatment. These observations propose that IL-17A helps to restore skin barrier function, but this action is antagonized by Th2 cytokines. This suggests that restoration of IL-17/IL-4 ratio in the skin of AD patients may improve barrier function and in so doing improve disease severity.

Keywords: atopic dermatitis; tight junctions; cytokines; claudin; STAT3

---

1. Introduction

Atopic dermatitis (AD) is a chronic, inflammatory skin disorder characterized by an overactive immune response to a host of environmental allergens and dry/itchy skin. Over the last decade, important discoveries have demonstrated that AD develops in part from genetic, and/or acquired defects in the skin barrier. The overall hypothesis is that skin barrier defects allow resident immune cells to gain access to microbes, allergens, and irritants, resulting in a robust type 2 immune response that can be characterized by the tissue recruitment of T helper (Th) 2 cells. Recently, other Th subtypes have been identified in acute (e.g., Th22/Th17) and chronic (e.g., Th22 and Th1) AD lesions, although the importance of these subsets in AD pathogenesis is still unclear [1–3].

Epidermal barrier is mediated by two structural components, the stratum corneum (SC) and tight junctions (TJs). TJs, found in the stratum granulosum, seal intercellular spaces and the 'tightness' of this structure constitutes a key physiological and dynamic function of the skin. The essential role that epidermal TJs play in maintaining the skin barrier was demonstrated by the claudin (CLDN) 1 deficient mouse, which suffers from extensive transepidermal water loss and succumbs shortly after birth to severe dehydration [4]. We and others have demonstrated functional defects in epidermal TJs from skin samples of AD subjects, which may in part be due to the reduced expression of two transmembrane TJ components, CLDN1 and CLDN23 [5–7]. Kast et al. reported that CLDN1 is the most abundant CLDN family member in the skin and in primary human keratinocytes (PHK), with lower levels of the CLDN family members 4 and 25 occurring, while CLDN 3, 5, 7, 8, 10, and 23 were only present in the skin (but not detected in primary culture models) [8].

Altered TJ protein expression has been observed in several human skin disorders including psoriasis, lichen planus, ichthyosis vulgaris, skin wounds, skin infections and in response to ultraviolet exposure; however, the functional consequences of these changes have not been fully evaluated yet [9]. Additionally, very little is known about what regulates epidermal TJ function and claudin expression in immune-mediated skin disorders. Several T-cell derived cytokines were found in AD skin (e.g., IL-4, IL-13, IL-17A, IL-22, IL-25, or TNF-α) and these cytokines were shown to reduce the expression of several SC barrier proteins such as filaggrin, loricrin, S100A11, and involucrin [10–12]. In one study, TNF-α was shown to alter TJ protein expression and localization in a skin explant model, as well as TJ function in cultured keratinocytes [9]. Another study showed contradictory results, with no effect observed with either TNF-α or IL-4 treatment, while IL-17 diminished and IL-22 enhanced TJ function in cultured keratinocytes, as measured by transepithelial electric resistance (TEER) [6]. Importantly, in vivo studies have strongly suggested that IL-17A has barrier promoting properties. This was highlighted by increased intestinal permeability in mice lacking IL-17A as well as spontaneous development of allergic skin inflammation in mice lacking the IL-17RA [13,14]. To begin to clarify the effects that AD related cytokines play in barrier function of the skin, we analyzed how the Th17 cytokine, IL-17A, influences TJ barrier properties in cultured keratinocytes and human skin explants and how Th2 cytokines may affect that response.

2. Results

2.1. IL-17A Enhances Epidermal TJ Barrier Integrity

PHK were exposed to cytokines throughout the differentiation process, thus modeling the effect they would have on TJ assembly during wound repair. In our submerged culture model, we observed that TJ barrier function, quantified by TEER, increases to a peak of approximately 2500 ohms × cm^2 around 120 ± 24 h after addition of differentiation media [15]. Notably, in our model we observed that as TEER increases, the immunoreactivity of TJ components changes from a dotted, disrupted pattern to a continuous "chicken wire" membranous staining typically observed in mature TJ of the human epidermis (Figure S1).

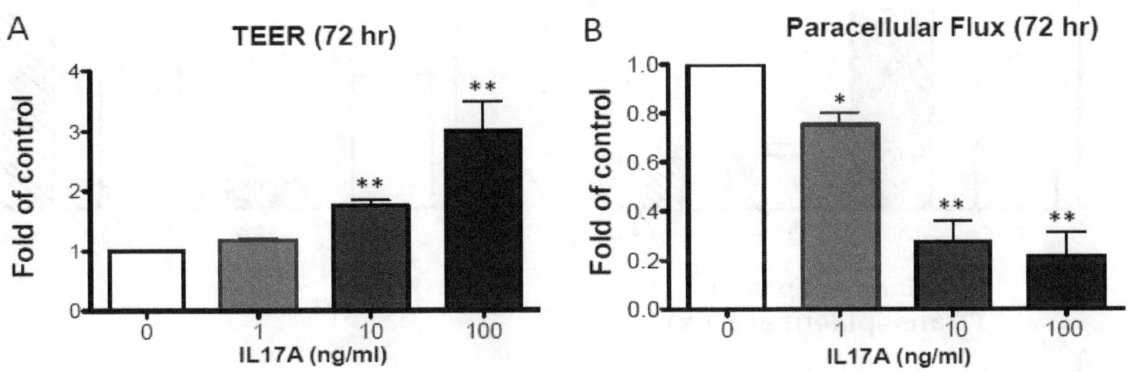

Figure 1. IL-17A enhances epidermal TJ barrier integrity. In PHK, IL-17A dose-dependently (**A**) enhanced TEER (n = 3–6) and (**B**) reduced paracellular flux of fluorescein (n = 3) 72 h after cytokine treatment. Data are shown as mean ± SEM fold of control. Significance was calculated compared to untreated controls. * $p \leq 0.05$, ** $p \leq 0.05$.

When PHK were differentiated in the presence of IL-17A, we observed a dose-dependent enhancement of the TJ barrier function by 72 h (Figure 1). This was demonstrated by a significant enhancement of TEER in IL-17A treated PHK (72 h, $p \leq 0.01$, n = 3–6; Figure 1A) and a reduction in the paracellular fluorescein flux (72 h, $p \leq 0.05$, n = 3; Figure 1B). To further validate our findings, we tested the effect of IL-17A on human epidermal samples in an ex vivo model. We tape stripped (15x) discarded healthy human skin, as a model of mechanical disruption and then isolated the epidermis and monitored epidermal TJ recovery with and without IL-17A stimulation. To do this we used a modified micro-Snapwell™ system, previously developed to study TJ function in intestinal epithelium [16], and adapted in our laboratory for epidermal sheets [17]. We observed a significant reduction in the transepidermal fluorescein flux ($p \leq 0.02$, $n = 3$; Figure 2C) after IL-17A treatment (24 h), thus confirming our barrier-enhancing findings in submerged PHK. Altogether, these observations demonstrate that IL-17A enhances the development of TJ epidermal barrier function.

2.2. IL-4 Inhibits IL-17A-Mediated TJ Barrier Enhancement

In contrast to the IL-17A effects on TEER in our model, IL-4 did not significantly alter TJ integrity of cultured PHK monolayers. Previous studies have demonstrated that Th2 cytokines antagonize IL-17A-induced production of antimicrobial peptides and S100A8 in human keratinocytes [18–20]. Therefore, we tested whether Th2 cytokines also inhibit IL-17A barrier-enhancing effects. When PHK were treated with both IL-17A and IL-4 we observed that the enhanced TEER and reduced paracellular fluorescein flux observed in response to IL-17A were completely inhibited by co-treatment with IL-4 (50 ng/mL; respectively $p = 0.024$ and $p = 0.002$, Figure 2A,B). Of note, we observed a slight, but not significant ($p = 0.08$) increase in paracellular fluorescein flux with IL-4 treatment at 72 h (Figure 2B), suggesting this cytokine might have an effect on TJ pore size. We again validated our finding in an ex vivo model, co-treatment with IL-4 was able to block the IL-17A induced increase in permeability flux of fluorescein (Figure 2C). To determine whether IL-4 blocks other known IL-17A mediated effects in our in vitro model we measured the expression of S100A7, a well-known downstream product of IL-17A signaling [21]. S100A7, also known as psoriasin, is a calcium-binding protein with chemotactic and antimicrobial properties that is expressed in AD lesions and even more so in psoriasis skin, a Th1/Th17 driven disease [22]. IL-17A-mediated S100A7 expression in PHK was blocked by co-stimulation with IL-4 (Figure S2).

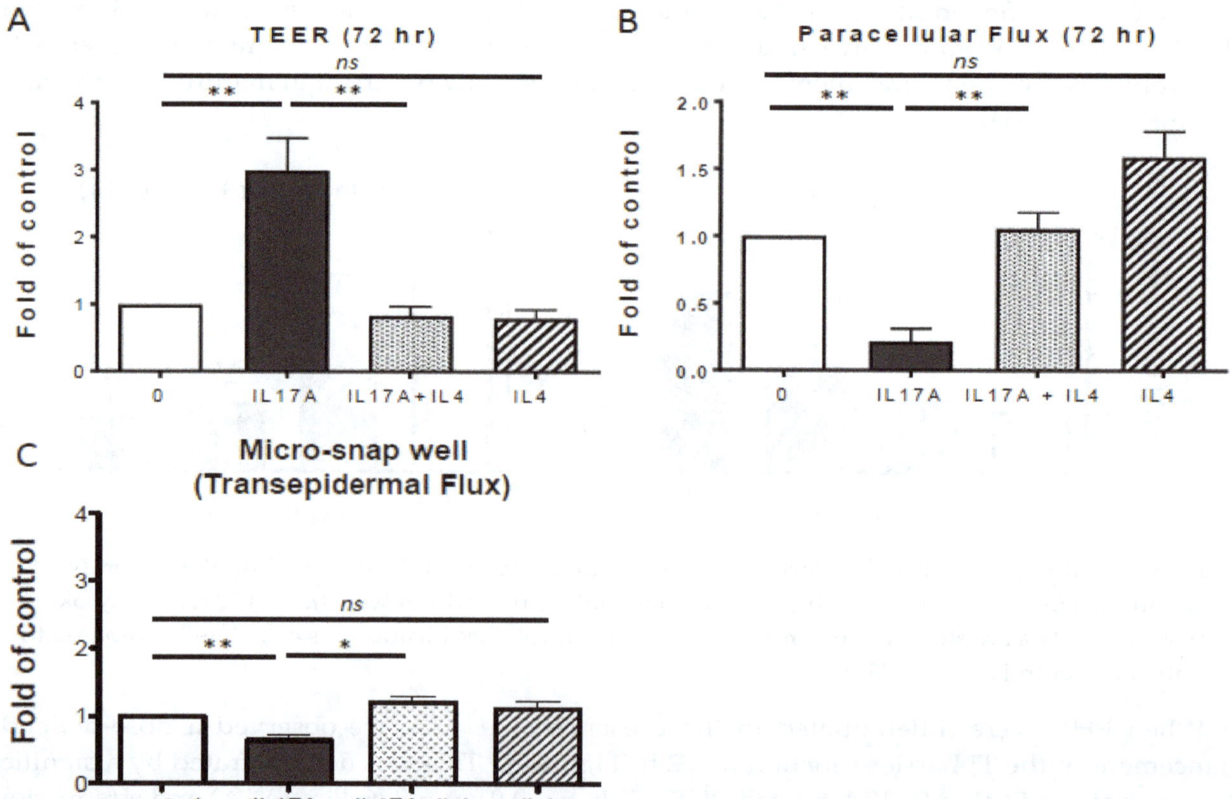

Figure 2. IL-4 inhibits IL-17A mediated barrier enhancement in PHK. Co-treatment with IL-4 (50 ng/mL) (**A**) inhibited IL-17A (100 ng/mL) increased TEER and (**B**) enhanced paracellular flux ($n = 3$–5). (**C**) IL-17A treatment reduced transepidermal flux of fluorescein (50 ng/mL 0.02% fluorescein; $n = 3$) in tape-stripped human skin samples. Data are shown as mean ± SEM fold of control. * $p \leq 0.05$, ** $p \leq 0.01$, ns: not significant.

2.3. *Inhibition of STAT3 Activation Reduces IL-17A-Induced TEER*

It has been suggested that Janus kinases (JAK) and mitogen-activated protein kinases (MAPK) are downstream signaling pathways activated by IL-17A [23]. Therefore, we examined if JAK and MAPK activation are involved in IL-17A mediated actions on epidermal TJ function. A major cytosolic target of JAK signaling is the phosphorylation and nuclear translocation of signal transducer and activator of transcription 3 (STAT3). In our model, we confirmed that IL-17A enhanced STAT3 activation in PHK, as demonstrated by increased STAT3 phosphorylation at amino acid Y705 (Figure 3A), suggesting the importance of the JAK-STAT3 pathway in the barrier-enhancing effect of IL-17A. Using a pan-JAK inhibitor (JAK_{Tot}) that blocks all four JAK isoforms (JAK inhibitor I; Calbiochem, 10 μM) a significant decrease in the IL-17A-dependent increase of TEER was observed in our PHK model (IL-17A vs JAK_{Tot} + IL-17A, $p = 0.025$; $n = 4$; Figure 3B). By contrast, no inhibition of IL-17A-enhanced TEER was observed from PHK treated with an upstream MAPK inhibitor (PD98056; data not shown). Treatment with JAK_{Tot} completely blocked both baseline and IL-17A-dependent STAT3 activation (Figure 3A). Notably, we observed that IL-4 reduced STAT3 phosphorylation both at baseline and in response to IL-17A treatment (Figure 3C), suggesting that IL-4's attenuation of the IL-17A barrier-enhancing effect may be mediated by the JAK-STAT3 pathway.

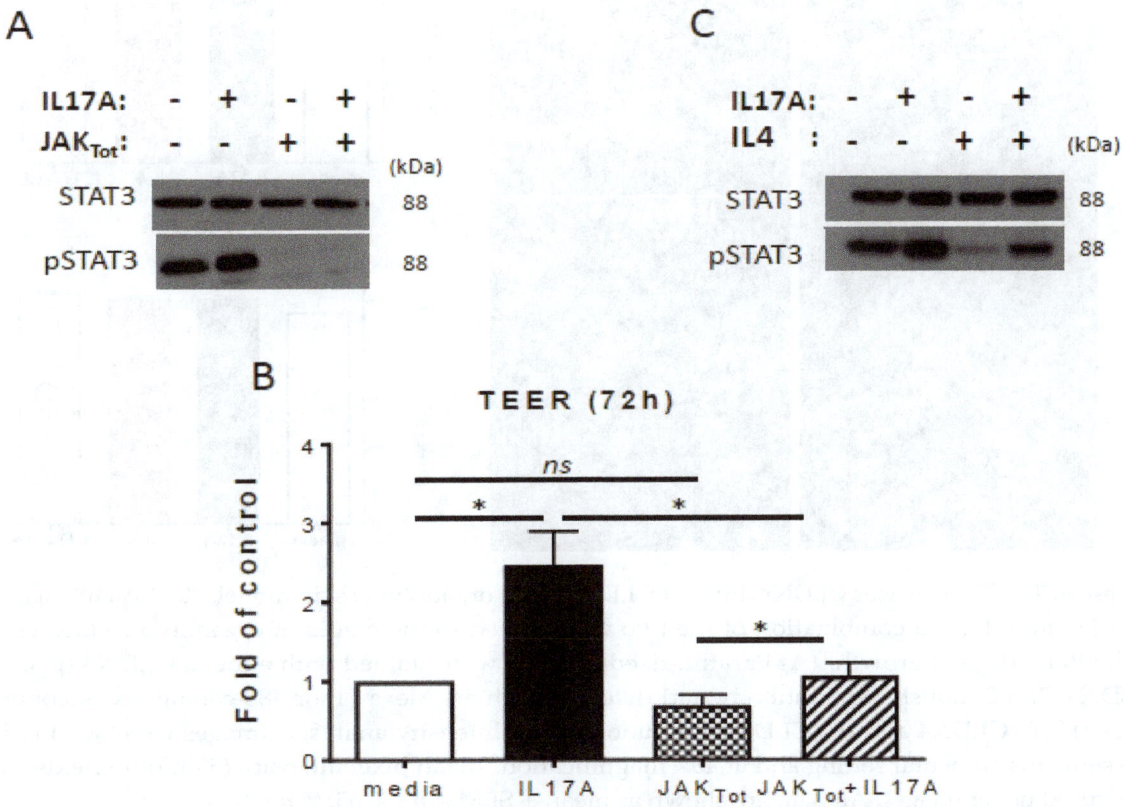

Figure 3. A pan-JAK inhibitor blocks IL-17A-induced TEER enhancement. (**A**) A representative Western blot demonstrates that pan-JAK inhibition blocks STAT3 activation (pSTAT3; EP2147Y), but has no effect on total STAT3 expression (STAAD22A) (**B**) Pretreatment with a pan-JAK inhibitor (JAKtot: 10 μM; $n = 4$) inhibited IL-17A (100 ng/mL) induced TEER in primary human keratinocytes. (**C**) IL-4 blocked STAT3 phosphorylation in PHK treated with IL-17A or media alone. Representative Western blot bands of $n = 3$ experiments are shown. All data are shown as mean ± SEM fold of control. * $p \leq 0.05$, ns: not significant.

2.4. *IL-17A Enhances Claudin-4 Expression*

To further investigate the mechanisms responsible for IL-17A mediated barrier enhancement, we quantified the expression of two highly expressed epidermal claudins, CLDN1 and CLDN4, in both submerged PHK and in the epidermal organotypic model. This latter model is three-dimensional and more closely mimics mature human epidermis with a well-differentiated, multilayered structure that includes both a stratum granulosum (where TJs reside) and SC layers. Using this model, we detected increased CLDN4 immunoreactivity in the presence of IL-17A (100 ng/mL; $p = 0.04$) (Figure 4A,B). IL-4 in contrast slightly, but significantly reduced CLDN4 expression compared to the media alone (50 ng/mL; $p = 0.04$). Additionally, IL-17A-induced CLDN4 expression was blocked by IL-4 co-treatment ($p = 0.009$; Figure 4A,B). The expression of CLDN1 was not affected by either cytokine alone, or in combination (Figure 4C). The TJ findings in the submerged PHK culture were in line with those observed in the organotypic model. IL-4 stimulation alone and in combination with IL-17A reduced CLDN4 expression as detected by Western blot analysis while CLDN1 expression was unaffected by cytokine treatment (Figure S3).

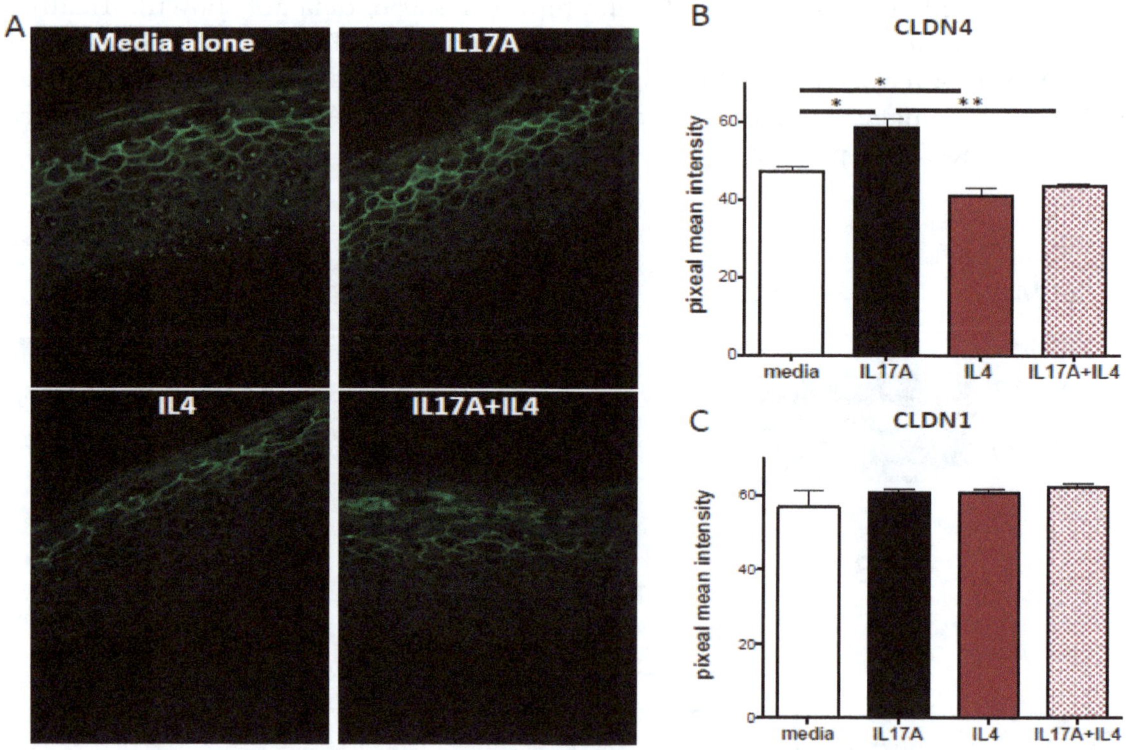

Figure 4. IL-17A increases CLDN4, but not CLDN1 in an organotypic skin model. IL-17A (100 ng/mL), IL-4 (50 ng/mL) or a combination of the two were added to the media of organotypic cultures for 72 h after 9 days of growth. (**A**) Paraffin-fixed sections were stained with either a CLDN4 (pAb) or CLDN1 (JAY.8; not shown) antibody and detected with an Alexa Fluor 488-conjugated secondary (green). (**B**) CLDN4 and (**C**) CLDN1 immunostaining intensity analysis. Images were acquired at the same instrumental setting and at 40× magnification. Mean pixel intensity (5 random fields) was measured using ImageJ. All data are shown as mean ± SEM. * $p \leq 0.05$, ** $p \leq 0.01$.

3. Discussion

There is no doubt that type 2 immunity plays a central role in atopic diseases, including AD. Less clear is the contribution of other cytokines, such as IL-17A, in AD pathogenesis. We have now unveiled a novel role for IL-17A in repairing and/or enhancing TJ epidermal barrier function. In this work, we demonstrated that IL-17A enhances TJ integrity at an early time point in differentiating PHK (Figure 1) and this was associated with enhanced expression of CLDN4 (Figure 4).

Notably, IL-17A effects may be inhibited in AD because the antagonistic effect of the Th2 cytokines in the tissue. Previous studies showed Th2 cytokines can reduce IL-17A expression by Th17 cells, as well as IL-17A-induced production of antimicrobial peptides, S100A7 and S100A8 by human keratinocytes [18–20,24]. We have added another example to strengthen the paradigm that it is the relative expression of IL-4 and IL-17A that determines the biological consequences of either of these cytokines (Figure 2). IL-17A's ability to induce antimicrobial peptides as well as recruit neutrophils strongly suggests that this cytokine plays a crucial protective role at epithelial sites. Clinical trials testing the effect of biological therapeutics targeting the Th1/Th17 pathways have been conducted in AD. Anecdotally, good responses in some AD patients were reported with Ustekinumab, an antibody that blocks Th1/Th23 signaling and thus reduces Th17 activation. However, in two independent, randomized, placebo-controlled, phase II studies Ustekinumab did not demonstrate meaningful efficacy in AD. Another trial with Secukinumab that targets the cytokine, IL-17A, was conducted in AD; however, results are not available yet. This trial will be extremely informative to better understand the role of specific cytokines in AD.

We demonstrated in this study that blocking STAT3 activation via a pan-JAK inhibitor was associated with loss of IL-17A-mediated TJ function enhancement (Figure 3). Remarkably, we observed that the IL-4 inhibitory effect on IL-17A's ability to enhance barrier was partly mediated by reduced STAT3 activation. Since another IL-17A downstream protein S100A7 (also STAT3 dependent) was inhibited by IL-4 co-stimulation, this observation further supports that IL-4 blocks IL-17A signaling at the level of transcriptional control. Activation of STAT3 has been shown to play a key role in many cellular processes such as cell growth and apoptosis. Indeed, STAT3 knockout mice are not viable [25]. K5-Cre/STAT3 transgenic mice, whose epidermal and follicular keratinocytes lack STAT3, were viable and the development of the epidermis and hair follicles appeared normal. However, hair cycle and wound healing processes were severely compromised [26]. When STAT3 is inactivated in intestinal epithelial cells through lentiviral transduction of a dominant negative STAT3 protein, barrier is severely impacted [27]. STAT3 activation was also shown to increase permeability in human retinal endothelial cells [28]. Contrasting with our data, another group showed that baseline STAT3 activation was not reduced when PHK were treated with IL-4 and -13 together [29]. This could be explained because IL-13 can also induce phosphorylation of STAT3 in PHK [30]. In humans, STAT3 mutations cause a form of the Hyper IgE syndrome (AD-HIES; OMIM: 147060). Severe AD-like phenotype, increased susceptibility to S. aureus infections and eosinophilia are some of the clinical features of this disease, in addition to reduced peripheral Th17 cells (STAT3 is involved in Th17 differentiation cells) [31]. However, involvement of the STAT3 pathway in TJ function and composition in the skin has not been directly evaluated in these patients.

An alternative and complementary, approach to investigate the effect of cytokines on TJ function is to directly look at the expression and localization of key TJ components. In this study, we investigated the effect of the IL-17A with or without IL-4 on the expression of CLDN1 and 4, the two most abundant CLDNs in human keratinocytes [8]. Our studies demonstrated that CLDN4, but not CLDN1 expression is enhanced by IL-17A, and this enhancement is prevented by co-treatment with IL-4 (Figure 4). Previous work by Kirschner et al. showed that silencing CLDN4 in PHK reduced TEER and enhanced permeability to small macromolecules (4 kDa dextran-FITC conjugated). CLDN23 is another TJ-associated molecule we hypothesized to be important in barrier function since it is down-regulated (mRNA) in the skin of AD patients [5]. Currently, the reagents for this protein are poorly validated in human skin, which is why it was not investigated. Future studies will have to clarify the specific effects of reduced expression of each of the claudins in AD subjects

Our findings in an epidermal in vitro model are in contrast with previous data showing IL-17A decreased TEER in ocular epithelial cells, suggesting impaired barrier function capabilities [32]. Gutowska-Owsiak D. et al. reported findings from a microarray study in HaCaT cells showing that IL-17A treatment (20 or 50 ng/mL for 24 h) reduced expression of several intercellular junction components, implying IL-17A could reduce keratinocyte barrier function at the level of the SC, but a link to TJs was not suggested [21]. It is important to point out that studying the mechanisms regulating epithelial TJ function and composition is made difficult by pleiotropic cytokine actions, in addition to temporal and dose-dependent variation in cell culture systems [33]. Additionally, the study of epidermal TJs is further complicated by different culture conditions employed, which also typically results in variable levels of differentiation. To minimize many of these sources of variability we chose to focus our studies on submerged keratinocyte cultures because they form robust barrier at the level of the TJ (Figure S1), which we believe more faithfully mimics the epithelium. Air liquid cultures cornify which is ideal for studying the SC, but importantly they do not form robust TJ, making this an unreliable model to study the effects of cytokines on barrier formation. This biological complexity might explain the discrepancy of data published so far in this field and as such requires further validation in vivo.

Dupilumab (Regeneron, Tarrytown, NY, USA & Sanofi, Paris, France) a fully human monoclonal antibody against the Th2 cytokine receptor IL-4Rα, is the first biologic medication approved worldwide for the treatment of AD. This has been a revolutionary treatment option for patients with AD, with a good safety profile [34]. This fully humanized monoclonal antibody targets the common receptor subunit (IL-4Rα) used by both IL-4 and IL-13 and thereby blocks signaling. We hypothesize that some of the clinical improvement and normalization of the epithelial transcriptome may be due to diminished antagonism of IL-17A. A better understanding of the complex interactions between relevant Th cytokines found in AD lesions and how they affect epidermal barrier function will be critical for a more thorough understanding of AD pathogenesis. The development and study of highly targeted AD therapies will likely provide an opportunity to look at this issue and allow us to tailor therapies to individual patient's tissue inflammation.

4. Methods

4.1. Isolation and Culture of Primary Human Foreskin Keratinocytes

Human keratinocytes were isolated from neonatal foreskin as previously described [35]. PHK were cultured in Keratinocyte-SFM (Thermo Fisher Scientific, Waltham, MA, USA) with 1% Pen/Strep, 0.2% Amphotericin B (Thermo Fisher Scientific, Waltham, MA, USA). To differentiate PHK, cells were grown in DMEM (11965092, Thermo Fisher Scientific, Waltham, MA, USA) with 1% Pen/Strep, 0.2% Amphotericin B. For cytokine experiments, the following reagents were added alone or in combination to the culture media from the time of differentiation and replaced with every 48-h media change: human IL-4 (5–50 ng/mL; R&D system, Minneapolis, MN, USA), human IL-13 (5–50 ng/mL; R&D system, Minneapolis, MN, USA), human IL-17A (1–100 ng/mL; R&D system), JAK inhibitor I (10 μM; Calbiochem, San Diego, CA, USA), and PD98056 (10 μM; Calbiochem, San Diego, CA, USA). For the epidermal organotypic model experiment, keratinocytes were grown as previously described [36]. For all cytokine treatment studies, keratinocytes were starved of growth factors for 24 h before stimulation with the indicated cytokines.

4.2. Transepithelial Electric Resistance (TEER) and Paracellular Flux

TJ integrity in cultured PHK was evaluated using two complementary assays, TEER and paracellular flux, as previously described [5]. Briefly, TEER was measured using an EVOMX voltohmmeter (World Precision Instruments, Sarasota, FL, USA). To evaluate the paracellular flux across PHK cultures, 0.02% Fluorescein (FITC) Sodium (Sigma, St. Louis, MO, USA) in HBSS was added to the upper chamber. Samples were collected from the lower chamber at different times (0.5–3 h). Paracellular permeability was presented as "Relative Fluorescence Flux" = experimental condition/media alone × 100.

We used the Micro-Snapwell™ barrier function assay (Corning Incorporated, Corning, NY, USA) to determine the permeability of ex vivo human skin. The protocol was adapted from previously described work [16]. Full thickness epidermis was isolated from discarded, deidentified skin samples. To disrupt skin barrier, the epidermis was taped stripped 15 times and then removed using a Weck blade (Goulian Skin Graft Knife Set, George Tiemann & Co. Hauppauge, NY, USA) and washed in PBS. Skin samples were mounted on filter supports (Whatman Nuclepore Track-Etch Membrane; GE Healthcare Biosciences, Pittsburgh, PA, USA). The epidermal side was faced upward and sandwiched between two custom made Plexiglas discs containing an opening of 3 mm. This sandwich was then placed in the modified Snapwell™ chambers. Samples were immersed in DMEM media and kept at

37 °C, 5% CO_2 for 30 min. Media was added to samples and to test the effect of IL-17A, 50 ng/mL of the cytokine was incorporated into the media, which was then added to both sides of the transwell. Paracellular flux of fluorescein into the lower levels of human epidermis was measured at 24 h later as described above. The Pathology Department at the University of Rochester Medical Center provided the discarded human skin, which was approved by the institutional Research Subject Review Board (RSRB 00042616; approved on 31 May 2012).

4.3. *Immunoblotting*

PHK were placed on ice in RIPA lysis buffer (20 mM Tris, 50mM NaCl, 2mM EDTA, 2mM EGTA, 1% sodium deoxycholate, 1% TX-100, and 2% SDS, pH 7.4) with a 100-fold dilution of both Protease Inhibitor Cocktail (Sigma, St. Louis, MO, USA) and Phosphatase Inhibitor Cocktail (Sigma, St. Louis, MO, USA) for 30 min. The samples were boiled for 10 min and then centrifuged for 15 min. Forty μg of protein, as determined by the BCA (Thermo Fisher Scientific, Waltham, MA, USA) protein quantification assay, were mixed in NuPage LDS Sample Buffer (Thermo Fisher Scientific, Waltham, MA, USA) and applied to 4-12% NuPage Bis-Tris gels (Thermo Fisher Scientific, Waltham, MA, USA). Electrophoresis was performed under reducing conditions with MES SDS Buffer (Thermo Fisher Scientific, Waltham, MA, USA). Membranes were incubated in blocking solution, which contained 5% non-fat dry milk + 0.05% Tween-20 in PBS, for 1 h at RT and then incubated with primary antibodies: claudin-4 (ab53156; Abcam, Cambridge, UK), claudin-1 (JAY.8; Thermo Fisher Scientific, Waltham, MA, USA) STAT3 (STAAD22A; Abcam, Cambridge, UK), pSTAT3 (EP2147Y, phospho-Y705; Abcam, Cambridge, UK), S100A7 (47c1068; Thermo Fisher Scientific, Waltham, MA, USA) and β-Actin (C4; Santa Cruz, Dallas, TX, USA). HRP-linked secondary antibody (GE Healthcare Biosciences, Pittsburgh, PA, USA) was used to visualize protein staining in tandem with ECL reagent (GE Healthcare Biosciences, Pittsburgh, PA, USA) by autoradiography with Kodak BioMax MR film (Kodak, Rochester, NY, USA).

4.4. *Immunostaining*

PHK were grown on a cover glass were fixed in methanol at −20 °C for 15 min, followed by blocking with 1% BSA in PBS and immunolabeled with the above pSTAT3 antibody for 2.5 h at 4 °C. This was followed by a 1-h incubation with secondary antibodies. A 1:1000 dilution of Alexa Fluor 568 donkey-anti-mouse IgG H+L (Thermo Fisher Scientific, Waltham, MA, USA) and 1:10,000 4′,6-diamidino-2-phenyl-indole, dihydrochloride (DAPI) (Molecular Probe, Eugene, OR, USA) was used. Fluorescent images were taken using an Olympus FV1000 laser scanning confocal microscope with the FV10-ASW 1.7 software (Olympus, Tokyo, Japan). The saturation level of fluorescence intensity was determined for the controls using the Hi-Lo feature of the Fluoview software. In each experiment, the images captured used identical settings during acquisition and no further manipulation of the images occurred before into importing into the FV1000 software. The organotypic samples were fixed in formalin then embedded in paraffin. Five μm sections were rehydrated after deparaffinization. Slides were incubated in EDTA containing solution, pH 8.0 at 95 °C for 10 min. Samples were incubated overnight at with the same primary Abs used for immunoblotting at 4 °C. After primary antibody binding a 1-h incubation with the secondary antibody Alexa Fluor 488 donkey-anti-rabbit IgG H+L (Thermo Fisher Scientific, Waltham, MA, USA) diluted 1000-fold was done. Fluorescent images were obtained with an Olympus B× 60 microscope

4.5. *Statistical Analysis*

Statistical significance was tested using either a pairwise analysis T test (Mann-Whitney) or the one-way analysis of variance (Kruskal-Wallis) to determine significance. Statistical analysis was accomplished using GraphPad Prism (GraphPad Software, Inc., San Diego, CA, USA). A value of $p < 0.05$ qualified as significant.

Author Contributions: Conceptualization, A.D.B, L.B. and M.B. Methodology, A.D.B, L.B., T.Y., F.I.K, S.F. Formal Analysis, L.B., A.D.B., T.Y., F.I.K, M.B. Writing—Original Draft Preparation, A.D.B, M.B. and L.B. Writing—Review & Editing, T.Y., F.I.K, M.B. and A.D.B.; Visualization, T.Y. Funding Acquisition, A.D.B and L.B.

Acknowledgments: The authors are thankful to S. Getsios at Northwestern University – Skin Disease Research Center Keratinocyte Core who assisted in the keratinocyte organotypic cultures with support from the National Institutes of Health (NIH) National Institute of Arthritis and Musculoskeletal and Skin (NIAMS) Diseases grant AR057216.

References

1. Chan, T.C.; Sanyal, R.D.; Pavel, A.B.; Glickman, J.; Zheng, X.; Xu, H.; Cho, Y.T.; Tsai, T.F.; Wen, H.C.; Peng, X.; et al. Atopic dermatitis in Chinese patients shows TH2/TH17 skewing with psoriasiform features. *J. Allergy Clin. Immunol.* **2018**, *142*, 1013–1017. [CrossRef] [PubMed]
2. Sanyal, R.D.; Pavel, A.B.; Glickman, J.; Chan, T.C.; Zheng, X.; Zhang, N.; Cueto, I.; Peng, X.; Estrada, Y.; Fuentes-Duculan, J.; et al. Atopic dermatitis in African American patients is TH2/TH22-skewed with TH1/TH17 attenuation. *Ann. Allergy Asthma Immunol.* **2019**, *122*, 99–110.
3. Gittler, J.K.; Shemer, A.; Suárez-Fariñas, M.; Fuentes-Duculan, J.; Gulewicz, K.J.; Wang, C.Q.; Mitsui, H.; Cardinale, I.; de Guzman Strong, C.; Krueger, J.G.; et al. Progressive activation of T(H)2/T(H)22 cytokines and selective epidermal proteins characterizes acute and chronic atopic dermatitis. *J. Allergy Clin. Immunol.* **2012**, *130*, 1344–1354. [CrossRef]
4. Furuse, M.; Hata, M.; Furuse, K.; Yoshida, Y.; Haratake, A.; Sugitani, Y.; Noda, T.; Kubo, A.; Tsukita, S. Claudin-based tight junctions are crucial for the mammalian epidermal barrier: A lesson from claudin-1-deficient mice. *J. Cell Biol.* **2002**, *156*, 1099–1111. [CrossRef]
5. De Benedetto, A.; Rafaels, N.M.; McGirt, L.Y.; Ivanov, A.I.; Georas, S.N.; Cheadle, C.; Berger, A.E.; Zhang, K.; Vidyasagar, S.; Yoshida, T.; et al. Tight junction defects in patients with atopic dermatitis. *J. Allergy Clin. Immunol.* **2011**, *127*, 773–786.
6. Yuki, T.; Tobiishi, M.; Kusaka-Kikushima, A.; Ota, Y.; Tokura, Y. Impaired Tight Junctions in Atopic Dermatitis Skin and in a Skin-Equivalent Model Treated with Interleukin-17. *PLoS ONE* **2016**, *11*, e0161759. [CrossRef] [PubMed]
7. Batista, D.I.S.; Perez, L.; Orfali, R.L.; Zaniboni, M.C.; Samorano, L.P.; Pereira, N.V.; Sotto, M.N.; Ishizaki, A.S.; Oliveira, L.M.S.; Sato, M.N.; et al. Profile of skin barrier proteins (filaggrin, claudins 1 and 4) and Th1/Th2/Th17 cytokines in adults with atopic dermatitis. *J. Eur. Acad. Dermatol. Venereol.* **2015**, *29*, 1091–1095. [PubMed]
8. Kast, J.I.; Wanke, K.; Soyka, M.B.; Wawrzyniak, P.; Akdis, D.; Kingo, K.; Rebane, A.; Akdis, C.A. The broad spectrum of interepithelial junctions in skin and lung. *J. Allergy Clin. Immunol.* **2012**, *130*, 544–547. [CrossRef]
9. Kirschner, N.; Poetzl, C.; von den Driesch, P.; Wladykowski, E.; Moll, I.; Behne, M.J.; Brandner, J.M. Alteration of tight junction proteins is an early event in psoriasis: Putative involvement of proinflammatory cytokines. *Am. J. Pathol.* **2009**, *175*, 1095–1106. [CrossRef] [PubMed]
10. Gutowska-Owsiak, D.; Schaupp, A.L.; Salimi, M.; Taylor, S.; Ogg, G.S. Interleukin-22 downregulates filaggrin expression and affects expression of profilaggrin processing enzymes. *Br. J. Dermatol.* **2011**, *165*, 492–498.
11. Hvid, M.; Vestergaard, C.; Kemp, K.; Christensen, G.B.; Deleuran, B.; Deleuran, M. IL-25 in atopic dermatitis: A possible link between inflammation and skin barrier dysfunction? *J. Investig. Dermatol.* **2011**, *131*, 150–157. [CrossRef] [PubMed]
12. Howell, M.D.; Fairchild, H.R.; Kim, B.E.; Bin, L.; Boguniewicz, M.; Redzic, J.S.; Hansen, K.C.; Leung, D.Y. Th2 cytokines act on S100/A11 to downregulate keratinocyte differentiation. *J. Investig. Dermatol.* **2008**, *128*, 2248–2258. [CrossRef] [PubMed]
13. Floudas, A.; Saunders, S.P.; Moran, T.; Schwartz, C.; Hams, E.; Fitzgerald, D.C.; Johnston, J.A.; Ogg, G.S.; McKenzie, A.N.; Walsh, P.T.; et al. IL-17 Receptor A Maintains and Protects the Skin Barrier to Prevent Allergic Skin Inflammation. *J. Immunol.* **2017**, *199*, 707–717. [CrossRef] [PubMed]
14. Lee, J.S.; Tato, C.M.; Joyce-Shaikh, B.; Gulen, M.F.; Cayatte, C.; Chen, Y.; Blumenschein, W.M.; Judo, M.; Ayanoglu, G.; McClanahan, T.K.; et al. Interleukin-23-Independent IL-17 Production Regulates Intestinal Epithelial Permeability. *Immunity* **2015**, *43*, 727–738. [CrossRef] [PubMed]
15. Nadeau, P.; Henehan, M.; De Benedetto, A. Activation of protease-activated receptor 2 leads to impairment of keratinocyte tight junction integrity. *J. Allergy Clin. Immunol.* **2018**, *142*, 281–284. [CrossRef] [PubMed]

16. El Asmar, R.; Panigrahi, P.; Bamford, P.; Berti, I.; Not, T.; Coppa, G.V.; Catassi, C.; Fasano, A. Host-dependent zonulin secretion causes the impairment of the small intestine barrier function after bacterial exposure. *Gastroenterology* **2002**, *123*, 1607–1615. [CrossRef]
17. Kuo, I.H.; Carpenter-Mendini, A.; Yoshida, T.; McGirt, L.Y.; Ivanov, A.I.; Barnes, K.C.; Gallo, R.L.; Borkowski, A.W.; Yamasaki, K.; Leung, D.Y.; et al. Activation of epidermal toll-like receptor 2 enhances tight junction function: Implications for atopic dermatitis and skin barrier repair. *J. Investig. Dermatol.* **2013**, *133*, 988–998. [PubMed]
18. Eyerich, K.; Pennino, D.; Scarponi, C.; Foerster, S.; Nasorri, F.; Behrendt, H.; Ring, J.; Traidl-Hoffmann, C.; Albanesi, C.; Cavani, A. IL-17 in atopic eczema: Linking allergen-specific adaptive and microbial-triggered innate immune response. *J. Allergy Clin. Immunol.* **2009**, *123*, 59–66. [CrossRef]
19. Nograles, K.E.; Suárez-Fariñas, M.; Shemer, A.; Fuentes-Duculan, J.; Chiricozzi, A.; Cardinale, I.; Zaba, L.C.; Kikuchi, T.; Ramon, M.; Bergman, R.; et al. Atopic dermatitis (AD) keratinocytes exhibit normal Th17 cytokine responses. *J. Investig. Dermatol.* **2010**, *125*, 744–746.
20. Bai, B.; Yamamoto, K.; Sato, H.; Sugiura, H.; Tanaka, T. Complex regulation of S100A8 by IL-17, dexamethasone, IL-4 and IL-13 in HaCat cells (human keratinocyte cell line). *J. Dermatol. Sci.* **2007**, *47*, 259–262. [CrossRef]
21. Gutowska-Owsiak, D.; Schaupp, A.L.; Salimi, M.; Selvakumar, T.A.; McPherson, T.; Taylor, S.; Ogg, G.S. IL-17 downregulates filaggrin and affects keratinocyte expression of genes associated with cellular adhesion. *Exp. Dermatol.* **2012**, *21*, 104–110. [CrossRef] [PubMed]
22. Nograles, K.E.; Zaba, L.C.; Guttman-Yassky, E.; Fuentes-Duculan, J.; Suárez-Fariñas, M.; Cardinale, I.; Khatcherian, A.; Gonzalez, J.; Pierson, K.C.; White, T.R.; et al. Th17 cytokines interleukin (IL)-17 and IL-22 modulate distinct inflammatory and keratinocyte-response pathways. *Br. J. Dermatol.* **2008**, *159*, 1092–1102. [CrossRef] [PubMed]
23. Huang, F.; Kao, C.Y.; Wachi, S.; Thai, P.; Ryu, J.; Wu, R. Requirement for both JAK-mediated PI3K signaling and ACT1/TRAF6/TAK1-dependent NF-kappaB activation by IL-17A in enhancing cytokine expression in human airway epithelial cells. *J. Immunol.* **2007**, *179*, 6504–6513. [CrossRef] [PubMed]
24. Newcomb, D.C.; Boswell, M.G.; Zhou, W.; Huckabee, M.M.; Goleniewska, K.; Sevin, C.M.; Hershey, G.K.K.; Kolls, J.K.; Peebles, R.S., Jr. Human TH17 cells express a functional IL-13 receptor and IL-13 attenuates IL-17A production. *J. Allergy Clin. Immunol.* **2011**, *127*, 1006–1013. [CrossRef] [PubMed]
25. Takeda, K.; Noguchi, K.; Shi, W.; Tanaka, T.; Matsumoto, M.; Yoshida, N.; Kishimoto, T.; Akira, S. Targeted disruption of the mouse Stat3 gene leads to early embryonic lethality. *Proc. Natl. Acad. Sci. USA* **1997**, *94*, 3801–3804. [CrossRef] [PubMed]
26. Sano, S.; Itami, S.; Takeda, K.; Tarutani, M.; Yamaguchi, Y.; Miura, H.; Yoshikawa, K.; Akira, S.; Takeda, J. Keratinocyte-specific ablation of Stat3 exhibits impaired skin remodeling, but does not affect skin morphogenesis. *EMBO J.* **1999**, *18*, 4657–4668. [CrossRef]
27. Wang, Z.; Li, R.; Tan, J.; Peng, L.; Wang, P.; Liu, J.; Xiong, H.; Jiang, B.; Chen, Y. Syndecan-1 Acts in Synergy with Tight Junction Through Stat3 Signaling to Maintain Intestinal Mucosal Barrier and Prevent Bacterial Translocation. *Inflamm. Bowel Dis.* **2015**, *21*, 1894–1907. [CrossRef]
28. Yun, J.H.; Park, S.W.; Kim, K.J.; Bae, J.S.; Lee, E.H.; Paek, S.H.; Kim, S.U.; Ye, S.; Kim, J.H.; Cho, C.H. Endothelial STAT3 Activation Increases Vascular Leakage Through Downregulating Tight Junction Proteins: Implications for Diabetic Retinopathy. *J. Cell. Physiol.* **2017**, *232*, 1123–1134. [CrossRef]
29. Amano, W.; Nakajima, S.; Kunugi, H.; Numata, Y.; Kitoh, A.; Egawa, G.; Dainichi, T.; Honda, T.; Otsuka, A.; Kimoto, Y.; et al. The Janus kinase inhibitor JTE-052 improves skin barrier function through suppressing signal transducer and activator of transcription 3 signaling. *J. Allergy Clin. Immunol.* **2015**, *136*, 667–677. [CrossRef]
30. Mitamura, Y.; Nunomura, S.; Nanri, Y.; Ogawa, M.; Yoshihara, T.; Masuoka, M.; Tsuji, G.; Nakahara, T.; Hashimoto-Hachiya, A.; Conway, S.J.; et al. The IL-13/periostin/IL-24 pathway causes epidermal barrier dysfunction in allergic skin inflammation. *Allergy* **2018**, *73*, 1881–1891. [CrossRef]
31. Ma, C.S.; Chew, G.Y.; Simpson, N.; Priyadarshi, A.; Wong, M.; Grimbacher, B.; Fulcher, D.A.; Tangye, S.G.; Cook, M.C. Deficiency of Th17 cells in hyper IgE syndrome due to mutations in STAT3. *J. Exp. Med.* **2008**, *205*, 1551–1557. [CrossRef]
32. Chen, Y.; Yang, P.; Li, F.; Kijlstra, A. The effects of Th17 cytokines on the inflammatory mediator production and barrier function of ARPE-19 cells. *PLoS ONE* **2011**, *6*, e18139. [CrossRef] [PubMed]
33. Capaldo, C.T.; Nusrat, A. Cytokine regulation of tight junctions. *Biochim. Biophys. Acta* **2009**, *1788*, 864–871.

[CrossRef] [PubMed]
34. Beck, L.A.; Thaçi, D.; Hamilton, J.D.; Graham, N.M.; Bieber, T.; Rocklin, R.; Ming, J.E.; Ren, H.; Kao, R.; Simpson, E.; et al. Dupilumab treatment in adults with moderate-to-severe atopic dermatitis. *N. Engl. J. Med.* **2014**, *371*, 130–139. [CrossRef] [PubMed]
35. Poumay, Y.; Roland, I.H.; Leclercq-Smekens, M.; Leloup, R. Basal detachment of the epidermis using dispase: Tissue spatial organization and fate of integrin alpha 6 beta 4 and hemidesmosomes. *J. Investig. Dermatol.* **1994**, *102*, 111–117. [CrossRef] [PubMed]
36. Gordon, K.; Kochkodan, J.J.; Blatt, H.; Lin, S.Y.; Kaplan, N.; Johnston, A.; Swindell, W.R.; Hoover, P.; Schlosser, B.J.; Elder, J.T.; et al. Alteration of the EphA2/Ephrin-A signaling axis in psoriatic epidermis. *J. Investig. Dermatol.* **2013**, *133*, 712–722. [CrossRef] [PubMed]

10

Brazilian Green Propolis Rescues Oxidative Stress-Induced Mislocalization of Claudin-1 in Human Keratinocyte-Derived HaCaT Cells

Kana Marunaka [1], Mao Kobayashi [1], Shokoku Shu [1], Toshiyuki Matsunaga [2] and Akira Ikari [1,*]

1 Laboratory of Biochemistry, Department of Biopharmaceutical Sciences, Gifu Pharmaceutical University, Gifu 501-1196, Japan
2 Education Center of Green Pharmaceutical Sciences, Gifu Pharmaceutical University, Gifu 502-8585, Japan
* Correspondence: ikari@gifu-pu.ac.jp.

Abstract: Claudin-1 (CLDN1) is expressed in the tight junction (TJ) of the skin granular layer and acts as a physiological barrier for the paracellular transport of ions and nonionic molecules. Ultraviolet (UV) and oxidative stress may disrupt the TJ barrier, but the mechanism of and protective agents against this effect have not been clarified. We found that UVB and hydrogen peroxide (H_2O_2) caused the internalization of CLDN1 and increased the paracellular permeability of lucifer yellow, a fluorescent marker, in human keratinocyte-derived HaCaT cells. Therefore, the mechanism of mislocalization of CLDN1 and the protective effect of an ethanol extract of Brazilian green propolis (EBGP) were investigated. The UVB- and H_2O_2-induced decreases in CLDN1 localization were rescued by EBGP. H_2O_2 decreased the phosphorylation level of CLDN1, which was also rescued by EBGP. Wild-type CLDN1 was distributed in the cytosol after treatment with H_2O_2, whereas T191E, its H_2O_2-insensitive phosphorylation-mimicking mutant, was localized at the TJ. Both protein kinase C activator and protein phosphatase 2A inhibitor rescued the H_2O_2-induced decrease in CLDN1 localization. The tight junctional localization of CLDN1 and paracellular permeability showed a negative correlation. Our results indicate that UVB and H_2O_2 could induce the elevation of paracellular permeability mediated by the dephosphorylation and mislocalization of CLDN1 in HaCaT cells, which was rescued by EBGP. EBGP and its components may be useful in preventing the destruction of the TJ barrier through UV and oxidative stress.

Keywords: claudin-1; hydrogen peroxide; phosphorylation

---

1. Introduction

The skin forms a barrier between the body and its external environment in order to prevent the intrusion of pathogens and the uncontrolled loss of internal water and solutes. The epidermis is the outermost layer of skin and provides the first line of defense against ultraviolet (UV) radiation and other environmental factors. UV is divided into three ranges based on wavelength: UVA (320–400 nm), UVB (280–320 nm), and UVC (100–290 nm). Among these, UVC is unlikely to be present in terrestrial sunlight because it is blocked by the ozone layer, whereas UVA and UVB can come into contact with the skin. The skin can be separated by the basement membrane into two layers, the dermis and epidermis. The mammalian epidermis is composed of four layers, including the basal, spinous, granular, and cornified layers [1]. UVA can penetrate the dermal layer of the skin and reach the capillaries, whereas UVB is blocked in the upper dermis. According to wavelength-dependent studies, UVB radiation has more cytotoxic and mutagenic effects than UVA does [2,3]. Long-term exposure to UVB induces damage to both the dermal and epidermal skin [4]. UVB-induced cell death is caused through directly induced DNA damage and indirect action mediated by the generation of reactive oxygen species (ROS) and nitric oxide [5]. The main causes of UV-induced DNA damage are 2,3-cyclobutane pyrimidine dimer and a pyrimidine (6-4) pyrimidone photoproduct [6]. In contrast, it is not fully understood what mechanisms are involved in damage to the skin barrier under low toxic conditions. An investigation of the effect of each toxic factor on the tight junction (TJ) barrier may be needed in order to clarify the molecular mechanism.

Human skin keratinocytes form a TJ at the most apical pole of the lateral membrane between neighboring cells. The TJ prevents the abnormal paracellular movement of water, solutes, and pathogens. Claudins (CLDNs) and occludin are integral membrane proteins in TJ, and they comprise a large family of 27 subtypes in mammals [7,8]. Among them, CLDN1 and CLDN4 are highly expressed in human keratinocytes [9]. Most CLDNs contain carboxyl-terminal PSD95/Dlg/ZO-1 (PDZ)-binding motifs that mediate interactions between CLDNs and scaffolding proteins such as ZO-1 and ZO-2. These scaffolding proteins indirectly link CLDNs to actin filaments. CLDN1-deficient mice develop aberrant stratum corneum barrier functions in the skin [10]. A premature stop codon of CLDN1, resulting in a lack of CLDN1 protein, has been identified in neonatal ichthyosis and sclerosing cholangitis syndrome, a disorder characterized by scalp hypotrichosis, ichthyosis, scarring alopecia, and sclerosing cholangitis [11]. Therefore, CLDN1 may have an important role in maintaining the barrier function in the skin.

An antioxidant effect is one of the key factors in reducing skin injury, aging, and cancer risk, and antioxidant activities have been reported in components extracted from plants, fruits, vegetables, and bee propolis [12,13]. Propolis contains many classes of compounds, including flavonoids, phenolic acids, and others. The components of propolis are different in each area [14]. The ethanol extract of Brazilian green propolis (EBGP) exerts strong antioxidant activity in mouse skin [15] and protects human keratinocytes against UV-induced apoptosis [16]. In addition, hydroalcoholic extracts of Brazilian propolis improve dermal burn healing [17]. EBGP may be useful in protecting skin damage from various external stimuli, but the effect of EBGP on the TJ barrier has not been examined.

Human keratinocyte-derived HaCaT cells can easily be plated as a monolayer and form the TJ [18]. Therefore, they may be useful in examining the function of TJ in the skin. In the present study, we investigated the effects of ROS and EBGP on the cellular localization and function of CLDN1 in HaCaT cells. In addition, the molecular mechanism of tight junctional localization of CLDN1 was assessed by immunoprecipitation, immunoblotting, and immunofluorescence measurements.

2. Results

2.1. *Effect of UVB Radiation on the Production of ROS*

The production of ROS and H_2O_2 in HaCaT cells was measured using 2′,7′-dichlorodihydrofluorescein diacetate (H_2DCFDA) and Hydrop, respectively. UVB radiation dose-dependently induced the fluorescence intensities of H_2DCFDA and Hydrop (Figure 1A,B). UVB radiation between 10 and 50 mJ/cm^2 may have increased the production of ROS and H_2O_2. Cell viability was dose-dependently decreased by exposure to UVB (Figure 1C). The effects were significant over 10 mJ/cm^2, but the percentage of cell toxicity was less than 40%. UVB radiation up to 50 mJ/cm^2 may have produced relatively low toxic conditions. To clarify the effects of ROS on viability, the cells were transiently exposed to H_2O_2 for 3 h. H_2O_2 dose-dependently decreased cell viability (Figure 1D). We decided to treat the cells with 50 mJ/ cm^2 UVB or 200 μM H_2O_2 in subsequent experiments.

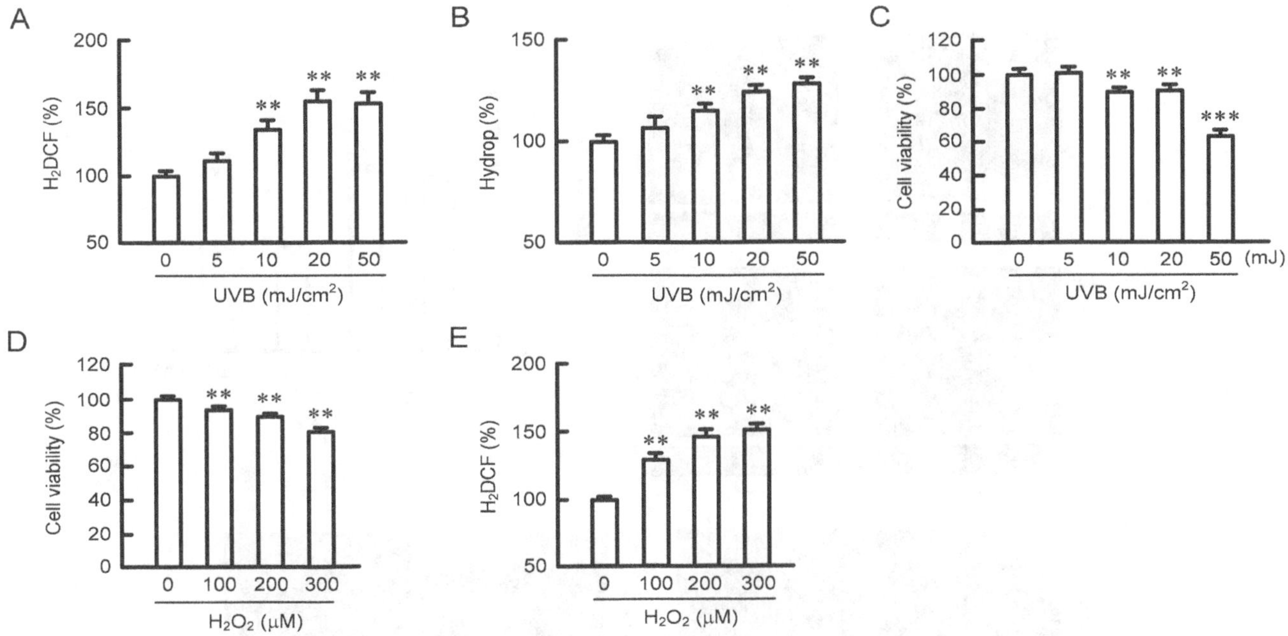

Figure 1. Effects of UVB and H_2O_2 on reactive oxygen species (ROS) production and cell toxicity in HaCaT cells. (**A**,**B**) Cells exposed to UVB were cultured for 3 h and then incubated with 2 ,7 -dichlorodihydrofluorescein diacetate (H_2DCFDA) (ROS) and Hydrop (H_2O_2) for 30 min. The florescence intensities of each dye were measured using an Infinite F200 Pro microplate reader and are shown relative to the values (0 mJ/cm^2). (**C**,**D**) Cells were exposed to UVB or transiently exposed to H_2O_2 for 3 h, and then they were cultured for 24 h. Cell viability was measured using a WST-1 assay. (**E**) Cells were exposed to H_2O_2 for 3 h, and then they were incubated with H_2DCFDA. The florescence intensities of H_2DCF are shown as relative to the values in 0 μM H_2O_2. n = 4–6; *** $p < 0.001$ and ** $p < 0.01$ significantly different from 0 mJ/cm^2 UVB or 0 μM H_2O_2.

2.2. *Effect of H_2O_2 on the Cell Localization of TJ Protein*

Immunofluorescence measurements showed that CLDN1, CLDN4, and ZO-1 were localized at the TJ under control conditions (Figure 2A,B). The fluorescence signal of CLDN1 at the TJ was not changed by treatment with 200 μM H_2O_2 for 3 h, but it became weaker at 6 h. In contrast, those of CLDN4 and ZO-1 were unchanged. CLDN1 was relocalized at the TJ after 48 h. The amount of CLDN1 protein was unchanged by H_2O_2 treatment, but that of CLDN4 transiently increased at 6 h (Figure 2C).

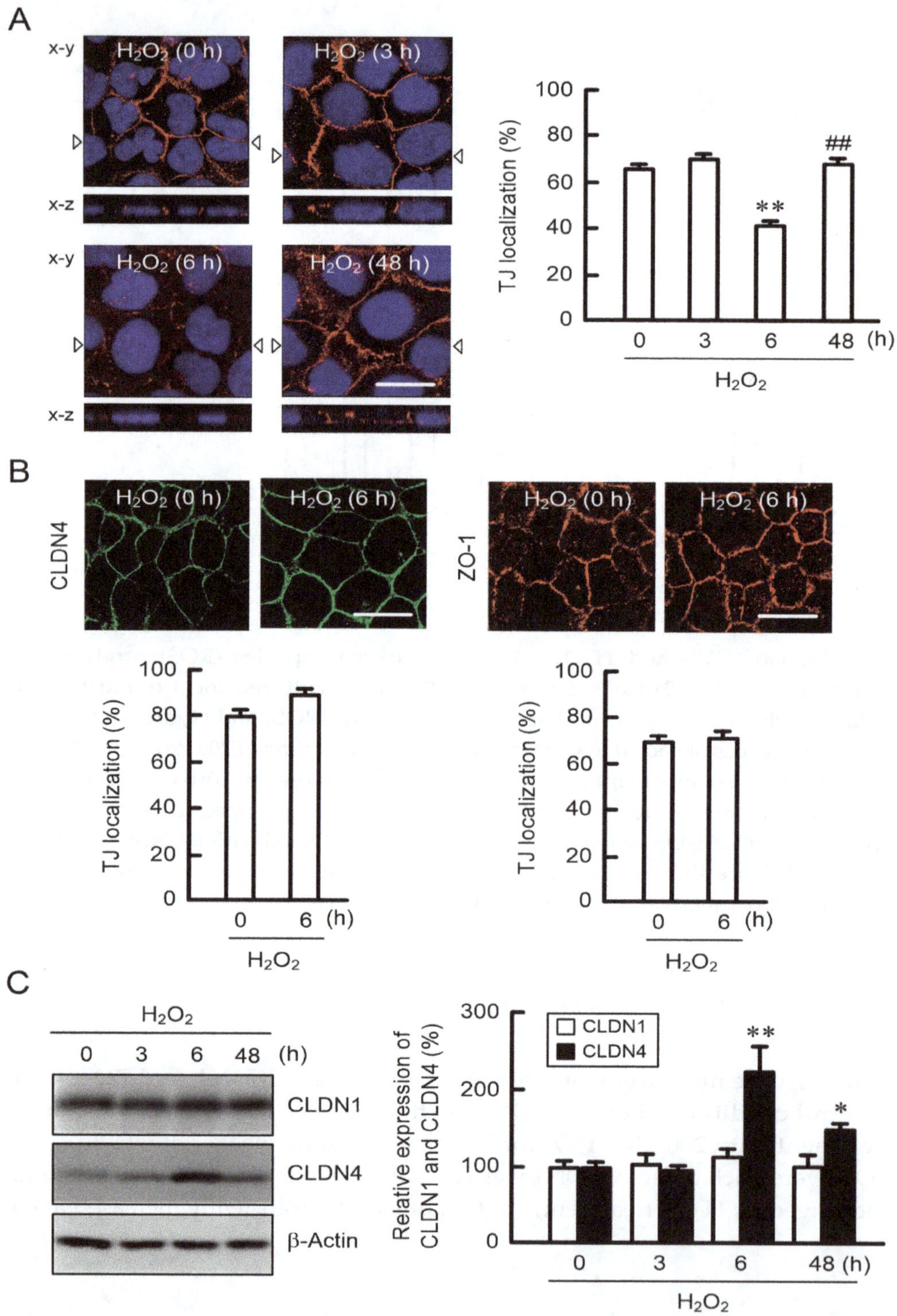

Figure 2. Effects of H_2O_2 on the localization and expression of claudins (CLDNs). Cells exposed to 200 μM H_2O_2 were cultured for 0, 3, 6, and 48 h. (**A**) The cells were immunostained with anti-CLDN1 (red) antibodies and DAPI (nuclear marker). The low panels (x-z) show the vertical sections indicated by the triangles of x-y images. Scale bars indicate 10 μm. The fluorescence values of CLDN1 at the tight junction (TJ) are shown as a percentage of the total fluorescence values. (**B**) The cells were immunostained with anti-CLDN4 (green) or anti-ZO-1 (red) antibodies. The fluorescence values of CLDN4 and ZO-1 at the TJ are shown as a percentage of the total fluorescence values. (**C**) Cell lysates were applied to 12.5% SDS-PAGE and blotted with anti-CLDN1, anti-CLDN4, and β-actin antibodies. The expression levels of CLDN1 and CLDN4 are represented relative to the values at 0 h. n = 3–8; ** $p < 0.01$ and * $p < 0.05$ significantly different from 0 h; ## $p < 0.01$ significantly different from 6 h.

2.3. *Effects of EBGP on the UVB- and H_2O_2-Induced Destruction of the TJ Barrier*

The function of the TJ barrier was evaluated using transepithelial electrical resistance (TER) and lucifer yellow (LY) flux with HaCaT cells cultured on Transwell inserts. TER decreased after 6 h of treatment with UVB or H_2O_2, then returned to the control level after 48 h (Figure 3A). EBGP-induced cell toxicity was examined using a 2-(4-Iodophenyl)-3-(4-nitrophenyl)-5-(2,4-disulfophenyl)-2*H*-tetrazolium (WST-1) assay. Cell viability was not significantly changed by EBGP at concentrations less than 20 μg/mL, but decreased at 50 μg/mL (Figure 3B). The H_2O_2-induced mislocalization of CLDN1 was significantly rescued by 10 μg/mL EBGP (Figure 3C), but not by 50 μg/mL EBGP. Similarly, the TJ localization of CLDN1 was decreased by UVB, which was rescued by 10 μg/mL EBGP. Both H_2O_2 and UVB caused a decrease in TER and an increase in LY flux, indicating that H_2O_2 and UVB induced the destruction of the TJ barrier. These effects were rescued by EBGP (Figure 3D). These results indicated that UVB and H_2O_2 may have induced the destruction of the TJ barrier mediated by the loss of CLDN1 in the TJ, which was rescued by EBGP.

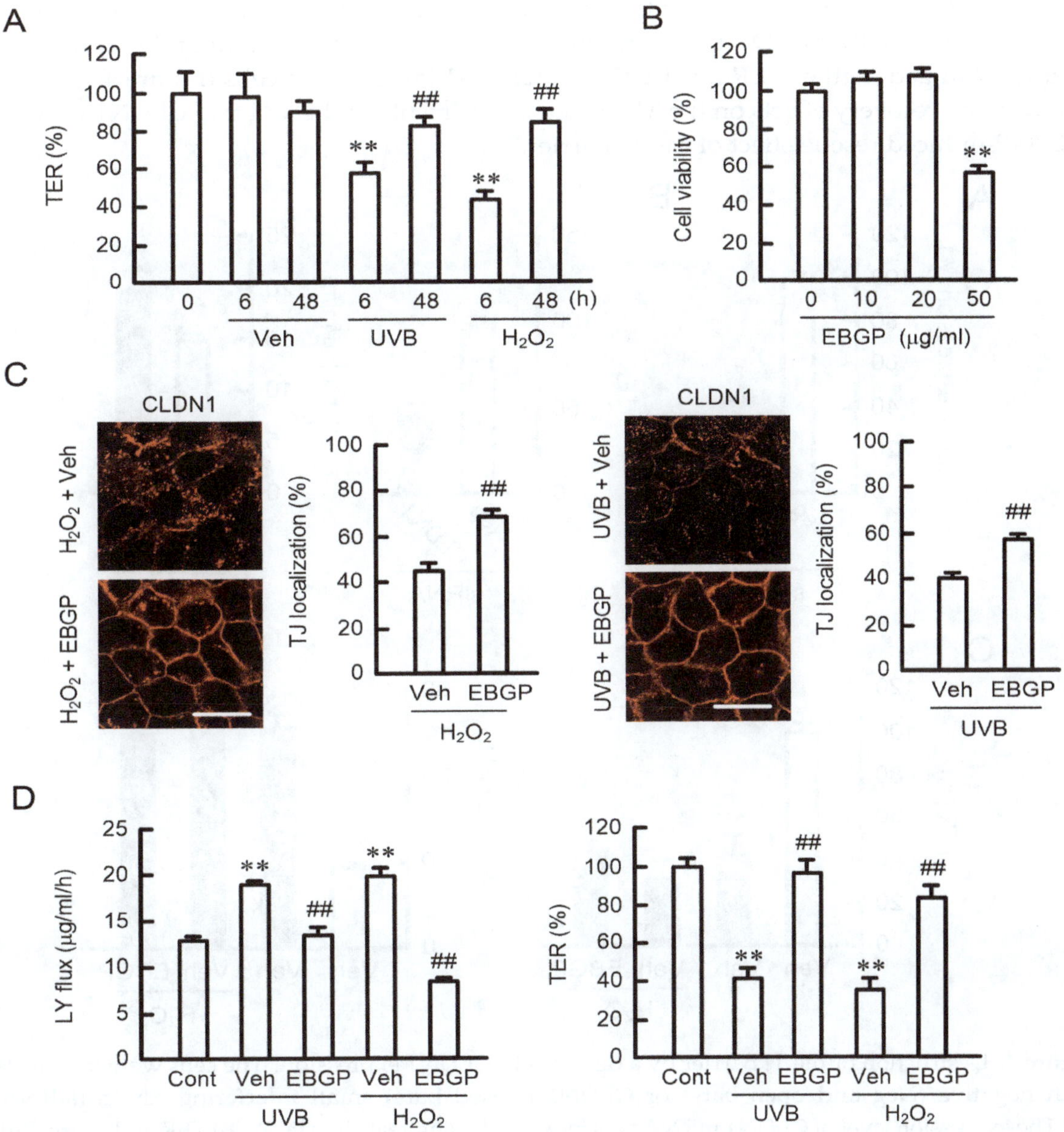

Figure 3. Effects of UVB, H_2O_2, and Brazilian green propolis (EBGP) on TJ permeability. (**A**) Cells plated on Transwell inserts were incubated with 50 mJ/cm^2 UVB or 200 μM H_2O_2 for the periods indicated. Transepithelial electrical resistance (TER) was measured using a volt-ohmmeter and is represented as a

percentage of the values at 0 h. (**B**) Cells were incubated in the presence or absence of EBGP for 24 h. Cell viability was measured using a WST-1 assay. (**C**) The cells were pre-incubated in the absence (Veh) or presence of 10 μg/mL EBGP for 30 min, followed by exposure to UVB or H_2O_2 for 6 h. The cells were immunostained with anti-CLDN1 antibodies. Scale bars indicate 10 μm. The TJ localization of CLDN1 is represented as the percentage of total amount. (**D**) The cells were pre-incubated in the absence (Veh) or presence of 10 μg/mL EBGP for 30 min, followed by exposure to UVB or H_2O_2 for 6 h. Control cells (Cont) were not treated with UVB or H_2O_2. TER was measured using a volt-ohmmeter and is represented as a percentage of the values in control cells. Paracellular lucifer yellow (LY) flux was analyzed using fluorescence spectrometry. $n = 4$; ** $p < 0.01$ significantly different from 0 h, 0 μg/mL, or Cont; ## $p < 0.01$ significantly different from 6 h of UVB and H_2O_2 or Veh of UVB and H_2O_2.

2.4. *Effect of CLDN1 Expression on the TJ Barrier*

To confirm the involvement of CLDN1, we examined the effect of knockdown of CLDN1 expression. The mRNA expression of CLDN1 was decreased by the introduction of CLDN1 small interfering RNAs (siRNAs) (Figure 4A). The knockdown of CLDN1 induced a decrease in TER and an increase in LY flux (Figure 4B), indicating that CLDN1 had an important role in the regulation of the TJ barrier. H_2O_2 significantly changed neither TER nor LY flux in CLDN1 knockdown cells (Figure 4C). In addition, EBGP showed no recovery effects on the TJ barrier. These results indicated that CLDN1 was necessary for the EBGP-induced rescue effect of the TJ barrier.

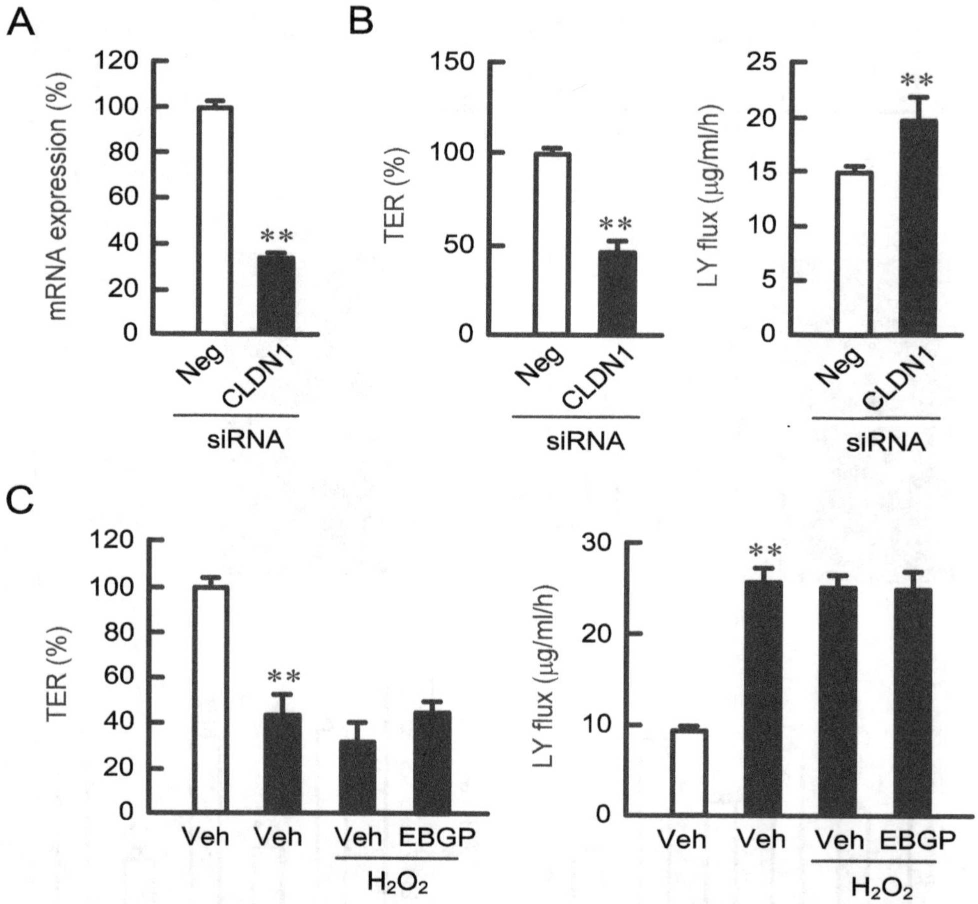

Figure 4. Destruction of the TJ barrier by a decrease in CLDN1 expression. The cells were transfected with negative (Neg and open bars) or CLDN1 (closed bars) small interfering RNAs (siRNAs). (**A**) The expression level of CLDN1 mRNA was measured using real-time PCR. (**B**) TER and paracellular LY flux were analyzed using a volt-ohmmeter and fluorescence spectrometry, respectively. (**C**) The cells were incubated in the absence (Veh) or presence of 200 μM H_2O_2 and 10 μg/mL EBGP for 6 h. TER and paracellular LY flux were analyzed. $n = 4$; ** $p < 0.01$ significantly different from Veh.

2.5. Effects of Endocytosis Inhibitors on the Localization of CLDN1

The localization of plasma membrane protein is regulated by endocytotic processes, including clathrin- and caveolae-dependent pathways. The H_2O_2-induced mislocalization of CLDN1 was inhibited by monodansylcadaverine (MDC), a clathrin-dependent endocytosis inhibitor, but not by methyl-β-cyclodextrin (MβCD), a caveolae-dependent endocytosis inhibitor (Figure 5A), indicating that H_2O_2 may increase the internalization of CLDN1 mediated by the clathrin-dependent endocytosis pathway. An increase in the paracellular permeability caused by H_2O_2 was rescued by MDC, but not by MβCD (Figure 5B). These results coincided with the effect of MDC on the H_2O_2-induced mislocalization of CLDN1 in the immunofluorescence measurements. The recovery effect of MDC was blocked by the knockdown of CLDN1 expression. These results support that the disappearance of CLDN1 in TJ may induce disruption of the TJ barrier. The localization of some CLDNs has been reported to change according to their phosphorylation status [19–21]. Therefore, we decided to investigate the effect of H_2O_2 on the phosphorylation of CLDN1.

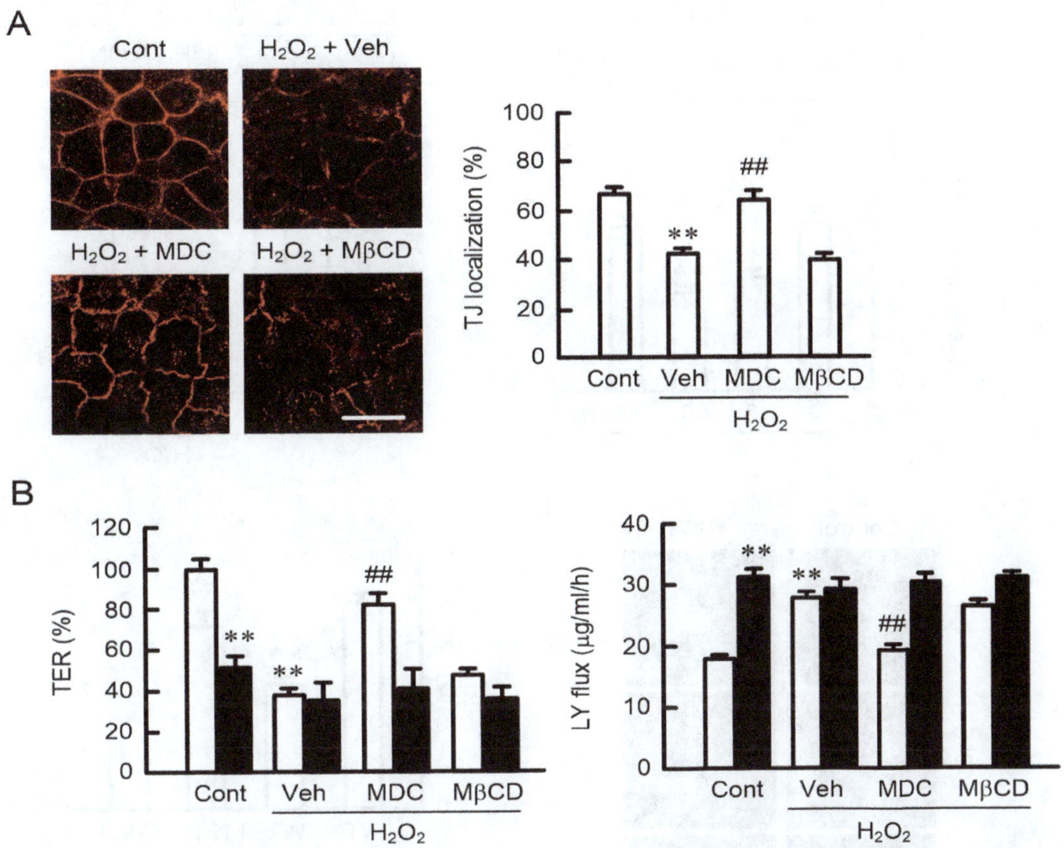

Figure 5. Rescue of the H_2O_2-induced mislocalization of CLDN1 by monodansylcadaverine (MDC). (**A**) Cells were pre-incubated in the absence (Veh) or presence of 5 μM MDC and 10 μM methyl-β-cyclodextrin (MβCD) for 30 min, followed by exposure to 200 μM H_2O_2. Control cells (Cont) were not treated with H_2O_2 and endocytosis inhibitors. The cells were immunostained with anti-CLDN1 antibodies. Scale bars indicate 10 μm. The TJ localization of CLDN1 is represented as the percentage of the total amount. (**B**) Negative (open bars) or CLDN1 siRNA-transfected cells (closed bars) were incubated with H_2O_2, MDC, and MβCD. TER and paracellular LY flux were analyzed using a volt-ohmmeter and fluorescence spectrometry, respectively. $n = 4–6$; ** $p < 0.01$ significantly different from Cont or Cont of negative siRNA; ## $p < 0.01$ significantly different from Veh of H_2O_2.

2.6. Effects of Phosphorylation Mimic Mutants of CLDN1 on the TJ Barrier

H_2O_2 transiently decreased the phosphothreonine (p-Thr) levels of CLDN1 after 3–6 h (Figure 6A). In contrast, phosphoserine (p-Ser) levels were unchanged by H_2O_2 (data not shown). The H_2O_2-induced dephosphorylation of CLDN1 was inhibited by EBGP (Figure 6B). It has been reported previously that CLDN1 may be phosphorylated at T191 and T195 [22]. To clarify the dephosphorylation site of CLDN1 associated with H_2O_2 treatment, we investigated the effects of phosphorylation mimic mutants of CLDN1, T191E, and T195E. In the absence of H_2O_2, the wild-type (WT), T191E, and T195E proteins were localized in the TJ area. H_2O_2 increased the intracellular distribution of WT and T195E, but T191E was still mainly localized in the TJ area (Figure 6C). Compared to WT in the absence of H_2O_2, T191E showed a similar TJ barrier function, but T195E did not (Figure 6D). These results indicated that the phosphorylation of T191 may be necessary for the TJ localization of CLDN1.

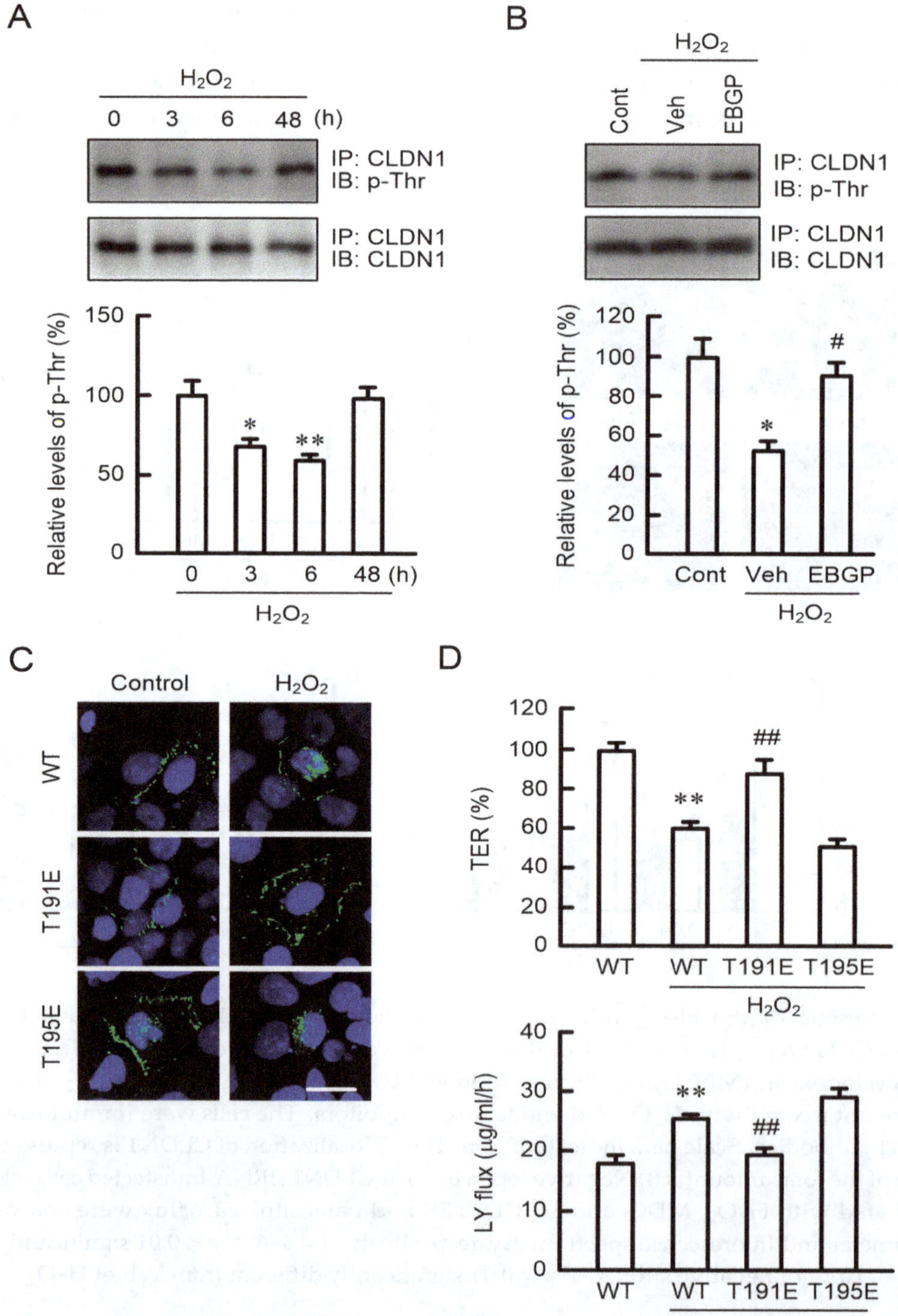

Figure 6. Effects of phosphorylation mimic mutants on H_2O_2-induced barrier destruction. (**A**) Cell lysates were prepared from the cells exposed to 200 μM H_2O_2 for the periods indicated. (**B**) Cell lysates

were prepared from the cells exposed to 200 μM H_2O_2 in the absence (Veh) or presence of 10 μg/mL EBGP for 6 h. After immunoprecipitation with anti-CLDN1 antibodies, the immunoprecipitants were applied to 12.5% SDS-PAGE and blotted with antiphosphothreonine (p-Thr) and anti-CLDN1 antibodies. The levels of p-Thr are represented relative to the values at 0 h or the control (Cont). (**C**,**D**) Cells were transfected with FLAG-tagged wild-type (WT), T191E, and T195E CLDN1. At 48 h after transfection, the cells were exposed to 200 μM H_2O_2 for 6 h. (**C**) The cells were immunostained with anti-FLAG antibodies (green) and DAPI (blue), a nuclear marker. Scale bars indicate 10 μm. (**D**) TER and paracellular LY flux were analyzed using a volt-ohmmeter and fluorescence spectrometry, respectively. $n = 3$–4; ** $p < 0.01$ and * $p < 0.05$ significantly different from 0 h, Cont, or WT alone; ## $p < 0.01$ and # $p < 0.05$ significantly different from Veh with H_2O_2 or WT with H_2O_2.

2.7. *Phosphorylation of CLDN1 by PKC*

The phosphorylation level of CLDN1 is regulated by atypical PKC and protein phosphatases (PPs) [23]. Go6976, a selective PKCα and β isoform inhibitor, decreased p-Thr levels and the TJ localization of CLDN1 (Figure 7A,B). H_2O_2 decreased PKC activity and increased PP activities, which were significantly inhibited by EBGP (Figure 7C). To support the involvement of PKC, we examined the effects of phorbol 12-myristate 13-acetate (PMA), a PKC activator, and cantharidin, a PP2A inhibitor, on the TJ barrier. The H_2O_2-induced mislocalization of CLDN1 was inhibited by both PMA and cantharidin (Figure 7D). Similarly, the reduction in the TJ barrier function by H_2O_2 was significantly rescued by both PMA and cantharidin (Figure 7E,F). The recovery effects of PMA and cantharidin were blocked by the knockdown of CLDN1 expression. These results indicated that PKC activator and PP inhibitor could rescue the H_2O_2-induced mislocalization of CLDN1.

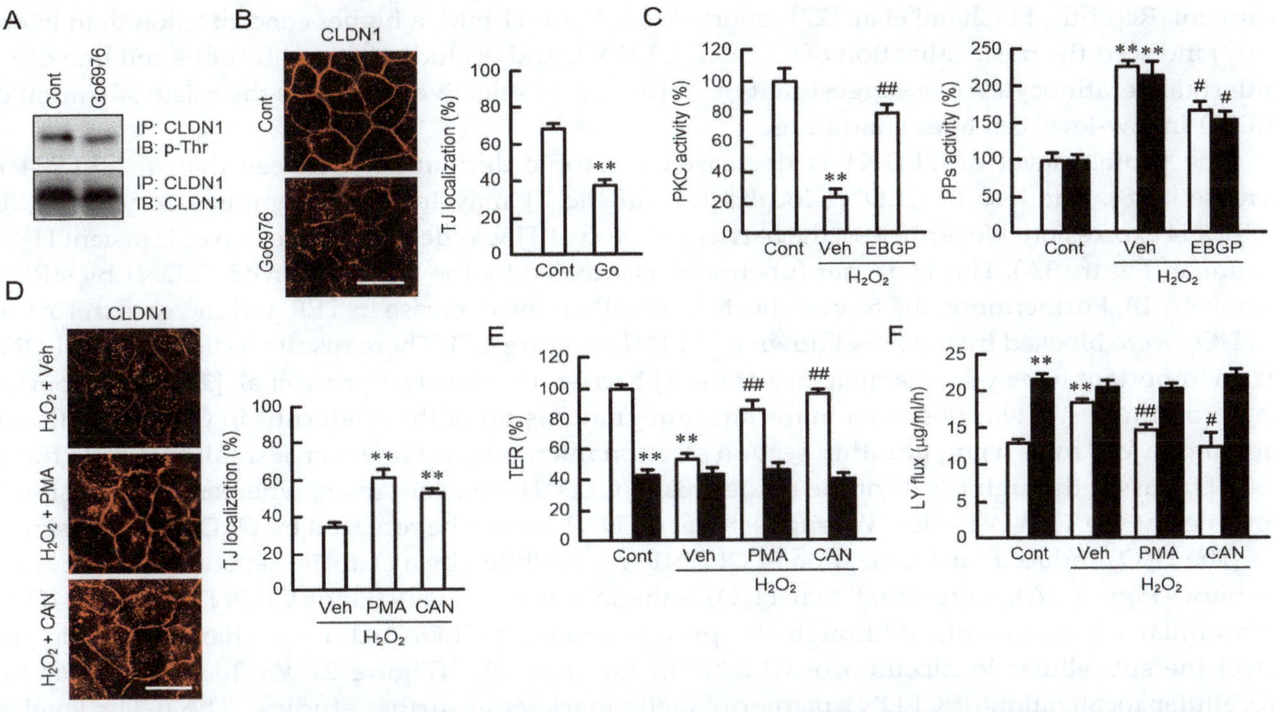

Figure 7. Upregulation of the TJ localization and phosphorylation of CLDN1 by PKC. (**A**,**B**) Cells were incubated in the absence (Cont) or presence of 10 μM Go6976 (Go) for 30 min (**A**) or 6 h (**B**). (**A**) After immunoprecipitation with anti-CLDN1 antibodies, the immunoprecipitants were blotted with anti-p-Thr and anti-CLDN1 antibodies. (**B**) The cells were immunostained with anti-CLDN1 antibodies. Scale bars indicate 10 μm. The TJ localization of CLDN1 is represented as the percentage of the total amount. (**C**) Cell lysates were prepared from the cells exposed to 200 μM H_2O_2 in the presence (EBGP) and absence (Veh) of 10 μg/mL EBGP. Control cells (Cont) were not treated with H_2O_2 and EBGP. PKC activity was measured using a CycLex PKC Super Family Kinase Assay Kit. Protein phosphatase (PP)

activities were measured using pNPP in neutral (open bars) and alkaline (closed bars) conditions. (**D**) The cells were pre-incubated with vehicle (Veh), 100 nM phorbol 12-myristate 13-acetate (PMA), or 1 μM cantharidin (CAN) for 30 min, followed by exposure to 200 μM H_2O_2 for 6 h. The cells were immunostained with anti-CLDN1 antibodies. The TJ localization of CLDN1 is represented as the percentage of the total amount. Scale bars indicate 10 μm. (**E,F**) Negative (open bars) or CLDN1 siRNA-transfected cells (closed bars) were incubated with H_2O_2, PMA, and CAN for 6 h. Control cells (Cont) were not treated with H_2O_2 and these drugs. TER and paracellular LY flux were analyzed using a volt-ohmmeter and fluorescence spectrometry, respectively. n = 3–8; ** $p < 0.01$ and * $p < 0.05$ significantly different from Cont, Cont of negative, or Veh with H_2O_2; ## $p < 0.01$ and # $p < 0.05$ significantly different from Veh with H_2O_2.

3. Discussion

Disruption of the TJ barrier in the skin is caused by UV exposure and oxidative stress. In previous research, TJ integrity has been examined using high doses of UV, which can induce noticeable cell damage [24]. The production of ROS and H_2O_2 was increased by UVB in a dose-dependent manner (Figure 1A,B). ROS production has been reported to be increased by UVB irradiation in a dose- and time-dependent manner [25]. Park et al. [26] reported that ROS production was increased 1.3-fold by UVB irradiation (10 mJ/cm^2) and incubation for 8 h in HaCaT cells, which is similar to our data. Transient H_2O_2 treatment increased the production of ROS and decreased cell viability in a dose-dependent manner (Figure 1D,E). The percentage of damaged cells was about 10%–20% in the present experimental conditions of transient H_2O_2 treatment. Transient H_2O_2 (200 μM) treatment induced the mislocalization of CLDN1 without affecting the amount of CLDN1 protein (Figure 2). There was no apparent change in CLDN4 localization, but the amount of CLDN4 protein increased transiently after 6 h of H_2O_2 treatment. Recently, El-Chami et al. [27] reported that H_2O_2 (1 mM, a higher concentration than in our study) induced the mislocalization of CLDN1, CLDN4, and occludin in a continuous cell line of rat epidermal keratinocytes. We suggest that oxidative stress selectively induces the mislocalization of CLDN1 in low-level cell toxic conditions.

The protein level of CLDN1 is decreased in atopic dermatitis, whereas that of CLDN4 is increased [28]. The loss of CLDN1 localization in the TJ may induce a compensatory elevation of CLDN4 expression. Nevertheless, the barrier function of TJ was decreased after 6 h of transient H_2O_2 treatment (Figure 3A). The TJ barrier function was reduced by the knockdown of CLDN1 by siRNA (Figure 4A,B). Furthermore, the rescue effects of EBGP on the decrease in TER and increase in LY flux by H_2O_2 were blocked by the knockdown of CLDN1 (Figure 4C). There results indicate that CLDN1 has an important role in the maintenance of the TJ barrier. Previously, Furuse et al. [29] have reported that a continuous TJ was observed in the stratum granulosum of the epidermis in CLDN1-deficient mice and in WT mice using ultrathin section electron microscopy. However, a small molecule tracer (~600 D) passes through the TJ of the epidermis in CLDN1-deficient mice, whereas the diffusion is prevented at the TJ in WT mice. We suggest that CLDN1 cannot be replaced by CLDN4 in the skin.

The H_2O_2-induced mislocalization of CLDN1 was inhibited by a clathrin-dependent endocytosis inhibitor (Figure 5A), suggesting that H_2O_2 enhances the endocytosis of CLDN1 from the TJ to intracellular compartments. Although the protein levels of CLDN1 did not change, we did not detect the subcellular localization of CLDN1 in the organelle (Figure 2). We have to clarify the subcellular localization of CLDN1 using organelle markers in further studies. The p-Thr level of CLDN1 was decreased by H_2O_2 (Figure 6A). The phosphorylation status of proteins is regulated by various protein kinases and PPs. Go6976 induced the dephosphorylation and mislocalization of CLDN1 (Figure 7). On the contrary, the H_2O_2-induced mislocalization of CLDN1 was rescued by the PKC activator PMA. These results suggest that the TJ localization of CLDN1 is upregulated by PKC-dependent phosphorylation. Dephosphorylation of CLDN1 is caused by the activation of PP2A [23]. H_2O_2 increased PP activities, and the mislocalization of CLDN1 was rescued by a PP inhibitor, cantharidin, indicating that PPs may also be involved in the dephosphorylation of CLDN1 by H_2O_2. CLDN1 is phosphorylated in both renal tubular MDCK I and colonic HT29 cells, but PKC-induced changes in the phosphorylation state were detected only in MDCK I cells [21]. The regulatory mechanism for the phosphorylation of CLDN1 may be different in each tissue.

Threonine phosphorylation sites of CLDN1 by PKC were predicted at T191 and T195 using the NetPhos 2.0 and Disphos 1.3 servers [22]. A phosphorylation mimic T191E mutant was localized to the TJ and maintained the TJ barrier in cells treated with H_2O_2, whereas WT and a T195E mutant lost their localization and barrier function in the TJ (Figure 6C,D). These results indicate that phosphorylation at T191 may be necessary for CLDN1 to localize to the TJ in HaCaT cells. The necessity of phosphorylation at T191 has been reported using human embryonic kidney 293 cells [30] and MDCK cells [31].

H_2O_2 caused a decrease in CLDN1 localization in TJ and an increase in paracellular permeability, which were rescued by EBGP (Figure 3C,D). Similar effects were observed in the UVB-treated cells. Although the effect of EBGP may be due to antioxidant activity, some cinnamic acid derivatives and flavonoids, which are contained in propolis, have beneficial effects on PKC. Artepillin C enhances adipocyte differentiation and glucose uptake mediated by the activation of PKC [32]. Chlorogenic acid prevents α-amino-hydroxy-5-methyl-isoxazole-4-propionate-mediated excitotoxicity in optic nerve oligodendrocytes through a PKC-dependent pathway [33]. Caffeic acid phenethyl ester inhibits the expression and activity of PP2A [34]. Our preliminary data indicate that kaempferide, which is abundant in EBGP [35], had lower antioxidant activity compared to EBGP, but it rescued the H_2O_2-induced mislocalization of CLDN1 and the reduction in the TJ barrier function. The components of propolis, including kaempferide, vary depending on the area [14]. A comparison of the effects of propolis produced in various places may be good for the identification of active components. Further studies are needed on which components of EBGP can rescue the H_2O_2-induced mislocalization of CLDN1 using human keratinocytes and what doses are effective. The identification of active components could lead to expanding the range of raw substances beyond green propolis, which could prove useful for the same functions.

In conclusion, we found that UVB irradiation increased ROS production, including H_2O_2, and both UVB and H_2O_2 caused the mislocalization of CLDN1 in HaCaT cells. The H_2O_2-induced mislocalization of CLDN1 was rescued by EBGP, PMA, and cantharidin. H_2O_2 decreased PKC activity and increased PP activities, which were inhibited by EBGP. Go6976 decreased p-Thr levels and the TJ localization of CLDN1. These results suggest that the TJ localization of CLDN1 is controlled by PKC. H_2O_2 decreased the p-Thr level of CLDN1, but this effect was blocked by EBGP. The T191E CLDN1 mutant was localized to the TJ after treatment with H_2O_2. We suggest that the phosphorylation of CLDN1 at T191 is necessary for its localization in the TJ. EBGP and its components may be useful in preventing the destruction of the TJ barrier by UVB and oxidative stress.

4. Materials and Methods

4.1. Materials

Rabbit anti-CLDN1, mouse anti-CLDN4, and rabbit anti-ZO-1 antibodies were obtained from Thermo Fisher Scientific (San Diego, CA, USA). Goat anti-β-actin and mouse anti-FLAG antibodies were from Santa Cruz Biotechnology (Santa Cruz, CA, USA) and Wako Pure Chemical (Osaka, Japan), respectively. Mouse anti-p-Ser and anti-p-Thr antibodies were from Sigma-Aldrich (Saint Louis, MO, USA). EBGP, LY, H_2DCFDA, and PMA were from Yamada Bee Company, Inc. (Lot: LY-009, Okayama, Japan), Biotium (Fremont, CA, USA), Thermo Fisher Scientific, and LC Laboratories (Woburn, MA, USA), respectively. OxiOrange and Hydrop were from Goryo Chemical (Hokkaido, Japan). All other reagents were of the highest grade of purity available.

4.2. Cell Culture

HaCaT cells, an immortalized nontumorigenic human keratinocyte-derived cell line [36], were grown in Dulbecco's Modified Eagle's Medium (Sigma-Aldrich) supplemented with 5% fetal bovine serum (FBS, Sigma-Aldrich), 0.07 mg/mL penicillin-G potassium, and 0.14 mg/mL streptomycin sulfate in a 5% CO_2 atmosphere at 37 °C. One day before experiments, cells were transferred to FBS-free medium. Cell viability was examined using a WST-1 assay.

4.3. UVB Irradiation

UVB irradiation was carried out using a UV Crosslinker CL-1000M (Analytik Jena, Upland, CA, USA), which emits most of its energy within the UVB range (peaking at 302 nm). HaCaT cells were irradiated at a dose of 5–50 mJ/cm^2 in Hank's balanced salt solution. After UVB radiation, cells were incubated in fresh medium until analysis.

4.4. Production of Reactive Oxygen Species

H_2DCFDA can detect several ROS, including H_2O_2, ·OH, and peroxy radical, whereas OxiOrange and Hydrop selectively detect ·OH and H_2O_2, respectively. HaCaT cells were incubated with these ROS-sensitive fluorescent probes for 30 min. The fluorescence intensity of each probe was detected using an Infinite F200 Pro microplate reader (Tecan, Mannedorf, Switzerland).

4.5. Confocal Microscopy

Cells were plated on cover glasses. After forming a confluent monolayer, the cells were incubated with cold methanol for 10 min at −30 °C and then permeabilized with 0.2% Triton X-100 for 10 min. Following permeabilization, the cells were blocked with 4% Block Ace (Dainippon Sumitomo Pharma, Osaka, Japan) for 30 min and incubated with anti-CLDN1, anti-CLDN4, anti-FLAG, or anti-ZO-1 antibodies (1:100 dilution) for 16 h at 4 °C, followed by incubation with Alexa Fluor 488- or 555-conjugated secondary antibodies (1:100 dilution). The fluorescence images were observed using an LSM 700 confocal microscope (Carl Zeiss, Jena, Germany).

4.6. SDS-Polyacrylamide Gel Electrophoresis and Immunoblotting

Cells were scraped into cold phosphate-buffered saline and precipitated by centrifugation. They were lysed in a RIPA buffer containing 150 mM NaCl, 0.5 mM EDTA, 1% Triton X-100, 0.1% SDS, 50 mM Tris-HCl (pH 8.0), and a protease inhibitor cocktail (Sigma-Aldrich) and were sonicated for 20 s. After centrifugation at 6000× *g* for 5 min, the supernatants were collected and used as cell lysates, which included membrane and cytoplasmic proteins. Samples were applied to SDS-polyacrylamide gel electrophoresis (SDS-PAGE) and blotted onto a polyvinylidene fluoride membrane. The membrane was then incubated with the respective primary antibody (1:1000 dilution) at 4 °C for 16 h, followed by a peroxidase-conjugated secondary antibody (1:3000 dilution) at room temperature for 1.5 h. Finally, the blots were incubated in EzWestLumi Plus (Atto, Tokyo, Japan) or ImmunoStar Basic (Wako Pure Chemical) and scanned with a C-DiGit Blot Scanner (LI-COR Biotechnology, Lincoln, NE, USA). The blots were stripped and reprobed with an anti-β-actin antibody. Band density was quantified using ImageJ software (National Institute of Health software). The signals were normalized using a β-actin loading control.

4.7. Measurement of Paracellular Permeability

Cells were plated on Transwell plates (0.4 μm pore size, Corning Inc., Corning, NY, USA). After forming a confluent monolayer, TER was measured using a Millicell-ERS epithelial volt-ohmmeter (Millipore, Billerica, MA, USA). TER values (ohms × cm^2) were normalized by the area of the monolayer and were calculated by subtracting the blank values from the filter and the bathing medium. The paracellular permeability to LY, a fluorescent paracellular flux marker, for 1 h from upper to lower compartments was measured with an Infinite F200 Pro microplate reader.

4.8. Isolation of Total RNA and Quantitative Real-Time Polymerase Chain Reaction

Total RNA was extracted using a TRI reagent (Sigma-Aldrich). Reverse transcription and quantitative real-time polymerase chain reaction (PCR) was performed as described previously [37]. The primer pairs used for PCR were human CLDN1 (sense: 5′-ATGAGGATGGCTGTCATTGG-3′; antisense: 5′-ATTGACTGGGGTCATAGGGT-3′) and human β-actin (sense: 5′-CCTGAGGCACTCTTCCAGCCTT-3′; antisense: 5′-TGCGGATGTCCACGTCACACTTC-3′).

4.9. PKC and Serine/Threonine Protein Phosphatase Activity Assays

Cells were harvested in 1× Passive Buffer (Promega, Madison, WI, USA). PKC activity was measured using a CycLex PKC Super Family Kinase Assay Kit (Medical & Biological Laboratories, Nagoya, Japan) in accordance with the manufacturer's protocol. Serine/threonine PP activities were investigated using paranitrophenylphosphate (pNPP) as a substrate at pH 7.5 (neutral condition) and pH 8.4 (alkaline condition). The hydrolysis of pNPP was measured by monitoring the absorbance at 405 nm.

4.10. Vector Construction and Transfection

A vector containing human CLDN1 cDNA was prepared as described previously [31] and was called CLDN1/pCMV, which encoded a FLAG tag at the amino terminus. Mutants of T191E and T195E were generated as described previously [31]: siRNAs against CLDN1 (SASI_Hs01_00211216) and a negative control (SIC-001) were purchased from Sigma-Aldrich. Plasmid vector and siRNAs were transfected into cells using Lipofectamine 2000, as recommended by the manufacturer.

4.11. Statistics

Results are presented as means ± S.E.M. Differences between groups were analyzed using one-way analysis of variance, and corrections for multiple comparison were made using Tukey's multiple comparison test. Comparisons between two groups were made using Student's *t*-test. Statistical analyses were performed using KaleidaGraph version 4.5.1 software (Synergy Software, Reading, PA, USA). Significant differences were assumed at $p < 0.05$.

Author Contributions: K.M., M.K., and S.S. performed the experiments and analyzed the data. T.M. contributed to the experiment plan and to the discussion of the manuscript. A.I. contributed to supervision of the project, interpretation of the data, and writing the paper.

Abbreviations

CLDN	Claudin
Ct	Threshold cycle
FBS	Fetal bovine serum
EBGP	Ethanol extract of Brazilian green propolis
H_2DCFDA	2′,7′-Dichlorodihydrofluorescein diacetate
H_2O_2	Hydrogen peroxide
LY	Lucifer yellow
MβCD	Methyl-β-cyclodextrin
MDC	Monodansylcadaverine
p-Ser	Phosphoserine
p-Thr	Phosphothreonine
PCR	Polymerase chain reaction
PDZ	PSD95/Dlg/ZO-1
PMA	Phorbol 12-myristate 13-acetate

pNPP	Paranitrophenylphosphate
PPs	Protein phosphatases
ROS	Reactive oxygen species
SDS-PAGE	SDS-polyacrylamide gel electrophoresis
siRNA	Small interfering RNA
TER	Transepithelial electrical resistance
TJ	Tight junction
UV	Ultraviolet
WST-1	2-(4-Iodophenyl)-3-(4-nitrophenyl)-5-(2,4-disulfophenyl)-2*H*-tetrazolium
WT	Wild-type

References

1. Coolen, N.A.; Schouten, K.C.; Middelkoop, E.; Ulrich, M.M. Comparison between human fetal and adult skin. *Arch. Dermatol. Res.* **2010**, *302*, 47–55. [CrossRef] [PubMed]
2. Young, A.R.; Chadwick, C.A.; Harrison, G.I.; Nikaido, O.; Ramsden, J.; Potten, C.S. The similarity of action spectra for thymine dimers in human epidermis and erythema suggests that DNA is the chromophore for erythema. *J. Investig. Dermatol.* **1998**, *111*, 982–988. [CrossRef] [PubMed]
3. De Gruijl, F.R. Action spectrum for photocarcinogenesis. *Recent Results Cancer Res.* **1995**, *139*, 21–30. [PubMed]
4. Moon, S.E.; Youn, J.I.; Kim, J.A. The effect of ultraviolet-B exposure scheduling on the photodamage of hairless mouse skin. *Photodermatol. Photoimmunol. Photomed.* **2000**, *16*, 74–77. [CrossRef] [PubMed]
5. Pustisek, N.; Situm, M. UV-radiation, apoptosis and skin. *Coll. Antropol.* **2011**, *35* (Suppl. 2), 339–341.
6. Ichihashi, M.; Ueda, M.; Budiyanto, A.; Bito, T.; Oka, M.; Fukunaga, M.; Tsuru, K.; Horikawa, T. UV-induced skin damage. *Toxicology* **2003**, *189*, 21–39. [CrossRef]
7. Mineta, K.; Yamamoto, Y.; Yamazaki, Y.; Tanaka, H.; Tada, Y.; Saito, K.; Tamura, A.; Igarashi, M.; Endo, T.; Takeuchi, K.; et al. Predicted expansion of the claudin multigene family. *FEBS Lett.* **2011**, *585*, 606–612. [CrossRef]
8. Turksen, K.; Troy, T.C. Barriers built on claudins. *J. Cell Sci.* **2004**, *117*, 2435–2447. [CrossRef]
9. Basler, K.; Bergmann, S.; Heisig, M.; Naegel, A.; Zorn-Kruppa, M.; Brandner, J.M. The role of tight junctions in skin barrier function and dermal absorption. *J. Control. Release* **2016**, *242*, 105–118. [CrossRef]
10. Sugawara, T.; Iwamoto, N.; Akashi, M.; Kojima, T.; Hisatsune, J.; Sugai, M.; Furuse, M. Tight junction dysfunction in the stratum granulosum leads to aberrant stratum corneum barrier function in claudin-1-deficient mice. *J. Dermatol. Sci.* **2013**, *70*, 12–18. [CrossRef]
11. Hadj-Rabia, S.; Baala, L.; Vabres, P.; Hamel-Teillac, D.; Jacquemin, E.; Fabre, M.; Lyonnet, S.; De Prost, Y.; Munnich, A.; Hadchouel, M.; et al. Claudin-1 gene mutations in neonatal sclerosing cholangitis associated with ichthyosis: A tight junction disease. *Gastroenterology* **2004**, *127*, 1386–1390. [CrossRef] [PubMed]
12. Zhou, Y.; Zheng, J.; Li, Y.; Xu, D.P.; Li, S.; Chen, Y.M.; Li, H.B. Natural Polyphenols for Prevention and Treatment of Cancer. *Nutrients* **2016**, *8*, 515. [CrossRef] [PubMed]
13. Sawicka, D.; Car, H.; Borawska, M.H.; Niklinski, J. The anticancer activity of propolis. *Folia Histochem. Cytobiol.* **2012**, *50*, 25–37. [CrossRef] [PubMed]
14. Toreti, V.C.; Sato, H.H.; Pastore, G.M.; Park, Y.K. Recent progress of propolis for its biological and chemical compositions and its botanical origin. *Evid. Based Complement. Alternat. Med.* **2013**, *2013*, 697390. [CrossRef] [PubMed]
15. Fonseca, Y.M.; Marquele-Oliveira, F.; Vicentini, F.T.; Furtado, N.A.; Sousa, J.P.; Lucisano-Valim, Y.M.; Fonseca, M.J. Evaluation of the Potential of Brazilian Propolis against UV-Induced Oxidative Stress. *Evid. Based Complement. Alternat. Med.* **2011**, *2011*, 863917. [CrossRef] [PubMed]
16. Kim, H.B.; Yoo, B.S. Propolis Inhibits UVA-Induced Apoptosis of Human Keratinocyte HaCaT Cells by Scavenging ROS. *Toxicol. Res.* **2016**, *32*, 345–351. [CrossRef]
17. De Almeida, E.B.; Cordeiro Cardoso, J.; Karla de Lima, A.; de Oliveira, N.L.; de Pontes-Filho, N.T.; Oliveira Lima, S.; Leal Souza, I.C.; de Albuquerque-Junior, R.L. The incorporation of Brazilian propolis into collagen-based dressing films improves dermal burn healing. *J. Ethnopharmacol.* **2013**, *147*, 419–425. [CrossRef] [PubMed]

18. Li, J.; Li, Q.; Geng, S. Alltrans retinoic acid alters the expression of the tight junction proteins Claudin1 and 4 and epidermal barrier functionassociated genes in the epidermis. *Int. J. Mol. Med.* **2019**, *43*, 1789–1805. [PubMed]
19. D'Souza, T.; Agarwal, R.; Morin, P.J. Phosphorylation of claudin-3 at threonine 192 by cAMP-dependent protein kinase regulates tight junction barrier function in ovarian cancer cells. *J. Biol. Chem.* **2005**, *280*, 26233–26240. [CrossRef]
20. Ikari, A.; Matsumoto, S.; Harada, H.; Takagi, K.; Hayashi, H.; Suzuki, Y.; Degawa, M.; Miwa, M. Phosphorylation of paracellin-1 at Ser217 by protein kinase A is essential for localization in tight junctions. *J. Cell Sci.* **2006**, *119*, 1781–1789. [CrossRef]
21. Sjo, A.; Magnusson, K.E.; Peterson, K.H. Protein kinase C activation has distinct effects on the localization, phosphorylation and detergent solubility of the claudin protein family in tight and leaky epithelial cells. *J. Membr. Biol.* **2010**, *236*, 181–189. [CrossRef] [PubMed]
22. Ahmad, W.; Shabbiri, K.; Ijaz, B.; Asad, S.; Sarwar, M.T.; Gull, S.; Kausar, H.; Fouzia, K.; Shahid, I.; Hassan, S. Claudin-1 required for HCV virus entry has high potential for phosphorylation and O-glycosylation. *Virol. J.* **2011**, *8*, 229. [CrossRef] [PubMed]
23. Nunbhakdi-Craig, V.; Machleidt, T.; Ogris, E.; Bellotto, D.; White, C.L.; Sontag, E. Protein phosphatase 2A associates with and regulates atypical PKC and the epithelial tight junction complex. *J. Cell Biol.* **2002**, *158*, 967–978. [CrossRef] [PubMed]
24. Alhasaniah, A.; Sherratt, M.J.; O'Neill, C.A. The Impact of Ultraviolet Radiation on Barrier Function in Human Skin: Molecular Mechanisms and Topical Therapeutics. *Curr. Med. Chem.* **2018**, *25*, 5503–5511. [CrossRef] [PubMed]
25. Hu, Y.; Ma, Y.; Wu, S.; Chen, T.; He, Y.; Sun, J.; Jiao, R.; Jiang, X.; Huang, Y.; Deng, L.; et al. Protective Effect of Cyanidin-3-O-Glucoside against Ultraviolet B Radiation-Induced Cell Damage in Human HaCaT Keratinocytes. *Front. Pharmacol.* **2016**, *7*, 301. [CrossRef] [PubMed]
26. Park, E.J.; Kim, Y.M.; Chang, K.C. Hemin Reduces HMGB1 Release by UVB in an AMPK/HO-1-dependent Pathway in Human Keratinocytes HaCaT Cells. *Arch. Med. Res.* **2017**, *48*, 423–431. [CrossRef] [PubMed]
27. El-Chami, C.; Haslam, I.S.; Steward, M.C.; O'Neill, C.A. Organic osmolytes preserve the function of the developing tight junction in ultraviolet B-irradiated rat epidermal keratinocytes. *Sci. Rep.* **2018**, *8*, 5167. [CrossRef]
28. Gruber, R.; Bornchen, C.; Rose, K.; Daubmann, A.; Volksdorf, T.; Wladykowski, E.; Vidal, Y.S.S.; Peters, E.M.; Danso, M.; Bouwstra, J.A.; et al. Diverse regulation of claudin-1 and claudin-4 in atopic dermatitis. *Am. J. Pathol.* **2015**, *185*, 2777–2789. [CrossRef]
29. Furuse, M.; Hata, M.; Furuse, K.; Yoshida, Y.; Haratake, A.; Sugitani, Y.; Noda, T.; Kubo, A.; Tsukita, S. Claudin-based tight junctions are crucial for the mammalian epidermal barrier: A lesson from claudin-1-deficient mice. *J. Cell Biol.* **2002**, *156*, 1099–1111. [CrossRef]
30. Shiomi, R.; Shigetomi, K.; Inai, T.; Sakai, M.; Ikenouchi, J. CaMKII regulates the strength of the epithelial barrier. *Sci. Rep.* **2015**, *5*, 13262. [CrossRef]
31. Fujii, N.; Matsuo, Y.; Matsunaga, T.; Endo, S.; Sakai, H.; Yamaguchi, M.; Yamazaki, Y.; Sugatani, J.; Ikari, A. Hypotonic stress-induced down-regulation of claudin-1 and -2 mediated by dephosphorylation and clathrin-dependent endocytosis in renal tubular epithelial cells. *J. Biol. Chem.* **2016**, *291*, 24787–24799. [CrossRef] [PubMed]
32. Choi, S.S.; Cha, B.Y.; Iida, K.; Lee, Y.S.; Yonezawa, T.; Teruya, T.; Nagai, K.; Woo, J.T. Artepillin C, as a PPARgamma ligand, enhances adipocyte differentiation and glucose uptake in 3T3-L1 cells. *Biochem. Pharmacol.* **2011**, *81*, 925–933. [CrossRef] [PubMed]
33. Rebai, O.; Amri, M. Chlorogenic Acid Prevents AMPA-Mediated Excitotoxicity in Optic Nerve Oligodendrocytes Through a PKC and Caspase-Dependent Pathways. *Neurotox. Res.* **2018**, *34*, 559–573. [CrossRef] [PubMed]
34. Avci, C.B.; Sahin, F.; Gunduz, C.; Selvi, N.; Aydin, H.H.; Oktem, G.; Topcuoglu, N.; Saydam, G. Protein phosphatase 2A (PP2A) has a potential role in CAPE-induced apoptosis of CCRF-CEM cells via effecting human telomerase reverse transcriptase activity. *Hematology* **2007**, *12*, 519–525. [CrossRef] [PubMed]
35. Andrade, J.K.S.; Denadai, M.; de Oliveira, C.S.; Nunes, M.L.; Narain, N. Evaluation of bioactive compounds potential and antioxidant activity of brown, green and red propolis from Brazilian northeast region. *Food Res. Int.* **2017**, *101*, 129–138. [CrossRef] [PubMed]

36. Boukamp, P.; Petrussevska, R.T.; Breitkreutz, D.; Hornung, J.; Markham, A.; Fusenig, N.E. Normal keratinization in a spontaneously immortalized aneuploid human keratinocyte cell line. *J. Cell Biol.* **1988**, *106*, 761–771. [CrossRef] [PubMed]
37. Sonoki, H.; Tanimae, A.; Endo, S.; Matsunaga, T.; Furuta, T.; Ichihara, K.; Ikari, A. Kaempherol and Luteolin Decrease Claudin-2 Expression Mediated by Inhibition of STAT3 in Lung Adenocarcinoma A549 Cells. *Nutrients* **2017**, *9*, 597. [CrossRef]

11

Curcumin Mitigates Immune-Induced Epithelial Barrier Dysfunction by *Campylobacter jejuni*

Fábia Daniela Lobo de Sá [1], Eduard Butkevych [1], Praveen Kumar Nattramilarasu [1], Anja Fromm [1], Soraya Mousavi [2], Verena Moos [3], Julia C. Golz [4], Kerstin Stingl [4], Sophie Kittler [5], Diana Seinige [5], Corinna Kehrenberg [6], Markus M. Heimesaat [2], Stefan Bereswill [2], Jörg-Dieter Schulzke [1,*] and Roland Bücker [1]

1. Institute of Clinical Physiology/Nutritional Medicine, Medical Department, Division of Gastroenterology, Infectiology, Rheumatology, Charité–Universitätsmedizin Berlin, 12203 Berlin, Germany
2. Institute of Microbiology, Infectious Diseases and Immunology, Charité–Universitätsmedizin Berlin, Campus Benjamin Franklin, 14195 Berlin, Germany
3. Medical Department, Division of Gastroenterology, Infectiology and Rheumatology, Charité–Universitätsmedizin Berlin, 12203 Berlin, Germany
4. German Federal Institute for Risk Assessment (BfR), Department of Biological Safety, National Reference Laboratory for Campylobacter, 12277 Berlin, Germany
5. University of Veterinary Medicine Hannover, Research Center for Emerging Infections and Zoonoses, 30559 Hannover, Germany
6. Institute for Veterinary Food Science, Justus-Liebig-University, 35392 Giessen, Germany

* Correspondence: joerg.schulzke@charite.de

Abstract: *Campylobacter jejuni* (*C. jejuni*) is the most common cause of foodborne gastroenteritis worldwide. The bacteria induce diarrhea and inflammation by invading the intestinal epithelium. Curcumin is a natural polyphenol from turmeric rhizome of *Curcuma longa*, a medical plant, and is commonly used in curry powder. The aim of this study was the investigation of the protective effects of curcumin against immune-induced epithelial barrier dysfunction in *C. jejuni* infection. The indirect *C. jejuni*-induced barrier defects and its protection by curcumin were analyzed in co-cultures with HT-29/B6-GR/MR epithelial cells together with differentiated THP-1 immune cells. Electrophysiological measurements revealed a reduction in transepithelial electrical resistance (TER) in infected co-cultures. An increase in fluorescein (332 Da) permeability in co-cultures as well as in the germ-free IL-$10^{-/-}$ mouse model after *C. jejuni* infection was shown. Curcumin treatment attenuated the *C. jejuni*-induced increase in fluorescein permeability in both models. Moreover, apoptosis induction, tight junction redistribution, and an increased inflammatory response—represented by TNF-α, IL-1β, and IL-6 secretion—was observed in co-cultures after infection and reversed by curcumin. In conclusion, curcumin protects against indirect *C. jejuni*-triggered immune-induced barrier defects and might be a therapeutic and protective agent in patients.

Keywords: *Campylobacter jejuni*; curcumin; tight junction; claudin; apoptosis; co-culture; mouse colon; cytokines; TNF; NFκB

1. Introduction

Campylobacter jejuni is the most prevalent pathogenic bacterium of zoonotic gastroenteritis [1]. Typical symptoms provoked by *C. jejuni* are watery to bloody diarrhea, abdominal pain, fever, and nausea [2]. This human pathogen is present in the intestinal microbiota of farm animals, especially poultry, which is the main source of infection for humans by ingestion of contaminated or undercooked food [1]. The bacteria adhere to the mucus and the surface of intestinal epithelial cells, invade the intestinal epithelium, while they pass the cells via the transcellular or paracellular route [3,4]. Consequently, direct epithelial barrier defects such as dysregulation of tight junction (TJ) proteins, induction of epithelial lesions or also indirect effects by inflammatory responses of epithelial or immune cells occur [5,6].

Epithelial lesions can go along with single-cell lesions by increased apoptosis, mid-sized leaks like focal leaks, as well as erosions or even ulcerations. These pathological findings could explain the type of diarrhea for the *Campylobacter* infection as leak flux pathomechanism, which is characterized by a loss of water and solutes from the organism into the intestinal lumen through a leaky epithelium [6,7]. For a physiological intestinal integrity, an intact epithelium is essential. The main components of intestinal epithelial barrier are the TJs. They are the apical components of the intercellular seal and divide the epithelial cell membrane into an apical and basolateral compartment (fence function) [8]. Tight junctions form a barrier for water, ions, and solutes (barrier function) [9] and consist of various molecular transmembrane proteins like claudins and occludin, junction adhesion molecules (JAM), tricellulin, and cytoplasmatic scaffolding proteins as zonula occludens protein-1 (ZO-1) [8,9].

Recently, the number of *C. jejuni* infections surpasses *Salmonella* infections as the leading cause of zoonotic bacterial gastroenteritis in the US and Europe. Moreover, high levels of fluoroquinolone resistance have been reported [10] and levels of multi drug resistances (MDR) are increasing [11]. In consequence, new therapeutic targets as well as innovative strategies to reduce the amount of *C. jejuni* in farm animals and environment are urgently required. A potent natural substance thought to interfere with *Campylobacter* infection is curcumin.

Curcumin is a polyphenolic compound found in the turmeric root of the *Curcuma longa* plant (commonly known as turmeric), a member of the ginger family [12–14]. It is used as coloring agent, cosmetics, and it is the principal ingredient in oriental food spice like curry powder [12,13]. For centuries, curcumin has been used in Asian traditional medicine as a common nontoxic and active agent against different gastrointestinal and digestive disorders [12,14,15]. In addition, curcumin is known to possess anti-inflammatory, antioxidative, antibacterial, anticarcinogenic [16], antiviral [15,17], antiapoptotic, and antiproliferative properties [12]. The antibacterial effect was shown against *Salmonella* spp. [18], *Helicobacter pylori* [16,19,20], methicillin-resistant *Staphylococcus aureus*, *Bacillus subtilis*, and *Escherichia coli* [21]. Various studies have shown that curcumin modulates inflammatory cytokine pathways like IL-1β, TNF-α, IL-6, IL-8, IL-10 [12,13,22,23], and inhibit signaling pathways like NFκB [12,17,19].

The aim of the presented study was to evaluate the barrier-protective and anti-inflammatory properties of curcumin on intestinal epithelial barrier function, which is compromised by the *C. jejuni*-mediated immune response. Therefore, a novel in vitro co-culture model consisting of intestinal epithelial HT-29/B6-GR/MR and immune THP-1 cells was established. This co-culture model enables the screening of potential barrier-protective and anti-inflammatory compounds in *C. jejuni* infection, as a result of which curcumin was identified as being effective.

2. Results

2.1. Establishment of a Co-Culture with Colon Epithelial Cells HT-29/B6-GR/MR and Immune THP-1 Cells

To analyze the impact of the immune response during a *Campylobacter* infection on epithelial barrier function, human colon epithelial cells and immune cells were co-cultured. Colonic epithelial cells (HT-29/B6-GR/MR) were seeded on cell culture filter supports (0.4 µm pore size) and represented the epithelial compartment. Under the membrane of the epithelial cell filter, THP-1 immune cells were seeded on the bottom of the cell culture well and represented the subepithelial compartment. Basal bacterial infection with *C. jejuni* decreased transepithelial electrical resistance (TER) when colonic epithelial and immune cells (differentiated THP-1 cells) were co-cultured (Figure 1). From the apical supernatant, no bacteria could be cultured after the experiment, indicating that the bacteria could not pass the filter membrane with 0.4 µm pore size. As a result of this, and in contrast to the results in the co-culture model, *C. jejuni* infection at the basal membrane had no effect on TER in HT-29/B6-GR/MR mono-culture (Figure 1).

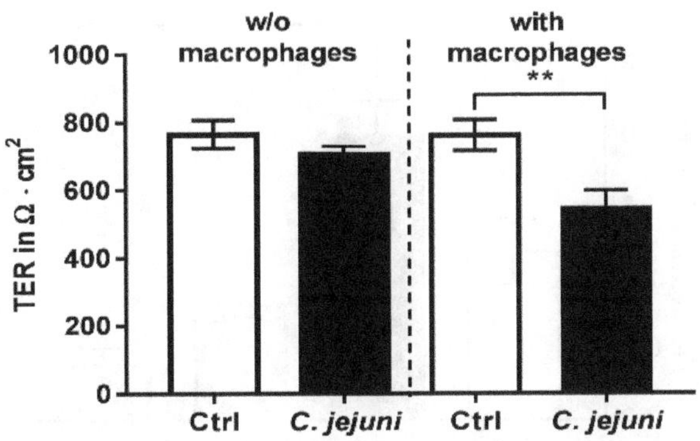

Figure 1. *C. jejuni* infection in human intestinal epithelial cells (HT-29/B6-GR/MR) or in co-culture together with human immune cells (differentiated THP-1 cells). Transepithelial electrical resistance (TER) in co-culture after *C. jejuni* infection from the basal side decreased 48 h post infection ($n = 7$–9, ** $p < 0.01$, unpaired Student's *t*-test).

2.2. *Curcumin Improves Disturbed Intestinal Barrier Function in the Co-Culture System*

Using mRNA sequencing and subsequent ingenuity pathway analysis (IPA) of human mucosa biopsies from *C. jejuni*-infected patients, inhibition of curcumin-dependent pathways indicated that the presence of curcumin might prevent the *C. jejuni*-induced changes in gene expression. Based on this bioinformatic prediction of curcumin's effect on host gene expression, the addition of curcumin should activate downstream pathways and might reduce the effects of *Campylobacter* infections (Supplementary Table S1). In order to test the barrier protective and anti-inflammatory properties of curcumin in *Campylobacter*-induced barrier dysfunction, a curcumin solution was added to the infected co-culture model. As shown in Figure 2, *C. jejuni* reduced TER significantly 48 h after infection, whereas this change was inhibited by the presence of curcumin.

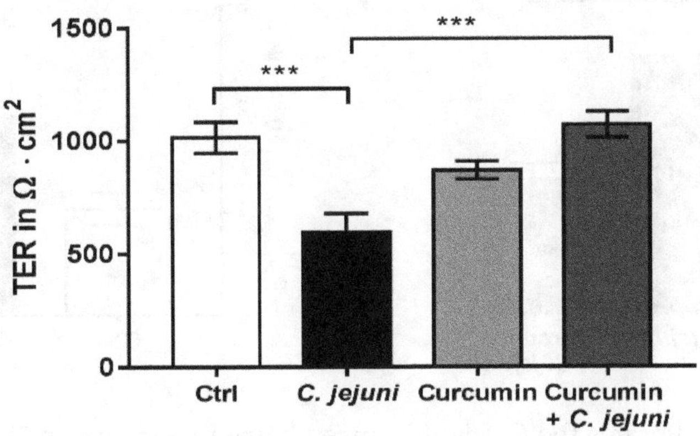

Figure 2. Curcumin protects against *C. jejuni*-induced barrier disruption in co-culture. Cells were treated with 50 μM curcumin and infected with *C. jejuni* for 48 h ($n = 13$, *** $p < 0.001$, one-way ANOVA with Bonferroni correction for multiple comparisons).

Permeability studies were performed with the paracellular flux marker fluorescein (332 Da). Compared to controls, *C. jejuni* infection caused an increase in permeability to fluorescein. Application of curcumin not only ameliorated the permeability increase for fluorescein, but even resulted in an improved barrier function for paracellular passage of macromolecules when exceeded the control level (Figure 3).

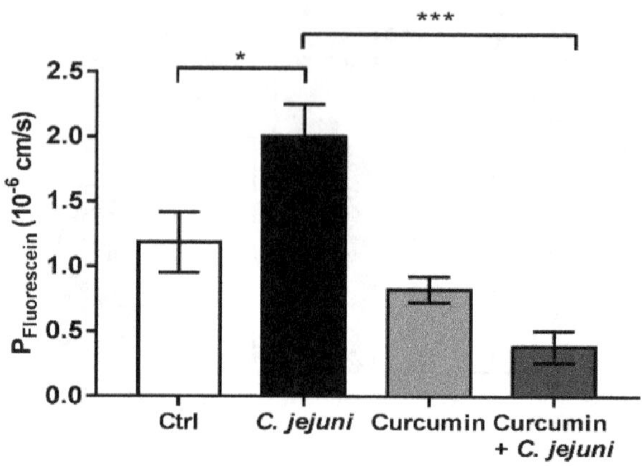

Figure 3. Curcumin improves the *C. jejuni*-induced increase in permeability to fluorescein (332 Da) in co-cultures with colon epithelial HT-29/B6-GR/MR and immune THP-1 cells. The co-cultures were infected with *C. jejuni* after incubation with 50 μM curcumin. Fluorescein permeabilities were measured 48 h post infection ($n = 7$–8, * $p < 0.05$, *** $p < 0.001$, one-way ANOVA with Bonferroni correction for multiple comparisons).

2.3. *Effect of Curcumin In Vivo on Intestinal Barrier Regulation in IL-10$^{-/-}$ Mice*

Secondary abiotic IL-10$^{-/-}$ mice were infected with *C. jejuni* and treated with 0.5 mg/mL of curcumin, starting 4 days prior infection until the end of the observation period (i.e., day 6 post infection). Epithelial barrier function of distal colon specimens from IL-10$^{-/-}$ mice was analyzed in Ussing chambers. Conductance (G, Figure 4A) and permeability to fluorescein (Figure 4B) was significant higher in *C. jejuni*-infected mice than in untreated infected mice. Mice receiving curcumin showed a significantly lower conductance and permeability to fluorescein than untreated infected mice. The electrophysiological results obtained in vivo (Figure 4) resemble the results from the in vitro co-culture system (Figure 3).

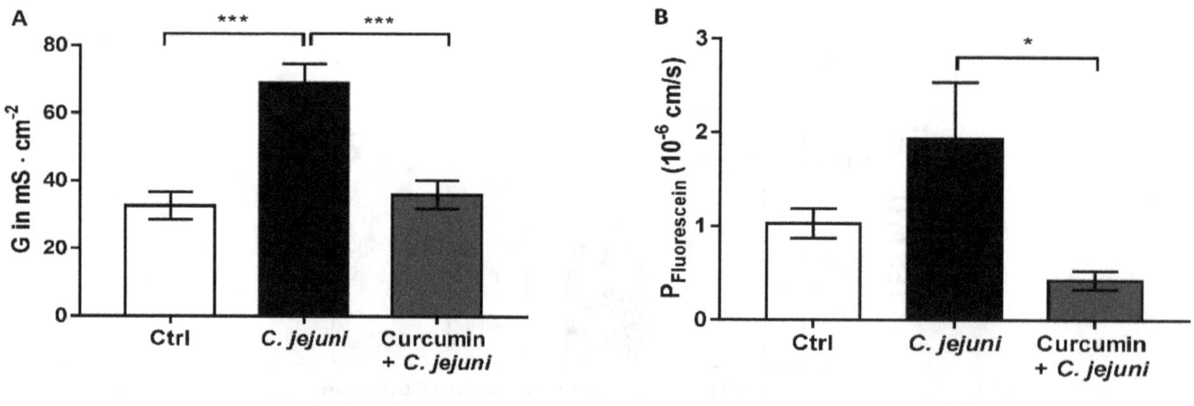

Figure 4. Curcumin ameliorates the *C. jejuni*-induced barrier dysfunction in colon of IL-10$^{-/-}$ mice. Colon specimens of IL-10$^{-/-}$ mice were mounted in Ussing chambers designed for impedance spectroscopy. (**A**) Epithelial conductance (G) ($n = 5$–18, *** $p <$ 0,001, Mann–Whitney *U*-test) and (**B**) fluorescein (332 Da) permeability ($n = 5$–14, * $p < 0.05$, Mann–Whitney *U*-test) of IL-10$^{-/-}$ mouse colon were measured in Ussing chambers. IL-10$^{-/-}$ mice were either infected with *C. jejuni* (per oral gavage with 10 ^{8} CFU in 0.3 mL) or infected and treated for 6 days with Curcumin (0.5 mg/mL).

2.4. *Effect of Curcumin on Bacterial Integrity*

To determine whether curcumin had a direct effect on the viability of *C. jejuni* in the concentration used in our experiments, the minimal inhibitory concentration (MIC) for curcumin was determined. The MIC of curcumin in *C. jejuni* 81–176 amounted to 87 μM under physiological conditions at a pH of 7.4, and is thus above our experimental curcumin concentrations in the co-culture. Moreover, if curcumin is used as treatment, it is necessary to know whether the substance exerts any stress to the bacterium, leading to enhanced adaption by, e.g., stimulation of natural transformation capacity. Therefore, competence development and DNA uptake of *C. jejuni* was investigated under the treatment of curcumin using a single cell assay. After curcumin treatment, no change in the ratio of competent bacteria was registered (control 26% ± 2% versus treated 25.5% ± 0.5% with 50 μM curcumin, $n = 3$, n.s., unpaired Student's t-test). In addition, also the DNA uptake processes in competent *C. jejuni* was unaffected after curcumin treatment (control 100% versus treated 86.9% ± 6.7%, $n = 3$, n.s., unpaired Student´s t-test).

2.5. Cytokine Secretion in C. jejuni-Infected Co-Cultures

To identify which cytokines were released by the immune cells and affected epithelial cells function, cytometric bead arrays (CBA) from supernatants of treated co-cultures were performed. TNF-α, IL-1β, and IL-6 were increased in *C. jejuni*-infected co-cultures, while this changes were prevented by curcumin treatment (Figure 5).

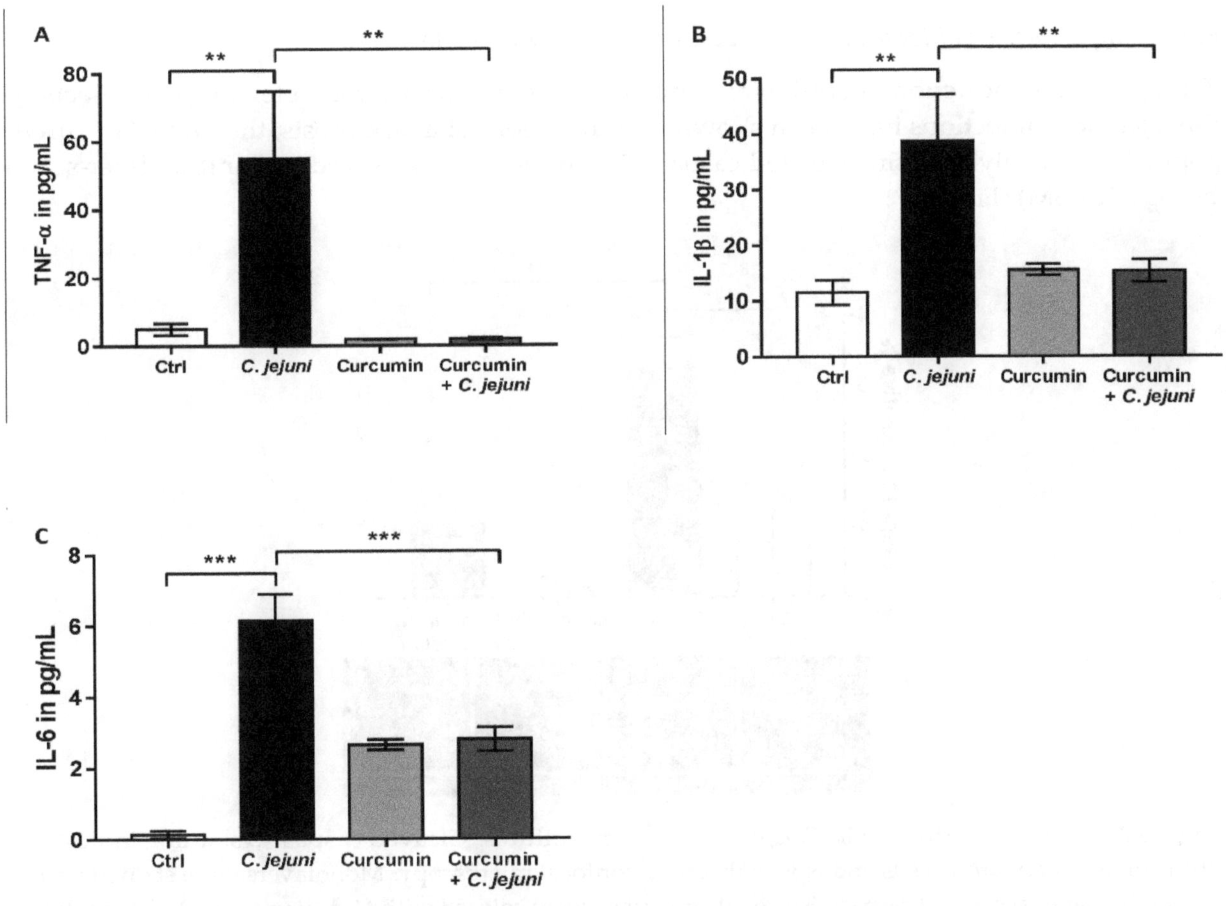

Figure 5. Cytokine release in co-cultures after *C. jejuni* infection. (**A**) Tumor necrosis factor-α (TNF-α), (**B**) Interleukin-1β (IL-1β), and (**C**) Interleukin-6 (IL-6) secretion after treatment with curcumin in *C. jejuni*-infected co-cultures ($n = 5–6$, * $p < 0.05$, ** $p < 0.01$, *** $p < 0.001$, one-way ANOVA with Bonferroni correction for multiple comparisons).

2.6. NFκB Signaling Pathway is Involved in C. jejuni-Induced Barrier Dysfunction

Furthermore, we evaluated the possible relevance of the NFκB pathway for barrier function in our cell model using the NFκB inhibitor BAY 11-7082 (BAY). BAY inhibited the *C. jejuni*-induced reduction in TER, similar to the effect of curcumin (Figure 6A). In addition, the increase in permeability to fluorescein could be reversed after BAY inhibition and almost reached control values (Figure 6B).

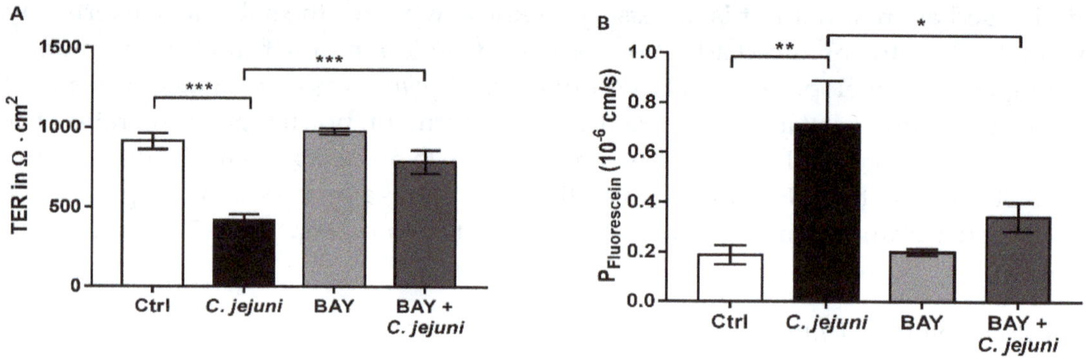

Figure 6. The NFκB pathway is barrier-relevant. Inhibitory effect of BAY 11-7082 in *C. jejuni*-infected co-cultures. (**A**) Transepithelial electrical resistance (TER) and (**B**) permeability to fluorescein (332 Da) after incubation with 10 µM BAY 11-7082 and infection with *C. jejuni* for 48 h ($n = 5$, * $p < 0.05$, ** $p < 0.01$, *** $p < 0.001$, one-way ANOVA with Bonferroni correction for multiple comparisons).

2.7. Epithelial Apoptosis is Blocked by Curcumin in the Co-Culture Model

Given that the induction of epithelial apoptosis constitutes a barrier-relevant pathomechanism, and *Campylobacter* infections have been shown to induce epithelial apoptosis, the impact of curcumin on apoptosis was analyzed using cleaved caspase-3 staining and visualized by confocal laser-scanning microscopy (CLSM) (Figure 7).

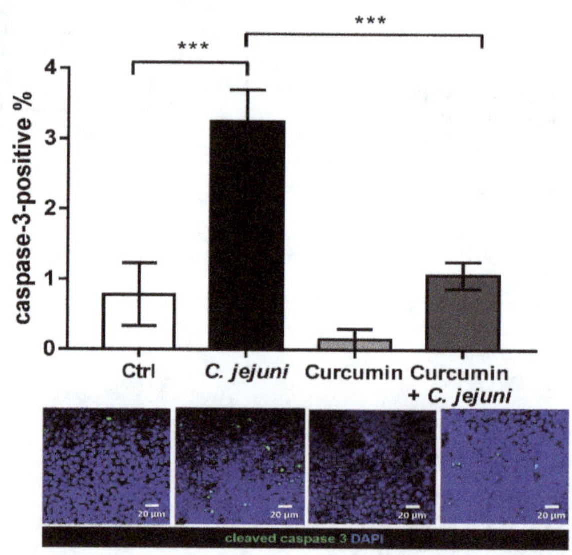

Figure 7. Apoptosis induction in *C. jejuni*-infected co-cultures. Cleaved caspase-3 staining indicates the number of apoptotic cells and is visualized by confocal microscopy. Monolayers were stained with antibodies against cleaved caspase-3 (green) and nuclei were colored with 4´-d-diamidino-2-phenylidole dihydrochloride (DAPI, blue). Signals of cleaved caspase-3-positive cells in CLSM were counted in the top view pictures and related to the number of DAPI stained nuclei (~300 nuclei/frame) to calculate the percentage of apoptotic cells in the monolayers ($n = 6$, *** $p < 0.001$, one-way ANOVA with Bonferroni correction for multiple comparisons). Representative pictures are shown for HT-29/B6-GR/MR cells, obtained from co-culture after treatment with 50 µM curcumin and infection with *C. jejuni*.

After infection with *C. jejuni*, the apoptotic ratio was increased 3-fold in comparison to control. Incubation with curcumin during the infection with *C. jejuni* reduced the apoptotic ratio to that of controls (Figure 7) (control 0.7% ± 0.4% versus infected 3.3% ± 0.4% versus treated 1.1% ± 0.2%, $n = 6$, ** $p < 0.01$, *** $p < 0.001$, one-way ANOVA with Bonferroni correction for multiple comparisons).

2.8. *Influence of* C. jejuni *and Curcumin on Tight Junction Protein Expression in the Co-Culture Model*

To define the *C. jejuni*-induced tight junction (TJ) expression changes, protein lysates were generated from colon epithelial cells of co-cultures after treatment with curcumin and infection with *C. jejuni* (Figure 8). The expression change was quantified by densitometric analysis with β-actin as loading control, which showed an increase in protein expression of claudin-1 protein expression after *C. jejuni* infection. Claudin-1 expression seemed to decrease again after curcumin treatment in the *C. jejuni* infected group. However, this did not reach statistical significance. No further expression changes of the tested TJ proteins could be observed. A reduction by trend after *C. jejuni* infection in occludin and claudin-8 was detectable, but without reaching statistical significance after multiple test correction (claudin-8 infected versus curcumin-treated $p = 0.046$, uncorrected Student´s t-test).

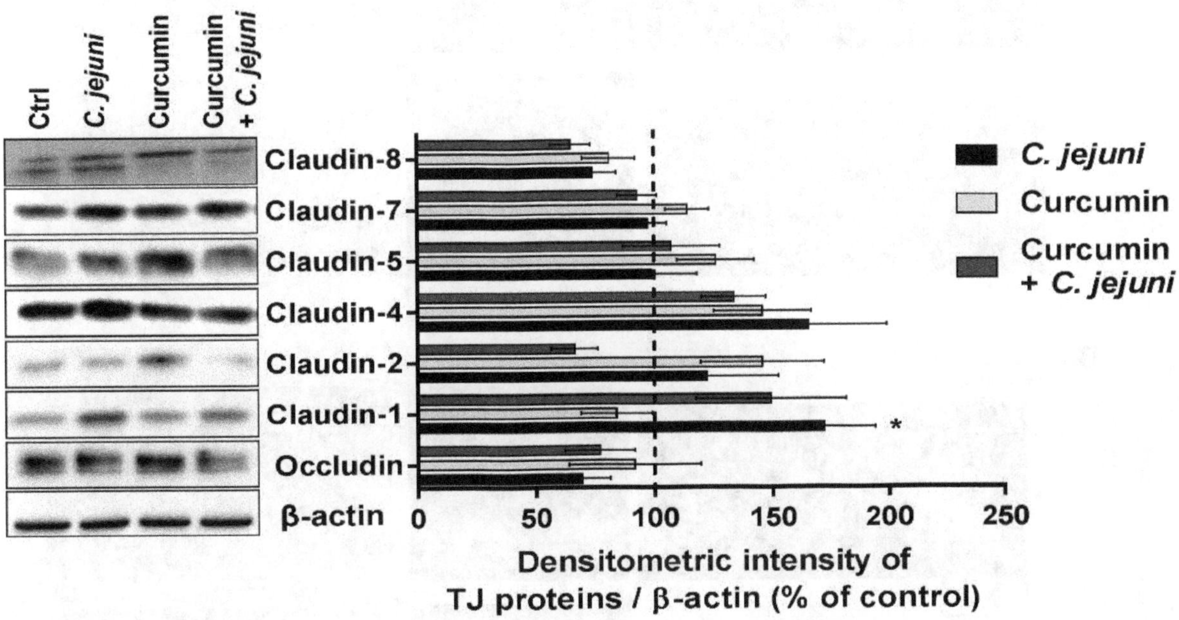

Figure 8. Tight junction (TJ) protein expression in western blot. Representative western blots and densitometry of TJ protein expression from HT-29/B6-GR/MR cells obtained from co-culture after incubation with 50 μM curcumin and infection with *C. jejuni*. TJ proteins were normalized to the level of β-actin ($n = 6$–9, * $p < 0.05$, two-way ANOVA with Bonferroni correction for multiple comparisons. Control value is marked with a dashed line.

2.9. *Influence of* C. jejuni *and Curcumin on Subcellular TJ Protein Distribution in Co-Cultures in Confocal Laser-Scanning Microscopy*

Since the composition and presence of TJ proteins in the TJ strands influence barrier function, the subcellular distribution of the TJs was analyzed with confocal laser-scanning microscopy (CLSM). Z-stacks were generated from different TJ proteins of the immunostained epithelial cell monolayers, and intensity-distance plots were generated. The TJ strands after *C. jejuni* infection indicated a redistribution of claudin-4 (Figure 9A) and claudin-8 (Figure 9B). Claudin-4 and claudin-8 signals were retracted from the TJ strands. Moreover, claudin-4 and claudin-8 were no longer co-localized with zonula occludens protein-1 (ZO-1) after *C. jejuni* infection. Intensity–distance plots with high background intensities indicate re-localization into the intracellular regions. After treatment with curcumin, the appearance of claudin-4 and claudin-8 improved. As a consequence, the infected and curcumin treated group showed co-localization (merging), with high peak intensities in the areas of TJ contact, comparable to the control. Curcumin control groups showed no difference compared to the untreated controls (Supplementary Figure S2).

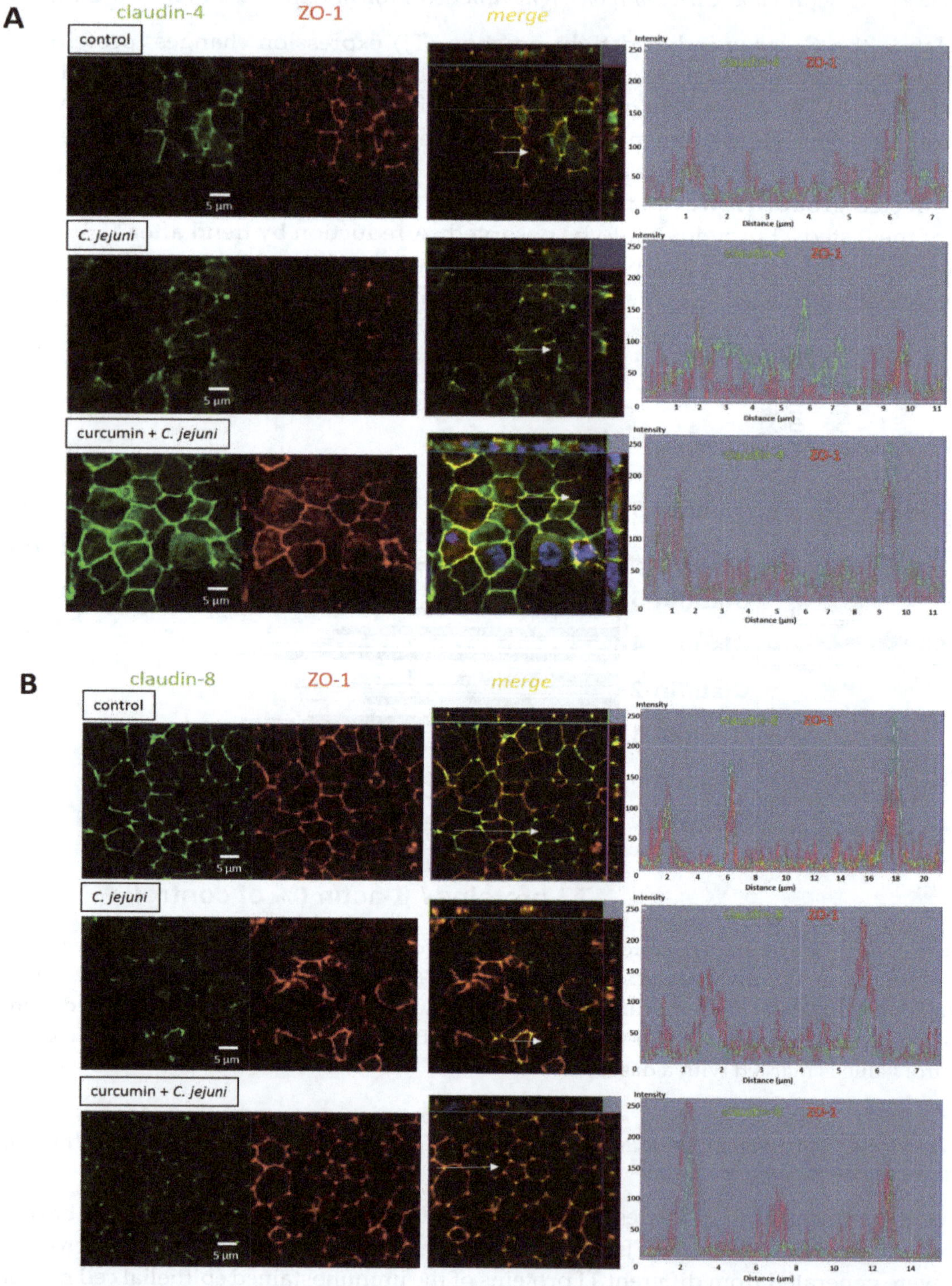

Figure 9. Tight junction distribution in *C. jejuni* infection after treatment with curcumin. Representative confocal laser-scanning microscopy pictures of HT-29/B6-GR/MR after co-culturing together with immune cells. (**A**) Claudin-4 (green) and zonula occludens protein-1 (ZO-1, red), and (**B**) claudin-8 (green) and ZO-1 (red). Nuclei are stained in blue with 4´-6-diamidino-2-phenylindole dihydrochloride (DAPI).

3. Discussion

It is important to contain the *Campylobacter jejuni* infection because of its impact on gastroenteritis worldwide and resulting sequelae, such as irritable bowel syndrome, reactive arthritis, or Guillian–Barré syndrome. Therefore, we aimed to study the underlying pathomechanisms for barrier dysfunction and the influence of the immune system in more detail. For this purpose, we used a new in vitro co-culture model and an in vivo mouse model to study the protective properties of the natural polyphenol curcumin.

Curcumin is a vitamin D receptor (VDR) ligand [24], and we found the downstream signaling pathway of curcumin to be inhibited in the mucosa of acutely infected *C. jejuni* patients by means of the bioinformatic prediction in Qiagen ingenuity pathway analysis (IPA) from RNA-sequencing (Supplementary Table S1). This prediction indicated that curcumin could be a potential therapeutic or preventive target against *C. jejuni* infections. Based on this bioinformatics prediction, we tested the effects of curcumin on *C. jejuni* infection in vitro and in vivo. The inhibitory effect of another top hit upstream regulator in IPA analysis on the mucosa of *C. jejuni*-infected patients was calcitriol (active form of vitamin D). Calcitriol exhibited a high significance value comparable to curcumin (Appendix A, Table S1). Annulment of *C. jejuni*-induced cytotoxicity after calcitriol treatment has been demonstrated in HT-29/B6 cells previously, supporting the view that the bacterium affects intestinal barrier function via a VDR-dependent pathway [6].

As predicted by the pathway analysis in acute human campylobacteriosis, curcumin was effective in our experiments. It could prevent the decrease in TER as well as the increase in fluorescein permeability in our co-culture model, indicating improved barrier function by curcumin in immune-mediated epithelial barrier dysfunction by *C. jejuni*.

Because we expected an essential contribution of the immune system, we established a co-culture model with basal infection. In this model, the *C. jejuni*-triggered immune response reduced the TER after 48 h. The apical bathing medium did not contain bacteria even after 48 h, which suggests that the effect on resistance was caused indirectly by the *C. jejuni*-induced immune response. The decrease in TER may therefore reflected an increase in paracellular permeability secondary to epithelial barrier defects, which would increase permeability to small molecules and toxins. Furthermore, this change in barrier function may increase the access of bacteria to the underlying tissue and potentiate the immune response, causing additional barrier disruption, with loss of cellular integrity [6,25]. Concomitantly, a leak-flux type of diarrhea is caused that represents a main diarrheal mechanism.

Depending on the cell type and the experimental infection conditions, *C. jejuni* was more or less able to modulate TER in previous studies. No effects on TER have been found in MKN-28 cells [26] or Caco-2 monolayers [4]. In contrast, exposure to *C. jejuni* has been reported to reduce TER in T84 and MDCK-I cells [27]. In our HT-29/B6-GR/MR cells, TER was also not different from control even after 48 h basal infection, although a decrease in TER after basolateral infection has been reported to be more pronounced than after apical infection in T84 cells [28]. That this was not seen in our HT-29/B6-GR/MR cells may reflect differences in cell type and/or the 0.4 μm filter pore size of our filters, limiting the bacterial access to the epithelial cells.

Within 6 days following peroral *C. jejuni* infection, secondary abiotic IL-$10^{-/-}$ mice develop acute ulcerative enterocolitis mimicking key features of severe campylobacteriosis in humans, with a specific cytokine response [29]. We were able to show for the first time, using impedance spectroscopy in Ussing chambers, compromised barrier integrity with an increased epithelial conductance and permeability to macromolecules in this mouse model. These data reflect the results of our co-culture model, in which the barrier defects could be similarly prevented by curcumin. Bücker et al. [6] incubated colonic mucosal biopsies from *C. jejuni*-infected patients in culture medium and quantified cytokine release into the supernatant. Secretion of TNF-α, IFN-γ, IL-1β, and IL-13 was enhanced in comparison to controls [6]. It was concluded that this mucosal cytokine storm led to disruption of intestinal TJs and the induction of epithelial apoptosis. In our infected co-cultures, release of cytokines was also mainly responsible for the barrier defects, even though THP-1 cells produce and secrete a lower amount and spectrum of cytokines in comparison to fresh isolated peripheral blood mononuclear cells (PBMC). After *C. jejuni* infection, levels of the pro-inflammatory cytokine TNF-α, IL-1β, and IL-6 increased remarkably, and these changes were largely prevented by curcumin. Curcumin not only reduced local inflammation in the intestine, it also reduced systemic inflammation triggered by LPS [23]. In another study, curcumin nanoparticles suppressed the expression of mRNAs encoding pro-inflammatory mediators, including TNF-α, IL-1β, and IL-6 [14].

C. jejuni was found to activate the NFκB signaling pathway [28], which was suppressed by curcumin [12,17,19]. To determine the role of the NFκB pathway in maintaining barrier function, inhibition studies with the NFκB-inhibitor BAY 11-7082 were performed in *C. jejuni*-infected co-cultures. BAY 11-7082 treatment inhibited both the *C. jejuni*-induced decrease in TER and the increase in fluorescein permeability. This points to the involvement of NFκB in the regulation of epithelial barrier function. Similar results were obtained by our group previously with the bioactive ginger ingredient 6-shogaol after TNF-α stimulation in Caco-2 and HT-29/B6 cells [30].

In our co-culture, *C. jejuni* provoked paradoxical upregulation of claudin-1 (Claudin-1 paradox); that is, upregulation of the barrier-forming TJ protein claudin-1 while epithelial resistance was reduced, which we explained by the re-distribution of claudin-1 off the tight junction domain [6]. Claudin-1 induction has also been observed before in human colon biopsies after *C. jejuni* infection [6], *C. fetus* or *C. coli* infection in HT-29/B6 cells [31], and pro-inflammatory cytokine stimulation in HT-29/B6 cells [32,33]. A correlation between increased claudin-1 protein expression and apoptosis induction also occurs in HT-29/B6 cells [34]. In addition, curcumin induced claudin-4 mRNA expression in Caco-2 cells [8]. However, in our co-culture model, curcumin did not induce claudin-4 protein expression. Cell type-dependent differences in TJ expression were also found for quercetin, another plant-derived polyphenol with barrier-improving properties, which increased claudin-4 protein expression in Caco-2, but not in HT-29/B6 cells [35,36].

In order to exclude antibacterial effects of curcumin in our study, the MIC was determined with 87 μM (at pH 7.4). This suggests that the protective effects of 50 μM curcumin were not based on its antibacterial properties. Thus, we conclude that curcumin had direct barrier-protective and anti-inflammatory properties in the host cells.

To further exclude interfering properties of curcumin with bacterial fitness and adaptive response, we investigated if curcumin changes natural transformation capacity. Therefore, we used a single cell DNA uptake assay and challenged the bacteria with 50 μM curcumin. *C. jejuni* is naturally competent for DNA uptake from the environment. Generally, this leads to genetic recombinations between bacterial strains, and as a consequence, to more diversity [37]. Curcumin did not influence the competence development and horizontal DNA uptake of *C. jejuni* in our experimental concentrations.

One of the major challenges to the development of therapeutic approaches is the remarkable diversity between different *Campylobacter* strains in animals, the environment, and food [2]. Consequently, the efficacy of curcumin should also be tested using other *C. jejuni* field strains and, subsequently, in other *Campylobacter* species such as *C. coli* or *C. concisus*. Curcumin is a natural agent, and should be more appropriate than synthetic drugs [12]. Curcumin has been shown to be beneficial in maintaining remission in ulcerative colitis patients when used as a complementary therapeutic substance together with the antiphlogistic compound mesalazine in several randomized clinical studies [38–40]. Since the immune response plays an important role in pathogenesis in IBD as well as in *C. jejuni* infection, curcumin may be equally effective in immune-induced barrier dysfunctions by *C. jejuni* as an add-on to other anti-inflammatory drugs. Indeed, multimodal therapies may have an even greater protective effect, and synergism has been reported between curcumin and quercetin [41], genistein [42], epigallocatechin-3-gallate [15,43], and antioxidants [12,44]. Thus, curcumin alone or in combination with other agents may be useful in preventing, treating, and combating *Campylobacter* infection in humans, *C. jejuni*-associated sequelae as well as colonization of farm animals and especially poultry.

Taken together, our results suggest that a significant part of the *C. jejuni*-induced barrier defect is mediated indirectly by the immune cells. Curcumin abolishes functional epithelial disorders as well as pro-inflammatory cytokine secretion. Further barrier-protective or anti-inflammatory compounds, as well as curcumin combinations with other therapies, should be tested in the co-culture model.

4. Materials and Methods

4.1. Epithelial Cell Culture and Differentiation of THP-1 Cells

The human colon carcinoma cell line HT-29/B6-GR/MR [45], a subclone of the HT-29/B6 line [46] was cultured in 25 cm^2 culture flasks in RPMI 1640 culture medium (Sigma Aldrich, St. Louis, MO, USA). HT-29/B6-GR/MR cells expressing the human glucocorticoid receptor α and the human mineralocorticoid receptor, were established earlier by stable transfection [45,47]. The culture medium was supplemented with 10% fetal calf serum (FCS; Gibco, Carlsbad, CA, USA), 1% penicillin/streptomycin (Corning, Wiesbaden, Germany), G418-BC (300 μg/mL; Invitrogen, Carlsbad, CA, USA) and hygromycin B (200 μg/mL; Biochrom GmbH, Berlin, Germany), and cells were passaged every week. Cells were seeded on Millicell PCF filters membranes (Merck Millipore, Billerica, MA, USA; 0.4 μm pore size and an effective growth area of 0.6 cm^2). The cell medium in the culture flask, as well as in the filters, were changed every other day. Experiments were performed between 7 and 9 days after seeding, when polarized cells formed confluent monolayers and the transepithelial electrical resistance was 600–900 $\Omega \cdot cm^2$. These cells were chosen because of their stable growth behavior and the high resistance. The human monocyte leukemia cell line THP-1 (ATCC TIB-202) was provided by Verena Moos. THP-1 cells were cultured in RPMI 1640 supplemented with 10% heat-inactivated FCS and 1% penicillin/streptomycin. Cultures were maintained by the addition of fresh medium and were subcultured before reaching a density of 8×10^5 cells/mL once a week. THP-1 cells were re-suspended in antibiotic-free medium supplemented with phorbol 12-myristate 13-acetate (PMA; Sigma Aldrich, St. Louis, MO, USA; solved in DMSO) at a final concentration of 100 nM, seeded with a density of 1.8×10^5 in 12-well plates, and incubated for 24 h for differentiation to macrophages [48]. After incubation, PMA-containing medium was removed and the differentiated and adherent macrophages-like cells were cultured together with epithelial monolayers grown on filters in co-culture (Figure 10). All cells were cultured at 37 °C in a humidified 5% CO_2 atmosphere.

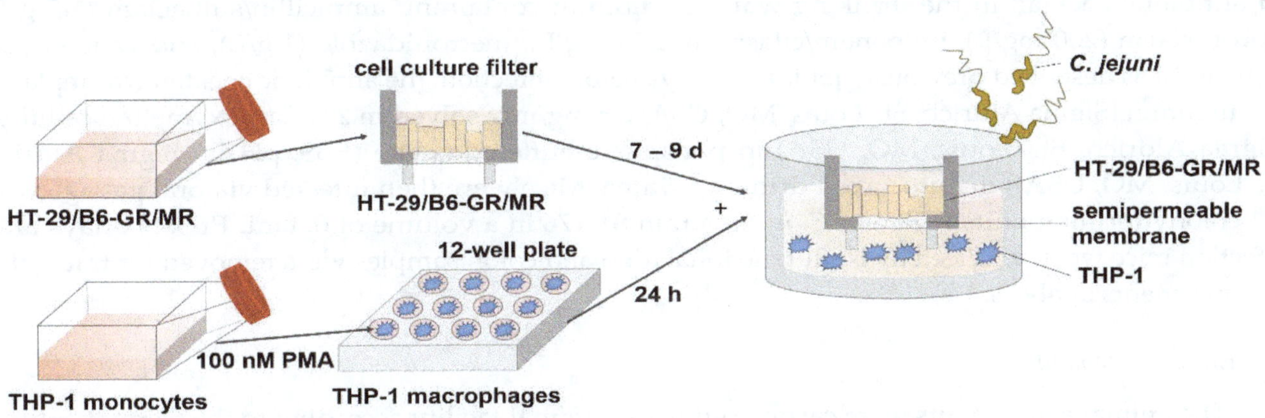

Figure 10. Experimental setting for co-culture of colon epithelial HT-29/B6-GR/MR and THP-1 immune cells. For the continuous culture of epithelial and immune cells, cells were cultured in culture flask. Epithelial cells were then seeded on filter membranes and differentiated over 7–9 days to a polarized monolayer with tight junctions. 24 h before co-culture started, immune THP-1 cells were stimulated with phorbol 12-myristate 13-acetate (PMA) to allow a differentiation to adherent macrophage-like immune cells. After the differentiation of epithelial and immune cells, filters were placed into the 12-well plate with epithelial cells in the apical and immune cells in the basal compartment. Infection with *C. jejuni* was performed in the basal compartment.

4.2. Growth Conditions of C. jejuni, Treatment and Infection Procedure In Vitro

Campylobater jejuni wildtype (wt) 81-176 reference strain was pre-cultured on blood agar plates (Oxoid, Thermo Scientific, Waltham, MA, USA) under microaerobic conditions (5–10% O_2, 10% CO_2, 85% N_2) at 37 °C and re-cultured a second time before infection experiments. Microaerobic conditions were generated in a plastic jar with CampyGen gaspacks from Oxoid (Oxoid, Thermo Scientific, Waltham, MA, USA). Bacteria were then cultured in Mueller–Hinton broth for at least 2.5 h at 37 °C under microaerobic conditions, centrifuged (2 min, 5000 *g*, 10 °C), and re-suspended in the cell culture medium. The optical density at OD_{600} was adjusted to 1 and the cells where infected from basal side with a multiplicity of infection (MOI) of 100. At least 1.5 h before the infection, cells were washed three times with antibiotic-free culture medium supplemented with 10% heat-inactivated FCS. To analyze the barrier-protective and anti-inflammatory properties of curcumin (Sigma Aldrich, St. Louis, MO, USA; final concentration 50 μM), different conditions were tested: control, *C. jejuni* infected, curcumin control (50 μM), and curcumin in combination with *C. jejuni*. Curcumin control showed no adverse effects on the cell viability of HT-29/B6-GR/MR monolayers at the concentration tested. Cells were pre-incubated with curcumin on both sides for 2 h, and infected with *C. jejuni* from the basal side for a direct immune cell infection and cytokine-induced barrier effect. Curcumin was added to the apical and basal side, since functional measurements depend on identical solutions on both the sides of the monolayer to avoid, e.g., potential osmotic effects. Moreover, curcumin readily accessed the basolateral compartment via *C. jejuni*-induced leaks. After infection, cells were incubated at 37 °C under microaerobic conditions favorable to the bacteria for 48 h.

4.3. Generation of Secondary Abiotic IL10$^{-/-}$ Mice, Treatment Infection

Permeability to fluorescein (332 Da; Sigma Aldrich, St. Louis, MO, USA) measurements were performed in Ussing chambers with colon of secondary abiotic IL-10$^{-/-}$ mice suffering from acute enterocolitis within 6 days following peroral *C. jejuni* infection [49]. IL-10$^{-/-}$ mice (in C57BL/6j background) were held under specific pathogen free (SPF) conditions in the animal facilities of the Forschungseinrichtung für Experimentelle Medizin (Charité–Universitätsmedizin Berlin). To remove the commensal gut microbiota, mice were transferred to sterile cages and treated for 8 weeks with an antibiotic cocktail in the drinking water ad libitum containing ampicillin/sulbactam (1.5 g/L), ciprofloxacin (200 mg/L), imipenem/cilastatin (250 mg/L), metronidazole (1 g/L), and vancomycin (500 mg/L) as described previously [50]. Four days before infection, the antibiotic cocktail was replaced by curcumin (Sigma Aldrich, St. Louis, MO, USA; 0.5 mg/mL; solved in 2% Carboxymethyl cellulose (Sigma Aldrich, St. Louis, MO, USA) in phosphate-buffered saline (PBS; pH7.4; Sigma Aldrich, St. Louis, MO, USA)) in autoclaved drinking water. Mice were then infected via oral gavage with 10^8 colony forming units (CFU) of *C. jejuni* strain 81-176 in a volume of 0.3 mL PBS. Six days after infection mice were sacrificed by isoflurane inhalation and colon samples were removed for tracer flux measurements analysis.

4.4. Ethics Statement

The animal experiments were carried out in our animal facility according to the German animal protection law (LaGeSo Berlin; approval number G0172/16, 13th Oct. 2016).

4.5. Electrophysiological Studies

Transepithelial electrical resistance (TER) was measured in vitro with a chopstick electrode pair under sterile conditions at 37 °C. The TER-values were corrected with the resistance of an empty cell filter and the bath solution. Colon samples from mice were mounted in modified miniaturized Ussing chambers (Institute of Clinical Physiology, Charité, Berlin, with an effective area of 0.049 cm^2) in modified Ringer´s solution. Ringer´s solution for the Ussing experiments contained (in mM): NaCl (113.6), $NaHCO_3$ (21), KCL (5.4), Na_2HPO_4 (2.4), $MgCl_2$ (1.2), $CaCl_2$ (1.2), NaH_2PO_4 (0.6), D(+)-glucose (10), D(+)-mannose (10) beta-hydroxybutyric acid (0.5), L-glutamine (2.5), and the antibiotics piperacillin (50 mg/L) and imipenem (4 mg/L), and was equilibrated for 15 min with carbogen gas to a pH of 7.4. One-path impedance spectroscopy measurements were performed as described previously [51] to delineate between epithelial and subepithelial conductance (G).

4.6. Epithelial Permeability

For permeability measurements, unidirectional flux studies were conducted with the paracellular marker fluorescein (332 Da; 100 μM) from the mucosal to the serosal compartment, either directly in 12-well plates or under short circuit current (I_{SC}) conditions in Ussing chambers, as described earlier [7]. I_{SC} was recorded under voltage clamp conditions by an automatic clamp device (CVC6, Fiebig Hard & Software, Berlin, Germany) at 37 °C over 1.5 h. Fluorescein was dissolved either in Ringer´s solution or in media. Samples were taken every 15 min for one hour from the basolateral side, and fluorescence was measured in a spectrophotometer (Tecan GmbH, Maennedorf, Switzerland). Permeability was calculated from flux over concentration difference.

4.7. Cytometric Bead Array

Supernatants of co-cultures were collected 48 h after infection and analyzed using the Cytometric Bead Assay (CBA; BD Biosciences, Franklin Lakes, NJ, USA) according to manufacturer´s instructions (human Th1, Th2, Th17 Kit, Flex Set IL-1β) to determine the secretion of cytokines TNF-α, IL-1β, IL-6. Flow cytometric measurement were performed with FACS CantoII (BD Biosciences; Franklin Lakes, NJ, USA) and analyzed with FACP Array^{TM} software v3.0 (BD Biosciences, Franklin Lakes, NJ, USA).

4.8. Western Blot Analysis

For protein quantification, epithelial cells were washed twice with ice-cold PBS. Whole cell lysates were extracted with ice-cold lysis buffer. Whole cell lysis buffer was prepared with 150 mM NaCl, 10 mM Tris buffer pH of 7.5, 0.5% Triton *X*-100, and 1% SDS. A volume of 10 mL lysis buffer was supplemented with one Complete Protease Inhibitor Cocktail tablet (Roche AG, Basel, Switzerland). After lysis, cells were scraped from the filters, incubated for 60 min on ice, and vortexed every 10 min. After centrifugation (30 min, 15,000× *g* at 4 °C), the supernatant was collected. Sonification of the lysate followed by further centrifugation. Total protein quantification was performed by Pierce BCA assay (Thermo Scientific, Waltham, MA, USA) according to the product instructions using a Tecan plate reader (Tecan GmbH, Maennedorf, Switzerland) at an absorbance of 562 nm. Protein samples (10–20 μg) were mixed with 5xLaemmli buffer and loaded on a SDS polyacrylamide gel (for claudins 12.5%, for occludin 10% polyacrylamide). After electrophoretic separation of the proteins and transfer to a nitrocellulose membrane, membranes were blocked for 2 h at room temperature with 1% PVP-40 (Polyvinylpyrrolidone; Sigma Aldrich, St. Louis, MO, USA) in TBST/0.05%Tween-20 buffer. Primary antibodies anti-occludin (1:100; Sigma Aldrich, St. Louis, MO, USA), anti-claudin-1, -2, -4, -5, -7, -8 (1:100; Invitrogen, Carlsbad, CA, USA), and anti-β-actin (1:5000; Sigma Aldrich, St. Louis, MO, USA) as internal loading control were incubated overnight at 4 °C. Peroxidase conjugated secondary antibodies goat anti-rabbit IgG or goat anti-mouse IgG (Jackson ImmunoResearch, Ely, UK) were incubated for 2 h at room temperature. For protein detection, SuperSignal West Pico PLUS Stable Peroxide Solution (Thermo Scientific, Waltham, MA, USA) was used and signals were detected with Fusion FX7 imaging system (Vilber Lourmat Deutschland GmbH, Eberhardzell, Germany). Densitometric quantification was performed using ImageJ software 1.48v/Java 1.6.0_20 (Rasband, W. S., ImageJ, NIH, Bethesda, MD, USA), and the values were normalized to β-actin.

4.9. Immunofluorescence Staining

Cells grown on filters were rinsed twice with PBS and fixed for 30 min in 2% paraformaldehyde (PFA; Electron Microscopy Sciences, Hatfield, PA, USA) for immunostaining and microscopic analysis. Afterwards, the cells were washed twice with PBS, permeabilized with 0.5% Triton *X*-100 (Sigma Aldrich, St. Louis, MO, USA) for 7 min, and blocked for 10 min with 1% goat serum (Gibco, Carlsbad, CA, USA). The cells were incubated with the primary antibodies anti-claudin-4 and -8 (1:100; Invitrogen, Carlsbad, CA, USA), anti-ZO-1 (1:100; BD Biosciences, Franklin Lakes, NJ, USA) or cleaved caspase-3 (1:100; Cell Signaling Technology, Cambridge, UK) for 1 h at room temperature, followed by the secondary anti-rabbit or anti-mouse antibody for 1 h (1:500; Invitrogen, Carlsbad, CA, USA). Secondary antibodies were conjugated to Alexa-Fluor 488 or 594. The nuclei were stained with 4′-6-diamidino-2-phenylindole dihydrochloride (DAPI, 1:1000, Roche AG, Basel, Switzerland). Subsequently, the cells on the filters were washed with water and ethanol, then embedded in ProTaq Mount Fluor (Biocyc, Luckenwalde, Germany). The subcellular distribution of the tight junctions was analyzed by confocal laser-scanning microscopy (CLSM, Zeiss LSM780, Jena, Germany). Stained apoptoses with cleaved caspase-3 were counted microscopically.

4.10. Determination of Minimal Inhibitory Concentration Values

Determinations of minimal inhibitory concentration values (MICs) of *Campylobacter* strain 81-176 were performed in a broth microdilution assay. The wells of microtiter plates contained 2-fold serial dilutions of curcumin with a test range of 1–1024 µg/mL. For MIC determinations, inoculum level, growth medium, incubation time, and conditions were performed in accordance with the recommendations given in the Clinical and Laboratory Standards Institute (CLSI) document VET01-A4. In brief, *Campylobacter* colonies were taken from a blood agar plate and transferred into a tube containing 5 mL of 0.85% saline. The suspension was adjusted to the turbidity equivalent of 0.5 McFarland standard, and diluted 1:100 with growth medium. Subsequently, 50 µL of the diluted inoculum was added to each test well (50 µL) in the dilution series (containing substances at double the desired final concentration) and mixed. The microtiter plates were incubated for 24 to 48 h at 42 °C under microaerobic conditions to obtain sufficient growth. They were analyzed visually, and the lowest concentration preventing visible growth of bacteria was defined as the MIC. All susceptibility tests were performed in growth medium adjusted to pH of 7.4. *Campylobacter jejuni* reference strain DSM 4688 was used for quality control purposes. The MIC values of the quality control strain were determined in advance in three independent experiments using the broth microdilution and macrodilution method.

4.11. DNA-Uptake Assay

C. jejuni strain 81-176 was streaked out from a −80 °C stock and subcultured on Columbia blood agar containing 5% sheep blood (Oxoid, Thermo Scientific, Waltham, MA, USA). Plates were incubated overnight under an atmosphere containing 3.5% H_2, 6% O_2, 7% CO_2, rest N_2 at 37 °C. Cells were inoculated at OD_{ini} ~0.3 in sterile-filtered brain heart infusion broth (BHI; Oxoid, Thermo Scientific, Waltham, MA, USA), and incubated under the same atmosphere at 37 °C for 6–9 h at 140 rpm. Liquid cultures were subcultured under the same conditions in 5 mL BHI with or without 50 µM curcumin. A suitable OD_{ini} was chosen to reach optical densities of 0.1–0.6 after 17 h (± 3 h). 500 µL (± 300 µL) of the cell suspension was harvested by centrifugation (~16000× *g* for 5 min). The pellet was resuspended in 100 µL BHI with or without 50 µM curcumin. DNA labeling and uptake analysis were performed as previously described for *H. pylori* [52]. In short, *C. jejuni* BfR-CA-14430 genomic DNA was extracted using the PureLink Kit (Life Technologies, Thermo Scientific, Waltham, MA, USA), and labeled with fluorescein in a 1:1 (volume:weight) ratio of Label IT reagent to nucleic acid according to the manufacturer's protocol (Mirus Label IT Fluorescein, Mirus Bio LLC, Madison, WI, USA). One µL of labeled DNA (100 ng/µL) was added to the cell suspension and incubated at 5% O_2, 10% CO_2, rest N_2 for 30 min at 37 °C. Subsequently, cells were centrifuged (~16000× *g* for 5 min) and resuspended in 15 µL BHI supplemented with at least 3 U DNaseI (Roche AG, Basel, Switzerland). DNaseI digestion was performed at 37 °C for 5–10 min. The fluorescence microscope Axio Observer ZI (Zeiss, Jena, Germany) with a plan apochromatic 63*x*/1.4 objective and differential interference contrast (DIC) was used for the analysis of *Campylobacter* cells immobilized on 1.5% agarose pads. A metal halide light source (HXP120C) and a filter set with excitation at 470 ± 20 nm and emission at 525 ± 25 nm were used to visualize fluorescein. Exposing times varied between 250 ms and 750 ms. Images were taken by the 12-bit monochromatic AxioCam MRm camera. Cells with at least one fluorescent focus were considered active for DNA uptake, i.e., competent for natural transformation. For each condition, the fraction of competent cells was calculated and data are presented from three independent experiments.

4.12. Cytotoxicity

The CCK-8 assay (Cell Counting Kit-8, Thermo Scientific, Waltham, MA, USA) was performed in 96-well plates with colon epithelial cells HT-29/B6-GR/MR 1 day after seeding in accordance with the manufactures' instructions. The cells were treated with different curcumin concentrations 2 h before infection. The absorption at 450 nm was measured by a spectrophotometer (Tecan GmbH, Maennedorf, Switzerland), and the percentage of viable cells calculated. In HT-29/B6-GR/MR cells, *C. jejuni* induced no significant loss in viability in a pre-test (control 100 ± 18% and DMSO control 81 ± 7% versus *C. jejuni*-infected 81 ± 4%, n = 3–4, *n.s.*, one-way ANOVA with Bonferroni´s multiple comparison). No effect on cell viability was also seen at curcumin concentrations 30 μM, 40 μM, and 50 μM (85 ± 8%, 64 ± 4%, and 66 ± 6%, n = 3–4, *n.s.*, one-way ANOVA with Bonferroni´s multiple comparison).

4.13. Ingenuity Pathways Analysis

Ingenuity Pathways Analysis (IPA, Qiagen Silicon Valley, Reswood, CA, USA) was used to evaluate curcumin-dependent expression data from a dataset from human colon biopsies, generated, and analyzed previously [6].

4.14. Statistical Analysis

All data are expressed as mean values ± standard error of the mean (SEM). Statistical analyses were performed with GraphPad Prism (version 7.0, GraphPad Software, Inc., San Diego, CA, USA) using one-way ANOVA with Bonferroni adjustment for multiple comparison or unpaired Student´s *t*-test. For data that were not normally distributed, the nonparametric Mann–Whitney *U*-test was used. $p < 0.05$ was considered to be statistically significant.

Author Contributions: Conceptualization, F.D.L.d.S., J.-D.S., and R.B.; Data curation, F.D.L.d.S., E.B., P.K.N., A.F., S.M., V.M., J.C.G., D.S., M.M.H., and S.B.; Formal analysis, F.D.L.d.S., E.B., P.K.N., A.F., S.M., V.M., J.C.G., K.S., S.K., D.S., C.K., M.M.H., and S.B.; Funding acquisition, J.-D.S. and R.B.; Investigation, F.D.L.d.S., J.-D.S., R.B.; Methodology, J.-D.S. and R.B.; Project administration, J.-D.S. and R.B.; Resources, V.M.; Supervision, J.-D.S. and R.B.; Writing—original draft, F.D.L.d.S.; Writing—review and editing, E.B., P.K.N., A.F., S.M., V.M., J.C.G., K.S., S.K., D.S., C.K., M.M.H., S.B., J.-D.S., and R.B.

Acknowledgments: In-Fah Maria Lee is gratefully acknowledged for her excellent technical support. We thank Geoffrey I. Sandle, (University of Leeds, Leeds, UK) for his critical discussion and support of the study.

Abbreviations

BHI	Brain heart infusion broth
CBA	Cytometric bead array
CFU	Colony forming units
CLSM	Confocal laser-scanning microscopy
DAPI	4´-6-diamidino-2-phenylindole dihydrochloride
IFN	Interferon
IL	Interleukin
LPS	Lipopolysaccharide
MIC	Minimal inhibitory concentration
PBS	Phosphate buffered saline
PMA	Phorbol 12-myristate 13-acetate
TER	Transepithelial electrical resistance
TJ	Tight junction
TNF	Tumor necrosis factor
ZO	Zonula occludens

Appendix A

Acute Campylobacteriosis Contains Activated Signaling Pathways for which Curcumin Has Counter-Regulatory Properties in IPA Analysis

From acute infected *C. jejuni* patients a RNA-Sequencing (RNA-Seq) analysis with concomitant ingenuity pathway analysis (IPA, Qiagen Silicon Valley) was performed. Once gene expression modifications were revealed, a bioinformatic prediction about possible inhibitors of the changed gene regulation could be carried out [6]. The hypothesis is, that upstream regulators, which have an inhibited activation pattern, may re-activate in *Campylobacter* infection, when the substance is applied during infection. Consequently, inhibited upstream regulators could be protective or therapeutic approaches in *Campylobacter* infection by activation of the corresponding downstream pathways. Different potential candidates were screened for barrier-protective and anti-inflammatory properties in *C. jejuni* infection. One promising predicted regulator candidate that might counter-regulate the *C. jejuni*-induced downstream pathways was curcumin. Curcumin showed a significant effect on downstream target genes, with a p-value of $2.06E^{-5}$ and an activation ^{z}-score of −3.489, and might therefore be another promising barrier-protecting or potential therapeutic substance in campylobacteriosis (Table S1). The *C. jejuni*-induced target genes in the dataset that could be counter-regulated by curcumin belong mainly to pro-inflammatory pathways, such as TNF-α or IL-1β. Another promising and studied candidate against *C. jejuni* infections is calcitriol (active vitamin D). Vitamin D shows in the RNA-Seq analysis with concomitant IPA analysis from patients in contrast to curcumin an even higher significance value (overlap p-value of $8.97E^{-25}$, z-score −6.25; with negative expression direction) [6].

References

1. Kaakoush, N.O.; Castaño-Rodríguez, N.; Mitchell, H.M.; Man, S.M. Global Epidemiology of *Campylobacter* Infection. *Clin. Microbiol. Rev.* **2015**, *28*, 687–720. [CrossRef] [PubMed]
2. Burnham, P.M.; Hendrixson, D.R. *Campylobacter jejuni*: Collective components promoting a successful enteric lifestyle. *Nat. Rev. Microbiol.* **2018**, *16*, 551–565. [CrossRef] [PubMed]
3. Konkel, M.E.; Mead, D.J.; Hayes, S.F.; Cieplak, W., Jr. Translocation of *Campylobacter jejuni* across human polarized epithelial cell monolayer cultures. *J. Infect. Dis.* **1992**, *166*, 308–315. [CrossRef] [PubMed]
4. Brás, A.M.; Ketley, J.M. Transcellular translocation of *Campylobacter jejuni* across polarised epithelial monolayers. *FEMS Microbiol. Lett.* **1999**, *179*, 209–215. [CrossRef] [PubMed]
5. Ketley, J.M. Pathogenesis of enteric infection by *Campylobacter*. *Microbiology* **1997**, *143*, 5–21. [CrossRef] [PubMed]
6. Bücker, R.; Krug, S.M.; Moss, V.; Bojarski, C.; Schweiger, M.R.; Kerick, M.; Fromm, A.; Janßen, S.; Fromm, M.; Hering, N.A.; et al. *Campylobacter jejuni* impairs sodium transport and epithelial barrier function via cytokine release in human colon. *Mucosal Immunol.* **2018**, *11*, 575–577. [CrossRef] [PubMed]
7. Bücker, R.; Schulz, E.; Günzel, D.; Bojarski, C.; Lee, I.F.; John, L.J.; Wiegand, S.; Janßen, T.; Wieler, L.H.; Dobrindt, U.; et al. α-Haemolysin of *Escherichia coli* in IBD: A potentiator of inflammatory activity in the colon. *Gut* **2014**, *63*, 1893–1901. [CrossRef] [PubMed]
8. Watari, A.; Yagi, K.; Kondoh, M. A simple reporter assay for screening Claudin-4 modulators. *Biochem. Biophys. Res. Commun.* **2012**, *426*, 454–460. [CrossRef]
9. Chiba, H.; Osanai, M.; Murata, M.; Kojima, T.; Sawada, N. Transmembrane proteins of tight junctions. *Biochim. Biophys. Acta* **2008**, *1778*, 588–600. [CrossRef]
10. Hakanen, A.; Jousimies-Somer, H.; Siitonen, A.; Huovinen, P.; Kotilainen, P. Fluoroquinolone resistance in *Campylobacter jejuni* isolates in travelers returning to Finland: Association of ciprofloxacin resistance to travel destination. *Emerg. Infect. Dis.* **2003**, *9*, 267–270. [CrossRef]
11. Hakanen, A.J.; Lehtopolku, M.; Siitonen, A.; Huovinen, P.; Kotilainen, P. Multidrug resistance in *Campylobacter jejuni* strains collected from Finnish patients during 1995–2000. *J. Antimicrob. Chemother.* **2003**, *52*, 1035–1039. [CrossRef] [PubMed]
12. Goel, A.; Kunnumakkara, A.B.; Aggarwal, B.B. Curcumin a "Curecumin": From kitchen to clinic. *Biochem. Pharmacol.* **2008**, *75*, 787–809. [CrossRef] [PubMed]
13. Cho, J.A.; Park, E. Curcumin utilizes the anti-inflammatory response pathway to protect the intestine against bacterial invasion. *Nutr. Res. Pract.* **2015**, *9*, 117–122. [CrossRef] [PubMed]
14. Masahi, O.; Nishida, A.; Sugitani, Y.; Nishino, K.; Inatomi, O.; Sugimoto, M.; Kawahara, M.; Andoh, A. Nanoparticle curcumin ameliorated experimental colitis via modulation of gut microbiota and induction of regulatory T cells. *PLoS ONE* **2017**, *12*, e0185999. [CrossRef]

15. Anggakusuma; Colpitts, C.C.; Schang, L.M.; Rachmawati, H.; Frentzen, A.; Pfaender, S.; Behrendt, P.; Brown, R.J.; Bankwitz, D.; Steinmann, J.; et al. Tumeric curcumin inhibits entry of all hepatitis C virus genotypes into human liver cells. *Gut* **2014**, *63*, 1137–1149. [CrossRef]
16. Mahady, G.B.; Pendland, S.L.; Yun, G.; Lu, Z.Z. Turmeric (*Curcuma longa*) and curcumin inhibit the growth of *Helicobacter pylori*, a group 1 carcinogen. *Anticancer Res.* **2002**, *22*, 4179–4181.
17. Chen, D.-Y.; Shien, J.-H.; Tiley, L.; Chiou, S.S.; Wang, S.Y.; Chang, T.J.; Lee, Y.J.; Chan, K.W.; Hsu, W.L. Curcumin inhibits influenza virus infection and haemagglutination activity. *Food Chem.* **2010**, *119*, 1346–1351. [CrossRef]
18. Marathe, S.A.; Balakrishnan, A.; Negi, V.D.; Sakorey, D.; Chandra, N.; Chakravortty, D. Curcumin reduces the motility of *Salmonella enterica* serovar typhymurium by binding to the flagella, thereby leading to flagellar fragility and shedding. *J. Bacteriol.* **2016**, *198*, 1798–1811. [CrossRef]
19. Moghadamtousi, S.Z.; Kadir, H.A.; Hassandarvish, P.; Tajik, H.; Abubakar, S.; Zandi, K. A review on antibacterial, antiviral, and antifungal activity of curcumin. *Biomed. Res. Int.* **2014**, *2014*, 186864. [CrossRef]
20. Santos, A.M.; Lopes, T.; Oleastro, M.; Gato, I.V.; Floch, P.; Benejat, L.; Chaves, P.; Pereira, T.; Seixas, E.; Machado, J.; et al. Curcumin inhibits gastric inflammation induced by *Helicobacter pylori* infection in a mouse model. *Nutrients* **2015**, *7*, 306–320. [CrossRef]
21. Gunes, H.; Gulen, D.; Mutlu, R.; Gumus, A.; Tas, T.; Topkaya, A.E. Antibacterial effects of curcumin: An in vitro minimum inhibitory concentration study. *Toxicol. Ind. Health* **2016**, *32*, 246–250. [CrossRef] [PubMed]
22. Sharma, R.A.; Geschner, A.J.; Steward, W.P. Curcumin: The story so far. *Eur J. Cancer* **2005**, *41*, 1955–1968. [CrossRef] [PubMed]
23. Wang, J.; Ghosh, S.S.; Ghosh, S. Curcumin improves intestinal barrier function: Modulation of intracellular signaling, and organization of tight junctions. *Am. J. Physiol.-Cell. Physiol.* **2017**, *312*, C438–C445. [CrossRef] [PubMed]
24. Bartik, L.; Whitfield, G.K.; Kaczmarska, M.; Lowmiller, C.L.; Moffet, E.W.; Furmick, J.K.; Hernandez, Z.; Haussler, C.A.; Haussler, M.R.; Jurutka, P.W. Curcumin: A novel nutritionally derived ligand of the vitamin D receptor with implications for colon cancer chemoprevention. *J. Nutr. Biochem.* **2010**, *21*, 1153–1161. [CrossRef] [PubMed]
25. Backert, S.; Boehm, M.; Wessler, S.; Tegtmeyer, N. Transmigration route of *Campylobacter jejuni* across polarized intestinal epithelial cells: Paracellular, transcellular or both? *Cell Commun. Signal.* **2013**, *11*, 72. [CrossRef] [PubMed]
26. Boehm, M.; Hoy, B.; Rohde, M.; Tegtmeyer, N.; Bæk, K.T.; Oyarzabal, O.A.; Brøndsted, L.; Wessler, S.; Backert, S. Rapid paracellular transmigration of *Campylobacter jejuni* across polarized epithelial cells without affecting TER: Role of proteolytic-active HtrA cleaving E-cadherin but not fibronectin. *Gut Pathog.* **2012**, *4*, 3. [CrossRef] [PubMed]
27. Wine, E.; Chan, V.L.; Sherman, P.M. *Campylobacter jejuni* Mediated Disruption of Polarized Epithelial Monolayers is Cell-Type Specific, Time Dependent, and Correlates with Bacterial Invasion. *Pediatric Res.* **2008**, *64*, 599–604. [CrossRef] [PubMed]
28. Chen, M.L.; Ge, Z.; Fox, J.G.; Schauer, D.B. Disruption of tight junctions and induction of proinflammatory cytokine responses in colonic epithelial cells by *Campylobacter jejuni*. *Infect. Immun.* **2006**, *74*, 6581–6589. [CrossRef]
29. Heimesaat, M.M.; Grundmann, U.; Alutis, M.E.; Fischer, A.; Bereswill, S. Small Intestinal Pro-Inflammatory Immune Responses Following *Campylobacter jejuni* Infection of Secondary Abiotic IL-10−/− Mice Lacking Nucleotide-Oligomerization-Domain-2. *Eur. J. Microbiol. Immunol.* **2017**, *7*, 138–145. [CrossRef]
30. Luettig, J.; Rosenthal, R.; Lee, I.-F.M.; Krug, S.M.; Schulzke, J.D. The ginger component 6-shogaol prevents TNF-α-induced barrier loss via inhibition of PI3K/Akt and NF-κB signaling. *Mol. Nutr. Food Res.* **2016**, *60*, 2576–2586. [CrossRef]
31. Bücker, R.; Krug, S.M.; Fromm, A.; Nielsen, H.L.; Fromm, M.; Nielsen, H.; Schulzke, J.-D. *Campylobacter fetus* impairs barrier function in HT-29/B6 cells through focal tight junction alterations and leaks. *Ann. N. Y. Acad. Sci.* **2017**, *1405*, 189–201. [CrossRef] [PubMed]
32. Amasheh, M.; Fromm, A.; Krug, S.M.; Amasheh, S.; Andres, S.; Zeitz, M.; Fromm, M.; Schulzke, J.-D. TNFalpha-induced and berberine-antagonized tight junction barrier impairment via tyrosine kinase, Akt and NFkappaB signaling. *J. Cell Sci.* **2010**, *123*, 4145–4155. [CrossRef]

33. Rosenthal, R.; Luettig, J.; Hering, N.A.; Krug, S.M.; Albrecht, U.; Fromm, M.; Schulzke, J.D. Myrrh exerts barrier-stabilising and –protective effects in HT-29/B6 and Caco-2 intestinal epithelial cells. *Int. J. Colorectal Dis.* **2017**, *32*, 623–634. [CrossRef]
34. Bojarski, C.; Weiske, J.; Schöneberg, T.; Schröder, W.; Mankertz, J.; Schulzke, J.-D.; Florian, P.; Fromm, M.; Tauber, R.; Huber, O. The specific fates of tight junction proteins in apoptotic epithelial cells. *J. Cell Sci.* **2004**, *117*, 2097–2107. [CrossRef]
35. Amasheh, M.; Schlichter, S.; Amasheh, S.; Mankertz, J.; Zeitz, M.; Fromm, M.; Schulzke, J.D. Quercetin enhances epithelial barrier function and increases claudin-4 expression in Caco-2 cells. *J. Nutr.* **2008**, *138*, 1067–1073. [CrossRef] [PubMed]
36. Amasheh, M.; Luettig, J.; Amasheh, S.; Zeitz, M.; Fromm, M.; Schulzke, J.-D. Effects of quercetin studied in colonic HT-29/B6 cells and rat intestine in vitro. *Ann. N. Y. Acad. Sci.* **2012**, *1258*, 100–107. [CrossRef] [PubMed]
37. Young, K.T.; Davis, L.M.; Dirita, V.J. *Campylobacter jejuni*: Molecular biology and pathogenesis. *Nat. Rev. Microbiol.* **2007**, *5*, 665–679. [CrossRef] [PubMed]
38. Hanai, H.; Iida, T.; Takeuchi, K.; Watanabe, F.; Maruyama, Y.; Andoh, A.; Tsujikawa, T.; Fujiyama, Y.; Mitsuyama, K.; Sata, M.; et al. Curcumin maintenance therapy for ulcerative colitis: Randomized, multicenter, double-blind, placebo-controlled trial. *Clin. Gastroenterol. Hepatol.* **2006**, *4*, 1502–1506. [CrossRef] [PubMed]
39. Singla, V.; Pratap Mouli, V.; Garg, S.K.; Rai, T.; Choudhury, B.N.; Verma, P.; Deb, R.; Tiwari, V.; Rohatgi, S.; Dhingra, R.; et al. Induction with NCB-02 (curcumin) enema for mild-to-moderate distal ulcerative colitis—A randomized, placebo-controlled, pilot study. *J. Crohn's Colitis* **2014**, *8*, 208–214. [CrossRef] [PubMed]
40. Lang, A.; Salomon, N.; Wu, J.C.; Kopylov, U.; Lahat, A.; Har-Noy, O.; Ching, J.Y.; Cheong, P.K.; Avidan, B.; Gamus, D.; et al. Curcumin in Combination With Mesalamine Induces Remission in Patients With Mild-to-Moderate Ulcerative Colitis in a Randomized Controlled Trial. *Clin. Gastroenterol. Hepatol.* **2015**, *13*, 1444–1449. [CrossRef] [PubMed]
41. Cruz-Correa, M.; Shoskes, D.A.; Sanchez, P.; Zhao, R.; Hylind, L.M.; Wexner, S.D.; Giardiello, F.M. Combination treatment with curcumin and quercetin of adenomas in familial adenomatous polyposis. *Clin. Gastroenterol. Hepatol.* **2006**, *4*, 1035–1038. [CrossRef] [PubMed]
42. Verma, S.P.; Salamone, E.; Goldin, B. Curcumin and genistein, plant natural products, show synergistic inhibitory effects on the growth of human breast cancer MCF-7 cells induced by estrogenic pesticides. *Biochem. Biophys. Res. Commun.* **1997**, *233*, 692–696. [CrossRef] [PubMed]
43. Balasubramanian, S.; Eckert, R.L. Green tea polyphenol and curcumin inversely regulate human involucrin promoter activity via opposing effects on CCAAT/enhancer-binding protein function. *J. Biol. Chem.* **2004**, *279*, 24007–24014. [CrossRef] [PubMed]
44. Wang, Y.J.; Pan, M.H.; Cheng, A.-L.; Lin, L.-I.; Ho, Y.-S.; Hsieh, C.-Y.; Lin, J.-K. Stability of curcumin in buffer solutions and characterization of its degradation products. *J. Pharm. Biomed. Anal.* **1997**, *15*, 1867–1876. [CrossRef]
45. Bergann, T.; Plöger, S.; Fromm, A.; Zeissig, S.; Borden, S.A.; Fromm, M.; Schulzke, J.D. A colonic mineralocorticoid receptor cell model expressing epithelial Na+ channels. *Biochem. Biophys. Res. Comm.* **2009**, *381*, 280–285. [CrossRef] [PubMed]
46. Kreusel, K.M.; Fromm, M.; Schulzke, J.D.; Hegel, U. Cl- secretion in epithelial monolayers of mucus-forming human colon cells (HT-29/B6). *Am. J. Physiol.* **1991**, *261*, C574–C582. [CrossRef] [PubMed]
47. Zeissig, S.; Bürgel, N.; Günzel, D.; Richter, J.; Mankertz, J.; Wahnschaffe, U.; Kroesen, A.J.; Zeitz, M.; Fromm, M.; Schulzke, J.D. Changes in expression and distribution of claudin 2, 5 and 8 lead to discontinuous tight junctions and barrier dysfunction in active Crohn's disease. *Gut* **2007**, *56*, 61–72. [CrossRef] [PubMed]
48. Kämpfer, A.A.M.; Urbán, P.; Gioria, S.; Kanase, N.; Stone, V.; Kinsner-Ovaskainen, A. Development of an in vitro co-culture model to mimic the human intestine in healthy and diseased state. *Toxicol. In Vitro* **2017**, *45*, 31–43. [CrossRef]
49. Haag, L.M.; Fischer, A.; Otto, B.; Plickert, R.; Kühl, A.A.; Göbel, U.B.; Bereswill, S.; Heimesaat, M.M. Intestinal microbiota shifts towards elevated commensal *Escherichia coli* loads abrogate colonization resistance against Campylobacter jejuni in mice. *PLoS ONE* **2012**, *7*, e35988. [CrossRef]
50. Bereswill, S.; Plickert, R.; Fischer, A.; Kühl, A.A.; Loddenkemper, C.; Batra, A.; Siegmund, B.; Gobel, U.B.; Heimesaat, M.M. What you eat is what you get: Novel *Campylobacter* models in the quadrangle relationship between nutrition, obesity, microbiota and susceptibility to infection. *Eur. J. Microbiol. Immunol.* **2011**, *1*,

237–248. [CrossRef]
51. Gitter, A.H.; Schulzke, J.-D.; Sorgenfrei, D.; Fromm, M. Ussing chamber for high-frequency transmural impedance analysis of epithelial tissues. *J. Biochem. Biophys. Methods* **1997**, *35*, 81–88. [CrossRef]
52. Krüger, N.J.; Knüver, M.T.; Zawilak-Pawlik, A.; Appel, B.; Stingl, K. Genetic Diversity as Consequence of a Microaerobic and Neutrophilic Lifestyle. *PLoS Pathog.* **2016**, *12*, e1005626. [CrossRef] [PubMed]

12

Tight Junctions of the Outer Blood Retina Barrier

Aisling Naylor, Alan Hopkins, Natalie Hudson and Matthew Campbell *

Smurfit Institute of Genetics, Trinity College Dublin, Dublin 2, Ireland; aisling.naylor@ucdconnect.ie (A.N.); ahopkin@tcd.ie (A.H.); natalie.hudson@tcd.ie (N.H.)

* Correspondence: campbem2@tcd.ie.

Abstract: The outer blood retina barrier (oBRB) formed by the retinal pigment epithelium (RPE) is critical for maintaining retinal homeostasis. Critical to this modified neuro-epithelial barrier is the presence of the tight junction structure that is formed at the apical periphery of contacting cells. This tight junction complex mediates size-selective passive diffusion of solutes to and from the outer segments of the retina. Unlike other epithelial cells, the apical surface of the RPE is in direct contact with neural tissue and it is centrally involved in the daily phagocytosis of the effete tips of photoreceptor cells. While much is known about the intracellular trafficking of material within the RPE, less is known about the role of the tight junction complexes in health and diseased states. Here, we provide a succinct overview of the molecular composition of the RPE tight junction complex in addition to highlighting some of the most common retinopathies that involve a dysregulation of RPE integrity

Keywords: retinal pigment epithelium; retinopathy; tight junction

1. Introduction

The blood retina barrier (BRB) is fundamental in establishing and maintaining a suitable environment for optimum retinal function [1]. While the inner blood retina barrier (iBRB) is composed of tight junctions (TJs) between retinal capillary endothelial cells, the so-called outer blood retina barrier (oBRB) is formed by the TJs between retinal pigment epithelial (RPE) cells. This oBRB acts to regulate and filter molecular movement of solutes and nutrients from the choroid to the sub-retinal space. Impairment of either of these barriers can lead to the accumulation of blood-borne proteins and other potentially toxic solutes within the retina [1] (differences between the oBRB and iBRB are summarised further in Table 1).

The RPE is composed of a single layer of epithelial cells separating the neural retinal from the underlying Bruch's membrane and fenestrated choriocapillaris. TJs connecting neighbouring RPE cells block the movement of plasma components and toxic molecules into the retina as well as allowing for a controlled flow of fluid and solutes across an osmotic gradient from the choroidal vasculature to the outer retina [2]. The retina has the highest oxygen consumption per weight of any tissue in the body and the BRB (both outer and inner) is essential in providing a distinct and regulated source of nutrients to facilitate for this high metabolic rate [3]. TJs at both the iBRB and oBRB are complex, dynamic structures [4] and in the context of these barriers, the integrity of these TJs is crucial to sight.

Specifically, the oBRB allows for the survival of the photoreceptors (PRs) by supporting essential functions including filtering and transport of nutrients and photoreceptor outer segment (POS) phagocytosis. Other important functions of the RPE include absorption of out of focus and scattered light, retinal adhesion, and vitamin A transport and processing, and re-isomerisation of all-trans-retinal to 11-cis retinal, which is crucial for the visual cycle [2,5]. The RPE is essential for visual function, a failure of any one of these functions can lead to degeneration of the retina, loss of visual function, and blindness [2]. The RPE is highly polarized and can be divided into apical and basolateral sides. The apical surface is in direct contact with the POS and the basolateral side acts as a barrier in its interaction with the highly permeable and highly perfused choriocapillaris of the choroid [6].

TJs allow for high degrees of selectivity in paracellular barrier function in both the iBRB and the oBRB. TJs are unique assemblies of transmembrane proteins and peripheral cytoplasmic proteins. Transmembrane proteins include the claudins, the MARVEL (Mal and related proteins for vesicle trafficking and membrane link) family and junctional adhesion molecules (JAMs), which span the plasma membrane. Peripheral cytoplasmic proteins such as zonula occludens-1, (ZO-1), -2 (ZO-2) and -3 (ZO-3) anchor these transmembrane proteins to the cytoskeleton and are vital in the initial formation and distinct organization of TJs [7,8].

In principle, the inner and outer BRBs are performing inherently similar roles of paracellular diffusion regulation; however, the molecular composition varies considerably in these junctions. In this review, we will discuss the structure and functions of the oBRB, and how its disruption contributes to the pathogenesis of a variety of ocular conditions including diabetic retinopathy (DR), age related macular degeneration (AMD), central serous chorioretinopathy (CSCR), Sorsby's fundus dystrophy, Retinitis Pigmentosa (RP), and conditions associated with mutations in CLDN-19, the gene encoding for the TJ protein claudin-19.

Table 1. A summary of the key differences between the outer blood retina barrier (oBRB) and inner blood retinal barrier (iBRB).

Outer Blood Retinal Barrier	Inner Blood Retinal Barrier
Formed by tight junctions (TJ) between neighbouring retinal pigment epithelium (RPE) cells [1]. Rests on underlying Bruch's membrane [1]	Formed by TJ between neighbouring retinal endothelial cells [1]. Rests on a basal lamina that is covered by the processes of astrocytes and Müller cells [1]
Regulates the paracellular movement of fluids and molecules between the choriocapillaris and the retina [1]	Regulates the paracellular movement of fluids and molecules across retinal capillaries [1]
Claudin-19 is the predominant claudin [9], claudin-3 and -10 are also expressed [10]	Claudin-5 is the most predominant claudin, claudin-1 and -2 are also expressed [11]
Plays a fundamental role in the microenvironment of the outer retina [1] including regulating access of nutrients from blood to photoreceptors (PRs), eliminating waste products, and maintaining retinal adhesion [1]	Plays a fundamental role in the microenvironment of the neural retina [1]
The relationship between the RPE apical villi and PR is considered to be crucial in maintaining visual function [1]	Regulatory signals of the retinal neuronal circuitry are transmitted by astrocytes, muller cells and pericytes thereby influencing the activity of the iBRB [1]

2. The Retinal Pigment Epithelium (RPE)

The RPE is composed of a single layer of cells joined laterally towards their apices by TJs between adjacent plasma membranes (Figure 1a demonstrates the topographic relationship of the RPE within the retina and Figure 1b demonstrates the structure and function of the RPE as described in this section). The apical membrane faces the photoreceptor outer segments (POS) and the basolateral membrane faces Bruch's membrane, thereby separating the RPE from the choroid [2]. In a planar view, RPE cells exhibit a hexagonal shape. Microvilli extend from the apical surface of the RPE and envelop both rod and cone POS. The apical microvilli cloak a greater length of rods than cones [12]. They increase the RPE surface area 30-fold and promote a unique metabolic and functional relationship between the RPE cells and PRs, which is considered to be critical for the maintenance of visual function [1 13].

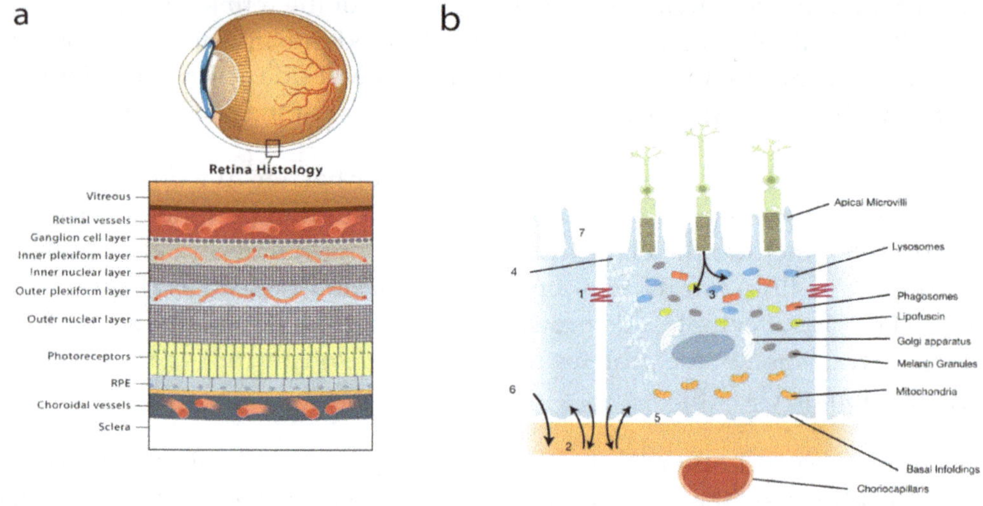

Figure 1. (**a**) Topographic relationship of the retinal pigment epithelium (RPE) within the retina. RPE anatomy and function. (**b**) The functions of the RPE are labelled as follows: (1) Tight junction (TJ) of the outer blood retina barrier (oBRB); (2) Transport of fluid, nutrients, and metabolites (paracellular and transcellular); (3) Phagocytosis of photoreceptor outer segments (POS); (4) Absorption of scattered light; (5) Retinal adhesion; (6) Paracrine secretion (including vascular endothelial growth factor (VEGF)); (7) Maintaining balance across the sub-retinal space.

Laterally, the RPE membrane is the site of cell communication and adhesion. Apically, the composite of gap junctions, adherens junctions (AJs), and TJs form a physical barrier and are involved in the maintenance of cell polarity and the prevention of intramembranous diffusion between the basolateral and apical membrane domains [4,14]. Unlike most epithelia however, the RPE is unusual as its apical surface abuts solid tissue rather than a lumen. This variance allows the neural retina to influence epithelial polarity and the structure and function of TJs [4].

The infoldings of the RPE basal plasma membrane is typical of cells adept in transport. Intracellularly, the organelles tend to display domain specific distribution, with melanosomes primarily in the apical cytoplasm and the mitochondria, Golgi, and nucleus located basally [13].

The RPE has many functions that are vital to the maintenance of normal ocular function and homeostasis. These include phagocytosis of shed POS, retinoid conversion and storage, absorption of scattered light, RPE to PR apposition, and ion and fluid transport [13]. With such a broad spectrum of essential activities involved in vision, it is clear that primary dysregulation of these cells can lead to serious clinical effects, and indeed, given its interaction in many pathways, they must also be vulnerable to secondary damage.

The role of RPE junctional complexes in preventing paracellular movement of molecules and ions make RPE cells the protectors and key effectors of the oBRB [4]. However, they do not form an absolute and persistent wall. In fact, the RPE is moderately leaky in order to support the unique needs of the PRs. This is demonstrated by assessing transepithelial electrical resistance (TEER), which is a measure of paracellular ion movement. By inhibiting the passage of certain ions, TJs allow the RPE to regulate transport and establish concentration gradients between the neural retina and Bruch's membrane [4].

In the context of the oBRB, TJs establish a barrier between the sub-retinal space and the choriocapillaris. RPE cells are responsible for the movement of nutrients and metabolic end products and serve to maintain the ion balance in the sub-retinal space [13]. Channels and pumps mediate the movement of ions and molecules via the transcellular pathway. In one direction, the RPE transports electrolytes and water from the sub-retinal space to the choroid, and in the other direction, the RPE transports glucose and other nutrients from the blood to the PRs [2]. Paracellular resistance is significantly higher than transcellular resistance, subsequently, water transport occurs mainly through transcellular pathways mediated by aquaporin-1 and -4 [2]. The Na^+/K^+-ATPase, which is located apically in RPE cells, provides the energy for the transport of electrolytes and water from the sub retinal space to the choroid [13]. A substantial amount of water is produced due to the large metabolic turnover of the retina. This constant elimination of water from the inner retina to the choroid produces an elimination force between the retina and RPE. This force is lost by the inhibition of the Na^+/K^+-ATPase. Proper anatomical apposition between RPE and PR cells is essential to the optics of the eye and health of PR cells [13].

Due to exposure to an intense level of light, toxic substances accumulate in PR cells daily [15]. In order for PRs to carry out their role of light transduction efficiently, the POS undergo a constant renewal process [15]. Each rod regenerates its outer segment in 7–12 days [13]. New POS are built at the cilium from the base of the outer segments. The highest concentration of radicals, photo damaged proteins, and lipids are located in the tips, which are shed from the PRs. Shed POS are phagocytosed by the RPE and the process of disk shedding and phagocytosis is tightly coordinated between the RPE and PR [15]. Through this coordinated POS tip shedding and the formation of new POS, a constant length of POS is maintained [15]. Lipofuscin compounds accumulate in the RPE as a consequence of the cells' role in phagocytosing the POS membrane [13]. The RPE facilitates the recycling of the digested shed POS, allowing for essential molecules to be recycled to PRs [2,15]. This process of POS shedding is under circadian control, with the major burst of phagocytosis taking place with the onset of light [16].

The RPE also produces a number of growth factors and other soluble proteins that are essential for the maintenance and structure of the retina and the choriocapillaris [6]. Vascular endothelial growth factor (VEGF) is secreted in low concentrations at the basolateral surface of the RPE in the healthy eye where it prevents endothelial cell apoptosis and is essential for maintaining an intact endothelium associated with the choriocapillaris. It is thought to act as a permeability factor stabilizing the fenestrations of the endothelium [2,13].

3. Tight Junctions (TJs)

TJs are composed of transmembrane proteins and peripheral membrane proteins that interact with each other to form a complex network. Transmembrane proteins extend into the paracellular space, creating a seal characteristic of TJs [3]. These proteins include the MARVEL and claudin family members and JAMs. Scaffolding proteins including ZO-1, ZO-2, and ZO-3 bind to transmembrane proteins, linking these to the cytoskeleton [14]. The TJs create a barrier to paracellular diffusion of solutes as well as the maintenance of cell polarity between the basolateral and apical plasma membrane domains, which are often referred to as the "barrier" and "fence" function, respectively [17].

Since their identification via electron microscopy in 1963, we have learnt a great deal more about the location and composition of TJs [18]. In ultra-thin sections of brain tissues, TJs appear like a sequence of fusions (or kisses), which are formed between two adjacent cells by the outer leaflets of the plasma membrane. At higher magnification, it becomes clear that the membranes are not fused, but rather, are in tight contact with each other [14].

Secretions of the neural retina regulate the assembly, maturation, and tissue-specific properties of these TJs [4]. Initially, adhesive membrane proteins of AJs and TJs form adhesion complexes at sites of cell-to-cell contact. Subsequently, they organise into zipper like structures by lateral adhesion along the cell border [7]. The intracellular partners of transmembrane adhesive proteins also vary during junction maturation and stabilization. Adhesions are in dynamic equilibrium, even after stable contacts have been formed and recycle continuously between the plasma membrane and cytoplasm [8]. In epithelial cells, TJs and AJs follow a well-defined spatial distribution along the cellular cleft with TJs located at the most apical regions and AJs below them [8]. TJs make the most significant contribution to the paracellular component of TEER [4].

The TAMP (TJ-associated MARVEL protein) family members include occludin (MarvelD1), tricellulin (MarvelD2), and MarvelD3 [19,20]. They contain a conserved four-transmembrane MARVEL (Mal and related proteins for vesicle trafficking and membrane link) domain [21]. Occludin, tricellulin, and marvelD3 have both redundant and unique contributions to epithelial function [19,20]. Occludin was the first transmembrane protein identified in the TJ. It has a molecular mass of 65 kDa, with two extracellular loops, aforementioned four transmembrane loops and its amino and carboxy termini localised intracytoplasmically [22]. Occludin binds directly with the three ZO proteins [8,23] and these interactions are required for the TJ localisation of occludin [20]. Occludin's two extracellular loops are involved in cell–cell adhesion and regulating paracellular permeability, respectively [3]. The C-terminal interacts with ZO-1 and ZO-2 and the last 150 amino acids interact with F-actin [3,22,23]. Expression of C-terminally truncated occludin in MDCKII cells resulted in increased paracellular permeability [24,25]. Over-expression of chicken occludin in MDCK cells leads to an increase in TEER, and therefore a decrease in paracellular permeability [3,26]. In combination, this suggests that occludin plays a potentially pivotal role in established paracellular permeability.

Occludin plays an essential regulatory role in the function of TJs, however, it is unnecessary for TJ formation [27–30]. Occludin associates with ZO-1 at the TJ upon recruitment by JAM-A [7]. This interaction is essential in modulating the function of occludin at the TJ [3]. Occludin has a half-life of approximately 1.5 h and rapidly dissociates from the TJ, indicating that it may facilitate TJs in adapting rapidly to physiologic changes [30,31]. Phosphorylation of occludin has been shown to act as an important regulatory mechanism. Post-translational phosphorylation status of occludin has been found to influence its location within the TJ and its regulation of paracellular permeability [26,32,33]. Threonine and serine phosphorylation of occludin appears to occur in conjunction with the preservation of TJ integrity [3]. Conversely, tyrosine phosphorylation of occludin appears to disrupt the association of ZO-1 and occludin, and demonstrates increased paracellular permeability [3]. In phosphorylated occludin, the binding of Z0-1, Z0-2, and Z0-3 to the C-terminal tail were decreased in comparison to non-phosphorylated occludin [3,34].

The Claudins are a family of transmembrane proteins of which there are more than 24 established members [14]. Like occludin, claudins have four membrane spanning regions, two extracellular loops, and two cytoplasmic termini/intracellular domains [14]. C-terminal amino acids encode PDZ-binding motifs, which are highly conserved throughout the claudin family. Through the C-terminal domain, claudins directly interact with peripheral PDZ-domain containing proteins including- ZO-1, ZO-2, and ZO-3 [14]. The first extracellular domain determines TEER and paracellular charge selectivity [35].

Claudin proteins mediate robust cell–cell adhesion and directly regulate permeability and selectivity [14,36]. They are almost certainly the main proteins important for TJ strand formation [14,30]. Claudins form pores facilitating the passive diffusion of molecules through the paracellular space [37]. Each claudin is thought to have a unique effect on selectivity and permeability [28,37] with the specific paracellular properties of different epithelia resulting from their individual pattern of claudin expression [36]. In the RPE, the claudins that are expressed vary among species [30]. In humans, claudin-3, claudin-10, and claudin-19 were detected in TJs [10] whereas in chicks, claudin-19 is not expressed and claudin-20 is the major claudin [9]. Some claudins have a limited distribution expressed in a tissue-specific manner, for example, claudin-5 appears to be confined to endothelial cells [8,38]. Claudin-1 to claudin-8 bind directly with ZO-1, ZO-2, and ZO-3 [8,39] via the cytoplasmic domain of claudins and the first PDZ domain of ZO proteins [39].

Claudin-19 is expressed and enriched in the RPE, where it is by far the predominant claudin [9]. High levels of claudin-19 are also found in the kidneys [9]. Claudin-19 determines permeability and semi-selectivity of the TJs in the RPE. By knocking-down claudin-19, the expressed claudin-3 was inadequate to form effective TJs. In contrast, the knockdown of claudin-3 demonstrated no effects [9]. Mutated claudin-19 affects multiple stages of RPE and retinal differentiation through its effects on multiple functions of the RPE [40,41]. Of note, claudin-10 was only expressed in a subset of cells [10].

The JAMs are a family of transmembrane proteins, of which four members have been identified: JAM-A, JAM-B, JAM-C, and JAM4/JAML [14]. JAMs are known to interact with many other proteins and may modulate TJ function by targeting other proteins to the TJ [8,14]. JAM-A can co-localize with occludin, ZO-1, and cingulin [7]. JAMs are members of the immunoglobulin superfamily and are expressed in epithelial cells along with endothelial cells, platelets, and leukocytes [14]. They are composed of a single transmembrane domain, and an extracellular domain containing two Ig-like motifs [14]. A PDZ-binding motif at the C terminus appears to be involved with the mediation of interaction with TJ scaffolding proteins, which appears to be important for the proper function of the TJ [42].

Peripheral membrane proteins anchor transmembrane proteins to the actin cytoskeleton and allow them to organize in the membrane and initiate cell signalling [3,14]. These include ZO-1, ZO-2, ZO-3, and cingulin [7].

The ZO proteins belong to the family of membrane associated guanylate kinases (MAGUK) that possesses a distinct molecular organisation. The core structure is composed of one or more PDZ domains, a Src homology 3 (SH3) domain, and a guanylate kinase (GUK) domain [14,43]. ZO-1 was the first TJ protein discovered in both epithelial and endothelial cells [44,45]. ZO-2 and ZO-3 were later discovered to localise to TJs with a similar sequence homology to ZO-1 [45]. ZO-1 has been shown to play a central role in the assembly and function of TJs [14]. PDZ domains facilitate the formation of specific multi-protein complexes including those necessary for TJ formation by recruiting downstream proteins in a signalling pathway [3]. At the TJ, the PDZ domain binds to the actin cytoskeleton through the C-terminal end and forms a bridge between the C-terminal sequences of occludin and β-actin [3]. It interacts with ZO-2 and ZO-3, its binding partners, through its second PDZ domain [11,43]. ZO proteins form a complex with AJ proteins in non-polarised cells where TJs have not formed, but upon polarisation, ZO proteins can separate from the AJ and concentrate in the TJ where they interact with TJ proteins such as claudins and occludin [46]. ZO-1 has also been found to participate in the regulation of gene expression via its binding in the nucleus to a transcription factor called Zonula-Occludens Associated Nucleic Acid-Binding protein (ZONAB). This interaction may allow TJs to regulate epithelial cell proliferation and cell density [47,48].

ZO-1 has been shown to be present in many ocular tissues including several retinal layers in the mammalian retina. This displays the role of ZO-1 in a wide portfolio of cellular functions in addition to mediating barrier function [3].

4. Role in Pathology

As we have established, the RPE has a wide ranging and diverse number of functions and the oBRB plays a crucial role in establishing the optimal conditions for the functioning of PRs and as such, also for visual function. In this section, we will discuss how the disruption of the oBRB contributes to the pathogenesis of a range of ocular pathologies namely DR, and diabetic macular oedema (DMO), AMD, CSCR, Sorsby's fundal dystrophy, and RP, and also examine conditions associated with mutations in CLDN-19. It should be noted that this is not a complete list of all ocular conditions caused by the disruption of the oBRB, however, for the purpose of the review, we will focus solely on these conditions.

5. Diabetic Retinopathy

Diabetic retinopathy (DR) is the leading cause of blindness among working aged individuals in developed countries. In type 1 diabetes, proliferative diabetic retinopathy (PDR) is the most common sight threatening lesion and is characterised by neovascularisation secondary to a hypoxic insult [2]. In type 2 diabetes, DMO is the primary cause of visual loss, resulting from vascular leakage due to the breakdown of the iBRB [2]. It has been well described previously that iBRB alteration from endothelial cell dysfunction leads to DMO and subsequently retinopathy progression [1]. However, less is known about the role of the oBRB in DR.

There is growing evidence for diabetes induced oBRB dysfunction as observed in both human and animal studies [49,50]. The oBRB specific leakage has been visualised in diabetic and ischemic rodents demonstrating diabetic and ischemia-induced breakdown of TJs in the RPE, with one study reporting that the oBRB contributed a third of the total vascular leakage in diabetic retinas [51]. Significant depletion of occludin in the RPE of ischemic and diabetic rodents was also observed, suggesting that this leakage is a consequence of the breakdown of TJs in the oBRB [51]. In rodents with early stages of diabetes (those without discernible vascular leakage or TJ breakdown), reduced RPE absorptive capacity occurred before the breakdown of TJ strands [52]. In examining ZO-1 staining in hyperglycaemic rats, the TJ appeared wider and demonstrated small holes [52]. It was found that as a consequence of these conditions, the retina becomes highly hypoxic, which causes the upregulation of hypoxia inducible factor-1 (HIF) alpha and VEGF [39]. Upregulation of the VEGF signalling pathway ultimately leads to the loss of barrier function and TJ-integrity in the oBRB [51,52]. The oBRB breakdown then results in the leakage of blood contents and influx of osmolytes, which precedes subretinal oedema and exudative retinal detachment [51]. Furthermore, hyperglycaemia induces a loss of Na^+/K^+-ATPase function, which could impair the transport of water from the sub retinal space to choriocapillaris and consequently might contribute to DMO development [2].

6. Age-Related Macular Degeneration (AMD)

Age-related macular degeneration (AMD) is the leading cause of central retinal vision loss in developed countries, with an estimated one in 10 people over the age of 55 showing signs of the condition [53]. It is a complex, multifactorial disease characterised by RPE dysfunction and macular PR loss [13]. Advanced AMD presents in two forms that are generally referred to as "dry" or "wet". Choroidal neovascularisation (CNV) is associated with wet AMD and occurs when new blood vessels sprout from the underlying choroidal vasculature and disrupt the integrity of the retina, leading to acute visual loss [54,55]. Geographic atrophy (GA) secondary to dry AMD occurs when the RPE begins to degenerate in the region of the macula leading to cone PR cell death and eventual central vision loss [55,56].

As discussed, the basal surface of the RPE rests on Bruch's membrane. In the aging eye, extracellular material is deposited in Bruch's membrane [13] and as a result, the thickness of Bruch's membrane increases and permeability decreases [56,57]. Aging involves an accumulation of oxidative insults and a concomitant decrease in protective mechanisms [58]. The accumulation of lipofuscin in the RPE has been suggested to act as a starting point [15]. The initial triggers for age-related degenerative diseases are thought to be oxidative damage [15,58]. In patients with AMD, the RPE's adaptive response to stress becomes dysregulated and an increasing imbalance of protective and toxic factors contributes to macular damage and the development of retinal lesions [15]. Reduced capacity of the RPE to absorb light energy is also an important factor in the cascade of events leading to AMD [15].

VEGF is secreted from the basolateral side of healthy RPE cells and is involved in the regulation of choroidal vasculature. Overexpression of VEGF is an important factor in the pathogenesis of CNV in wet AMD [6]. An imbalance between pro-angiogenic and anti-angiogenic factor in favour of the formation of new vessels [59] is likely to occur as a result of local hypoxia, inflammation, and dysregulated wound healing secondary to RPE cell loss [15]. It is the integrity of the oBRB that keeps the choroidal vascular response from invading the retina and changing dry into wet AMD [1] and the loss of this RPE–RPE attachment can induce VEGF overexpression [60]. RPE degeneration, tears, drusen formation, or apoptosis associated with aging can cause a loss of the RPE–RPE attachment [60]. Intravitreal agents that block VEGF have revolutionised the care of patients with wet AMD, decreasing growth, and leakage from CNV lesions, and preventing moderate and severe visual loss [59].

7. Central Serous Chorioretinopathy (CSCR)

Central serous chorioretinopathy (CSCR) is a disease of the retina characterised by serous detachment of the neurosensory retina secondary to focal lesions of the RPE. The mechanism whereby sub retinal fluid is produced is poorly understood [61]. It is thought that the choroidal vascular hyperpermeability occurs in CSCR, possibly as a result of stasis, ischaemia, or inflammation. Once the hydrostatic pressure is sufficiently high, the RPE is pushed forward, leading to discontinuing of the barrier and promoting pigment epithelial detachments. This overwhelms the barrier function of the RPE and leads to fluid accumulation within the RPE [62].

8. Sorsby's Fundus Dystrophy

Sorsby's fundus dystrophy is an autosomal dominant condition of the central retina. Similar to AMD, there is an accumulation of drusen deposits in Bruch's membrane [13]. It is caused by variants in the gene for tissue inhibitor of metalloproteinases-3 (TIMP-3), which has been found in the drusen-like deposits that occur in patients with the condition [63]. TIMP-3 functions to regulate the thickness of Bruch's membrane by inhibiting matrix metalloproteinases and also inhibits VEGF signalling and therefore angiogenesis [63]. Sorsby's fundus dystrophy leads to eventual CNV and PR cell loss [63].

9. Retinitis Pigmentosa

Retinitis Pigmentosa (RP) is the term given to a set of hereditary retinal diseases that feature degeneration of rod and cone PRs and characteristic retinal deposits [64]. In most cases, it is due to the primary degeneration of rods with secondary cone degeneration [65]. In the first instance, RP patients typically lose night vision, followed by peripheral vision, and eventually central vision later in life [64,66]. Rhodopsin was the first gene identified and encodes rod visual pigment. The gene products localised in rods being involved in several visual pathways including a protein of rod visual transduction, cytoskeleton proteins, trafficking proteins, and PR differentiation [65]. There are at least 200 different mutations in more than 60 genes that are linked to RP [67]. Many of the pathways vary dramatically ranging from visual pigments to visual cycle to metabolism and RNA splicing [65]. In RHO(-/-) knockout mice (mice that carry a targeted disruption of the rhodopsin gene and present with rapid PR degeneration) upregulation of ZO-1 has been demonstrated when compared to wild type [66]. This finding may represent a physiological response to PR cell degeneration and appears to be associated with the retinal vasculature of the RHO (−/−) mice. In RHO (−/−) mice, blood vessels appear more clustered and thicker in appearance compared to the wild-type [66]. In contrast, a study looking at retinal degeneration 9 (Rd9) mice (mice that represent a rare model of X-linked RP) demonstrate a decrease in the intensity of ZO-1 staining [67].

In addition, RP may be complicated by cystoid macular oedema (CMO), which represents an important and treatable cause of central visual loss in RP [68]. The underlying pathogenesis remains unclear, however, breakdown of the BRB is a proposed mechanism for the development of CMO in RP [68]. The release of toxic products from the degenerating retina or RPE may cause CMO by disrupting the BRB. The iBRB is the primary site of vascular leakage in RP complicated by CMO in both patients with early and severe RP. Interestingly, patients with more extensive RP demonstrated greater leakage through the RPE compared to those with less extensive RP [69]. In eyes with RP alone, albumin leakage was greatest from the iBRB, however, in RP-associated with other complications such as aphakia and glaucoma leakage varied between iBRB and oBRB [70].

Mutations in CLDN-19

As discussed earlier, claudin-19 is the predominant claudin in the human RPE [9]. Mutations in CLDN-19 cause the renal disease familial hypomagnesemia with hypercalciuria and nephrocalcinosis (FHHNC); ocular involvement (FHHNCOI) results in a varied array of ocular defects demonstrating incomplete penetrance [40,41]. In severe cases, symptoms appear in infants as young as 4–9 months old including bilateral macular coloboma, chorioretinal degeneration, nystagmus, strabismus, and visual loss [41]. Another study has reported that patients affected this mutation present with disrupted optic disc development resulting in near blindness and horizontal nystagmus [40]. The pathogenesis of such defects in FHHNCOI is yet to be determined.

10. Conclusions

In conclusion, the RPE is a single layer of cells that separates the neural retina from the underlying Bruch's membrane and choroid. The TJs between the RPE form the oBRB, which is essential for the maintenance of visual function. In this review, we discussed the RPE and the oBRB in detail and examined the effect of its dysfunction, which can lead to an array of pathological ocular conditions. Understanding the complex molecular function of the oBRB will inform our understanding of a range of retinal diseases and could ultimately lead to the development of therapies aimed at restoring oBRB function and integrity.

Author Contributions: Conceptualisation, A.N.; A.H.; N.H.; and M.C.; Writing—original draft preparation, A.N.; and A.H; Writing—review and editing, A.N.; N.H.; and M.C.; supervision, N.H.; and M.C. All authors have read and agreed to the published version of the manuscript.

References

1. Cunha-Vaz, J.; Bernardes, R.; Lobo, C. Blood-retinal barrier. *Eur. J. Ophthalmol.* **2011**, *21*, 3–9. [CrossRef]
2. Simó, R.; Villarroel, M.; Corraliza, L.; Hernández, C.; Garcia-Ramírez, M. The Retinal pigment epithelium: Something more than a constituent of the blood-retinal barrier—Implications for the pathogenesis of diabetic retinopathy. *J. Biomed. Biotechnol.* **2010**, *2010*, 190724. [CrossRef] [PubMed]
3. Campbell, M.; Humphries, P. The blood-retina barrier. In *Biology and Regulation of Blood-Tissue Barriers*; Cheng, C.Y., Ed.; Advances in Experimental Medicine and Biology; Springer: New York, NY, USA, 2013; pp. 70–84, ISBN 978-1-4614-4711-5.
4. Rizzolo, L.J. Development and role of tight junctions in the retinal pigment epithelium. *Int. Rev. Cytol.* **2007**, *258*, 195–234. [CrossRef] [PubMed]
5. Marmorstein, A.D.; Finnemann, S.C.; Bonhila, V.L.; Rodriguez-Boulan, E. Morphogenesis of the retinal pigment epithelium: Toward understanding retinal degenerative diseases. *Ann. N.Y. Acad. Sci.* **1998**, *857*, 1–12. [CrossRef] [PubMed]
6. Marneros, A.G.; Fan, J.; Yokoyama, Y.; Gerber, H.P.; Ferrara, N.; Crouch, R.K.; Olsen, B.R. Vascular endothelial growth factor expression in the retinal pigment epithelium Is essential for choriocapillaris development and visual function. *Am. J. Pathol.* **2005**, *167*, 1451–1459. [CrossRef]
7. Bazzoni, G.; Martínez-Estrada, O.M.; Orsenigo, F.; Cordenonsi, M.; Citi, S.; Dejana, E. Interaction of Junctional Adhesion Molecule with the Tight Junction Components ZO-1, Cingulin, and Occludin. *J. Biol. Chem.* **2000**, *275*, 20520–20526. [CrossRef]
8. Bazzoni, G.; Dejana, E. Endothelial cell-to-cell junctions: Molecular organization and role in vascular homeostasis. *Physiol. Rev.* **2004**, *84*, 869–901. [CrossRef]
9. Peng, S.; Rao, V.S.; Adelman, R.A.; Rizzolo, L.J. Claudin-19 and the barrier properties of the human retinal pigment epithelium. *Investig. Ophthalmol. Vis. Sci.* **2011**, *52*, 1392–1403. [CrossRef]
10. Peng, S.; Adelman, R.A.; Rizzolo, L.J. Minimal effects of VEGF and anti-VEGF drugs on the permeability or selectivity of RPE tight junctions. *Investig. Ophthalmol. Vis. Sci.* **2010**, *51*, 3216–3225. [CrossRef]
11. Luo, Y.; Xiao, W.; Zhu, X.; Mao, Y.; Liu, X.; Chen, X.; Huang, J.; Tang, S.; Rizzollo, L.W. Differential expression of claudins in retinas during normal development and the angiogenesis of oxygen induced retinopathy. *Investig. Ophthalmol. Vis. Sci.* **2011**, *52*, 7556–7564. [CrossRef]
12. Anderson, D.H.; Fisher, S.K. The relationship of primate foveal cones to the pigment epithelium. *J. Ultrastruct. Res.* **1979**, *67*, 23–32. [CrossRef]
13. Sparrow, J.R.; Hicks, D.; Hamel, C.P. The retinal pigment epithelium in health and disease. *Curr. Mol. Med.* **2010**, *10*, 802–823. [CrossRef] [PubMed]
14. Shin, K.; Fogg, V.C.; Margolis, B. Tight junctions and cell polarity. *Annu. Rev. Cell Dev. Biol.* **2006**, *22*, 207–235. [CrossRef] [PubMed]
15. Strauss, O. The retinal pigment epithelium in visual function. *Physiol. Rev.* **2005**, *85*, 845–881. [CrossRef]
16. Besharse, J.C.; Hollyfield, J.G.; Raybron, M.E. Photoreceptor outer segments: Accelerated membrane renewal in rods after exposure to light. *Science* **1997**, *196*, 536–538. [CrossRef]
17. Sakakibara, A.; Furuse, M.; Saitou, M.; Ando-Akatsuka, Y.; Tsukita, S. Possible involvement of phosphorylation of occludin in tight junction formation. *J. Cell Biol.* **1997**, *137*, 1393–1401. [CrossRef]
18. Farquhar, M.G.; Palade, G.E. Junctional complexes in various epithelia. *J. Cell Biol.* **1963**, *17*, 375–412. [CrossRef]
19. Diaz-Coranguez, M.; Liu, X.; Antonetti, D.A. Tight junctions in cell proliferation. *Int. J. Mol. Sci.* **2019**, *20*, 5972. [CrossRef]
20. Raleigh, D.R.; Marchiando, A.M.; Zhang, Y.; Shen, L.; Sasaki, H.; Wang, Y.; Long, M.; Turner, L.R. Tight

junction-associated MARVEL proteins marvelD3, tricellulin and occludin have distinct but overlapping functions. *Mol. Biol. Cell* **2010**, *21*, 1200–1213. [CrossRef]
21. Mariano, C.; Sasaki, H.; Brites, D.; Brito, M.A. A look at tricellulin and its role in tight junction formation and maintenance. *Eur. J. Cell Biol.* **2011**, *90*, 787–796. [CrossRef]
22. Furuse, M.; Hirase, T.; Itoh, M.; Nagafuchi, A.; Yonemura, S.; Tsukita, S.; Tsukita, S. Occludin: A novel integral membrane protein localizing at tight junctions. *J. Cell Biol.* **1993**, *123*, 1777–1788. [CrossRef]
23. Furuse, M.; Itoh, T.; Hirase, A.; Nagafuchi, S.; Yonemura, S.; Tsukita, S.; Tsukita, S. Direct association of occludin with ZO-1 and its possible involvement in the localization of occludin at tight junctions. *J. Cell Biol.* **1994**, *127*, 1617–1626. [CrossRef]
24. Balda, M.S.; Whitney, J.A.; Flores, C.; Gonzalez, S.; Cereijido, M.; Matter, K. Functional dissociation of paracellular permeability and transepithelial electrical resistance and disruption of the apical-basolateral intramembrane diffusion barrier by expression of a mutant tight junction membrane protein. *J. Cell Biol.* **1996**, *134*, 1031–1049. [CrossRef] [PubMed]
25. Chen, Y.; Merzdorf, C.; Paul, D.L.; Goodenough, D.A. COOH terminus of occludin is required for tight junction barrier function in early Xenopus embryos. *J. Cell Biol.* **1997**, *138*, 891–899. [CrossRef] [PubMed]
26. Feldman, G.J.; Mullin, J.M.; Ryan, M.P. Occludin: Structure, function and regulation. *Adv. Drug Deliv. Rev.* **2005**, *57*, 883–917. [CrossRef] [PubMed]
27. Saitou, M.; Fujimoto, K.; Doi, Y.; Fujimoto, T.; Furuse, M.; Takano, H.; Noda, T.; Tsukita, S. Occludin-deficient embryonic stem cells can differentiate into polarized epithelial cells bearing tight junctions. *J. Cell Biol.* **1998**, *141*, 397–408. [CrossRef] [PubMed]
28. Saitou, M.; Furuse, M.; Sasaki, H.; Schulzke, J.D.; Fromm, M.; Takano, H.; Noda, T.; Tsukita, S. Complex phenotype of mice lacking occludin, a component of tight junction strands. *Mol. Biol. Cell* **2000**, *11*, 4131–4142. [CrossRef]
29. Wong, V.; Gumbiner, B.M. A synthetic peptide corresponding to the extracellular domain of occludin perturbs the tight junction permeability barrier. *J. Cell Biol.* **1997**, *136*, 399–409. [CrossRef]
30. Rizzolo, L.; Peng, S.; Luo, Y.; Xiao, W. Integration of tight junctions and claudins with the barrier functions of the retinal pigment epithelium. *Prog. Retin. Eye Res.* **2011**, *30*, 296–323. [CrossRef]
31. Traweger, A.; Fang, D.; Liu, Y.C.; Stelzhammer, W.; Krizbai, I.A.; Bauer, H.C.; Bauer, H. The Tight-Junction Specific Protein Occludin is a Functional Target of the E3 Ubiquitin-Protein Ligase Itch. *J. Biol. Chem.* **2002**, *277*, 10201–10208. [CrossRef]
32. Wong, V. Phosphorylation of occludin correlates with occludin localization and function at the tight junction. *Am. J. Physiol.* **1997**, *273*, 1859–1867. [CrossRef] [PubMed]
33. Staddon, J.M.; Smales, C.; Schulze, C.; Esch, F.S.; Rubin, L.L. p120, a p120-related protein (p100), and the cadherin/catenin complex. *J. Cell Biol.* **1995**, *130*, 369–381. [CrossRef] [PubMed]
34. Kale, G.; Naren, A.P.; Sheth, P.; Rao, R.K. Tyrosine phosphorylation of occludin attenuates its interactions with ZO-1, ZO-2, and ZO-3. *Biochem. Biophys. Res. Commun.* **2003**, *302*, 324–329. [CrossRef]
35. Colegio, O.R.; Van Itallie, C.; Rahner, C.; Anderson, J.M. Claudin extracellular domains determine paracellular charge selectivity and resistance but not tight junction fibril architecture. *Am. J. Physiol. Cell Physiol.* **2003**, *284*, 1346–1354. [CrossRef]
36. Colegio, O.R.; Van Itallie, C.; McCrea, H.J. Claudins create charge-selective channels in the paracellular pathway between epithelial cells. *Am. J. Physiol. Cell Physiol.* **2002**, *282*, 142–147. [CrossRef]
37. Van Italie, C.M.; Anderson, J.M. Claudins and paracellular transport. *Annu. Rev. Physiol.* **2006**, *68*, 403–429. [CrossRef]
38. Morita, K.; Sasaki, H.; Furuse, M.; Tsukita, S. Endothelial claudin: Claudin-5/TMVCF constitutes tight junction strands in endothelial cells. *J. Cell Biol.* **1999**, *147*, 185–194. [CrossRef]
39. Itoh, M.; Furuse, M.; Morita, K.; Kubota, K.; Saitou, M.; Tsukita, S. Direct binding of three tight junction-associated MAGUKs, ZO-1, ZO-2, and ZO-3, with the COOH termini of claudins. *J. Cell Biol.* **1999**, *147*, 1351–1363. [CrossRef]
40. Konrad, M.; Schaller, A.; Seelow, D.; Pandey, A.V.; Waldegger, S.; Lesslauer, A.; Vitzthum, H.; Suzuki, Y.; Luk, J.M.; Becker, C.; et al. Mutations in the tight-junction gene claudin 19 (CLDN19) are associated with renal magnesium wasting, renal failure, and severe ocular involvement. *Am. J. Hum. Genet.* **2006**, *79*, 949–957. [CrossRef]
41. Wang, S.B.; Xu, T.; Peng, S.; Singh, D.; Ghiassi-Nejed, M.; Adelman, R.A.; Rizzolo, L.J. Disease-associated

mutations of claudin-19 disrupt retinal neurogenesis and visual function. *Commun. Biol.* **2019**, *2*, 113. [CrossRef]

42. Mandell, K.J.; Parkos, C.A. The JAM family of proteins. *Adv. Drug Deliv. Rev.* **2005**, *57*, 857–867. [CrossRef] [PubMed]
43. Gonzalez-Mariscal, L.; Betanzos, A.; Avila-Flores, A. MAGUK proteins: Structure and role in the tight junction. *Semin. Cell. Dev. Biol.* **2000**, *11*, 315–324. [CrossRef] [PubMed]
44. Stevenson, B.R.; Siliciano, J.D.; Mooseker, M.S.; Goodenough, D.A. Identification of ZO-1: A high molecular weight polypeptide associated with the tight junction (zonula occludens) in a variety of epithelia. *J. Cell Biol.* **1986**, *103*, 755–766. [CrossRef] [PubMed]
45. Tsukita, S.; Furuse, M.; Itoh, M. Multifunctional strands in tight junctions. *Nat. Rev. Mol. Cell Biol.* **2001**, *2*, 285–293. [CrossRef]
46. Muller, S.L.; Portwich, M.; Schmidt, A.; Utepbergenov, D.I.; Huber, O.; Blasig, I.E.; Krause, G. The tight junction protein occludin and the adherens junction protein α-catenin share a common interaction mechanism with ZO-1. *J. Biol. Chem.* **2005**, *280*, 3747–3756. [CrossRef]
47. Balda, M.S.; Matter, K. The tight junction protein ZO-1 and an interacting transcription factor regulate ErbB-2 expression. *EMBO. J.* **2000**, *19*, 2024–2033. [CrossRef]
48. Balda, M.S.; Garrett, M.D.; Matter, K. The ZO-1-associated Y-box factor ZONAB regulates epithelial cell proliferation and cell density. *J. Cell Biol.* **2003**, *160*, 423–432. [CrossRef]
49. Vinores, S.A.; Gadegbeku, C.; Campochiaro, P.A.; Green, W.R. Immunohistochemical localization of blood-retinal barrier breakdown in human diabetics. *Am. J. Pathol.* **1989**, *134*, 231–235. [CrossRef]
50. Xia, T.; Rizzolo, L.J. Effects of diabetic retinopathy on the barrier functions of the retinal pigment epothelium. *Vis. Res.* **2017**, *129*, 72–81. [CrossRef]
51. Xu, H.Z.; Le, Y.Z. Significant of outer blood-retina breakdown in diabetes and ischaemia. *Investig. Ophthalmol. Vis. Sci.* **2011**, *53*, 2160–2164. [CrossRef]
52. Desjardins, D.M.; Yates, P.W.; Dahrouj, M.; Yiu, Y.; Crosson, C.E.; Ablonczy, Z. Progressive early breakdown of retinal pigment epithelium function in hyperglycaemic rats. *Investig. Ophthalmol. Vis. Sci.* **2016**, *57*, 2706–2713. [CrossRef] [PubMed]
53. Wong, W.L.; Su, X.; Li, X.; Cheung, C.M.; Klein, R.; Cheng, C.Y.; Wong, T.Y. Global prevalence of age-related macular degeneration and disease burden projection for 2020 and 2040: A systematic review and meta-analysis. *Lancet Glob. Health* **2014**, *2*, 106–116. [CrossRef]
54. Chakravarthy, U.; Evans, J.; Rosenfeld, P.J. Age related macular degeneration. *BMJ* **2010**, *340*, 981. [CrossRef] [PubMed]
55. Cook, H.L.; Pater, P.J.; Tufail, A. Age-related macular degeneration: Diagnosis and management. *Br. Med. Bull.* **2008**, *85*, 127–149. [CrossRef]
56. Fleckenstein, M.; Mitchell, P.; Freund, K.B.; Sadda, S.; Holz, F.G.; Brittain, C.; Henry, E.C.; Ferrara, D. The progression of geographic atrophy secondary to age-related macular degeneration. *Ophthalmology* **2018**, *125*, 369–390. [CrossRef]
57. Ramrattan, R.S.; van der Schaft, T.L.; Mooy, C.M.; de Bruijn, W.C.; Mulder, P.G.; de Jong, P.T. Morphometric analysis of Bruch's membrane, the choriocapillaris, and the choroid in aging. *Investig. Ophthalmol. Vis. Sci.* **1994**, *35*, 2857–2864.
58. Chen, M.; Xu, H. Parainflammation, chronic inflammation, and age related macular degeneration. *J. Leukoc. Biol.* **2015**, *98*, 713–723. [CrossRef]
59. Brassler, S.B. Introduction: Understanding the role of angiogenesis and antiangiogenic agents in age-related macular degeneration. *Ophthalmology* **2009**, *116*, S1–S7. [CrossRef]
60. Farjood, F.; Vargis, E. Physical disruption of cell-cell contact induced VEGF expression in RPE cells. *Mol. Vis.* **2017**, *23*, 431–446, pmc:5524271.
61. Wang, M.; Munch, I.C.; Hasler, P.W.; Prunte, C.; Larsen, M. Central serous retinopathy. *Acta. Ophthalmol. Scand.* **2008**, *86*, 126–145. [CrossRef]
62. Nicholson, B.; Noble, J.; Forooghian, F.; Meyerle, C. Central serous chorioretinopathy: Update on pathophysiology and treatment. *Surv. Ophthalmol.* **2014**, *58*, 103–126. [CrossRef] [PubMed]
63. Christensen, D.R.G.; Brown, F.E.; Cree, A.J.; Ratnayaka, J.A.; Lotery, A.J. Sorsby fundus dystrophy—A review of pathology and disease mechanisms. *Exp. Eye Res.* **2017**, *165*, 35–46. [CrossRef] [PubMed]
64. Hartong, D.T.; Berson, E.L.; Dryja, T.P. Retinitis Pigmentosa. *Lancet* **2006**, *368*, 1795–1809. [CrossRef]

65. Hamel, C. Retinitis Pigmentosa. *Orphanet J. Rare Dis.* **2006**, *1*, 40. [CrossRef]
66. Campbell, M.; Humphries, M.; Kennan, A.; Kenna, P.; Humphries, P.; Brankin, B. Aberrant retinal tight junction and adherens junction protein expression in an animal model of autosomal recessive Retinitis pigmentosa: The Rho (−/−) mouse. *Exp. Eye Res.* **2006**, *83*, 484–492. [CrossRef]
67. Falasconi, F.; Biagioni, M.; Novelli, E.; Piabo, I.; Gargini, C.; Strettoi, E. Retinal phenotype in the rd9 mutant mouse, a model of X-linked RP. *Front. Neurosci.* **2019**, *13*, 991. [CrossRef]
68. Wang, W.; Kini, A.; Wang, Y.; Liu, T.; Chen, Y.; Vukmanic, E.; Emery, D.; Liu, Y.; Jin, L.; Lee, S.J.; et al. Metabolic deregulation of the blood-outer retinal barrier in retinitis pigmentosa. *Cell Rep.* **2019**, *28*, 1323–1334. [CrossRef]
69. Vinores, S.A.; Kuchle, M.; Derevjanik, N.L.; Henderer, J.D.; Mahlow, J.; Green, W.R.; Campochiaro, P.A. Blood-retinal barrier breakdown in retinitis pigmentosa: Light and electron microscopic immunolocalization. *Histol. Histopathol.* **1995**, *10*, 913–923.
70. Strong, S.; Liew, G.; Michaelides, M. Retinitis pigmentosa-associated cystoid macular oedema: Pathogenesis and avenues of intervention. *Br. J. Ophthalmol.* **2017**, *101*, 31–37. [CrossRef]

13

Use of Modified *Clostridium perfringens* Enterotoxin Fragments for Claudin Targeting in Liver and Skin Cells

Laura-Sophie Beier [1], Jan Rossa [2], Stephen Woodhouse [3], Sophia Bergmann [4], Holger B. Kramer [5,†] Jonas Protze [2], Miriam Eichner [1], Anna Piontek [2], Sabine Vidal-y-Sy [4], Johanna M. Brandner [4], Gerd Krause [2], Nicole Zitzmann [3] and Jörg Piontek [1,*]

1 Charité-Universitätsmedizin Berlin, Institute of Clinical Physiology, Hindenburgdamm 30, 12203 Berlin, Germany; Laura-Sophie.Beier@charite.de (L.-S.B.); miriam.eichner@googlemail.com (M.E.)
2 Leibniz-Forschungsinstitut für Molekulare Pharmakologie (FMP), Robert-Rössle-Straße 10, 13125 Berlin, Germany; JanRossa@gmx.de (J.R.); Protze@fmp-berlin.de (J.P.); dr.anna.piontek.ext@bayer.com (A.P.); GKrause@fmp-berlin.de (G.K.)
3 Oxford Glycobiology Institute, Department of Biochemistry, University of Oxford, South Parks Road, Oxford OX1 3QU, UK; Stephen.woodhouse@cardiov.ox.ac.uk (S.W.); nicole.zitzmann@bioch.ox.ac.uk (N.Z.)
4 Department of Dermatology and Venerology, University Hospital Hamburg Eppendorf, 20246 Hamburg, Germany; s.bergmann@t-online.de (S.B.); s.vidal-y-sy@uke.de (S.V.-y.-S.); brandner@uke.de (J.M.B.)
5 Oxford Centre for Gene Function, Department of Physiology, Anatomy and Genetics, University of Oxford, Oxford OX3 7DQ, UK; h.kramer@lms.mrc.ac.uk
* Correspondence: joerg.piontek@charité.de
† Current Address: London Institute of Medical Sciences, Biological Mass Spectrometry and Proteomics Facility, London W12 0NN, UK.

Abstract: Claudins regulate paracellular permeability in different tissues. The claudin-binding domain of *Clostridium perfringens* enterotoxin (cCPE) is a known modulator of a claudin subset. However, it does not efficiently bind to claudin-1 (Cldn1). Cldn1 is a pharmacological target since it is (i) an essential co-receptor for hepatitis C virus (HCV) infections and (ii) a key element of the epidermal barrier limiting drug delivery. In this study, we investigated the potential of a Cldn1-binding cCPE mutant (i) to inhibit HCV entry into hepatocytes and (ii) to open the epidermal barrier. Inhibition of HCV infection by blocking of Cldn1 with cCPE variants was analyzed in the Huh7.5 hepatoma cell line. A model of reconstructed human epidermis was used to investigate modulation of the epidermal barrier by cCPE variants. In contrast to cCPEwt, the Cldn1-binding cCPE-S305P/S307R/S313H inhibited infection of Huh7.5 cells with HCV in a dose-dependent manner. In addition, TJ modulation by cCPE variant-mediated targeting of Cldn1 and Cldn4 opened the epidermal barrier in reconstructed human epidermis. cCPE variants are potent claudin modulators. They can be applied for mechanistic *in vitro* studies and might also be used as biologics for therapeutic claudin targeting including HCV treatment (host-targeting antivirals) and improvement of drug delivery.

Keywords: *Clostridium perfringens* enterotoxin; claudin-1; Hepatitis C Virus; viral entry; epidermal barrier; reconstructed human epidermis; claudin targeting

---

1. Introduction

Claudins, tetraspan transmembrane proteins, are key structural and sealing components of tight junctions (TJ). In endo- and epithelia, they seal the paracellular space and regulate the paracellular ion permselectivity of the tissue and contribute to membrane polarity [1]. Besides their role in the apical TJ, claudins are also localized outside of the TJ complex, indicating that claudins might function apart from their classical role regulating paracellular permeability. Claudins have been shown to be important factors in cell signaling, cell adhesion, epithelial to mesenchymal transition, cell migration, mucosal homeostasis as well as disease pathogenesis (for review see [2]).

Various claudins serve as cell-surface receptors for pathogens, e.g., the *Clostridium perfringens* enterotoxin. The C-terminal domain of this toxin (cCPE) interacts with a subgroup of the claudin protein family (CPE-receptor claudins, e.g., claudin-3 (Cldn3), -4, -6, -9) [3]. cCPE was used in several studies to modulate the claudin-based barrier function of TJs [4]. To overcome the limitation of targeting CPE-receptor claudins only, claudin-subtype specificity of cCPE has been adapted by structure-guided mutagenesis and baculoviral display system. This enabled targeting of Cldn1-9 with the broad specificity binder cCPE-S305P/S307R/S313H (cCPE-SSS) [5,6], preferential binding to Cldn4 with cCPE-L254A/S256A/I258A/D284A [7] or preferential binding to Cldn5 with cCPE-Y306W/S313H and cCPE-N218Q/Y306W/S313H [6,8]. These variants bind with nanomolar affinities to their respective receptors [4,7,9,10]. The high affinity for their receptors as well as the claudin-specific binding renders cCPE variants appropriate tools for claudin targeting. The comparatively low production cost of the small recombinant protein and the concomitant targeting of a defined set of claudins with one molecule are convenient advantages over monoclonal antibodies against claudins.

Similar to other junctional proteins [11,12], claudins are employed as virus receptors. Cldn1 is an essential co-factor in hepatitis C virus (HCV) entry into hepatocytes [13]. Blocking of viral entry is a potent target for anti-HCV therapies with pan-genomic activity against multiple HCV subtypes; however, appropriate agents are not yet clinically available [14]. The multi-claudin binder cCPE-SSS might be a suitable agent to inhibit HCV receptor formation by binding Cldn1 and thereby blocking HCV entry into hepatocytes. HCV is a hepatotropic virus that has a high propensity to establish a persistent infection in human liver. HCV primarily infects human hepatocytes, which over time leads to chronic inflammation, progressive fibrosis, and development of hepatocellular carcinoma [15].

According to the World Health Organization, 3% of the world's population is infected with HCV [16]. About 15 to 45% of infected people spontaneously clear the virus within six months. The remaining 60 to 80%, currently ~71 million people worldwide, develop a chronic HCV infection, which, in a significant number of cases, will lead to cirrhosis or liver cancer. Each year, around 400,000 people die from hepatitis C-related cirrhosis or hepatocellular carcinoma [17]. Thus, even with the high success-rates of new direct-acting antivirals, HCV remains a major burden on global health.

HCV entry into hepatocytes is the first step of the viral life cycle resulting in productive viral infection [13,18]. The entry pathway consists of three steps: (1) viral attachment to hepatocytes; (2) receptor-mediated endocytosis of viral particle; and (3) endosomal fusion [19]. Viral entry is mediated by the viral envelope glycoproteins E1 and E2 and several host entry factors. These include heparan sulfate, tetraspanin cluster of differentiation 81 (CD81), scavenger receptor class B type I (SR-BI), and the TJ proteins Cldn1 [20] and occludin [21,22]. Subsequent studies demonstrated that Cldn6 and -9, displaying high levels of homology with Cldn1, are also able to mediate HCV entry into non-permissive cell lines [23,24]. However, in hepatoma cells, Cldn6 and Cldn9 appeared to function weaker than Cldn1 [24]. In addition, low expression levels of Cldn6 and Cldn9 in human hepatocytes seem to impede their efficient use as HCV entry factors [25].

It was suggested that Cldn1 plays a dual role in HCV entry [25]: (i) formation of a HCV co-receptor complex with CD81 [26] and (ii) direct interaction with the virus envelope E1E2 proteins [27]. While predominantly localized at the TJ, Cldn1 also localizes on the basolateral membrane of hepatocytes. Notably, this pool of Cldn1 outside of the TJ co-localizes with CD81 and allows the formation of the CD81-Cldn1 co-receptor complex for HCV entry [26,28].

HCV entry is a major target of the host neutralizing response [15,19,29] and a target for antiviral immunopreventive and therapeutic strategies (for review see [13,19,30,31]). In particular, use of blocking anti-Cldn1 antibodies showed promising results in blocking viral cell entry and cell-to-cell transmission in vitro and in vivo [14,30,32]. Treatment with an anti-Cldn1 monoclonal antibody cured chronically infected human liver chimeric mice [33]. As an alternative HCV treatment, in this study, we investigated the potential of claudin-binding cCPE variants to block HCV infection.

Cldn1 is also one of the main sealing claudins in human epidermis [34,35], a stratified epithelium, where multiple layers of cells protect the body against water loss, ultraviolet radiation and penetration of harmful xenobiotics [36]. There are two physiologic barriers serving this function: dead keratinocytes containing lipids and proteins, i.e., the *stratum corneum* (SC), and the TJ between the viable cells of the *stratum granulosum* (SG) [36,37]. Claudin targeting by cCPE variants might modify the latter barrier [37–39]. Thus, we used a skin model in addition to the HCV-hepatoma model to test cCPE variants in two different Cdn1-dependent cellular processes.

We utilized models of reconstructed human epidermis (RHE), which recreate all epidermal layers including the TJ-containing SG and the biologically dead yet highly dynamic SC. In order to develop a functional SC, the apical side of the keratinocytes is exposed to air, creating an air-liquid interface (ALI) [40]. RHE enables the analysis of keratinocyte-specific responses within the epidermis, making it an ideal model for this study, in which the capability of the cCPE variants to open the epidermal barrier was tested.

In this study, we investigated the potential of cCPE variants to target processes that depend on extra-junctional or junctional claudins. Targeting of extra-junctional Cldn1 in hepatocytes, which serves as a co-receptor for HCV infection, by cCPE-SSS efficiently inhibited HCV infection. In RHE, on the other hand, we were able to modify the junctional epidermal barrier by targeting claudins in the SG. Thus, we showed that cCPE variants are suitable for directed claudin targeting.

2. Results

2.1. Structure Homology Modeling and FRET Analyses Suggest that cCPE-Cldn1 Interaction Interferes with Hepatitis C Virus Entry into Host Cells

HCV entry into hepatocytes depends on Cldn1 as a HCV co-receptor. We hypothesized that HCV infection might be inhibited by a Cldn1-binding cCPE variant since its interaction with Cldn1 on the cell surface might block the virus entry. The interaction mechanism of cCPE with defined claudin subtypes was well characterized by binding, mutagenesis and modeling studies. The motif (N/D(P-1) P(P) L/M/V(P+1) V/T(P+2) P/A/D(P+3)) in the extracellular segment 2 (ECS2) of claudins binds to the claudin-binding pocket of cCPE consisting of a triple tyrosine and a triple leucine pit [6,7,10,41]. This mechanism was confirmed by complex structures of cCPE bound to Cldn19 or Cldn4 [42,43]. Using these structures, a homology model of a complex between the Cldn1-binding cCPE-SSS and Cldn1 was generated (Figure 1). The model clearly indicates that binding of cCPE-SSS to the extracellular domains of Cldn1 buries most of the residues, which were shown to be involved in Cldn1-CD81 HCV receptor complex formation [20,26]. In particular, this is the case for D38, E48, T42, N72 in ECS1 and M 152 and V155 in ECS2. Other residues involved in Cldn1-CD81 association include W30, G49 and W51, which are part of the claudin consensus sequence and presumably involved in the general fold of claudins [44]. In summary, the modeling suggests that cCPE binding to Cldn1 interferes with formation of the HCV receptor/entry factor complex which includes Cldn1 and CD81. Thus, cCPE-SSS might block HCV entry.

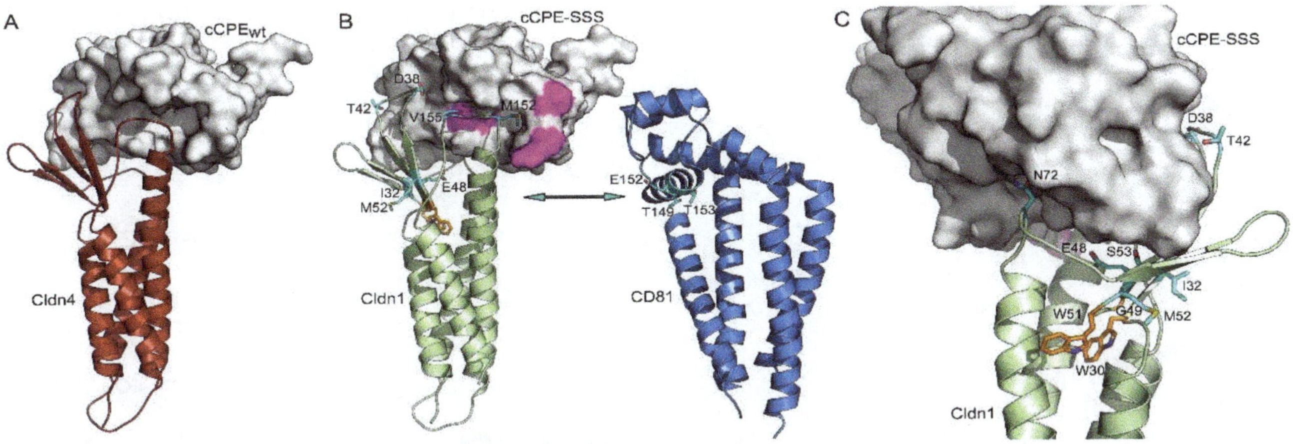

Figure 1. Homology modeling indicates that cCPE-Cldn1 interaction interferes with formation of HCV receptor complex. (**A**) X-ray structure (PDB ID: 5B2G, [43]) of Cldn4 (cartoon, red) interacting with cCPEwt (surface, gray) (**B**) Binding of cCPE-SSS to Cldn1 likely buries large parts of the extracellular domain of Cldn1 preventing Cldn1-CD81 interaction (arrow). Homology model of Cldn1 in complex with cCPE-SSS (template PDB ID: 5B2G) and CD81 (PDB ID: 5TCX). CD81 (blue) and Cldn1 (green) shown as cartoon; cCPE-SSS shown as the surface in gray and the substitutions (S305P, S307R, S313H) enhancing Cldn1 binding in magenta. Cldn1-CD81 interaction-relevant residues shown as cyan sticks (**C**) Other perspective highlighting Cldn1 residues involved in HCV entry [20,26] as cyan sticks. D38, T42, E48, S53, N72, M152, V155, reported to be involved in HCV infection are masked by cCPE. Involvement of W30, G49, W51 (sticks, orange) in HCV infection is likely to be due to a role in claudin folding [44]. I32 and M52 reported to be HCV relevant are not covered by cCPE.

Cldn1 and CD81 interact with each other in the plasma membrane outside of TJ. This interaction can be detected by fluorescence resonance energy transfer (FRET) [28]. Using human embryonic kidney cells (HEK-293), that were co-transfected with cyan fluorescent protein-fused Cldn1 and yellow fluorescent protein-fused CD81, we could show that FRET efficiency was significantly decreased after treatment with 10 μg/mL cCPE-SSS compared to the untreated control (Figure S1), indicating a reduced Cldn1-CD81 interaction. These results suggested that cCPE-SSS indeed interferes with Cldn1-CD81 co-receptor formation.

2.2. *S305P/S307R/S313H Substitution in cCPE Improves Its Binding to Cldn1 of Huh7 Hepatoma Cells*

The cCPE-variant cCPE-S305P/S307R/S313H (cCPE-SSS) binds more strongly to Cldn1 than cCPEwt [5,6]. To confirm this interaction in a hepatocyte-derived cell culture model, we first used human hepatocarcinoma Huh7 cells [23]. Binding of cCPE variants to Cldn1 endogenously expressed in Huh7 cells was analyzed by means of a pull down assay using GST-cCPE and cell lysates [7]. cCPEwt bound weakly yet specifically to Cldn1 whereas negative control cCPE-Y306A/L315A (cCPE-YALA), which does not interact with claudins [6,45], showed no binding (Figure 2). In contrast, cCPE-SSS showed strong binding to Cldn1 (Figure 2). These data demonstrate that cCPE-SSS binds efficiently to Cldn1 endogenously expressed in hepatocyte model cells.

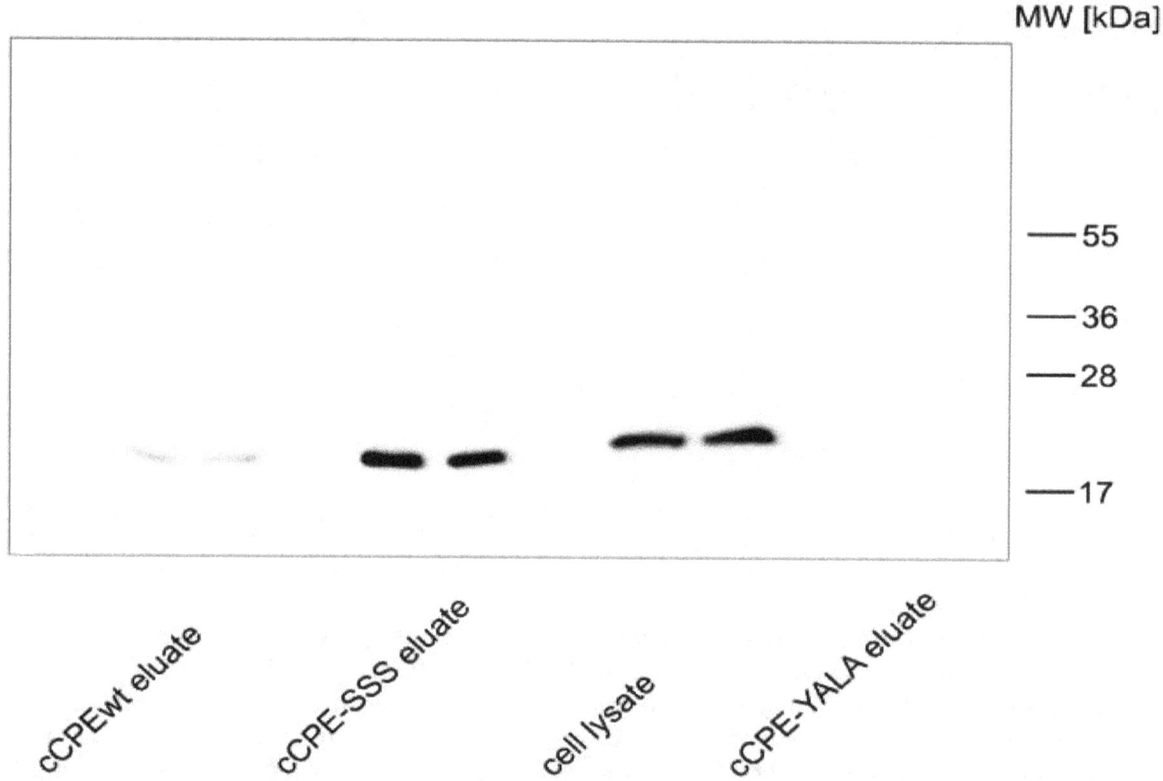

Figure 2. Compared to cCPEwt, cCPE-SSS interacts more strongly with Cldn1 expressed in Huh7 cells. Pull down assays using Huh7 lysates and GST-cCPEwt, -cCPE-SSS or -cCPE-YALA as negative control. Eluates and cell lysate were analyzed in duplicate by SDS-PAGE and Western blot using anti-Cldn1 antibodies. Strong immunoreactive bands in the expected range of ~ 20 kDa were detected for pull down with cCPE-SSS, whereas for cCPEwt, only weak bands and for cCPE-YALA, no bands were detected. Representative results are shown.

2.3. cCPE-S305P/S307R/S313H Specifically Inhibits HCV infection Of Huh7.5 Cells in a Dose-dependent Manner

Next, we tested whether cCPE-SSS is able to block Cldn1 as a HCV co-receptor and thus to inhibit HCV infection. To this end, infection of human hepatocarcinoma Huh7.5 cells with cell culture-derived HCV particles (HCVcc) was used as an established in vitro HCV infection model [46]. Compared to the parental cell line Huh7, the clone Huh7.5 is more permissive to HCV replication [47]. The infection assay was performed in the absence (control) or presence of cCPE variants. Strikingly, cCPE-SSS inhibited HCV infection of Huh7.5 cells in a dose-dependent manner (0.5–50 μg/mL), whereas the non-binding negative control cCPE-YALA had no inhibitory effect (Figure 3A). Surprisingly, CPEwt significantly enhanced HCV infection at low concentrations (0.5 and 2.5 μg/mL). Increased concentrations of cCPEwt (10 and 50 μg/mL) abolished this enhancing effect but never had an inhibiting effect (Figure 3A).

The inhibitory effect of cCPE-SSS on HCV infection was even more pronounced after extension of HCV incubation of the cells from 2 h to 26 h. Under these conditions, 10 μg/mL cCPE-SSS reduced the amount of HCV-infected cells to 0.095 ± 0.016-fold of the cells which were infected in the non-treated control. In contrast, at this concentration, neither cCPEwt nor cCPE-YALA had a significant effect (Figure 3B). Furthermore, 50 μg/mL cCPE-SSS inhibited the infection of HCV almost completely (0.013 ± 0.002-fold compared to the untreated control).

To exclude that these effects were caused by a reduced cell viability, Huh7.5 cells were treated with different concentrations of cCPEwt or –SSS and cell viability was analyzed with a tetrazolium salt-based colorimetric assay (MTS assay). Neither cCPEwt nor –SSS had adverse effects on Huh7.5 cell viability (Figure S2). This is in accordance with similar results previously obtained with claudin-transfected HEK-293 cells and brain endothelial cells, which were treated with different cCPE variants [8].

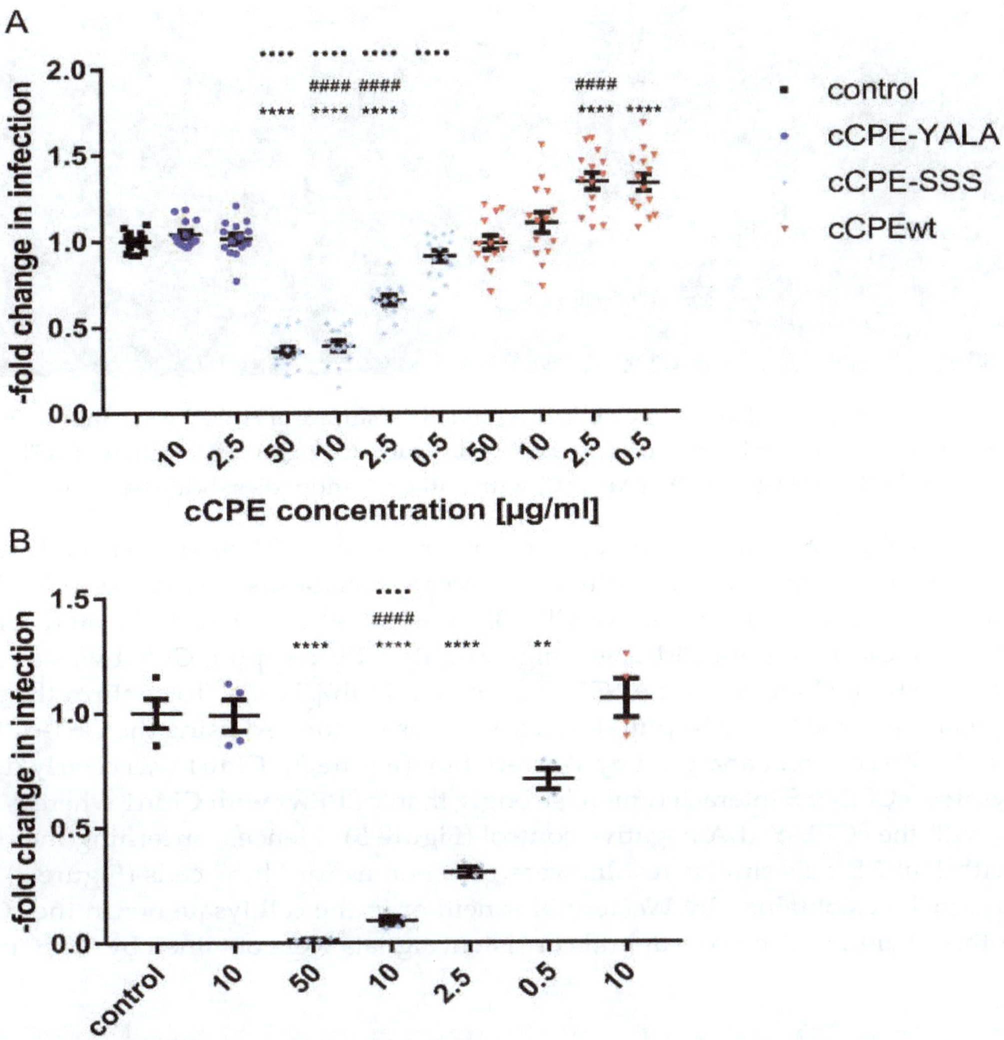

Figure 3. HCVcc infection of Huh7.5 cells in the presence of cCPE variants. Huh7.5 cells were pre-incubated for 2 h at 37 °C with cCPEwt, cCPE-SSS, cCPE–YALA (negative binding control) or not treated prior to infection (control). (**A**) HCVcc was added for 2 h, the medium was changed and 24 h later, HCV infection was analyzed by quantification of infection-positive cells after staining with an anti HCV NS5A antibody. Results are expressed as a percentage of infected cells compared to the control. Results of four experiments with total $n = 14$, for wt 50 $n = 13$. Results shown as means ± SEM. (**B**) HCVcc was added for 26 h. After this, HCV infection was analyzed by quantification of infection-positive cells after staining with an anti HCV NS5A antibody. Results are expressed as a percentage of infected cells compared to control. $n = 4$ Results shown as means ± SEM. One-way ANOVA with post-hoc Sidak's test was performed to analyze differences between treatment groups. ** $p \leq 0.01$ vs. control; **** $p \leq 0.0001$ vs. control; #### $p \leq 0.0001$ vs. cCPE-YALA; ●●●● $p < 0{,}0001$ vs. cCPEwt. cCPE-SSS strongly and specifically inhibits Cldn1-mediated HCV infection of Huh7.5 cells in a dose-dependent manner.

Together, the data demonstrate that cCPE-SSS specifically inhibits infection of Huh7.5 cells with HCV.

2.4. Both cCPE-SSS and cCPEwt Bind to Huh7.5 Cells

To further investigate the differential effects on HCV infection caused by cCPEwt and cCPE-SSS, the capability of cCPE variants to bind to Huh7.5 cells was analyzed. Confocal microscopy after immunostaining revealed binding of cCPEwt and cCPE-SSS but not of the cCPE-YALA control on the surface of Huh7.5 cells (Figure 4).

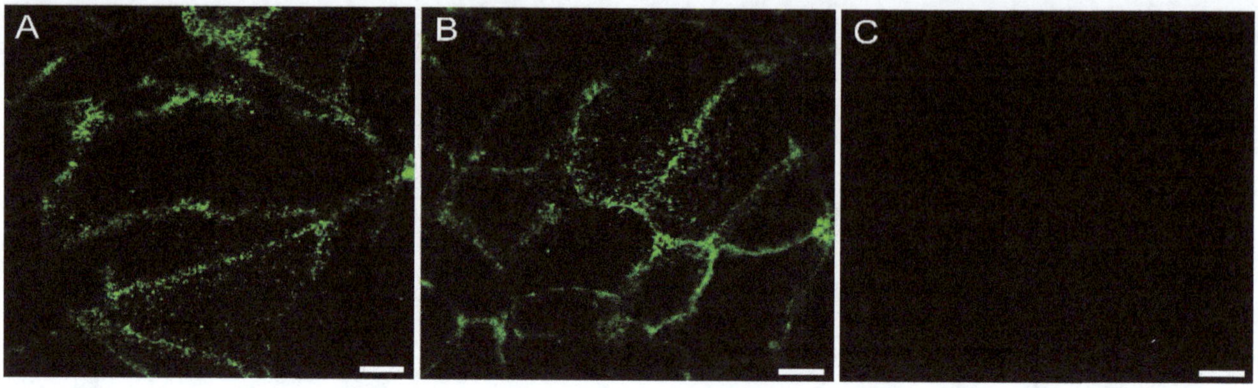

Figure 4. cCPEwt and cCPE-SSS but not cCPE-YALA bind to the surface of Huh7.5 cells. Immunostaining of Huh7.5 cells after 30 min of incubation with cCPE-constructs (10 μg/mL). Detection of GST-cCPEwt (**A**), GST-cCPE-SSS (**B**) and GST-cCPE-YALA (**C**) with anti-GST antibodies. Scale bar 10 μm.

cCPEwt and cCPE–SSS differ in their affinity towards Cldn1. Their similar binding to Huh7.5 cells indicated that, in addition to Cldn1, other CPE-receptor claudins are expressed in Huh7.5 cells. Expression of Cldn1, -2, and -4, but not of Cldn3, -6, -7, or -9 detected by Western blotting was reported [24]. Since Cldn4 but not Cldn2 is a high affinity CPE-receptor, Cldn4 was a candidate to contribute strongly to the binding of the cCPE variants to Huh7.5 cells. To confirm the presence of cCPE-binding claudins in these cells, pull down assay was performed using the GST-cCPE variants and lysates of Huh7.5 cells and analyzed by Western blot (Figure 5). Cldn1 was clearly detectable in the Huh7.5 lysates. cCPE-SSS interacted more strongly than cCPEwt with Cldn1 whereas no binding was detected with the cCPE-YALA negative control (Figure 5). Hence, concerning the cCPE-Cldn1 interaction with Huh7.5 cells similar results were obtained as for Huh7 cells (Figure 2). However, for Cldn4 no signal was obtained by Western blot neither in the cell lysate nor in the cCPE-bound fraction; for other claudins, also no clear and consistent signals were obtained by Western blot (data not shown).

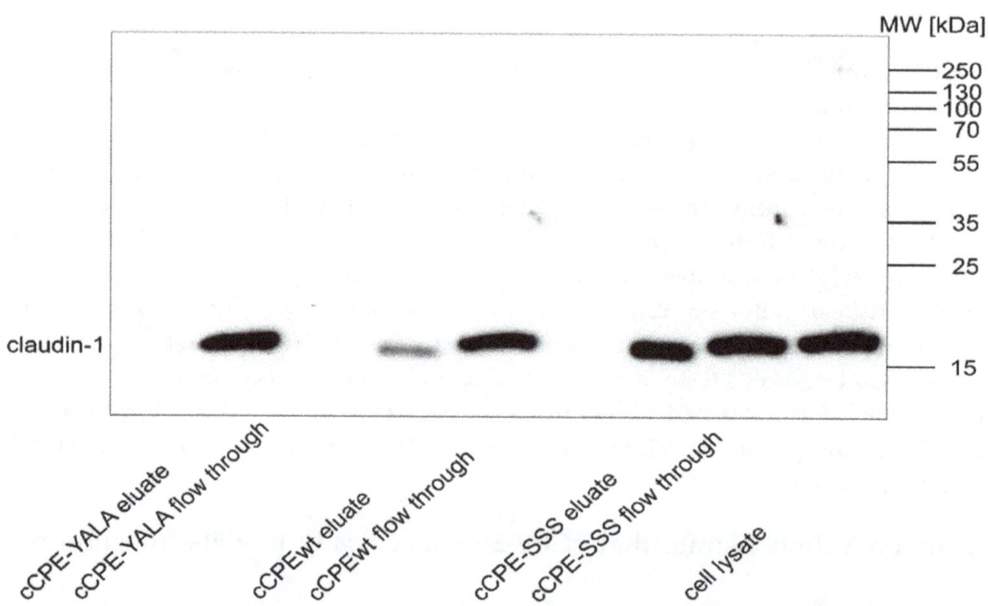

Figure 5. cCPE variants bind to Cldn1 of Huh7.5 cells. Pull down assays using GST-cCPE variants and lysates of Huh7.5 cells were performed. Eluate and flow through were separated by SDS-PAGE and analyzed by Western blot. A strong anti-Cldn1 immunoreactive band in the expected range of ~20 kDa was detected in the pull-down eluates for cCPE-SSS, whereas for cCPEwt, only a weak band and for cCPE-YALA, no band was detected. Representative results are shown.

2.5. Huh7.5 Cells Express Cldn1 and Cldn6 as Receptors for cCPE Variants

To overcome the limitations connected to sensitivity, specificity and cross-reactivity of anti-claudin antibodies, mass spectrometry was used to identify claudins which are expressed in Huh7.5 cells and able to bind to cCPEwt. A pull-down assay was performed with GST-cCPEwt and lysates of Huh7.5 cells, the cCPEwt-bound fraction as well as the non-bound fraction separated by SDS-PAGE, gel slices corresponding to areas between 15–25 kDa cut and processed for mass spectrometry. Consistent with the Western blots, Cldn1 was detected in the cCPEwt-bound and in the non-bound fraction (Table 1). In contrast, Cldn4 was not found in either fraction. However, Cldn6 was detected in the cCPE-bound fraction (Table 1). Subsequently, literature search-based purchase of another anti-Cldn6 antibody enabled confirmation of Cldn6 expression also by Western blot (Figure 6A). Cldn6 is known to be a CPE-receptor [6,10,48] and a weak entry factor for HCV [23,24]. Expression of Cldn6 mRNA in Huh7.5 cells was previously reported [29]. However, reports on Cldn6 protein expression were not completely consistent. Here, we clearly verify the expression of Cldn6 protein in Huh7.5 cells. Hence, cCPEwt can use Cldn6 and Cldn1 as receptors in Huh7.5 cells. We cannot entirely exclude that other claudins were not detected by mass spectrometry, due to methodological reasons, e.g., unsuitable peptide fragments. However, using mass spectrometry we were previously able to detect claudins such as Cldn1,-3,-4,-7 in liver cells [49].

Table 1. Mass spectrometry confirmed Cldn1 and Cldn6 protein expression in Huh7.5 cells. Pull down assay with GST-cCPEwt and lysates of Huh7.5 cells. cCPEwt-bound and non-bound fractions were separated by SDS-PAGE; gel slices corresponding to areas between 15–25 kDa were cut and processed for mass spectrometry.

Fraction	Cldn	Mascot Score	Sequence Cover-Age	Peptides	Position	Ions Score	Mass Determined	Mass Expected
E1: ~15kD	6	96	12%	K.VYDSLLALPQDLQAAR.A	66–81	46	1771.8914	1771.9468
				R.DFYNPLVAEAQK.R	146–157	90	1393.6274	1393.6878
E2: ~17KD	1	144	14%	K.VFDSLLNLSSTLQATR.A	66–81	91	1763.9194	1763.9418
				R.IVQEFYDPMTPVNAR.Y	144–158	56	1778.7754	1778.8662
				R.IVQEFYDPMTPVNAR.Y Oxidation M	144–158	80	1794.7974	1794.8611
E2: ~17KD	6	111	17%	K.VYDSLLALPQDLQAAR.A	66–81	77	1771.8914	1771.9468
				R.DFYNPLVAEAQK.R	146–157	71	1393.6614	1393.6878
				R.YSTSAPAISR.G	200–209	43	1051.4934	1051.5298
E3: ~18KD	6	88	17%	K.VYDSLLALPQDLQAAR.A	66–81	57	1771.8774	1771.9468
				R.DFYNPLVAEAQK.R	146–157	66	1393.6734	1393.6878
				R.YSTSAPAISR.G	200–209	32	1051.5194	1051.5298
FT9: ~17KD	1	97	14%	K.VFDSLLNLSSTLQATR.A	66–81	82	1763.9034	1763.9418
				R.IVQEFYDPMTPVNAR.Y	144–158	55	1778.7994	1778.8662

cCPEwt binds much more strongly to Cldn6 than to Cldn1 [6,9,10]. cCPE-SSS has a much higher affinity for Cldn1 than cCPEwt (Figure 6C) [5,6]. In contrast to cCPEwt, cCPE-SSS binds with high affinity (K_d~10 nM) both Cldn1 and to Cldn6 (Figure 6B,C). Taken together, the data indicate that high affinity cCPE-SSS binding to Cldn1 is efficiently inhibiting HCV infection of Huh7.5 cells whereas high affinity cCPEwt binding to Cldn6 is not sufficient to inhibit the infection efficiently.

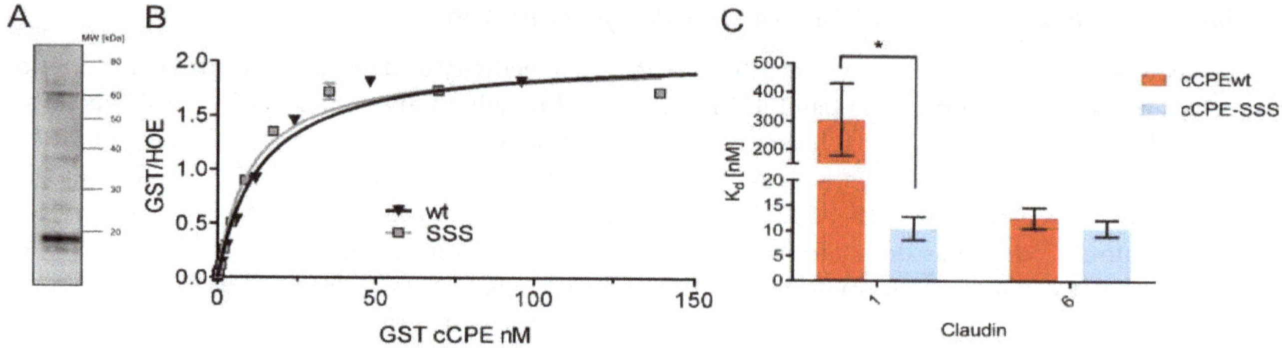

Figure 6. (**A**) Detection of Cldn6 expression in Huh7.5 cells by Western Blot. Huh7.5 cell pellets were lysed, the membrane protein fraction extracted, separated by SDS-PAGE and analyzed by Western blot using anti-Cldn6 antibodies. Cldn6 immunoreactive bands were observed in the expected range (~18 kDa). (**B**) cCPEwt and cCPE–SSS bind with similar affinities to Cldn6 on the cell surface. Binding of GST-cCPEwt and –cCPE-SSS to Cldn6. HEK-293 cells were transfected with Cldn6. Binding of GST-cCPE variants was analyzed using a cellular binding assay with varying cCPE concentrations, fixation, staining with phycoerythrin-coupled anti-GST antibody and Hoechst 33342 (nuclei) for normalization. (**C**) cCPE-SSS binds with similar high affinities to Cldn1 and -6, whereas cCPEwt has a very low affinity for Cldn1. The derived respective affinities of cCPEwt and cCPE–SSS for Cldn6 are not significantly different, but cCPE-SSS binds with a significantly stronger affinity to Cldn1 than cCPEwt does. Dissociation constants (K_d) were calculated from the binding assays using a non-linear regression analysis in GraphPad Prism 6 (model: Saturation binding, equation: one site—specific binding) [6]. Results are shown as means ± SEM. Student's t-Test, data from 3 (Cldn1) and 5 to 6 (Cldn6) independent K_d calculation experiments, *, $p < 0.05$.

2.6. *Treatment of Reconstructed Human Epidermis with cCPE Variants Leads to Barrier Opening*

To test claudin modulation by cCPE variants in another Cldn1-expressing tissue we analyzed TJ-modulation by cCPE variants in a human in vitro skin model. RHE expressing Cldn1 and Cldn4 [50] was treated with 50 µg/mL of the cCPE variants for 48 h at two different time points (day 4 and day 8). Subsequently, the barrier to ions (trans-epithelial resistance, TER) and the permeability for the tracer molecule Lucifer Yellow (LY) were measured to analyze the barrier function of the RHE.

The treatment with cCPEwt (binding strongly to Cldn4) led to a significant decrease in TER to 0.55 ± 0.07-fold ($p < 0.0001$ compared to untreated RHE and cCPE-YALA, Figure 7A) and a significant increase in LY permeability to 3.42 ± 0.24-fold ($p < 0.0001$ compared to the negative control and $p < 0.001$ compared to cCPE-YALA) on day four. However, on day eight, only a slight reduction in TER (0.753 ± 0.09-fold, Figure 7A, $p < 0.01$ compared to to cCPE-YALA) and a slight increase in LY permeability (1.97 ± 0.27-fold), could be observed (Figure 7B).

On day four, compared to cCPEwt, cCPE-SSS (binding strongly to Cldn1 and Cldn4) had a similar effect on TER (0.49 ± 0.04-fold, $p < 0.0001$ compared to untreated control and cCPE-YALA, Figure 7A) and a significantly greater effect on LY permeability (5.23 ± 0.44-fold, $p < 0.0001$ compared to untreated control and cCPE-YALA, $p < 0.001$ compared to cCPEwt, Figure 7B). On day eight, TER was significantly decreased after treatment with cCPE-SSS (0.66 ± 0.08-fold, $p = 0.001$ compared to untreated control, Figure 7A) and there was a slight tendency for an increased outside-to-inside passage of LY (1.78 ± 0.32%, Figure 7B) for cCPE-SSS.

The different cCPE variants had no effect on the expression of the relevant claudins in the RHE, (Cldn1, -3, -4 and -7), on day 4 or on day 8 (data not shown).

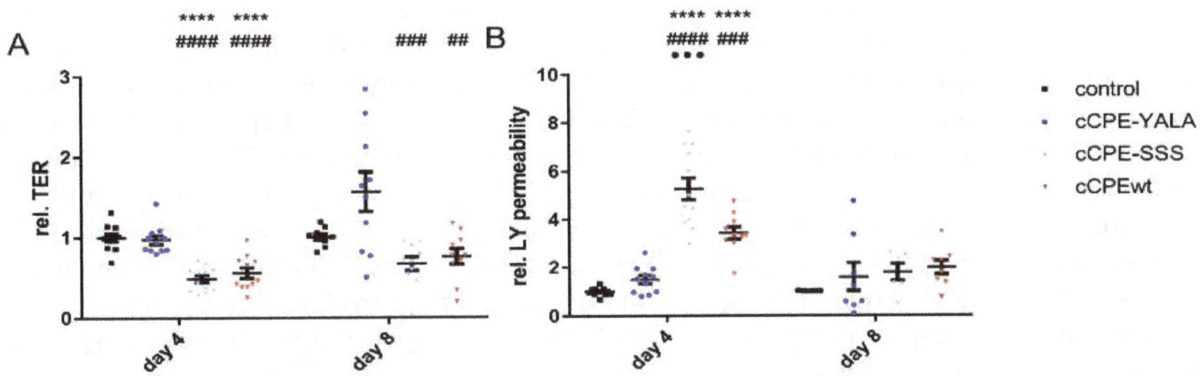

Figure 7. Effect of cCPE variants on (**A**) TER and (**B**) Lucifer Yellow permeability in reconstructed human epidermis. Average relative trans-epithelial resistance (**A**) and relative LY permeability (**B**) after 48 h incubation with 50 μg/mL of either cCPEwt, cCPE-SSS, cCPE-YALA or untreated control on days 4 and 8 after a change to ALI. Results were standardized to respective untreated control. Results are shown as means ± SEM. Statistical significances of one-way ANOVA: * compared to control, # compared to cCPE-YALA, ● compared to cCPEwt. ## $p < 0.01$; ###, ●●● $p < 0.001$; ****, #### $p < 0.0001$. $n = 5–12$.

In summary, the data showed that cCPE variants can be used as TJ modulators in the skin and that removal of Cldn1 and Cldn4 from the TJ impairs the barrier properties of the human epidermis.

3. Discussion

In this study, we showed that Cldn1 expressed in liver and skin can be efficiently targeted with the broad-specificity claudin binder cCPE-SSS. In contrast to cCPEwt, cCPE-SSS strongly inhibited Cldn1-dependent HCV infection in a hepatocyte in vitro model. In addition, cCPE-SSS improved paracellular permeability for small molecules in an in vitro human skin model to a larger extend than cCPEwt.

Claudins have been suggested as targets for diverse therapeutic application throughout numerous studies, including targeting of claudin-overexpressing cancer entities [51,52], visualization of upregulated claudin expression in colon polyps [53], opening of tissue barriers, e.g., the blood-brain-barrier [8] or intestinal barriers [54,55].

Use of monoclonal antibodies for claudin targeting led to very promising in vivo results; in particular with anti-Cldn18 or -Cldn1 antibodies for treatment of gastric cancer or HCV infections, respectively [14,18,32,33,56,57]. As an alternative for claudin targeting, application of CPE or cCPE variants has been suggested [8,58–60]. Not only is the bacterial production of these proteins very cost-effective compared to antibody production, but cCPE can also be adjusted to target several claudin subtypes. The latter is advantageous e.g., in viral infections like hepatitis C, where a pan-genomic treatment and prevention of resistance toward treatment (e.g., due to viral use of multiple claudins as receptors) is desirable. However, as claudins are expressed throughout the entire body, a thorough evaluation of possible adverse effects of claudin-targeting agents is necessary. Full-length CPE is known to have a high systemic toxicity. In contrast, more than ten weekly intravenous administration of 5 mg/kg cCPE did not increase biochemical markers of kidney and liver injury in mice [61]. These findings suggest that cCPE variants have limited and controllable adverse side effects even when applied systemically. Nevertheless, in vivo effects on tissue barriers have to be investigated in more detail. At least, blocking of extra-junctional HCV co-receptors might be achieved at much milder cCPE doses than opening of preformed junctional barrier, as indicated by the data (Figure 3, Figure 7).

Host-targeting agents are promising therapeutic approaches for novel anti-viral treatments for patients that do not respond to conventional therapies of direct-acting antivirals, or combination of ribavirin and pegylated interferon. So far, antibodies recognizing the extracellular domains of various cellular HCV receptors, for example Cldn1 [14,32,33], CD81 [62] and SR-BI [63] have been successfully used in vitro and in vivo to prevent HCV infection.

We showed that the broad-specificity claudin binder cCPE-SSS can effectively inhibit HCV infection in the hepatoma cell line Huh7.5. The substitution S305P/S307R/S313H improved cCPE binding to Cldn1 of Huh7 cells significantly compared to cCPEwt, providing a tool to target hepatoma cells potently. In contrast to cCPEwt, cCPE-SSS inhibited HCVcc infection of hepatoma Huh7.5 cells in a dose-dependent manner (Figure 3). Surprisingly, low concentrations of cCPEwt even aggravated HCV infection. Due to its much lower affinity to Cldn1 than to Cldn6, cCPEwt should bind mainly to Cldn6 under these conditions. We assume that Cldn1 and Cldn6 compete with each other for CD81 association, that cCPEwt binding to Cldn6 interferes with Cldn6-CD81 interaction (similar as for cCPE-SSS – Cldn1, indicated by homology modeling) and that in turn formation of the more potent Cldn1-CD81 HCV co-receptor complex is favored. Such a competition mechanism might explain aggravation of infection by cCPEwt.

Cldn1-specific neutralizing antibodies only partially inhibited infection of Huh7.5 cells by HCV strains with broad claudin tropisms, suggesting a possible escape from Cldn1-specific targeting molecules through Cldn6 or Cldn9 in cell culture [64]. In addition, a previously Cldn1-dependent HCV strain was shown to evolve the capacity to use Cldn6 as a host entry factor via a single amino acid mutation in the HCV E1 protein [65]. cCPE-SSS concomitantly targets Cldn1 and Cldn6 on the surface of Huh7.5 with nanomolar affinities (Figures 5 and 6). Hence, use of cCPE-SSS as a broad spectrum claudin binder might avoid the escape from Cldn1-specific antibodies, making it a more powerful agent in HCV treatment.

In summary, we showed that cCPE can be modified and used to prevent infection of hepatoma cells with HCV by binding to claudins including Cldn1, thus inhibiting formation of the viral entry complex.

Additionally, we analyzed the potential of cCPE variants to modulate the epidermal barrier by targeting claudins of the SG in RHE models. A purposeful and directed opening of the epidermal barrier can be useful to improve transdermal drug delivery [61,66].

Transdermal drug delivery is a desirable method of application. Its numerous benefits include a non-invasive technique, avoidance of first-pass metabolism by digestive enzymes and thus a better bioavailability of poorly orally bioavailable drugs, a constant plasma drug concentration as well as enhancement of patient compliance [61,66]. Treatment of RHE with cCPE variants prevents consecutive incorporation of claudins in TJs and allowed us to study the effects of TJ rearrangements on the epidermal barrier.

cCPEwt removes Cldn3, -4, -6, -7, -8 and -14 from the TJ. However, its affinity for Cldn1 is very low. We show here that cCPEwt resulted in an impaired epidermal barrier to small molecule tracers (LY) in the RHE at day 4, i.e., during formation of the SC. Impairment of the epidermal barrier to molecular tracers after cCPEwt treatment during formation of SC has been described earlier [39], and is in accordance with our results. In the study by Yuki et al., 10 or 50 µg/mL cCPEwt was added to the culture medium at the beginning of SC formation. After a four-day incubation period, the human skin equivalent models showed a significant increase in LY permeability, indicating a weakened epidermal barrier. [39]. In addition, we demonstrate that the ion barrier was weakened by cCPEwt on day 4. On day 8, when SC was more mature, cCPEwt did not have a significant effect even though there was a trend.

cCPE-SSS has a higher affinity for Cldn1-9 [5,6], which should result in a greater barrier impairment compared to cCPEwt because Cldn-1 plays a major role in epidermal barrier function [35]. Indeed, in the four-day model, the cCPE-SSS treatment led to a significant decrease in TER and a significant increase in LY permeability. This effect was greater than the effect of cCPEwt.

In the eight-day model, the TER was significantly lower after cCPE-SSS treatment compared to the other groups. This indicates that loss of Cldn1 from the TJ weakens the ionic barrier even in a well-developed SC.

All observed effects were weaker in the eight-day compared to the four-day model although the cCPE variants were applied from the basolateral side and thus were not held up by the SC. This finding might indicate a reduced accessibility of claudins in the more mature epithelium. If the extracellular domains of claudins are involved in trans-interactions, cCPE cannot bind anymore [67]. Likewise, the contribution of the TJ to epidermal barrier sealing might be less strong in the presence of a mature and intact SC.

The most effective approach to improve transdermal drug delivery should modify both epidermal barriers, the SC as well as the TJ. Hence, cCPE-based biologics might be used in combination with a chemical absorption enhancer, like surfactants (surface-active agents), fatty-acid esters, terpenes and solvents. Chemical enhancers increase skin permeability by various means, including solubilization of lipids and denaturing of keratin as well as fluidizing the crystalline structure of the SC [68]. Combining these methods with the directed claudin-targeting by cCPE could have a therapeutically significant effect on the epidermal barrier.

In conclusion, cCPE variants are promising tools for claudin targeting in diverse tissues including liver and skin. Their claudin binding properties can be adjusted to fit a certain demand, their production is very cost-effective compared to monoclonal antibodies and their systemic administration so far has been without fatal adverse side effects.

4. Materials and Methods

4.1. *Structural Bioinformatics and Molecular Modeling*

Homology model of the Cldn1-cCPE-SSS complex was built based on the Cldn4-cCPE complex (PDB ID: 5B2G; [43]). All manipulations, optimizations of the models, as well as calculations of hydrophobic and electrostatic potentials on the molecular surfaces, were performed with Sybyl X2.1.1 (Certara USA Inc., St. Louis, MO, USA). Models were energetically minimized using the AMBER7 FF99 force field. Structure images were generated with PyMOL Molecular Graphics System 1.8.4.1 (Schrödinger LLC, 2017).

4.2. *Chemicals and Solutions*

Antibodies: Phycolink® anti-GST R-phycoerythrin-conjugated antibodies (Europa Bioproducts Ltd., Cambridge, UK), rabbit anti-Cldn1 and anti-Cldn4, (Invitrogen, Carlsbad, USA), rabbit polyclonal anti-Cldn6 (Immuno-Biological Laboratories, Fujioka, JP), mouse anti-GST (Sigma-Aldrich, Steinheim, DE), mouse anti-NS5a (ABCAM, Cambridge, UK).

Chemicals: Lucifer Yellow: Sigma-Aldrich Chemie GmbH (München, DE).

4.3. *Cells and Tissues*

4.3.1. Cultivation of Cells

Primary human keratinocytes were isolated from juvenile foreskin as described previously [50]. The use of these samples was in accordance with approval of the ethics committee of the Ärztekammer Hamburg (WF 61/12). Samples were used anonymously and all investigations were conducted in accordance to the principles of the Declaration of Helsinki.

Human Huh7 cells [23] kindly provided by Dr. Nora Gehne (FMP), were cultured in Dulbecco's modified Eagle's medium supplemented with 10% (v/v) FBS and 1% (v/v) penicillin/streptomycin solution (10,000 U/mL penicillin, 10,000 µg/mL streptomycin, Gibco™, Thermo Fisher Scientific, Waltham, MA, USA). Huh7.5. cells (Apath, LLC, St. Louis, MO, USA) were cultured as described previously [69].

HEK-293 cells (CRL-1573, A.T.C.C., Manassas, VA, USA), [70] were kindly provided by Professor Otmar Huber (Friedrich Schiller University of Jena), cultivated in Minimal Essentail Medium with Earle's salts and L-glutamine (Merck KGaA, Darmstadt, DE) supplemented with 10% (v/v) FBS and 1% (v/v) penicillin/streptomycin solution (10000 U/mL penicillin, 10,000 µg/mL streptomycin, Gibco™, Thermo Fisher Scientific, Waltham, MA, USA). For transfection, cells were seeded onto coverslips with a diameter of 32 mm (Menzel Gläser, Thermo Fisher Scientific, Waltham, MA, USA). At 80% confluency, cells were transfected with 2.9 ng plasmid encoding N-terminally YFP-fused human CD81 and 1.1 ng plasmid encoding N-terminally CFP-fused Cldn1 using polyethlenimine (PEI MAX 40K, Polysciences, Inc., Warrington, PA, USA). Cells were further incubated at 37 °C for 2 days before measuring FRET efficiency. In the case of cCPE-SSS treatment, cells were incubated with 10 μg/mL of the protein for 30 min prior to starting the measurement.

4.3.2. HCVcc Infection Assay

Two to three days after plating on 48 well plates, Huh7.5 cells were incubated for 2 h with GST-cCPEwt, GST-cCPE-SSS or GST-cCPE-YALA at concentrations between 0.5–50 μg/mL at 37 °C or not further treated (control). After incubation the cells were infected with Gt2a HCV J6CF-JFH1 (Jc1) at a multiplicity of infection (moi) of 1.0 at 37 °C. The medium was changed 2 h after infection and Huh7.5 cells were incubated at 37 °C for further 24 h or Huh7.5 cells were incubated at 37 °C for 26 h without medium exchange.

Huh7.5 cells were fixed with acetone/methanol (1:1 ratio), incubated for 1 h at room temperature with anti NS5a antibodies, and incubated for one hour at room temp with Alexa Fluor 488-conjugated goat anti-mouse secondary antibody (washed with PBS between steps) and counterstained with 4′,6-diamidino-2-phenylindole (DAPI). Infection levels were determined using a Nikon Eclipse TE2000-U microscope (Nikon Corporation, Tokyo, Japan), and expressed as a fold change of the cCPE untreated control [46,69].

4.3.3. MTS Cell Proliferation Assay

Cellular toxicity was measured using a Cell Titer 96 aqueous nonradioactive cell proliferation assay kit, according to the manufacturer's instructions (Promega, Madison, WI, USA) as described previously [71]. Briefly, 2×10^4 Huh7.5 cells were seeded in a 96 well plate. After cultivation for 72 h, cells were incubated with cCPEwt and cCPE-SSS at concentrations ranging from 0 to 100 ng/μL for 2 h. 3-(4,5-dimethylthiazol-2-yl)-5-(3-carboxymethoxyphenyl)-2-(4-sulfophenyl)-2*H*-tetrazolium) (MTS)-phenazine methosulfate solution (40 μL) was added to each well and the samples were incubated at 37 °C in a humidified 5% CO_2 atmosphere for 3 h. Absorbance was measured at 490 nm with a UVmax plate reader (Molecular Devices, San José, CA, USA). Each sample was analyzed in triplicate.

4.3.4. Live Cell Imaging, Acceptor Photobleaching and FRET Data Analysis

Live cell imaging and photo bleaching were performed using the laser scanning microscope LSM 780 (Carl Zeiss Microscopy GmbH, Jena, DE). Experiments were performed as described previously [72].

FRET efficiency as calculated using the following formula: $E_{FRET} = (CFP_A - CFP_B)/CFP_A$; CFP_A and CFP_B referring to CFP fluorescence intensity after and before photobleaching, respectively.

4.3.5. Cultivation of Keratinocytes and Creation of Reconstructed Human Epidermis

Reconstructed human epidermis was generated from primary human keratinocytes as already described [50]. Briefly, the 3×10^5 cells were seeded in 500 μL EpiLife medium with 1.5 mM $CaCl_2$ onto Millicell cell culture inserts (Merck Millipore, Tullagreen, Ireland) and 2.5 mL cell culture medium was applied to the basal compartment. The cells were cultivated at 37 °C in humidified atmosphere for 30 h before lifting to the air-liquid interface (ALI). Concurrently, the basal medium was exchanged with 1.5 mL of EpiLife medium containing 1.5 mM $CaCl_2$, 92 μg/mL ascorbic acid (Sigma, Darmstadt, DE) and 10 ng/mL recombinant human keratinocyte growth factor (R&D Systems, Wiesbaden-Nordenstadt, DE). Cells derived from an individual donor were cultured as duplicates.

4.3.6. Treatment of Reconstructed Human Epidermis with cCPE and Barrier Assays

For cCPE treatment, 50 μg/mL of the respective cCPE mutants were added to the basal compartment of the RHE models at day 2 or 6 after transitioning to ALI. After 48 h of incubation at 37 °C in a humidified atmosphere, the transepidermal resistance as a measure for the epidermal ion barrier was measured using chopstick electrodes. Relative TER values were used due to intrinsic variation in starting absolute TER values between the different RHE cultures since they were derived from different human donors. The absolute TER values of the control RHE cultures were between 0.53 and 2.26 kOhm × cm^2 at day 4 and 0.71 and 2.92 kOhm × cm^2 at day 8.

To examine the outside-to-inside epidermal barrier for small molecules, the RHE models were transferred to 6-well plates with 1.5 mL of medium. Then, 200 µL of a 1 mM Lucifer Yellow solution was added to the apical side. After 6 h incubation at 37 °C, 100 µL of the basal medium was transferred into a microtiter plate. The LY fluorescence was measured with the Infinite M2000 fluorescence reader (Tecan Group Ltd., Männedorf, Switzerland) using the following settings: excitation wavelength 425 nm, emissions wavelength 550 nm, 10 flashes, 50-fold gain. By measuring the fluorescence of LY standard samples of known concentration, the concentration of the basal media samples could be calculated. With this, the LY permeability was calculated using the following formula: $P_{APP}\left[\frac{cm}{s}\right] = \frac{flux\left[\frac{\mu g}{s \cdot cm^2}\right]}{LY\ concentration\left[\frac{\mu g}{ml}\right]}$.

4.4. Plasmids

Plasmid encoding GST-CPE194-319wt (GST-cCPE), GST-cCPE-S305P/S307R/S313H and GST-cCPE–Y306A/L315A fusion protein were reported previously [6], as well as plasmid encoding N-terminally CFP-tagged Cldn1 [72]. pEZ-M15-NYFP/huCD81 was obtained from GeneCopoeia™ (Rockville, MD, USA).

4.5. Expression and Purification of cCPE-Constructs

CPE194-319 (cCPE) with N-terminal GST-fusions as well as GST (control) were expressed in *E. coli* BL21 as reported previously [6]. Briefly, bacteria were grown to A_{600} = 0.6–0.8, expression induced by addition of 1 mM isopropyl-β-D-thiogalactopyranoside. 3 h after induction, bacteria were harvested (10 min, 20,000× *g*, 4 °C) and lysed in PBS (phosphate buffered saline) with 1% (v/v) Triton X-100, 0.1 mM PMSF, 1 mM EDTA, protease inhibitor cocktail (Sigma-Aldrich, Steinheim, DE) and sonicated with Vibra Cell™ Model 72434 (BioBlock Scientific, Strasbourg, France) by 10 × 1 s pulses. Insoluble cell debris was removed by centrifugation (20,000× *g* for 1 h at 4 °C). Proteins were purified from supernatants using glutathione-agarose (Sigma-Aldrich, Steinheim, DE) and dialyzed against PBS. Protein concentration was determined with BCA Protein Kit (Thermo-Scientific, Rockford, USA).

4.6. Pull-down Assay

Pull-down assay was performed as described previously [10]. Briefly, Huh7 and Huh7.5 cells were lysed with 1% Triton X-100, EDTA-free protease inhibitor cocktail (Roche, Mannheim, DE) in PBS. The 10,000× *g* supernatant was incubated with GST-cCPE constructs bound to Glutathione Sepharose beads (GE Healthcare, Uppsala, Sweden) for 2 h on a shaker at 4 °C. Beads were washed 3 times with PBS containing 0.5% Triton X-100, bound proteins eluted with Laemmli buffer and analyzed by SDS-PAGE and Western blot with anti-Cldn1, or anti-Cldn4 antibodies. After stripping, membranes were incubated with anti-GST antibodies to verify that similar amounts of GST-cCPE were bound to the beads.

4.7. Immunostaining

Immunocytochemistry was performed as described [10]. Briefly, 2–3 days after plating, Huh7.5 cells were incubated with 10 µg/mL GST-cCPEwt, GST-cCPE-SSS or GST-cCPE-YALA for 30 min at 37 °C and washed with PBS. Cells were fixed with methanol/acetone 1:1 ratio, blocked with 1% BSA in PBS, incubated with anti-GST antibodies, washed, and incubated with Alexa Fluor 488-conjugated goat anti-mouse secondary antibodies. Cells were analyzed with the LSM 510 META confocal microscope (Zeiss, Jena, DE).

4.8. Mass Spectrometry

4.8.1. In-gel Trypsin Digestion

Gel bands of interest were excised after Coomassie blue staining and cut into 1–2 mm^3 gel pieces, which were placed into 1.5 mL sample tubes. Gel pieces were rinsed twice with wash solution for 18 h in total (200 µL, 50% methanol, 5% acetic acid). The solutions were removed and gel pieces were dehydrated in acetonitrile (200 µL, 5 min). Supernatants were removed and gel pieces were dried in a vacuum centrifuge for 3 min. Disulfide reduction was performed with 10 mM DTT (30 µL) for 0.5 h, followed by alkylation with 100 mM iodoacetamide (30 µL) for 0.5 h. Supernatants were removed from the gel samples and dehydration with acetonitrile and evaporation performed as described above. Gel pieces were washed with 100 mM ammonium bicarbonate (200 µL, 10 min). Supernatants were removed and dehydration performed with acetonitrile and evaporation as above. The gel samples were then rehydrated on ice with freshly prepared trypsin solution (30 µL, 20 ng/µL sequencing grade trypsin (Promega, Madison, WI, USA) in 50 mM ammonium bicarbonate). After rehydration, excess trypsin solution was removed and 50 mM ammonium bicarbonate (10 µL) was added to prevent dehydration of gel pieces. Gel samples were digested at 37 °C for 18 h. The gel pieces were then extracted sequentially with 50 mM ammonium bicarbonate (60 µL), 50% acetonitrile, 5% formic acid (60 µL) and 85% acetonitrile, 5% formic acid (60 µL). The combined extracts were evaporated in a vacuum centrifuge and were re-dissolved in 5% acetonitrile, 0.1% formic acid (20 µL) on an ultrasonic bath and transferred into LC-MS sample vials.

4.8.2. LC-MS/MS Analysis

For the analysis of in-gel digested protein material, liquid chromatography was performed using an Ultimate 3000 nano-HPLC system (Dionex, Sunnyvale, CA, USA) comprising a WPS-3000 micro auto sampler, a FLM-3000 flow manager and column compartment, a UVD-3000 UV detector, an LPG-3600 dual-gradient micro-pump, and an SRD-3600 solvent rack controlled by Hystar (Bruker Daltonics, Billerica, MA, USA) and DCMSLink 2.0 software (Dionex, Sunnyvale, CA, USA). Samples were concentrated on a trapping column Dionex (Sunnyvale, CA, USA), 300 µm i.d., 0.1 cm) at a flow rate of 20 µL/min. For the separation with a C18 Pepmap column (75 µm i.d., 15 cm, Dionex), a flow rate of 250 nL/min was used as generated by a cap-flow splitter cartridge (1/1000). Peptides were eluted by the application of a 30 min multi-step gradient using solvents A (98% H_2O, 2% acetonitrile, 0.1% formic acid) and B (80% acetonitrile, 20% H_2O, 0.1% formic acid) (Table 2):

Table 2. Elution protocol of peptides. A C18 Pepmap column (75 µm i.d., 15 cm) was used for separation of peptide fragments. Solvent A (98% H_2O, 2% acetonitrile, 0.1% formic acid) and solvent B (80% acetonitrile, 20% H_2O, 0.1% formic acid) were used in the ratio indicated.

Composition (% Solvent B)	Run Time (min)
2–10	0–3
10–25	3–18
25–50	18–30
50–90	30–30.2

The liquid chromatography was interfaced directly with a 3D high capacity ion trap mass spectrometer (amaZon; Bruker Daltonics, Billerica, MA, USA) utilizing 10 µm i.d. distal coated SilicaTips (New Objective, Woburn, MA, USA) and nano-ESI mode. SPS parameter settings on the ion trap were tuned for a target mass of 850 *m/z*, compound stability 100% and a smart ICC target of 250,000. MS/MS analysis was initiated on a contact closure signal triggered by HyStar software (version 3.2).

Up to five precursor ions were selected per cycle with active exclusion (0.5 min) in collision-induced dissociation (CID) mode. CID fragmentation was achieved using helium gas and a 30–200% collision energy sweep with amplitude 1.0 (ions are ejected from the trap as soon as they fragment).

4.8.3. Data Processing and Database Searching

Raw LC-MS/MS data were processed and Mascot compatible files were created using DataAnalysis 4.0 software (Bruker Daltonics, Billerica, MA, USA). Database searches were performed using the Mascot algorithm (version 2.4) and the UniProt_SwissProt database with mammalian taxonomy restriction (v2012.09.17, number of entries 537,505, after taxonomy filter: 66,032). The following parameters were applied: 2+, 3+ and 4+ ions, peptide mass tolerance 0.3 Da, $^{13}C = 2$, fragment mass tolerance 0.6 Da, number of missed cleavages: two, instrument type: ESI-TRAP, fixed modifications: Carbamidomethylation (Cys), variable modifications: Oxidation (Met).

4.9. Statistical Analysis

Results are expressed as means ± standard error of the mean (SEM). Gaussian distribution was tested using the D´Agostino-Pearson omnibus test. Statistical analyses were performed using one-way ANOVA with post-hoc Sidak's test or Student´s t-test (in case of failed normality test: Mann-Whitney test), as indicated. $p < 0.05$ was considered statistically significant.

Author Contributions: Conceptualization, J.R., N.Z. and J.P.; Data curation, L.-S.B. and J.R.; Formal analysis, N.Z. and J.P.; Funding acquisition, J.M.B., G.K., N.Z. and J.P.; Investigation, L.-S.B., J.R., S.W., S.B., H.B.K., J.P., M.E., A.P. and S.V.-y.-S.; Methodology, L.-S.B., J.R., S.W., S.B., H.B.K., J.P., M.E., A.P., N.Z. and J.P.; Project administration, N.Z. and J.P.; Resources, H.B.K., M.E., A.P., J.M.B., G.K., N.Z. and J.P.; Software, H.B.K., J.P., N.Z. and J.P.; Supervision, A.P., J.M.B., G.K., N.Z. and J.P.; Validation, L.-S.B., J.R., A.P., J.M.B., N.Z. and J.P.; Visualization, H.B.K., J.P., N.Z. and J.P.; Writing—original draft, L.-S.B., J.R., S.W., S.B., H.B.K., J.P., A.P., J.M.B., G.K., N.Z. and J.P.; Writing—review & editing, L.-S.B., J.R., S.W., S.B., H.B.K., J.P., A.P., J.M.B., G.K., N.Z. and J.P.

Abbreviations

ALI	air-liquid interface
ANOVA	analysis of variance
CD81	cluster of differentiation 81
CFP	cyan fluorescent protein
Cldn	claudin
cCPE	amino acids 194–319 of *Clostridium perfringens* enterotoxin
cCPE-SSS	cCPE-S305P/S307R/S313H
cCPEwt	wildtype cCPE
cCPE-YALA	cCPE-Y306A/L315A
CPE	*Clostridium perfringens* enterotoxin
ECS	extracellular segment
FBS	fetal bovine serum
FRET	fluorescence resonance energy transfer
GST	Glutathione S-transferase
HCV	hepatitis C virus
HCVcc	cell culture-derived hepatitis C virus particles
LC/MS-MS	liquid chromatography - tandem mass spectrometry
LY	Lucifer yellow
MTS	3-[4,5,dimethylthiazol-2-yl]-5-[3-carboxymethoxy-phenyl]-2-[4-sulfophenyl]-2H -tetrazolium
RHE	reconstructed human epidermis
SC	*stratum corneum*
SEM	standard error of the mean
SG	*stratum granulosum*

SR-BI	scavenger receptor class B type I
TER	trans-epithelial resistance
TJ	tight junction
YFP	yellow fluorescent protein

References

1. Krause, G.; Winkler, L.; Mueller, S.L.; Haseloff, R.F.; Piontek, J.; Blasig, I.E. Structure and function of claudins. *Biochim. Biophys. Acta.* **2008**, *1778*, 631–645. [CrossRef] [PubMed]
2. Hagen, S.J. Non-canonical functions of claudin proteins: Beyond the regulation of cell-cell adhesions. *Tissue Barriers* **2017**, *5*, e1327839. [CrossRef] [PubMed]
3. Freedman, J.C.; Shrestha, A.; McClane, B.A. Clostridium perfringens Enterotoxin: Action, Genetics, and Translational Applications. *Toxins* **2016**, *8*, 73. [CrossRef] [PubMed]
4. Veshnyakova, A.; Protze, J.; Rossa, J.; Blasig, I.E.; Krause, G.; Piontek, J. On the interaction of Clostridium perfringens enterotoxin with claudins. *Toxins* **2010**, *2*, 1336–1356. [CrossRef] [PubMed]
5. Takahashi, A.; Saito, Y.; Kondoh, M.; Matsushita, K.; Krug, S.M.; Suzuki, H.; Tsujino, H.; Li, X.; Aoyama, H.; Matsuhisa, K.; et al. Creation and biochemical analysis of a broad-specific claudin binder. *Biomaterials* **2012**, *33*, 3464–3474. [CrossRef] [PubMed]
6. Protze, J.; Eichner, M.; Piontek, A.; Dinter, S.; Rossa, J.; Blecharz, K.G.; Vajkoczy, P.; Piontek, J.; Krause, G. Directed structural modification of Clostridium perfringens enterotoxin to enhance binding to claudin-5. *Cell Mol. Life Sci.* **2015**, *72*, 1417–1432. [CrossRef] [PubMed]
7. Veshnyakova, A.; Piontek, J.; Protze, J.; Waziri, N.; Heise, I.; Krause, G. Mechanism of Clostridium perfringens enterotoxin interaction with claudin-3/-4 protein suggests structural modifications of the toxin to target specific claudins. *J. Biol. Chem.* **2012**, *287*, 1698–1708. [CrossRef] [PubMed]
8. Neuhaus, W.; Piontek, A.; Protze, J.; Eichner, M.; Mahringer, A.; Subileau, E.A.; Lee, I.M.; Schulzke, J.D.; Krause, G.; Piontek, J. Reversible opening of the blood-brain barrier by claudin-5-binding variants of Clostridium perfringens enterotoxin's claudin-binding domain. *Biomaterials* **2018**, *161*, 129–143. [CrossRef] [PubMed]
9. Fujita, K.; Katahira, J.; Horiguchi, Y.; Sonoda, N.; Furuse, M.; Tsukita, S. Clostridium perfringens enterotoxin binds to the second extracellular loop of claudin-3, a tight junction integral membrane protein. *FEBS Lett.* **2000**, *476*, 258–261. [CrossRef]
10. Winkler, L.; Gehring, C.; Wenzel, A.; Muller, S.L.; Piehl, C.; Krause, G.; Blasig, I.E.; Piontek, J. Molecular determinants of the interaction between Clostridium perfringens enterotoxin fragments and claudin-3. *J. Biol. Chem.* **2009**, *284*, 18863–18872. [CrossRef] [PubMed]
11. Petermann, P.; Rahn, E.; Thier, K.; Hsu, M.J.; Rixon, F.J.; Kopp, S.J.; Knebel-Morsdorf, D. Role of Nectin-1 and Herpesvirus Entry Mediator as Cellular Receptors for Herpes Simplex Virus 1 on Primary Murine Dermal Fibroblasts. *J. Virol.* **2015**, *89*, 9407–9416. [CrossRef]
12. Rahn, E.; Thier, K.; Petermann, P.; Rubsam, M.; Staeheli, P.; Iden, S.; Niessen, C.M.; Knebel-Morsdorf, D. Epithelial Barriers in Murine Skin during Herpes Simplex Virus 1 Infection: The Role of Tight Junction Formation. *J. Invest Dermatol.* **2017**, *137*, 884–893. [CrossRef] [PubMed]
13. Evans, M.J.; von Hahn, T.; Tscherne, D.M.; Syder, A.J.; Panis, M.; Wolk, B.; Hatziioannou, T.; McKeating, J.A.; Bieniasz, P.D.; Rice, C.M. Claudin-1 is a hepatitis C virus co-receptor required for a late step in entry. *Nature* **2007**, *446*, 801–805. [CrossRef] [PubMed]
14. Fukasawa, M.; Nagase, S.; Shirasago, Y.; Iida, M.; Yamashita, M.; Endo, K.; Yagi, K.; Suzuki, T.; Wakita, T.; Hanada, K.; et al. Monoclonal antibodies against extracellular domains of claudin-1 block hepatitis C virus infection in a mouse model. *J. Virol.* **2015**, *89*, 4866–4879. [CrossRef] [PubMed]
15. Wahid, A.; Helle, F.; Descamps, V.; Duverlie, G.; Penin, F.; Dubuisson, J. Disulfide bonds in hepatitis C virus glycoprotein E1 control the assembly and entry functions of E2 glycoprotein. *J. Virol.* **2013**, *87*, 1605–1617. [CrossRef] [PubMed]
16. Horsley-Silva, J.L.; Vargas, H.E. New Therapies for Hepatitis C Virus Infection. *Gastroenterol Hepatol. (N Y)* **2017**, *13*, 22–31.
17. Organization, W.H.; Hepatits, C. Available online: https://www.who.int/news-room/fact-sheets/detail/hepatitis-c (accessed on 1 May 2019).

18. Krieger, S.E.; Zeisel, M.B.; Davis, C.; Thumann, C.; Harris, H.J.; Schnober, E.K.; Mee, C.; Soulier, E.; Royer, C.; Lambotin, M.; et al. Inhibition of hepatitis C virus infection by anti-claudin-1 antibodies is mediated by neutralization of E2-CD81-claudin-1 associations. *Hepatology* **2010**, *51*, 1144–1157. [CrossRef] [PubMed]
19. Zhu, Y.Z.; Qian, X.J.; Zhao, P.; Qi, Z.T. How hepatitis C virus invades hepatocytes: The mystery of viral entry. *World J. Gastroenterol* **2014**, *20*, 3457–3467. [CrossRef] [PubMed]
20. Cukierman, L.; Meertens, L.; Bertaux, C.; Kajumo, F.; Dragic, T. Residues in a highly conserved claudin-1 motif are required for hepatitis C virus entry and mediate the formation of cell-cell contacts. *J. Virol.* **2009**, *83*, 5477–5484. [CrossRef]
21. Sarhan, M.A.; Pham, T.N.; Chen, A.Y.; Michalak, T.I. Hepatitis C virus infection of human T lymphocytes is mediated by CD5. *J. Virol.* **2012**, *86*, 3723–3735. [CrossRef]
22. Zona, L.; Tawar, R.G.; Zeisel, M.B.; Xiao, F.; Schuster, C.; Lupberger, J.; Baumert, T.F. CD81-receptor associations–impact for hepatitis C virus entry and antiviral therapies. *Viruses* **2014**, *6*, 875–892. [CrossRef] [PubMed]
23. Zheng, A.; Yuan, F.; Li, Y.; Zhu, F.; Hou, P.; Li, J.; Song, X.; Ding, M.; Deng, H. Claudin-6 and claudin-9 function as additional coreceptors for hepatitis C virus. *J. Virol.* **2007**, *81*, 12465–12471. [CrossRef] [PubMed]
24. Meertens, L.; Bertaux, C.; Cukierman, L.; Cormier, E.; Lavillette, D.; Cosset, F.L.; Dragic, T. The tight junction proteins claudin-1, -6, and -9 are entry cofactors for hepatitis C virus. *J. Virol.* **2008**, *82*, 3555–3560. [CrossRef] [PubMed]
25. Tawar, R.G.; Colpitts, C.C.; Lupberger, J.; El-Saghire, H.; Zeisel, M.B.; Baumert, T.F. Claudins and pathogenesis of viral infection. *Semin. Cell Dev. Biol.* **2015**, *42*, 39–46. [CrossRef] [PubMed]
26. Harris, H.J.; Davis, C.; Mullins, J.G.; Hu, K.; Goodall, M.; Farquhar, M.J.; Mee, C.J.; McCaffrey, K.; Young, S.; Drummer, H.; et al. Claudin association with CD81 defines hepatitis C virus entry. *J. Biol. Chem.* **2010**, *285*, 21092–21102. [CrossRef] [PubMed]
27. Douam, F.; Dao Thi, V.L.; Maurin, G.; Fresquet, J.; Mompelat, D.; Zeisel, M.B.; Baumert, T.F.; Cosset, F.L.; Lavillette, D. Critical interaction between E1 and E2 glycoproteins determines binding and fusion properties of hepatitis C virus during cell entry. *Hepatology* **2014**, *59*, 776–788. [CrossRef]
28. Harris, H.J.; Farquhar, M.J.; Mee, C.J.; Davis, C.; Reynolds, G.M.; Jennings, A.; Hu, K.; Yuan, F.; Deng, H.; Hubscher, S.G.; et al. CD81 and claudin 1 coreceptor association: Role in hepatitis C virus entry. *J. Virol.* **2008**, *82*, 5007–5020. [CrossRef] [PubMed]
29. Fofana, I.; Zona, L.; Thumann, C.; Heydmann, L.; Durand, S.C.; Lupberger, J.; Blum, H.E.; Pessaux, P.; Gondeau, C.; Reynolds, G.M.; et al. Functional analysis of claudin-6 and claudin-9 as entry factors for hepatitis C virus infection of human hepatocytes by using monoclonal antibodies. *J. Virol.* **2013**, *87*, 10405–10410. [CrossRef]
30. Timpe, J.M.; Stamataki, Z.; Jennings, A.; Hu, K.; Farquhar, M.J.; Harris, H.J.; Schwarz, A.; Desombere, I.; Roels, G.L.; Balfe, P.; et al. Hepatitis C virus cell-cell transmission in hepatoma cells in the presence of neutralizing antibodies. *Hepatology* **2008**, *47*, 17–24. [CrossRef]
31. Zeisel, M.B.; Fofana, I.; Fafi-Kremer, S.; Baumert, T.F. Hepatitis C virus entry into hepatocytes: Molecular mechanisms and targets for antiviral therapies. *J. Hepatol.* **2011**, *54*, 566–576. [CrossRef]
32. Fofana, I.; Krieger, S.E.; Grunert, F.; Glauben, S.; Xiao, F.; Fafi-Kremer, S.; Soulier, E.; Royer, C.; Thumann, C.; Mee, C.J.; et al. Monoclonal anti-claudin 1 antibodies prevent hepatitis C virus infection of primary human hepatocytes. *Gastroenterology* **2010**, *139*, 953–964. [CrossRef]
33. Mailly, L.; Xiao, F.; Lupberger, J.; Wilson, G.K.; Aubert, P.; Duong, F.H.T.; Calabrese, D.; Leboeuf, C.; Fofana, I.; Thumann, C.; et al. Clearance of persistent hepatitis C virus infection in humanized mice using a claudin-1-targeting monoclonal antibody. *Nat. Biotechnol.* **2015**, *33*, 549–554. [CrossRef] [PubMed]
34. Brandner, J.M.; Kief, S.; Grund, C.; Rendl, M.; Houdek, P.; Kuhn, C.; Tschachler, E.; Franke, W.W.; Moll, I. Organization and formation of the tight junction system in human epidermis and cultured keratinocytes. *Eur. J. Cell Biol.* **2002**, *81*, 253–263. [CrossRef] [PubMed]
35. Furuse, M.; Hata, M.; Furuse, K.; Yoshida, Y.; Haratake, A.; Sugitani, Y.; Noda, T.; Kubo, A.; Tsukita, S. Claudin-based tight junctions are crucial for the mammalian epidermal barrier: A lesson from claudin-1-deficient mice. *J. Cell Biol.* **2002**, *156*, 1099–1111. [CrossRef] [PubMed]
36. Basler, K.; Bergmann, S.; Heisig, M.; Naegel, A.; Zorn-Kruppa, M.; Brandner, J.M. The role of tight junctions in skin barrier function and dermal absorption. *J. Control Release* **2016**, *242*, 105–118. [CrossRef] [PubMed]

37. Nakajima, M.; Nagase, S.; Iida, M.; Takeda, S.; Yamashita, M.; Watari, A.; Shirasago, Y.; Fukasawa, M.; Takeda, H.; Sawasaki, T.; et al. Claudin-1 Binder Enhances Epidermal Permeability in a Human Keratinocyte Model. *J. Pharmacol. Exp. Ther.* **2015**, *354*, 440–447. [CrossRef]
38. Zhang, J.; Ni, C.; Yang, Z.; Piontek, A.; Chen, H.; Wang, S.; Fan, Y.; Qin, Z.; Piontek, J. Specific binding of Clostridium perfringens enterotoxin fragment to Claudin-b and modulation of zebrafish epidermal barrier *Exp. Dermatol.* **2015**, *24*, 605–610. [CrossRef]
39. Yuki, T.; Komiya, A.; Kusaka, A.; Kuze, T.; Sugiyama, Y.; Inoue, S. Impaired tight junctions obstruct stratum corneum formation by altering polar lipid and profilaggrin processing. *J. Dermatol. Sci.* **2013**, *69*, 148–158. [CrossRef]
40. Poumay, Y.; Dupont, F.; Marcoux, S.; Leclercq-Smekens, M.; Herin, M.; Coquette, A. A simple reconstructed human epidermis: Preparation of the culture model and utilization in in vitro studies. *Arch. Dermatol. Res.* **2004**, *296*, 203–211. [CrossRef]
41. Robertson, S.L.; Smedley, J.G., 3rd; McClane, B.A. Identification of a claudin-4 residue important for mediating the host cell binding and action of Clostridium perfringens enterotoxin. *Infect. Immun.* **2010**, *78*, 505–517. [CrossRef]
42. Saitoh, Y.; Suzuki, H.; Tani, K.; Nishikawa, K.; Irie, K.; Ogura, Y.; Tamura, A.; Tsukita, S.; Fujiyoshi, Y. Tight junctions. Structural insight into tight junction disassembly by Clostridium perfringens enterotoxin. *Science* **2015**, *347*, 775–778. [CrossRef] [PubMed]
43. Shinoda, T.; Shinya, N.; Ito, K.; Ohsawa, N.; Terada, T.; Hirata, K.; Kawano, Y.; Yamamoto, M.; Kimura-Someya, T.; Yokoyama, S.; et al. Structural basis for disruption of claudin assembly in tight junctions by an enterotoxin. *Sci. Rep.* **2016**, *6*, 33632. [CrossRef] [PubMed]
44. Suzuki, H.; Nishizawa, T.; Tani, K.; Yamazaki, Y.; Tamura, A.; Ishitani, R.; Dohmae, N.; Tsukita, S.; Nureki, O.; Fujiyoshi, Y. Crystal structure of a claudin provides insight into the architecture of tight junctions. *Science* **2014**, *344*, 304–307. [CrossRef] [PubMed]
45. Takahashi, A.; Komiya, E.; Kakutani, H.; Yoshida, T.; Fujii, M.; Horiguchi, Y.; Mizuguchi, H.; Tsutsumi, Y.; Tsunoda, S.; Koizumi, N.; et al. Domain mapping of a claudin-4 modulator, the C-terminal region of C-terminal fragment of Clostridium perfringens enterotoxin, by site-directed mutagenesis. *Biochem. Pharmacol.* **2008**, *75*, 1639–1648. [CrossRef] [PubMed]
46. Pollock, S.; Nichita, N.B.; Bohmer, A.; Radulescu, C.; Dwek, R.A.; Zitzmann, N. Polyunsaturated liposomes are antiviral against hepatitis B and C viruses and HIV by decreasing cholesterol levels in infected cells. *Proc. Natl. Acad. Sci. USA* **2010**, *107*, 17176–17181. [CrossRef] [PubMed]
47. Feigelstock, D.A.; Mihalik, K.B.; Kaplan, G.; Feinstone, S.M. Increased susceptibility of Huh7 cells to HCV replication does not require mutations in RIG-I. *Virol. J.* **2010**, *7*, 44. [CrossRef] [PubMed]
48. Lal-Nag, M.; Battis, M.; Santin, A.D.; Morin, P.J. Claudin-6: A novel receptor for CPE-mediated cytotoxicity in ovarian cancer. *Oncogenesis* **2012**, *1*, e33. [CrossRef]
49. Lohrberg, D.; Krause, E.; Schumann, M.; Piontek, J.; Winkler, L.; Blasig, I.E.; Haseloff, R.F. A strategy for enrichment of claudins based on their affinity to Clostridium perfringens enterotoxin. *BMC Mol. Biol.* **2009**, *10*, 61. [CrossRef]
50. Basler, K.; Galliano, M.F.; Bergmann, S.; Rohde, H.; Wladykowski, E.; Vidal, Y.S.S.; Guiraud, B.; Houdek, P.; Schuring, G.; Volksdorf, T.; et al. Biphasic influence of Staphylococcus aureus on human epidermal tight junctions. *Ann. N Y Acad. Sci.* **2017**, *1405*, 53–70. [CrossRef]
51. English, D.P.; Santin, A.D. Claudins overexpression in ovarian cancer: Potential targets for Clostridium Perfringens Enterotoxin (CPE) based diagnosis and therapy. *Int. J. Mol. Sci.* **2013**, *14*, 10412–10437. [CrossRef]
52. Walther, W.; Petkov, S.; Kuvardina, O.N.; Aumann, J.; Kobelt, D.; Fichtner, I.; Lemm, M.; Piontek, J.; Blasig, I.E.; Stein, U.; et al. Novel Clostridium perfringens enterotoxin suicide gene therapy for selective treatment of claudin-3- and -4-overexpressing tumors. *Gene Ther.* **2012**, *19*, 494–503. [CrossRef] [PubMed]
53. Rabinsky, E.F.; Joshi, B.P.; Pant, A.; Zhou, J.; Duan, X.; Smith, A.; Kuick, R.; Fan, S.; Nusrat, A.; Owens, S.R.; et al. Overexpressed Claudin-1 Can Be Visualized Endoscopically in Colonic Adenomas In Vivo. *Cell Mol. Gastroenterol Hepatol.* **2016**, *2*, 222–237. [CrossRef] [PubMed]
54. Krug, S.M.; Amasheh, M.; Dittmann, I.; Christoffel, I.; Fromm, M.; Amasheh, S. Sodium caprate as an enhancer of macromolecule permeation across tricellular tight junctions of intestinal cells. *Biomaterials* **2013**, *34*, 275–282. [CrossRef] [PubMed]

55. Radloff, J.; Cornelius, V.; Markov, A.G.; Amasheh, S. Caprate Modulates Intestinal Barrier Function in Porcine Peyer's Patch Follicle-Associated Epithelium. *Int. J. Mol. Sci.* **2019**, *20*, 1418. [CrossRef] [PubMed]
56. Singh, P.; Toom, S.; Huang, Y. Anti-claudin 18.2 antibody as new targeted therapy for advanced gastric cancer. *J. Hematol. Oncol.* **2017**, *10*, 105. [CrossRef] [PubMed]
57. Colpitts, C.C.; Tawar, R.G.; Mailly, L.; Thumann, C.; Heydmann, L.; Durand, S.C.; Xiao, F.; Robinet, E.; Pessaux, P.; Zeisel, M.B.; et al. Humanisation of a claudin-1-specific monoclonal antibody for clinical prevention and cure of HCV infection without escape. *Gut* **2018**, *67*, 736–745. [CrossRef] [PubMed]
58. Hashimoto, Y.; Yagi, K.; Kondoh, M. Roles of the first-generation claudin binder, Clostridium perfringens enterotoxin, in the diagnosis and claudin-targeted treatment of epithelium-derived cancers. *Pflugers. Arch.* **2017**, *469*, 45–53. [CrossRef] [PubMed]
59. Gao, Z.; McClane, B.A. Use of Clostridium perfringens Enterotoxin and the Enterotoxin Receptor-Binding Domain (C-CPE) for Cancer Treatment: Opportunities and Challenges. *J. Toxicol.* **2012**, *2012*, 981626. [CrossRef]
60. Black, J.D.; Lopez, S.; Cocco, E.; Schwab, C.L.; English, D.P.; Santin, A.D. Clostridium perfringens enterotoxin (CPE) and CPE-binding domain (c-CPE) for the detection and treatment of gynecologic cancers. *Toxins* **2015**, *7*, 1116–1125. [CrossRef]
61. Suzuki, H.; Kondoh, M.; Li, X.; Takahashi, A.; Matsuhisa, K.; Matsushita, K.; Kakamu, Y.; Yamane, S.; Kodaka, M.; Isoda, K.; et al. A toxicological evaluation of a claudin modulator, the C-terminal fragment of Clostridium perfringens enterotoxin, in mice. *Pharmazie* **2011**, *66*, 543–546.
62. Meuleman, P.; Hesselgesser, J.; Paulson, M.; Vanwolleghem, T.; Desombere, I.; Reiser, H.; Leroux-Roels, G. Anti-CD81 antibodies can prevent a hepatitis C virus infection in vivo. *Hepatology* **2008**, *48*, 1761–1768. [CrossRef] [PubMed]
63. Meuleman, P.; Catanese, M.T.; Verhoye, L.; Desombere, I.; Farhoudi, A.; Jones, C.T.; Sheahan, T.; Grzyb, K.; Cortese, R.; Rice, C.M.; et al. A human monoclonal antibody targeting scavenger receptor class B type I precludes hepatitis C virus infection and viral spread in vitro and in vivo. *Hepatology* **2012**, *55*, 364–372. [CrossRef] [PubMed]
64. Haid, S.; Grethe, C.; Dill, M.T.; Heim, M.; Kaderali, L.; Pietschmann, T. Isolate-dependent use of claudins for cell entry by hepatitis C virus. *Hepatology* **2014**, *59*, 24–34. [CrossRef] [PubMed]
65. Hopcraft, S.E.; Evans, M.J. Selection of a hepatitis C virus with altered entry factor requirements reveals a genetic interaction between the E1 glycoprotein and claudins. *Hepatology* **2015**, *62*, 1059–1069. [CrossRef] [PubMed]
66. Iqbal, B.; Ali, J.; Baboota, S. Recent advances and development in epidermal and dermal drug deposition enhancement technology. *Int. J. Dermatol.* **2018**, *57*, 646–660. [CrossRef] [PubMed]
67. Krause, G.; Protze, J.; Piontek, J. Assembly and function of claudins: Structure-function relationships based on homology models and crystal structures. *Semin. Cell Dev. Biol.* **2015**, *42*, 3–12. [CrossRef] [PubMed]
68. Prausnitz, M.R.; Mitragotri, S.; Langer, R. Current status and future potential of transdermal drug delivery. *Nat. Rev. Drug Discov.* **2004**, *3*, 115–124. [CrossRef] [PubMed]
69. Pollock, S.; Antrobus, R.; Newton, L.; Kampa, B.; Rossa, J.; Latham, S.; Nichita, N.B.; Dwek, R.A.; Zitzmann, N. Uptake and trafficking of liposomes to the endoplasmic reticulum. *FASEB J.* **2010**, *24*, 1866–1878. [CrossRef]
70. Piontek, J.; Winkler, L.; Wolburg, H.; Muller, S.L.; Zuleger, N.; Piehl, C.; Wiesner, B.; Krause, G.; Blasig, I.E. Formation of tight junction: Determinants of homophilic interaction between classic claudins. *FASEB J.* **2008**, *22*, 146–158. [CrossRef] [PubMed]
71. Woodhouse, S.D.; Smith, C.; Michelet, M.; Branza-Nichita, N.; Hussey, M.; Dwek, R.A.; Zitzmann, N. Iminosugars in combination with interferon and ribavirin permanently eradicate noncytopathic bovine viral diarrhea virus from persistently infected cells. *Antimicrob. Agents Chemother.* **2008**, *52*, 1820–1828. [CrossRef]
72. Milatz, S.; Piontek, J.; Schulzke, J.D.; Blasig, I.E.; Fromm, M.; Gunzel, D. Probing the cis-arrangement of prototype tight junction proteins claudin-1 and claudin-3. *Biochem. J.* **2015**, *468*, 449–458. [CrossRef]

14

The Blood–Brain Barrier and its Intercellular Junctions in Age-Related Brain Disorders

Laura Costea [1], Ádám Mészáros [2,3], Hannelore Bauer [4], Hans-Christian Bauer [5], Andreas Traweger [5], Imola Wilhelm [1,2], Attila E. Farkas [2,6,*,†] and István A. Krizbai [1,2,*,†]

1 Institute of Life Sciences, Vasile Goldiş Western University of Arad, 310414 Arad, Romania; laura.m.costea@gmail.com (L.C.); wilhelm.imola@brc.hu (I.W.)
2 Institute of Biophysics, Biological Research Centre, 6726 Szeged, Hungary; meszaros.adam@brc.hu
3 Doctoral School of Biology, University of Szeged, 6726 Szeged, Hungary
4 Department of Biological Sciences, University of Salzburg, 5020 Salzburg, Austria; hannelore.bauer@sbg.ac.at
5 Institute of Tendon and Bone Regeneration, Paracelsus Medical University—Spinal Cord Injury and Tissue Regeneration Center Salzburg, 5020 Salzburg, Austria; hans.bauer@pmu.ac.at (H.-C.B.); andreas.traweger@pmu.ac.at (A.T.)
6 Department of Physiology, Anatomy and Neuroscience, University of Szeged, 6726 Szeged, Hungary
* Correspondence: farkas.attilae@brc.hu (A.E.F.); krizbai.istvan@brc.hu (I.A.K.);

† These authors contributed equally to this study.

Abstract: With age, our cognitive skills and abilities decline. Maybe starting as an annoyance, this decline can become a major impediment to normal daily life. Recent research shows that the neurodegenerative disorders responsible for age associated cognitive dysfunction are mechanistically linked to the state of the microvasculature in the brain. When the microvasculature does not function properly, ischemia, hypoxia, oxidative stress and related pathologic processes ensue, further damaging vascular and neural function. One of the most important and specialized functions of the brain microvasculature is the blood–brain barrier (BBB), which controls the movement of molecules between blood circulation and the brain parenchyma. In this review, we are focusing on tight junctions (TJs), the multiprotein complexes that play an important role in establishing and maintaining barrier function. After a short introduction of the cell types that modulate barrier function via intercellular communication, we examine how age, age related pathologies and the aging of the immune system affects TJs. Then, we review how the TJs are affected in age associated neurodegenerative disorders: Alzheimer's disease and Parkinson's disease. Lastly, we summarize the TJ aspects of Huntington's disease and schizophrenia. Barrier dysfunction appears to be a common denominator in neurological disorders, warranting detailed research into the molecular mechanisms behind it. Learning the commonalities and differences in the pathomechanism of the BBB injury in different neurological disorders will predictably lead to development of new therapeutics that improve our life as we age.

Keywords: aging; blood–brain barrier; tight junction

1. Introduction

Aging and age-related co-morbidities are rapidly increasing unresolved health and socio-economic problems in developed countries. Decline of cognitive brain functions represents one of the main health challenges of aging and includes vascular and neurodegenerative dementias such as Alzheimer's or Parkinson's disease.

The functional state of the central nervous system (CNS) is greatly dependent on the quality of the vasculature. As the centuries old saying goes: "A man is as old as his arteries". Today, especially for the brain, this concept should be redefined: You are as old as your microvessels and capillaries [1]. There is increasing evidence that the cerebral microvasculature and the neurovascular unit play a critical role in age-related brain dysfunctions. The multitude of brain microvascular changes accompanied by aging includes endothelial dysfunction, blood–brain barrier (BBB) breakdown, decrease in blood flow, microhemorrhages, vessel rarefication and neurovascular uncoupling. In this review, our main focus is the breakdown of the paracellular barrier and tight junctions.

2. Cells of the Neurovascular Unit (NVU)

The vasculature in the brain forms a functional unit with the surrounding neural tissue, thus the term neurovascular unit was coined [2]. A functionally intact neurovascular unit (NVU) is a prerequisite for the proper function of the CNS. The most important cellular components of the NVU are cerebral endothelial cells, pericytes, astrocytic endfeet and neurons; however, other cellular elements like microglia may also play a modulatory role. The main role of the NVU besides neurovascular coupling is the formation of the BBB.

Cerebral endothelial cells (CECs) lining brain capillaries are considered the principal barrier forming endothelial cells. They are interconnected by a continuous line of tight junctions and characterized by a high number of mitochondria and low number of caveolae [3–5]. These characteristics contribute to the formation of a paracellular and transcellular barrier.

Pericytes are localized in the duplication of the basement membrane covering the basal surface of the endothelium. The estimates of pericyte coverage show large variations in the literature [6,7]. Pericytes can secrete a large number of substances that may influence endothelial function including TGFβ, angiopoetin-1 or VEGF. It seems that the differentiation stage of pericytes determines their effect on the endothelium as well [8]. The role of pericytes in the formation of the BBB is supported by the finding that absence of pericytes leads to endothelial hyperplasia, abnormal vasculogenesis and an increased BBB permeability [9,10].

Although the role of astrocytes in the formation of the physical barrier is limited, due to their influence on cerebral endothelial cells they play an important role in the maintenance of the BBB [11]. The astrocytic endfeet ensheath the brain vasculature almost completely [12] and express the water channel protein, aquaporin 4, which is suggested to play a crucial role in creating a bulk flow in the brain parenchyma from arterioles towards venules. This flow was shown to contribute to the clearance of extracellular proteins and metabolic waste products through the newly rediscovered glymphatic system [13,14]. Despite much interest in the glymphatic system, some experimental results do not support or even contradict its function [15]. Thus further refining of the glymphatic hypothesis is likely necessary [16]. Nevertheless, the aquaporin 4 in astrocytic endfeet is indispensable to BBB function as its knockout results in altered brain microvasculature and decreased water exchange through the BBB [17], furthermore its subcellular distribution shows an age dependent depolarization accompanied by decreased protein clearance [18].

The BBB is in the forefront of the defense line of the CNS and restricts the free movement of solutes and cellular elements between the systemic circulation and neuronal tissue. The BBB is involved in the pathogenesis of a large number of CNS disorders [19].

Cell types comprising the NVU are in close communication in order to maintain physiologic function and react to pathologies. Ligand-receptor type intercellular interactions and ion channels were described early as pathways that coordinate the function of the cell types constituting the NVU [20] An example of bidirectional information exchange is the role of CD146 in coordinating the development of pericyte coverage on brain vasculature during early ontogenesis. At first CD146 is expressed by endothelial cells but as pericyte coverage increases, CD146 expression shifts to pericytes where it acts as a co-receptor of PDGFRβ. Endothelia attached pericytes down-regulate endothelial CD146 via TGFβ1 secretion, promoting BBB maturation [21]. Recently a growing body of evidence suggests that extracellular vesicle—mainly exosome—mediated bidirectional communication coordinates key functions of the NVU at the local and systemic level as well [22–25]. In a mouse model of spinal cord injury, pericyte-derived exosomes improved microcirculation and protected barrier function [26]. In response to traumatic brain injury, the loss of pericytes and consequent impairment of crosstalk among NVU cells causes barrier dysfunction, brain edema and leakage of cerebral vasculature. A recent example of pericyte to endothelial cell communication in the retina is the circular RNA cPWWP2A that is synthesized in pericytes and downregulates angiopoietin 1, occludin and sirtuin-1expression in endothelial cells by sequestering miRNA-579 [27].

3. Brain Capillaries in Aging

With aging, the density of brain vasculature is decreased and cerebrovascular dysfunction appears to precede and accompany cognitive dysfunction and neurodegeneration. Cerebrovascular angiogenesis is decreased and cerebral blood flow is inhibited by anomalous blood vessels such as tortuous arterioles and thick collagen deposits in the walls of veins and venules [28].

In most mammals, the capacity of CECs to divide is limited and endothelial cells are prone to be senescent. Aging is associated with endothelial dysfunction, arterial stiffening and remodeling, impaired angiogenesis, defective vascular repair and with an increasing prevalence of atherosclerosis [29]. In the aging brain cerebral blood flow declines and perfusion pressure either is constant or increases. In Brown-Norway and Fisher 344/Brown-Norway rats that maintain a relatively consistent cortical volume throughout life the densities of arterioles and arteriole-to-arteriole anastomoses on the cortical surface was found to be decreased with age [30]. In a spontaneously hypertensive rat model, long term hypertension was found to gradually destroy BBB, resulting in white matter lesions, one of the most important pathological changes in vascular dementia [31]. At the capillary level, increased capillary diameters and decreased capillary density paired with increased red blood cell velocities were observed [28,32]. Some capillary density measurements in humans contradict these observations, as no changes were observed in the intervascular distance on CD31 stained brain sections [33].

4. Junctional Proteins in Aging and Related Disorders

The paracellular barrier properties of CECs are determined by the tight junction (TJ), which are composed of transmembrane proteins that control the transport across the intercellular space between adjacent cells and cytoplasmic plaque. Claudin 5, member of the 27 strong claudin family, is primarily responsible for controlling the paracellular transport of water and small molecules through capillary vessel walls [34]. Other claudins may contribute to BBB function as well, though the isoforms present in human and murine brain vessels is still under investigation [35,36]. Like claudins, MARVEL (MAL and related proteins for vesicle trafficking and membrane link) proteins especially occludin play a role in restricting paracellular transport and also regulate homeostasis and TJ organization [37–40]. The third group of integral membrane proteins of the TJ belongs to the CTX (cortical thymocyte marker in *Xenopus*) family within the immunoglobulin superfamily such as junctional adhesion molecules (JAM) and coxsackie and adenovirus receptor (CAR). Integral membrane proteins of the TJ are coupled to cytoplasmic plaque proteins that provide a platform for anchoring the junction to the cytoskeleton and for interactions with signaling molecules. The main components of the cytoplasmic plaque are zonula occludens (ZO) proteins [41–43].

Junctional proteins are not restricted to the endothelial cells of the NVU. Recently, pericytic occludin was described in a new role as a NADH oxidase enzyme and an important player in BBB pericyte metabolism including the modulation of pericyte energy sharing with other NVU components via intercellular transport of mitochondria and glucose [44]. Astrocytic occludin was described two decades ago [45], and a recent report proposed a new role for it in RNA metabolism [46]. Astrocytic TJ, composed of claudin 1, 4 and JAM may also form in inflammation to control lymphocyte segregation [47].

4.1. *TJ in Aging and Aging Models*

Limited data is available on what changes develop in the function of the BBB and the composition and structure of endothelial TJs in the healthy aging human brain. In a meta-analysis of BBB permeability studies, the barrier function was negatively impacted by age. Though there were some discrepancies, paracellular permeability was generally increased in the aged human brain [48]. Multiple accelerated aging animal models are available to complement this data, which we will discuss mainly together with the appropriate aging associated disorders in later paragraphs. Literature data presented in this chapter are summarized in Table 1. In the review we will include large molecule (e.g., albumin or Evans blue albumin) permeability and blood component extravasation as well even though these are not markers of TJ dysfunction but rather indicate transcellular transport or disruption of the vascular wall respectively.

Table 1. Blood–brain barrier (BBB) function and junctional proteins in aging.

Experimental System	Claudin 5	Occludin	ZO-1	Permeability	Other	Authors
BubR1 accelerated aging, cell culture	Discontinuous immunofluorescence for all three.					Baker et al., 2011 [49]
aged mouse and human brain tissue sirtuin-1 overexpression sirtuin-1 knockdown	fragmented, weak immunofluorescence, reduced protein level. rescued by sirtuin-1 overexpression reduced claudin 5 mRNA and protein		fragmented, weak immunofluorescence	increased 3kDa dextran permeability rescued by sirtuin-1 overexpression. increased		Stamatovic et al., 2019 [50]
senescent female rats				Evans blue extravasation, hippocampus, olfactory bulb		Bake and Sohrabji 2004 [51]
senescent rats postmenopausal women	weaker and discontinuous immunofluorescence in hippocampal vessels in senescent compared to young female rats. No change in middle aged vs. young male rats altered distribution	no change in reproductive senescent vs. young female rats			age and gender dependent IgG extravasation, e.g., increased IgG extravasation in senescent female rat thalamus	Bake et al., 2009 [52]
ovariectomized rats +estrogen replacement			unaffected	increased rescued	connexin43 redistribution rescued	Wilson et al., 2008 [53]
Pdgfrβ+/−, pericyte deficient mice	All three decreased.					Bell et al., 2010 [54]
24-month-old rats, compared to 12-months-old 24-month-old rats, compared to 3-months-old		protein decreased, mRNA unaffected protein decreased, mRNA unaffected	mRNA increased, protein unaffected protein unaffected			Mooradian et al., 2003 [55]
mouse primary brain microvessel endothelial culture. disrupted endothelial integrin/basal lamina interaction	decreased			increased		Osada et al., 2011 [56]
integrin α5 knockout mice	increased mRNA			improved barrier	resistant to MCAO induced stroke	Roberts et al., 2017 [57]

Experimental results linking BBB dysfunction to declining cognitive function prompted the use of modern in vivo imaging such as dynamic contrast enhanced magnetic resonance imaging to study cerebrovascular permeability. A recent work revealed an age dependent BBB permeability increase in the hippocampus of individuals without any cognitive impairment, which was exacerbated in the case of mild cognitive impairment of Alzheimer's, Huntington's and multiple sclerosis patients [58].

Permeability changes are likely the result of decreased expression and disorganized localization of TJ proteins. As shown in the BubR1 hypomorphic murine model of accelerated aging, claudin 5, occludin and ZO-1 expression of cortical and striatal microvessels were decreased in aged mice, with reactive astrogliosis also apparent at the vessels. BubR1 is a key member of the mitotic checkpoint that ensures accurate chromosome segregation [49]. In an experiment using cells isolated from BubR1 hypomorphic mice, aged CECs (co-cultured with pericytes) showed a discontinuous junctional staining for claudin 5, occludin and ZO-1 and the presence of abnormal cells with smaller cytoplasm that are highly immune-reactive for TJ proteins [59]. The BubR1 hypomorphic CECs also showed increased number of spikes in the claudin 5 staining, reminiscent to those previously described as budding and docking sites for claudin bearing vesicles [60].

Aged mouse and human brains both showed fragmented and weak claudin 5 and ZO-1 staining in microvessels compared to healthy young controls, with an accompanying reduction of claudin 5 protein expression. BBB permeability to 3 kDa dextran was also increased in aged mice. Both claudin 5 levels and BBB function were rescued in transgenic mice overexpressing sirtuin-1, a NAD-dependent deacetylase, while inducible, endothelial-specific knockdown of sirtuin-1 increased barrier permeability and reduced claudin 5 mRNA and protein levels [50]. These results add a novel mode in which sirtuin-1 affects longevity and neurological disorders by maintaining BBB integrity [61,62].

Barrier function of the cerebral vasculature was shown to be under the control of estrogen [63]. Thus, a potential influence of the age-dependent decline of estrogen levels on BBB integrity in postmenopausal women or reproductive senescent female animals appeared plausible. In this context, Bake and Sohrabji [51] showed enhanced extravasation of Evans Blue-albumin in the hippocampus and olfactory bulb of reproductive senescent female rats. Moreover, the localization of the tight junction protein claudin 5 in hippocampal microvessels was affected in senescent female rats compared to young adult ones. In case of human brain microvessels, disrupted distribution and poor junctional localization of claudin 5 was altered in postmenopausal women, in contrast to premenopausal women [52]. Further confirmation to suggest that reproductive hormones regulate BBB permeability came from studies with ovariectomized rats [53]. While expression of ZO-1 was unaffected, redistribution and increased expression of connexin-43 was observed following ovariectomy. Estrogen replacement to young (4 months old) ovariectomized rats restored BBB function, while reproductive senescent animals did not benefit from estrogen treatment or even showed increased BBB permeability in the hippocampus [53].

Further evidence of age-related junctional alterations at the BBB came from a study showing that the cerebral occludin protein content in total cerebral tissue extracts from 24-month-old rats was significantly reduced compared to 12-month-old or 3-month-old male rats without a corresponding change in occludin mRNA. ZO-1 protein levels were found to be unchanged while ZO-1 mRNA was significantly increased in 24-month-old rats compared to 3-month-olds [55].

By using transmission electron microscopy, a decline in the pericytic coverage of aged capillaries has been described over thirty years ago [64]. Beyond the assumption that the age-dependent loss of pericytes decreases the ability of the BBB to compensate for transient leaks, there is evidence that pericytes regulate the expression of TJ/BBB-specific proteins in microvascular endothelial cells. A comparison of claudin 5, ZO-1 and occludin expression in 6–8 month old Pdgfrβ+/− mice has revealed significant reductions of protein expression in pericyte-deficient mice compared to controls [54]. Therefore, it is likely that aging-associated TJ dysregulation may be partly the consequence of pericyte loss.

Integrins are heterodimeric proteins that anchor cells to the extracellular matrix and are involved in ligand dependent bidirectional signaling between the cell and its environment thus playing important roles in all cell functions. Researchers have found connections between integrins and aging [65,66] and neurological disorders [67]. Disruption of cerebral endothelial integrin interaction with the basal lamina lead to decreased claudin 5 levels and BBB integrity [56]. Recently, integrin α5 knockout mice were found resistant to transient middle cerebral artery occlusion (MCAO) induced ischemic stroke with a small initial infarct volume that disappeared within 3 days. The α5 knockout animals showed improved barrier function and higher claudin 5 mRNA levels compared to controls [57].

CTX family proteins have not been directly linked to aging; however, some play important roles in the brain development and function or tumorigenesis and invasiveness. JAM-A in CECs undergoes subcellular relocalization in response to inflammatory stimuli [68] and its levels are reduced in response to brain injury [20]. JAM3 was found to be essential for BBB integrity and for normal lens development in humans. Homozygous mutations of JAM3 in humans led to brain hemorrhage and brain calcification [69], the latter being somewhat similar to calcification in the basal ganglia in occludin knockout mice [70]. Brain endothelial function of CAR is scarcely known. However, it was found to play key roles in neurons in synapse development in the postnatal and adult brain and its loss contributes to cognitive defects [71,72].

4.2. Brain Endothelial TJ in Ischemic Injury

In the healthy brain, practically every neuron has a capillary running next to it. With the rarefaction of capillaries in aging brain, the ability to maintain homogeneous flow during periods of localized ischemia is reduced. Hypoxia and reoxygenation are important components of many disorders that affect the central nervous system, including stroke and dementia [73–75]. The incidence of stroke is increased with age and the prognosis of elderly stroke patients is unfavorable compared to young adults [76,77]. Literature data regarding TJs presented in this chapter are summarized in Table 2.

Table 2. BBB function and junctional proteins in age related disorders, ischemic injury and immune system.

Experimental System	Claudin 5	Occludin	ZO-1	Permeability	Other	Authors
hypoxia, 8% O_2, mouse, brain lysate	unaffected	decreased protein levels disrupted immunofluorescence for both	unchanged protein levels	Na-fluorescein leakage increased	MMP and VEGF dependence	Bauer et al., 2010 [78]
hypoxia in primary bovine brain microvessel endothelial culture		membrane localization perturbed	ZO2 membrane localization perturbed, ZO1 and ZO2 diffuse cytosolic staining	sucrose permeability increased	actin stress fibers, increased actin protein	Mark and Davis 2002 [79]
H_2O_2 ROS +GM6001	unaffected	phosphorylation change, degradation rescues degradation			TEER increase does not rescue	Lischper et al., 2010 [80]
hypoxia/reoxygenation modeled by DMNQ treatment of primary mouse brain microvessel endothelial culture		protein decreased				Krizbai et al., 2005 [81]
hypoxia in bEND.3	altered localization, decreased protein			increased		Koto et al., 2007 [82]
reactive oxygen species	altered localization, decreased protein for both				xanthine oxidase and 100 μM hypoxanthine	Schreibelt et al., 2007. [83]
MCAO, 12 months vs. 2 months old mice	decrease in both age groups, after 3 days less protein in old compared to young		decrease in both age groups, after 1 and 3 days less protein in old	larger Evans blue extravasation in old with stroke		Shen et al., 2019 [84]
young adult rats hypoxia/reoxygenation		phosphorylation	no change in protein	increased		Witt et al., 2003 [85]
young adult rats hypoxia/reoxygenation as above + free radical scavenger tempol		disrupted rescued				Lochhead et al., 2010 [73]
young adult rats hypoxia/reoxygenation	disrupted localization, slightly increase membrane fraction	disrupted localization, slightly increase membrane fraction	disrupted localization	PKC-dependent focal BBB leakage		Willis et al., 2010 [86]
EGFP-Claudin 5 transgenic mouse tMCAO after 24–30 hours after 48–58 hours	eGFP-claudin 5 no change eGFP-claudin 5 non membrane	no change reduced	no change reduced		Claudin 5 transgene protein expression is double of wild type.	Knowland et al., 2014 [87]

Table 2. *Cont.*

Experimental System	Claudin 5	Occludin	ZO-1	Permeability	Other	Authors
600–950 g rats compared to 200–250 g rats		degradation				Kaur et al., 2011 [88]
MCAO, both ages	slightly increased phosphorylation					
tPA after reperfusion, only in old	greatly increased phosphorylation	increased degradation				
rat tMCAO +PP2 Src kinase inhibitor, brain tissue	time dependent decrease of claudin 5 protein and mRNA was rescued by PP2			brain lysate protein decrease was significant from 24, mRNA from 72 hours and persisted 7 days later, fibrinogen leakage from brain coincided claudin 5 decrease and vas rescued by PP2		Bai et al., 2014, [89]
rat microsphere embolism, capillary lysate		decreased amount, increased phosphorylation 6–72 h after microsphere injection	decreased amount	albumin leakage starting from 48 hours		Kago et al., 2006 [90]
rat microsphere embolism, capillary lysate, +PP2		decreased amount and increased phosphorylation rescued by PP2		albumin leakage 6-24h, rescued by PP2		Takenaga et al., 2009 [91]
spontaneously hypertensive rats tMCAO, brain sections	rapid reorganization in localization 3 h after reperfusion + MMP2 activity at vessels. Protein appeared in astrocytes.			Increased gelatinase activity 3 h after reperfusion, rescued by MMP2 inhibition.		Yang et al., 2007, Yang and Rosenberg 2011 [92,93]
brain lysate	decreased mRNA, degraded protein rescued by BB-1101 MMP inhibitor					
spontaneously hypertensive rats tMCAO+GM6001 MMP inhibitor	protein levels were rescued for all three					Yang et al., 2013 [94]
Primary human brain microvascular endothelial cells+PARP inhibition	increased protein levels for both					Rom et al., 2015 [95]

Table 2. *Cont.*

Experimental System	Claudin 5	Occludin	ZO-1	Permeability	Other	Authors
spontaneously hypertensive rats tMCAO + mynocicline		increased protein levels for all three				Yang et al., 2015 [96]
SRF, MRTF-A/-B knockdown hemorrhagic stroke model, whole brain tissue	claudin1, 3, 5 and 12 mRNA and claudin5 protein downregulated		ZO-2 and ZO-3 mRNA downregulated, ZO-1 unaffected			Weinl et al., 2015 [97]
bovine brain microvessel endothelial cells +VEGF		distribution changed from membrane to cytoplasm, protein levels increased		permeability increase, resistance decrease		Wang et al., 2001 [98]
adenoviral expression of IL-1 in VEGFA knockout mouse	VEGFA KO rescued protein expression of brain microvessels in inflammatory lesions for both					Argaw et al., 2012 [99]
EAE in VEGFA KO mouse	VEGFA KO rescued protein expression of brain microvessels in inflammatory lesions					
VEGFR2 blocking antibodies in Cortical VEGF injected wild type mice	rescued protein expression for both					
eNOS inhibition or silencing in human brain microvessel endothelial cells	rescued protein expression					
24 months old mice compared to 3 months old, via fluorescence activated cell sorting	significantly decreased		decreased	IgG leakage	neurovascular inflammation and neuronal stress	Elahy et al., 2015 [100]
SAMP8 mice, 12 vs. 4 months old, permeability				no change in albumin permeability and insulin uptake		Banks et al., 2000 [101]

In vitro studies of hypoxia and hypoxia followed by reoxygenation have revealed much of the cellular mechanisms affected by ischemia in CECs. In oxidative stress, the interactions of occludin with claudins or proteins of the ZO family are affected, directly influencing the formation and function of the TJ. Oxidative stress downregulates occludin, reduces its specific membrane localization and regulatory contribution to barrier tightness via multiple signaling pathways [102].

Significant disruptions were seen in the distribution pattern of occludin after hypoxia [78] These changes were less apparent in posthypoxic reoxygenation, correlating to the increased protein expression [79]. MCAO caused NADPH oxidase upregulation, ROS (reactive oxygen species) generation, matrix metalloproteinase (MMP)-9 activation, and edema formation [103]. MMP activity in oxidative stress increases paracellular permeability by occludin cleavage. However, applying the same level of MMP activity without oxidative damage, neither is occludin cleaved nor BBB permeability is increased [80]. In an opposite approach, normobaric hyperoxia protected the BBB, and the expression and distribution of occludin against MMP-9–mediated effects in cerebral ischemia [103]. Thus occludin plays a key role in the response of cellular barriers to redox changes and could be a redox sensor at the TJs [104,105]. In a brain endothelial cell culture model of hypoxia/reoxygenation, ERK1/2 was activated and the amount of occludin and barrier function decreased which was exacerbated by glucose deprivation [81]. Hypoxia (1% O_2) altered the location of claudin 5 in the plasma membrane and the level of claudin 5 protein in bEND.3 cells, and these changes were accompanied by an increase in BBB permeability. In vivo, claudin 5 was also significantly reduced under hypoxic conditions [82] ROS-induced BBB disruption via the redistribution and decrease of claudin 5 and occludin levels was paralleled by cytoskeleton rearrangements. This rearrangement was mediated by RhoA, PI3 kinase and protein kinase B (PKB/Akt), and specific inhibitors prevented ROS-induced monocyte migration across an in vitro model of the BBB [83].

The effect of age on the outcome of ischemic stroke was studied in murine distal MCAO experiments, demonstrating an increased inflammatory reaction in 12 month old mice compared to 2 month olds based on IL-6 and IL-1β levels in the ischemic tissue. At the same time, there was an increased number of CD68 immunoreactive cells, denoting phagocytosis and microglia activation in peri-infarct tissue and the older mice suffered more severe BBB damage as well [84]. Similar BBB permeability increase was observed in young adult rats after one hour of hypoxia (6% O_2) followed by 10 minutes of reoxygenation, which also lead to increased occludin phosphorylation without significant changes in ZO-1 expression [85]. Disrupted occludin staining of brain endothelial cells in rats was reversible by administration of ROS scavengers [73]. Another study under similar conditions showed PKC-dependent focal BBB leakage and disrupted claudin 5, occludin and ZO-1 organization at endothelial cell borders while the protein levels were slightly increased in the membrane fraction for claudin 5 and occludin [86].

Functional changes of the BBB appear shortly after a hypoxic event, but structural TJ changes develop much slower. Using an EGFP-claudin 5 transgenic mouse model, researchers demonstrated a stepwise alteration of the BBB, leading to an increase in transcytosis followed by alterations of TJ proteins, which undergo significant ultrastructural remodeling and localization changes starting at 48–58 h post tMCAO (transient middle cerebral artery occlusion) [87]. Early permeability increase elicited by tissue plasminogen activator (tPA) during reperfusion in elderly rats was accompanied by greatly increased claudin 5 phosphorylation while unphosphorylated claudin 5 decreased compared to the contralateral side. MCAO treatment itself increased claudin 5 phosphorylation slightly. Occludin protein level was decreased in elderly compared to young rats and further degradation was observed in response to MCAO and tPA. These age-dependent differences in claudin 5 phosphorylation and occludin degradation were inferred from redistribution of western blot bands [88].

In a multifocal cerebral ischemia model, Src kinase dependent tyrosine phosphorylation of occludin was found to contribute to barrier disruption. Systemic inhibition of Src resulted in decreased infarct size, decreased VEGF-A, rescuing claudin 5, occludin expression and barrier function [89–91]. Reperfusion experiments following 90-minute hypoxia in spontaneously hypertensive rats revealed

MMP dependence of BBB leakage and the disruption of claudin 5 and occludin at the TJ in vivo. Interestingly the TJ proteins from disrupted endothelial barrier appeared in adjacent astrocytes underlining intercellular communication [92,93]. In vivo use of the MMP inhibitor GM6001 in rats as an early treatment given at the time of MCAO increased the number of vessels and improved ZO-1, occludin and claudin 5 expression in the infarct region 3 weeks after the ischemic injury [94].

In a sterile inflammation model induced by TNF-α in human primary brain microvascular endothelial cells, PARP inhibition promoted BBB barrier function. Furthermore, PARP inhibition in primary endothelial cells improved TEER and increased occludin and claudin 5 protein levels [95]. In spontaneously hypertensive rats that underwent MCAO, minocycline, an inhibitor of PARP-1 and anti-inflammatory, anti-apoptotic and neuroprotective drug, significantly reduced the infarct size and prevented tissue loss, improved perfusion, reduced BBB permeability and increased ZO-1, occludin and claudin 5 protein levels. At the same time, increased MMP-2 and -3 were detected at four weeks, at which time peri-infarct microglia showed M2 activation [96].

In a recent genetic model of intracerebral hemorrhagic stroke and vascular dementia, cerebral small vessel disease symptoms were elicited by inducible knockdown of the transcription factor, serum response factor (SRF) or its cofactors Myocardin Related Transcription Factor (MRTF-A/-B). The microhemorrhagic phenotype observed in this model is attributed to loss of TJ components and is specific for brain tissue. The authors demonstrated mRNA level decrease of claudin 1, 3, 5 and 12, as well as ZO-2 and ZO-3, and significant loss of claudin 5 protein in isolated brain endothelial cells [97].

Exposing mice to normobaric hypoxia led to an increase in brain vascular permeability associated with diminished expression of occludin and inhibition of VEGF attenuated vascular leakage [78]. This confirms the results of studies in primary brain endothelial cells in which VEGF decreased occludin expression [98]. Astrocyte derived VEGF-A acts directly on brain microvasculature to downregulate occludin protein and mRNA in vitro and in mouse models of experimental autoimmune encephalomyelitis (EAE), inducing BBB permeability. In vivo inactivation of astrocytic VEGF-A expression or systemic use of the eNOS inhibitor cavtratin rescued BBB function [99].

4.3. *Involvement of Brain Barriers in Aging-Associated Alterations of Immune Functions*

Brain barriers are important interfaces for neuroimmune communication and any disturbance may inevitably result in CNS dysfunctions. Aging-associated alterations at the BBB and brain cerebrospinal fluid (BCSF) barrier are thus considered to be one trigger for the development of cognitive impairment(s) in the aged brain (reviewed in [106,107]). Literature data concerning TJs presented in this chapter can be found at the end of Table 2.

While it is well established that migration of immune cells from the blood into the CNS parenchyma is a process occurring in inflammatory and neurodegenerative state, accumulating evidence suggests that immune cells also infiltrate the perivascular space of non-diseased brain, though at very low levels, maintaining and contributing to the surveillance and homeostasis of the CNS (reviewed in [108–110]).

Less information exists concerning age-related immune cell trafficking across the BBB and BCSF, obviously due to the fact that normal aging of the brain is usually accompanied by at least low level inflammatory activity [111]. Leakage of IgG into the cerebral cortex and hippocampus of healthy aged (24 months old) but not young (3 months old) mice were reported [100]. This leakage was accompanied by a decrease in occludin and ZO-1 expression in cerebral endothelial cells but no signs of leukocyte transmigration into the CNS.

A progressive age-related enhancement of TNFα-elicited T cell infiltration in brain parenchyma and choroid plexus of mice was demonstrated by Xu et al. [112]. Further, Stichel and Luebbert [113] reported on the presence of T cells and dendritic-like cells in mouse brain parenchyma at about 12 months of age, with numbers increasing during aging. T cells from aged mice were found to exhibit increased expression of adhesion molecules, such as CD11a and CD49d, facilitating diapedesis [114] and elevated ICAM1 expression was found in brain vasculature of aged human and mouse brains [115].

The role of aging-associated T cell infiltration of the non-diseased brain is still unclear. Results from single-cell RNA-sequencing and immunofluorescence staining of the subventricular zone from young (3 month old) and old (28–29 months) mice suggest that T cells, expressing interferon-γ, were clonally expanded and markedly enriched in the subventricular zone from old mice [116]. In vitro studies have shown that T cells can inhibit the proliferation of neural stem cells in co-cultures in an interferon-γ-dependent mechanism [116]. Thus, it may be assumed that interaction of T cells with neural stem cells in neurogenic niches of old animals may contribute to the proliferative decline of neural stem cells during aging. A distinct role of immune cells was demonstrated in increasing cerebral vascular permeability [117]. Using a mouse model of CD8 T cell-mediated CNS vascular permeability it was shown that CD8 T cells are capable of initiating BBB disruption in a non-apoptotic perforin-dependent manner.

4.4. Alzheimer's Disease (AD)

The blood–brain barrier contributes to the pathology and may play a causative role in the development of the most common neurodegenerative disease afflicting the elderly population: Alzheimer's disease (AD). The involvement of the BBB happens through multiple mechanisms such as barrier disruption, transporter dysfunction and secretion of neurotoxic substances [118,119]. Amyloid-β peptide (Aβ), the main component of senile plaques in the AD influences the expression and localization of TJ proteins [120–122]. The extracellular matrix and TJ proteins are substrates of MMPs, whose activity is also increased in AD and after ischemic injury, which results in lower TJ proteins [123]. Data concerning TJ changes in Alzheimer's, Parkinson's and Huntington's diseases, as well as schizophrenia are summarized in Table 3.

Table 3. BBB function and junctional proteins in age related disorders, neurodegenerative diseases.

Experimental System	Claudin 5	Occludin	ZO-1	Permeability	Other	Authors
Triple-transgenic mice (3 × Tg-AD), 10–11 vs. 4 month old				no change in [3H]-inulin and [14C]-sucrose uptake	Age dependent reduction in cerebrovascular volume, thickened basal membrane	Bourasset et al., 2009, Mehta et al., 2013 [124]
18–20 month old 3 × Tg-AD vs. wild type mice				no change in [3H]-inulin and [14C]-sucrose uptake decreased uptake of passively diffusing markers: [3H] diazepam and [3H] propranolol in AD mice	thickened basal membrane	
rat brain microvessel endothelial cells +Aβ1-42 for 24–72 h, immunofluorescence	disrupted localization after 24 h, increased fluorescence of Claudin 5 and 1.	no change in localization. Decreased fluorescence at 24 hours.	ZO-2 diffuse localization. ZO-2 increased fluorescence at 24 h and decreased at 72 h.			Marco and Skaper 2006 [120]
capillary cerebral amyloid angiopathy (capCAA) patient brain sections	loss of protein in capillaries with amyloid β deposits					Carrano et al., 2011, 2012 [125,126]
hCMEC/D3 cell culture + Aβ1-42 for 24 h		dose dependent, ROS mediated decrease of mRNA for both				
ARPE-19 cell culture + oligomeric Aβ1-42		disrupted membrane staining, significantly decreased mRNA		increased 40 kDa FITC-dextran permeability		Bruban et al., 2009 [127]
subretinal Aβ1-42 injection in C57BL/6 J mice		Similarly disrupted membrane staining of both in young, adult and aged mice.				
bEnd.3+Aβ1-42	decreased protein level		decreased protein levels rescued by inhibition of either RAGE, calcineurin or MMP	increased 40 kDa FITC-dextran permeability rescued by inhibition of calcineurin or MMP		Kook et al., 2012 [128]
8 months old transgenic mice with five familial AD mutations (5XFAD) compared to wild type	TJ cross section length significantly decreased, by transmission electronmicroscopy				RAGE expression was increased in brain capillaries	
bEnd.3	Oligomeric, but not monomeric or fibrillar Aβ1-42 dose dependently decreased protein levels. Protein levels were rescued by silencing RAGE with siRNA or blocking RAGE with a polyclonal antibody.					Wan et al., 2015 [129]

Table 3. *Cont.*

Experimental System	Claudin 5	Occludin	ZO-1	Permeability	Other	Authors
bEnd.3 + oligomeric Aβ1–42	decreased protein levels and increased permeability were rescued by EGb761 in a dose dependent manner				Aβ1–42-Oligo induced expression of MMP-2 and MMP-9 was rescued by blocking RAGE via antibody.	Wan, Cao, et al., 2014 [130]
bEnd.3 cells+ Aβ1-42 and isolated capillaries from 5XFAD mice	exogenous ANXA1 rescued Aβ1-42 induced decrease in protein levels by blocking Rho-ROCK signaling		exogenous ANXA1 rescued Aβ1-42 induced decrease in protein levels by blocking Rho-ROCK signaling			Park et al., 2017 [131]
APOE4 transgene expressing Apoe-/- mice	protein levels were decreased dependent on the pericytic cyclophilin A/NF-κB/MMP9 dependent pathway					Bell et al., 2010 [54]
bEnd.3, CD-1 or PARP-1 KO mouse primary glial culture +Aβ1–42		endothelial protein expression reduced by activated microglia was rescued by PARP inhibition.				Mehrabadi et al., 2017 [132]
MPTP mouse model +caffeine		MPTP treated animals showed decreased protein levels, which was rescued by caffeine.		Caffeine rescued MPTP induced Evans blue leakage in striatum.	Caffeine blocked MPTP-induced increases in MMP9 activity	Chen et al., 2008 [133]
MPTP mouse model +trehalose		immunofluorescent signal was greatly diminished in MPTP animals, which was rescued by trehalose.				Sarkar et al., 2014 [134]
unilateral striatal 6OHDA injection in rats				FITC-albumin or horseradish peroxidase leakage in striatum and substantia nigra in response to 6OHDA		Carvey et al., 2005 [135]
unilateral striatal 6OHDA injection in rats	immunohistochemistry signal decreased	protein levels decreased for both				Huang et al., 2016 [136]
hCMEC/D3 (co-cultured with mouse primary astrocytes) + preformed α-synuclein fibrils	decreased protein levels					Kuan et al., 2016 [137]

Table 3. *Cont.*

Experimental System	Claudin 5	Occludin	ZO-1	Permeability	Other	Authors
Parkinson's Disease patient tissue sections compared to age matched and deep brain stimulated patient tissue	decreased immunofluorescent staining compared to control, rescued by deep brain stimulation.			IgG extravasation		Pienaar et al., 2015 [138]
Parkinson's disease patient tissue sections				blood extravasation		Gray and Woulfe 2015 [139]
Huntington's disease patient putamen samples compared to control	decreased protein levels			extravascular fibrin deposition., increased leakage in caudate and putamen by DCE MRI	abnormal vessel morphology	Drouin-Ouellet et al., 2015 [140]
and R6/2 mouse model, striatal samples	decreased protein levels			increased pinocytic activity and increased albumin extravasation	morphological changes in vessel walls	
iPSC derived brain microvascular endothelial cells from HD patients	miss-localization to the cytoplasm			decreased TEER		Lim et al., 2017 [141]
schizophrenic patient brain tissue	discontinuous localization			anti-psychotic medications dose dependently increased claudin 5 expression		Greene et al., 2018. [142]

Similar to data on permeability in the aging brain, the BBB permeability data of AD patients was heterogeneous. This could partly be ascribed to difficulties in separating different types of dementia and co-morbidities confounding results. In a meta analysis by Farrall and Wardlaw some results show increased BBB permeability in AD compared to controls but the permeability increase is higher and more clear cut for vascular dementia and mixed dementias compared to AD [48]. Murine models of AD allow for more precise control over experiments and make it possible to concentrate on separate aspects of the disease; however, available results are still conflicting. A number of studies showed that the paracellular route is unaffected in murine models of AD. There was a substantially decreased occludin expression in APP/PS1 mice AD model compared to wild-type controls [143]. In some cases, when using an APP/PS1 mouse model there was an unaltered BBB permeability to ^{131}I-albumin [144] and to sodium fluorescein [145]. Even though earlier work showed small localized disruption in Senescence Accelerated Mouse-Prone 8 (SAMP8), no global BBB permeability change was observed when comparing 12 and 4 months old animals [101]. Comparing 11 and 4 months old 3 × Tg-AD mice (triple transgenic mice for familial Alzheimer's disease mutations), Bourasset et al. found no changes in the distribution of vascular space markers ^{3}H-inulin and ^{14}C-sucrose, but found a global reduction in cerebrovascular volume that was most significant in the hippocampus [124]. In a comparison of 18–20 month old 3×Tg-AD and wild type mice researchers also did not find changes in ^{3}H-inulin and ^{14}C-sucrose uptake, while the uptake of passively diffusing markers ^{3}H-diazepam and ^{3}H-propranolol was less in AD mice [146]. These results suggest that in murine AD models global BBB breakdown is not detectable and research probably should focus on local barrier disruption, as data from multiple research groups point to the importance of cerebral amyloid angiopathy in AD and there is ample amount of research revealing the molecular mechanisms behind it.

Regarding TJ proteins in post-mortem human brain, a dramatic loss of claudin 5, occludin and ZO-1 was demonstrated in Aβ-laden capillaries surrounded by NADPH oxidase-2 (NOX-2)-positive activated microglia. Further results show that Aβ is toxic to endothelial cells in the human brain via binding to the receptor for advanced glycation end-products (RAGE) and induction of ROS production, which ultimately leads to disruption of TJs and loss of BBB integrity [125,126]. In a model of age-related macular degeneration, increased formation of ROS was found after administration of the oligomeric form of Aβ1–42 to retinal pigmented epithelial cells [127], which was accompanied by a decrease in the expression of occludin. The Aβ1-42 peptide accumulation around brain microvessels was found to be mediated by transport via RAGE in 5XFAD mice, increasing permeability, disrupting claudin 5, ZO-1 and occludin. This MMP and Ca^{2+}-calcineurin dependent BBB disruption could be prevented in bEnd.3 cells by neutralizing antibodies against RAGE [128], small interfering RNA knockdown of RAGE [129] or *Ginkgo biloba* extract EGb761 [130].

An in vitro study on Aβ1-42-treated bEnd.3 cells showed that AnnexinA1 (ANXA1), an anti-inflammatory messenger, significantly rescued the expression of claudin 5 and ZO-1 and barrier function in Aβ1-42 -treated bEnd.3 cells. Aβ1–42 reduced ANXA1 bEnd.3 cells, and also had reduced expression in capillaries of 5XFAD mice, and the human serum of patients with AD. ANXA1 acted via the inhibition of RhoA-ROCK signaling. In co-culture experiments, pericyte secreted ANXA1 attenuated the Aβ1–42-induced disruption of the tight junction [131].

Pericytes influence the BBB by promoting TJ protein expression in endothelial cells [54] and helping the alignment of TJs [147]. It has also been shown that a loss of pericytes plays a role in AD development and is followed by a decreased expression of TJ proteins [148,149].

The strongest genetic risk factor for late onset AD is Apolipoprotein E4 (ApoE4). ApoE4 and its receptors are expressed throughout the NVU and are linked to many aspects of cerebrovascular dysfunction [150,151]. The molecular mechanisms of ApoE4 mediated neurovascular injury demonstrate how the coordinated effort of multiple cell types maintains a functional NVU. Bell et al. revealed using multiple transgenic mice that the ApoE induced degradation of TJ proteins claudin 5, ZO-1 and occludin and basal membrane protein collagen IV is the result of an intercellular communication error. ApoE4 secreted by astrocytes is unable to bind LRP1 on pericytes and thus does not block the pericytic cyclophilin A/NF-κB/MMP9 pathway, which results in vascular dysfunction [152].

Soluble Aβ is also known to induce secretion of proinflammatory cytokines (TNF and IL-6) and chemokines, which stimulate the production of MMP-2 and MMP-9 and it also activates the production of ROS [153]. Furthermore, experimental data confirms the involvement of all cells of the NVU in the effect of Aβ. Microglia activated by Aβ treatment shows reduced expression of trophic factors that are responsible for inflammatory resolution and increased pro inflammatory NO and TNFα release. This affects both astrocytes and capillary endothelium leading to reduced BBB integrity and function [132].

AD is also accompanied by tau protein accumulation and hyperphosphorylation, which was shown to promote BBB dysfunction in AD and other tauopathies [154–156].

4.5. *Parkinson's Disease (PD)*

The BBB is also involved in the progression of the second most common neurodegenerative disorder: Parkinson's disease (PD). The contribution of BBB disruption to PD is not widely studied despite the implication of known BBB damaging mechanisms and agents such as oxidative stress and MMPs in the pathomechanism of the disease [157,158]. Thickened basement membrane in the cingulate cortex and degeneration of the brain microvasculature in PD was reported by Farkas et al. and Guan et al. [159,160]. Gray and Woulfe published the first report of BBB disruption evidenced by blood extravasation in striatal PD tissue in 2015 [139]. Motor function and regional blood flow can be improved in PD patients by deep brain stimulation [161], which is possibly a result of normalizing aberrant microvasculature in PD. In a study regarding the effects of deep brain stimulation of the subthalamic nucleus in PD patients, the decreased immunofluorescence signal of claudin 5, occludin and ZO-1 of PD samples could be rescued by deep brain stimulation treatments [138]. This is in line with multiple studies using experimental parkinsonism models. Significant decrease was detected in the amount of occludin in the striatum, which was associated with increased BBB leakage, in the 1-methyl-4-phenyl-1,2,3,6-tetrahydropyridine (MPTP)-treated mouse model. Increased striatal MMP-9 activity was also detected in the MPTP model with a possible role in TJ opening [133,134]. Increased permeability of the BBB was also observed in the striatum in rats after treatment with unilateral injections with 6-Hydroxydopamine (6OHDA) [135] that leads to downregulation of occludin [136]. Similar to amyloid oligomers in AD the Lewy body component α-synuclein was found to damage BBB. An in vitro study showed that preformed α-synuclein fibrils lead to a decrease in the expression of occludin and ZO-1, but not tricellulin or marveld3, in human brain microvascular endothelial cells [137].

4.6. *Huntington's Disease (HD)*

Huntington's disease (HD) is a dominantly inherited autosomal neurodegenerative disorder that is not usually considered aging related as it is typically diagnosed at the age of 40, and the onset of the disease can vary from earlier than ten to over eighty years of age. However, recent research indicates that epigenetic age acceleration in specific brain regions is associated with HD [162].

Recently, morphological changes of blood vessels and BBB leakage in the caudate and putamen were observed in HD patients using magnetic resonance imaging. In putamen samples of HD patients and in striatal samples of the HD model R6/2 mice, occludin and claudin 5 protein levels were decreased and evidence of increased BBB permeability was found [140]. More recently, these results were expanded in induced pluripotent stemm cell (iPSC) derived brain microvascular endothelial cells. While in this system no significant changes in TJ protein levels were observed, claudin 5 was found to mislocalize to the cytoplasm of cells derived from HD patients; however, trans endothelial electric resistance (TEER) and MDR1 function was significantly decreased compared to control cell lines. Some of the changes in cell lines derived from HD patients could be attributed to aberrant Wnt signaling [141].

4.7. *Schizophrenia*

Some aspects of schizophrenia raise the possibility that accelerated aging of the brain can be responsible for some of its symptoms [163]. Individuals with the chromosomal abnormality: 22q11 deletion syndrome (22q11DS) have an increased risk of developing schizophrenia, in fact it occurs in 30% of such individuals. Interestingly, 22q11DS are haploinsufficient for claudin 5. Recent experiments on AAV mediated claudin 5 suppression in mice and inducible claudin 5 knockdown uncovered focal BBB breakdown and behavioral changes. Anti-psychotic medications were found to dose dependently increase claudin 5 expression. Furthermore post mortem analysis of schizophrenic brains revealed aberrant, discontinuous expression of claudin 5 [142].

5. Closing Remarks

Aging comes with the deterioration of all bodily functions of which the loss of cognitive function is possibly the most serious in our heavily information dependent society. With the building evidence that dysfunction of the microvasculature is not just coincident but is part of the underlying mechanisms of aging and associated neurovascular and neurological disorders, new therapeutic possibilities are opened. The significant heterogeneity of BBB disruption data in studies using aging postmortem brain tissue suggests that more data is necessary to clearly understand the role of BBB disruption and to see whether it is a symptom or a cause [48]. It is not surprising that human BBB data before the mid-1990s is mainly regarding permeability; however, even since the discovery of occludin, zonula occludens proteins and claudins, BBB TJ data in the aging brain is scarce and this deficit is just partly made up for by data from aging associated disorders. Furthermore many studies disregard that claudin 5, occludin or ZO1 does not constitute the paracellular barrier by itself and all three should be studied along with permeability to different molecular size tracers at the same time to get a picture of barrier status Worsening the situation is that the reliability of existing TJ data has been questioned recently [35] Thus further comprehensive BBB TJ and permeability studies are needed in the field of aging and aging associated disorders. Another focus needs to be the study of both classical and novel intercellular communication pathways between brain capillary endothelium, pericytes, astrocytes, microglia and neurons as the concerted activity of all these cell types is necessary for proper neural function. We are just beginning to understand the depth and mechanisms of intercellular information exchange that makes the NVU a functional unit. The knowledge of how the interdependent functions of the cell types constituting the NVU are affected by the process of aging can lead to the alleviation of age related impairments and a better quality of life.

Author Contributions: All authors participated in researching and discussing literature and writing parts of the manuscripts. L.C., I.A.K. and A.E.F. wrote the paper with input from all authors. All authors approved the final version.

Abbreviations

BBB	blood–brain barrier
TJ	tight-junction
CNS	central nervous system
NVU	neurovascular unit
CEC	Cerebral endothelial cells
ZO-1	zonula occludens 1 protein
MCAO	middle cerebral artery occlusion
Aβ	amyloid-beta protein
Aβ1-42	amyloid-beta peptide containing amino acids 1-42
AD	Alzheimer's disease
MMP	matrix metalloproteinase
PD	Parkinson's disease
HD	Huntington's disease

References

1. Wilhelm, I.; Nyúl-Tóth, Á.; Kozma, M.; Farkas, A.E.; Krizbai, I.A. Role of pattern recognition receptors of the neurovascular unit in inflamm-aging. *Am. J. Physiol. Heart Circ. Physiol.* **2017**, *313*, H1000–H1012. [CrossRef] [PubMed]
2. Iadecola, C. The Neurovascular Unit Coming of Age: A Journey through Neurovascular Coupling in Health and Disease. *Neuron* **2017**, *96*, 17–42. [CrossRef] [PubMed]
3. Brightman, M.W.; Reese, T.S. Junctions between intimately apposed cell membranes in the vertebrate brain. *J. Cell Biol.* **1969**, *40*, 648–677. [CrossRef] [PubMed]
4. Oldendorf, W.H.; Cornford, M.E.; Brown, W.J. The large apparent work capability of the blood-brain barrier: A study of the mitochondrial content of capillary endothelial cells in brain and other tissues of the rat. *Ann. Neurol.* **1977**, *1*, 409–417. [CrossRef]
5. Nag, S. Morphology and molecular properties of cellular components of normal cerebral vessels. *Methods Mol. Med.* **2003**, *89*, 3–36. [CrossRef]
6. Hill, J.; Rom, S.; Ramirez, S.H.; Persidsky, Y. Emerging roles of pericytes in the regulation of the neurovascular unit in health and disease. *J. Neuroimmune Pharm.* **2014**, *9*, 591–605. [CrossRef]
7. Winkler, E.A.; Sengillo, J.D.; Bell, R.D.; Wang, J.; Zlokovic, B.V. Blood-spinal cord barrier pericyte reductions contribute to increased capillary permeability. *Br. J. Cereb. Blood Flow Metab.* **2012**, *32*, 1841–1852. [CrossRef]
8. Thanabalasundaram, G.; Schneidewind, J.; Pieper, C.; Galla, H.J. The impact of pericytes on the blood-brain barrier integrity depends critically on the pericyte differentiation stage. *Int. J. Biochem. Cell Biol.* **2011**, *43*, 1284–1293. [CrossRef]
9. Hellström, M.; Gerhardt, H.; Kalén, M.; Li, X.; Eriksson, U.; Wolburg, H.; Betsholtz, C. Lack of pericytes leads to endothelial hyperplasia and abnormal vascular morphogenesis. *J. Cell Biol.* **2001**, *153*, 543–553. [CrossRef]
10. Armulik, A.; Genové, G.; Mäe, M.; Nisancioglu, M.H.; Wallgard, E.; Niaudet, C.; He, L.; Norlin, J.; Lindblom, P.; Strittmatter, K.; et al. Pericytes regulate the blood-brain barrier. *Nature* **2010**, *468*, 557–561. [CrossRef]
11. Abbott, N.J.; Rönnbäck, L.; Hansson, E. Astrocyte-endothelial interactions at the blood-brain barrier. *Nat. Rev. Neurosci.* **2006**, *7*, 41–53. [CrossRef] [PubMed]
12. Delaney, C.; Campbell, M. The blood brain barrier: Insights from development and ageing. *Tissue Barriers* **2017**, *5*, e1373897. [CrossRef] [PubMed]
13. Verheggen, I.C.M.; Van Boxtel, M.P.J.; Verhey, F.R.J.; Jansen, J.F.A.; Backes, W.H. Interaction between blood-brain barrier and glymphatic system in solute clearance. *Neurosci. Biobehav. Rev.* **2018**, *90*, 26–33. [CrossRef] [PubMed]
14. Iliff, J.J.; Wang, M.; Liao, Y.; Plogg, B.A.; Peng, W.; Gundersen, G.A.; Benveniste, H.; Vates, G.E.; Deane, R.; Goldman, S.A.; et al. A paravascular pathway facilitates CSF flow through the brain parenchyma and the clearance of interstitial solutes, including amyloid beta. *Sci. Transl. Med.* **2012**, *4*, 147ra111. [CrossRef] [PubMed]
15. Smith, A.J.; Verkman, A.S. The "glymphatic" mechanism for solute clearance in Alzheimer's disease: Game changer or unproven speculation? *FASEB J.* **2018**, *32*, 543–551. [CrossRef] [PubMed]
16. Abbott, N.J.; Pizzo, M.E.; Preston, J.E.; Janigro, D.; Thorne, R.G. The role of brain barriers in fluid movement in the CNS: Is there a 'glymphatic' system? *Acta Neuropathol.* **2018**, *135*, 387–407. [CrossRef] [PubMed]
17. Zhang, Y.; Xu, K.; Liu, Y.; Erokwu, B.O.; Zhao, P.; Flask, C.A.; Ramos-Estebanez, C.; Farr, G.W.; LaManna, J.C.; Boron, W.F.; et al. Increased cerebral vascularization and decreased water exchange across the blood-brain barrier in aquaporin-4 knockout mice. *PLoS ONE* **2019**, *14*, e0218415. [CrossRef] [PubMed]
18. Kress, B.T.; Iliff, J.J.; Xia, M.; Wang, M.; Wei, H.S.; Zeppenfeld, D.; Xie, L.; Kang, H.; Xu, Q.; Liew, J.A.; et al. Impairment of paravascular clearance pathways in the aging brain. *Ann. Neurol.* **2014**, *76*, 845–861. [CrossRef]
19. Neuwelt, E.; Abbott, N.J.; Abrey, L.; Banks, W.A.; Blakley, B.; Davis, T.; Engelhardt, B.; Grammas, P.; Nedergaard, M.; Nutt, J.; et al. Strategies to advance translational research into brain barriers. *Lancet Neurol.* **2008**, *7*, 84–96. [CrossRef]
20. Bhowmick, S.; D'Mello, V.; Caruso, D.; Wallerstein, A.; Abdul-Muneer, P.M. Impairment of pericyte-endothelium crosstalk leads to blood-brain barrier dysfunction following traumatic brain injury. *Exp. Neurol.* **2019**, *317*, 260–270. [CrossRef]

21. Chen, J.; Luo, Y.; Hui, H.; Cai, T.; Huang, H.; Yang, F.; Feng, J.; Zhang, J.; Yan, X. CD146 coordinates brain endothelial cell-pericyte communication for blood-brain barrier development. *Proc. Natl. Acad. Sci. USA* **2017**, *114*, E7622–E7631. [CrossRef] [PubMed]
22. Zagrean, A.M.; Hermann, D.M.; Opris, I.; Zagrean, L.; Popa-Wagner, A. Multicellular Crosstalk Between Exosomes and the Neurovascular Unit After Cerebral Ischemia. Therapeutic Implications. *Front. Neurosci.* **2018**, *12*, 811. [CrossRef] [PubMed]
23. Ramirez, S.H.; Andrews, A.M.; Paul, D.; Pachter, J.S. Extracellular vesicles: Mediators and biomarkers of pathology along CNS barriers. *Fluids Barriers CNS* **2018**, *15*, 19. [CrossRef] [PubMed]
24. András, I.E.; Toborek, M. Extracellular vesicles of the blood-brain barrier. *Tissue Barriers* **2016**, *4*, e1131804. [CrossRef]
25. Saeedi, S.; Israel, S.; Nagy, C.; Turecki, G. The emerging role of exosomes in mental disorders. *Transl. Psychiatry* **2019**, *9*, 122. [CrossRef]
26. Yuan, X.; Wu, Q.; Wang, P.; Jing, Y.; Yao, H.; Tang, Y.; Li, Z.; Zhang, H.; Xiu, R. Exosomes Derived from Pericytes Improve Microcirculation and Protect Blood-Spinal Cord Barrier After Spinal Cord Injury in Mice. *Front. Mol. Neurosci.* **2019**, *13*, 319. [CrossRef]
27. Liu, C.; Ge, H.M.; Liu, B.H.; Dong, R.; Shan, K.; Chen, X.; Yao, M.D.; Li, X.M.; Yao, J.; Zhou, R.M.; et al. Targeting pericyte-endothelial cell crosstalk by circular RNA-cPWWP2A inhibition aggravates diabetes-induced microvascular dysfunction. *Proc. Natl. Acad. Sci. USA* **2019**, *116*, 7455–7464. [CrossRef]
28. Brown, W.R.; Thore, C.R. Review: Cerebral microvascular pathology in ageing and neurodegeneration. *Neuropathol. Appl. Neurobiol.* **2011**, *37*, 56–74. [CrossRef]
29. Erusalimsky, J.D. Vascular endothelial senescence: From mechanisms to pathophysiology. *J. Appl. Physiol.* **2009**, *106*, 326–332. [CrossRef]
30. Sonntag, W.E.; Lynch, C.D.; Cooney, P.T.; Hutchins, P.M. Decreases in cerebral microvasculature with age are associated with the decline in growth hormone and insulin-like growth factor 1. *Endocrinology* **1997**, *138*, 3515–3520. [CrossRef]
31. Fan, Y.; Yang, X.; Tao, Y.; Lan, L.; Zheng, L.; Sun, J. Tight junction disruption of blood-brain barrier in white matter lesions in chronic hypertensive rats. *Neuroreport* **2015**, *26*, 1039–1043. [CrossRef] [PubMed]
32. Desjardins, M.; Berti, R.; Lefebvre, J.; Dubeau, S.; Lesage, F. Aging-related differences in cerebral capillary blood flow in anesthetized rats. *Neurobiol. Aging* **2014**, *35*, 1947–1955. [CrossRef] [PubMed]
33. Goodall, E.F.; Wang, C.; Simpson, J.E.; Baker, D.J.; Drew, D.R.; Heath, P.R.; Saffrey, M.J.; Romero, I.A.; Wharton, S.B. Age-associated changes in the blood-brain barrier: Comparative studies in human and mouse. *Neuropathol. Appl. Neurobiol.* **2018**, *44*, 328–340. [CrossRef] [PubMed]
34. Nitta, T.; Hata, M.; Gotoh, S.; Seo, Y.; Sasaki, H.; Hashimoto, N.; Furuse, M.; Tsukita, S. Size-selective loosening of the blood-brain barrier in claudin-5–deficient mice. *J. Cell Biol.* **2003**, *161*, 653–660. [CrossRef] [PubMed]
35. Castro Dias, M.; Coisne, C.; Lazarevic, I.; Baden, P.; Hata, M.; Iwamoto, N.; Francisco, D.M.F.; Vanlandewijck, M.; He, L.; Baier, F.A.; et al. Claudin-3-deficient C57BL/6J mice display intact brain barriers. *Sci. Rep.* **2019**, *9*, 203. [CrossRef]
36. Berndt, P.; Winkler, L.; Cording, J.; Breitkreuz-Korff, O.; Rex, A.; Dithmer, S.; Rausch, V.; Blasig, R.; Richter, M.; Sporbert, A.; et al. Tight junction proteins at the blood-brain barrier: Far more than claudin-5. *Cell Mol. Life Sci.* **2019**, *76*, 1987–2002. [CrossRef]
37. Dörfel, M.J.; Westphal, J.K.; Bellmann, C.; Krug, S.M.; Cording, J.; Mittag, S.; Tauber, R.; Fromm, M.; Blasig, I.E.; Huber, O. CK2-dependent phosphorylation of occludin regulates the interaction with ZO-proteins and tight junction integrity. *Cell Commun. Signal.* **2013**, *11*, 40. [CrossRef]
38. Raleigh, D.R.; Boe, D.M.; Yu, D.; Weber, C.R.; Marchiando, A.M.; Bradford, E.M.; Wang, Y.; Wu, L.; Schneeberger, E.E.; Shen, L.; et al. Occludin S408 phosphorylation regulates tight junction protein interactions and barrier function. *J. Cell Biol.* **2011**, *193*, 565–582. [CrossRef]
39. Runkle, E.A.; Sundstrom, J.M.; Runkle, K.B.; Liu, X.; Antonetti, D.A. Occludin localizes to centrosomes and modifies mitotic entry. *J. Biol. Chem.* **2011**, *286*, 30847–30858. [CrossRef]
40. Furuse, M.; Hirase, T.; Itoh, M.; Nagafuchi, A.; Yonemura, S.; Tsukita, S. Occludin: A novel integral membrane protein localizing at tight junctions. *J. Cell Biol.* **1993**, *12 Pt 2*, 1777–1788. [CrossRef]
41. Furuse, M.; Itoh, M.; Hirase, T.; Nagafuchi, A.; Yonemura, S.; Tsukita, S. Direct association of occludin with ZO-1 and its possible involvement in the localization of occludin at tight junctions. *J. Cell Biol.* **1994**, *127 Pt 1*

1617–1626. [CrossRef]
42. Haskins, J.; Gu, L.; Wittchen, E.S.; Hibbard, J.; Stevenson, B.R. ZO-3, a novel member of the MAGUK protein family found at the tight junction, interacts with ZO-1 and occludin. *J. Cell Biol.* **1998**, *141*, 199–208. [CrossRef] [PubMed]
43. Itoh, M.; Furuse, M.; Morita, K.; Kubota, K.; Saitou, M.; Tsukita, S. Direct binding of three tight junction-associated MAGUKs, ZO-1, ZO-2, and ZO-3, with the COOH termini of claudins. *J. Cell Biol.* **1999**, *147*, 1351–1363. [CrossRef] [PubMed]
44. Castro, V.; Skowronska, M.; Lombardi, J.; He, J.; Seth, N.; Velichkovska, M.; Toborek, M. Occludin regulates glucose uptake and ATP production in pericytes by influencing AMP-activated protein kinase activity. *J. Cereb. Blood Flow Metab.* **2018**, *38*, 317–332. [CrossRef] [PubMed]
45. Bauer, H.; Stelzhammer, W.; Fuchs, R.; Weiger, T.M.; Danninger, C.; Probst, G.; Krizbai, I.A. Astrocytes and neurons express the tight junction-specific protein occludin in vitro. *Exp. Cell Res.* **1999**, *250*, 434–438. [CrossRef] [PubMed]
46. Morgan, S.V.; Garwood, C.J.; Jennings, L.; Simpson, J.E.; Castelli, L.M.; Heath, P.R.; Mihaylov, S.R.; Vaquéz-Villaseñor, I.; Minshull, T.C.; Ince, P.G.; et al. Proteomic and cellular localisation studies suggest non-tight junction cytoplasmic and nuclear roles for occludin in astrocytes. *Eur. J. Neurosci.* **2018**, *47*, 1444–1456. [CrossRef] [PubMed]
47. Horng, S.; Therattil, A.; Moyon, S.; Gordon, A.; Kim, K.; Argaw, A.T.; Hara, Y.; Mariani, J.N.; Sawai, S.; Flodby, P.; et al. Astrocytic tight junctions control inflammatory CNS lesion pathogenesis. *J. Clin. Investig.* **2017**, *127*, 3136–3151. [CrossRef] [PubMed]
48. Farrall, A.J.; Wardlaw, J.M. Blood-brain barrier: Ageing and microvascular disease-systematic review and meta-analysis. *Neurobiol. Aging* **2009**, *30*, 337–352. [CrossRef]
49. Baker, D.J.; Wijshake, T.; Tchkonia, T.; LeBrasseur, N.K.; Childs, B.G.; van de Sluis, B.; Kirkland, J.L.; van Deursen, J.M. Clearance of p16Ink4a-positive senescent cells delays ageing-associated disorders. *Nature* **2011**, *479*, 232–236. [CrossRef]
50. Stamatovic, S.M.; Martinez-Revollar, G.; Hu, A.; Choi, J.; Keep, R.F.; Andjelkovic, A.V. Decline in Sirtuin-1 expression and activity plays a critical role in blood-brain barrier permeability in aging. *Neurobiol. Dis.* **2019**, *126*, 105–116. [CrossRef]
51. Bake, S.; Sohrabji, F. 17beta-estradiol differentially regulates blood-brain barrier permeability in young and aging female rats. *Endocrinology* **2004**, *145*, 5471–5475. [CrossRef] [PubMed]
52. Bake, S.; Friedman, J.A.; Sohrabji, F. Reproductive age-related changes in the blood brain barrier: Expression of IgG and tight junction proteins. *Microvasc. Res.* **2009**, *78*, 413–424. [CrossRef] [PubMed]
53. Wilson, A.C.; Clemente, L.; Liu, T.; Bowen, R.L.; Meethal, S.V.; Atwood, C.S. Reproductive hormones regulate the selective permeability of the blood-brain barrier. *Biochim. Biophys. Acta (BBA) Mol. Basis Dis.* **2008**, *1782*, 401–407. [CrossRef] [PubMed]
54. Bell, R.D.; Winkler, E.A.; Sagare, A.P.; Singh, I.; LaRue, B.; Deane, R.; Zlokovic, B.V. Pericytes control key neurovascular functions and neuronal phenotype in the adult brain and during brain aging. *Neuron* **2010**, *68*, 409–427. [CrossRef] [PubMed]
55. Mooradian, A.D.; Haas, M.J.; Chehade, J.M. Age-related changes in rat cerebral occludin and zonula occludens-1 (ZO-1). *Mech. Ageing Dev.* **2003**, *124*, 143–146. [CrossRef]
56. Osada, T.; Gu, Y.H.; Kanazawa, M.; Tsubota, Y.; Hawkins, B.T.; Spatz, M.; Milner, R.; del Zoppo, G.J. Interendothelial claudin-5 expression depends on cerebral endothelial cell-matrix adhesion by β (1)-integrins. *J. Cereb. Blood Flow Metab.* **2011**, *31*, 1972–1985. [CrossRef] [PubMed]
57. Roberts, J.; de Hoog, L.; Bix, G.J. Mice deficient in endothelial α5 integrin are profoundly resistant to experimental ischemic stroke. *J. Cereb. Blood Flow Metab.* **2017**, *37*, 85–96. [CrossRef] [PubMed]
58. Montagne, A.; Barnes, S.R.; Sweeney, M.D.; Halliday, M.R.; Sagare, A.P.; Zhao, Z.; Toga, A.W.; Jacobs, R.E.; Liu, C.Y.; Amezcua, L.; et al. Blood-brain barrier breakdown in the aging human hippocampus. *Neuron* **2015**, *85*, 296–302. [CrossRef]
59. Yamazaki, Y.; Baker, D.J.; Tachibana, M.; Liu, C.C.; van Deursen, J.M.; Brott, T.G.; Bu, G.; Kanekiyo, T. Vascular Cell Senescence Contributes to Blood-Brain Barrier Breakdown. *Stroke* **2016**, *47*, 1068–1077. [CrossRef]
60. Schlingmann, B.; Overgaard, C.E.; Molina, S.A.; Lynn, K.S.; Mitchell, L.A.; Dorsainvil White, S.; Mattheyses, A.L.; Guidot, D.M.; Capaldo, C.T.; Koval, M. Regulation of claudin/zonula occludens-1

complexes by hetero-claudin interactions. *Nat. Commun.* **2016**, *7*, 12276. [CrossRef]
61. Lee, S.H.; Lee, J.H.; Lee, H.Y.; Min, K.J. Sirtuin signaling in cellular senescence and aging. *BMB Rep* **2019**, *52*, 24–34. [CrossRef] [PubMed]
62. Herskovits, A.Z.; Guarente, L. SIRT1 in neurodevelopment and brain senescence. *Neuron* **2014**, *81*, 471–483 [CrossRef] [PubMed]
63. Stirone, C.; Duckles, S.P.; Krause, D.N. Multiple forms of estrogen receptor-alpha in cerebral blood vessels Regulation by estrogen. *Am. J. Physiol. Endocrinol Metab.* **2003**, *284*, E184–E192. [CrossRef] [PubMed]
64. Stewart, P.A.; Magliocco, M.; Hayakawa, K.; Farrell, C.L.; Del Maestro, R.F.; Girvin, J.; Kaufmann, J.C. Vinters, H.V.; Gilbert, J. A quantitative analysis of blood-brain barrier ultrastructure in the aging human *Microvasc. Res.* **1987**, *33*, 270–282. [CrossRef]
65. Borghesan, M.; O'Loghlen, A. Integrins in senescence and aging. *Cell Cycle* **2017**, *16*, 909–910. [CrossRef]
66. Labat-Robert, J. Cell-matrix interactions, alteration with aging and age associated diseases. A review. *Pathol Biol.* **2001**, *49*, 349–352. [CrossRef]
67. Wu, X.; Reddy, D.S. Integrins as receptor targets for neurological disorders. *Pharmacol. Ther.* **2012**, *134*, 68–81 [CrossRef]
68. Stamatovic, S.M.; Sladojevic, N.; Keep, R.F.; Andjelkovic, A.V. Relocalization of junctional adhesion molecule A during inflammatory stimulation of brain endothelial cells. *Mol. Cells Biol.* **2012**, *32*, 3414–3427. [CrossRef]
69. Mochida, G.H.; Ganesh, V.S.; Felie, J.M.; Gleason, D.; Hill, R.S.; Clapham, K.R.; Rakiec, D.; Tan, W.H. Akawi, N.; Al-Saffar, M.; et al. A homozygous mutation in the tight-junction protein JAM3 causes hemorrhagic destruction of the brain, subependymal calcification, and congenital cataracts. *Am. J. Hum. Genet.* **2010**, *87*, 882–889. [CrossRef]
70. Saitou, M.; Furuse, M.; Sasaki, H.; Schulzke, J.D.; Fromm, M.; Takano, H.; Noda, T.; Tsukita, S. Complex phenotype of mice lacking occludin, a component of tight junction strands. *Mol. Biol. Cell* **2000**, *11*, 4131–4142 [CrossRef]
71. Gonzalez-Lozano, M.A.; Klemmer, P.; Gebuis, T.; Hassan, C.; van Nierop, P.; van Kesteren, R.E.; Smit, A.B. Li, K.W. Dynamics of the mouse brain cortical synaptic proteome during postnatal brain development. *Sci. Rep.* **2016**, *6*, 35456. [CrossRef] [PubMed]
72. Zussy, C.; Loustalot, F.; Junyent, F.; Gardoni, F.; Bories, C.; Valero, J.; Desarménien, M.G.; Bernex, F.; Henaff, D. Bayo-Puxan, N.; et al. Coxsackievirus Adenovirus Receptor Loss Impairs Adult Neurogenesis, Synapse Content, and Hippocampus Plasticity. *J. Neurosci.* **2016**, *36*, 9558–9571. [CrossRef] [PubMed]
73. Lochhead, J.J.; McCaffrey, G.; Quigley, C.E.; Finch, J.; DeMarco, K.M.; Nametz, N.; Davis, T.P. Oxidative stress increases blood-brain barrier permeability and induces alterations in occludin during hypoxia-reoxygenation *J. Cereb. Blood Flow Metab.* **2010**, *30*, 1625–1636. [CrossRef] [PubMed]
74. Raz, L.; Knoefel, J.; Bhaskar, K. The neuropathology and cerebrovascular mechanisms of dementia. *J. Cereb. Blood Flow Metab.* **2016**, *36*, 172–186. [CrossRef]
75. Sugawara, T.; Fujimura, M.; Noshita, N.; Kim, G.W.; Saito, A.; Hayashi, T.; Narasimhan, P.; Maier, C.M. Chan, P.H. Neuronal death/survival signaling pathways in cerebral ischemia. *NeuroRx* **2004**, *1*, 17–25 [CrossRef]
76. Roy-O'Reilly, M.; McCullough, L.D. Age and Sex Are Critical Factors in Ischemic Stroke Pathology *Endocrinology* **2018**, *159*, 3120–3131. [CrossRef]
77. Lutski, M.; Zucker, I.; Shohat, T.; Tanne, D. Characteristics and Outcomes of Young Patients with First-Ever Ischemic Stroke Compared to Older Patients: The National Acute Stroke ISraeli Registry. *Front. Neurol.* **2017**, *8*, 421. [CrossRef]
78. Bauer, A.T.; Bürgers, H.F.; Rabie, T.; Marti, H.H. Matrix metalloproteinase-9 mediates hypoxia-induced vascular leakage in the brain via tight junction rearrangement. *J. Cereb. Blood Flow Metab.* **2010**, *30*, 837–848 [CrossRef]
79. Mark, K.S.; Davis, T.P. Cerebral microvascular changes in permeability and tight junctions induced by hypoxia-reoxygenation. *Am. J. Physiol. Heart Circ. Physiol.* **2002**, *282*, H1485–H1494. [CrossRef]
80. Lischper, M.; Beuck, S.; Thanabalasundaram, G.; Pieper, C.; Galla, H.J. Metalloproteinase mediated occludin cleavage in the cerebral microcapillary endothelium under pathological conditions. *Brain Res.* **2010**, *1326*, 114–127. [CrossRef]
81. Krizbai, I.; Bauer, H.; Bresgen, N.; Eckl, P.; Farkas, A.; Szatmari, E.; Traweger, A.; Wejksza, K.; Bauer, H. Effect of oxidative stress on the junctional proteins of cultured cerebral endothelial cells. *Cell. Mol. Neurobiol.*

2005, *25*, 129–139. [CrossRef] [PubMed]

82. Koto, T.; Takubo, K.; Ishida, S.; Shinoda, H.; Inoue, M.; Tsubota, K.; Okada, Y.; Ikeda, E. Hypoxia disrupts the barrier function of neural blood vessels through changes in the expression of claudin-5 in endothelial cells. *Am. J. Pathol.* **2007**, *170*, 1389–1397. [CrossRef] [PubMed]
83. Schreibelt, G.; Kooij, G.; Reijerkerk, A.; van Doorn, R.; Gringhuis, S.I.; van der Pol, S.; Weksler, B.B.; Romero, I.A.; Couraud, P.O.; Piontek, J.; et al. Reactive oxygen species alter brain endothelial tight junction dynamics via RhoA, PI3 kinase, and PKB signaling. *Faseb. J.* **2007**, *21*, 3666–3676. [CrossRef] [PubMed]
84. Shen, F.; Jiang, L.; Han, F.; Degos, V.; Chen, S.; Su, H. Increased Inflammatory Response in Old Mice is Associated with More Severe Neuronal Injury at the Acute Stage of Ischemic Stroke. *Aging Dis.* **2019**, *10*, 12–22. [CrossRef] [PubMed]
85. Witt, K.A.; Mark, K.S.; Hom, S.; Davis, T.P. Effects of hypoxia-reoxygenation on rat blood-brain barrier permeability and tight junctional protein expression. *Am. J. Physiol. Heart Circ. Physiol.* **2003**, *285*, H2820–H2831. [CrossRef] [PubMed]
86. Willis, C.L.; Meske, D.S.; Davis, T.P. Protein kinase C activation modulates reversible increase in cortical blood-brain barrier permeability and tight junction protein expression during hypoxia and posthypoxic reoxygenation. *J. Cereb. Blood Flow Metab.* **2010**, *30*, 1847–1859. [CrossRef]
87. Knowland, D.; Arac, A.; Sekiguchi, K.J.; Hsu, M.; Lutz, S.E.; Perrino, J.; Steinberg, G.K.; Barres, B.A.; Nimmerjahn, A.; Agalliu, D. Stepwise recruitment of transcellular and paracellular pathways underlies blood-brain barrier breakdown in stroke. *Neuron* **2014**, *82*, 603–617. [CrossRef]
88. Kaur, J.; Tuor, U.I.; Zhao, Z.; Barber, P.A. Quantitative MRI reveals the elderly ischemic brain is susceptible to increased early blood-brain barrier permeability following tissue plasminogen activator related to claudin 5 and occludin disassembly. *J. Cereb. Blood Flow Metab.* **2011**, *31*, 1874–1885. [CrossRef]
89. Bai, Y.; Xu, G.; Xu, M.; Li, Q.; Qin, X. Inhibition of Src phosphorylation reduces damage to the blood-brain barrier following transient focal cerebral ischemia in rats. *Int. J. Mol. Med.* **2014**, *34*, 1473–1482. [CrossRef]
90. Kago, T.; Takagi, N.; Date, I.; Takenaga, Y.; Takagi, K.; Takeo, S. Cerebral ischemia enhances tyrosine phosphorylation of occludin in brain capillaries. *Biochem. Biophys. Res. Commun.* **2006**, *339*, 1197–1203. [CrossRef]
91. Takenaga, Y.; Takagi, N.; Murotomi, K.; Tanonaka, K.; Takeo, S. Inhibition of Src activity decreases tyrosine phosphorylation of occludin in brain capillaries and attenuates increase in permeability of the blood-brain barrier after transient focal cerebral ischemia. *J. Cereb. Blood Flow Metab.* **2009**, *29*, 1099–1108. [CrossRef] [PubMed]
92. Yang, Y.; Estrada, E.Y.; Thompson, J.F.; Liu, W.; Rosenberg, G.A. Matrix metalloproteinase-mediated disruption of tight junction proteins in cerebral vessels is reversed by synthetic matrix metalloproteinase inhibitor in focal ischemia in rat. *J. Cereb. Blood Flow Metab.* **2007**, *27*, 697–709. [CrossRef] [PubMed]
93. Yang, Y.; Rosenberg, G.A. MMP-mediated disruption of claudin-5 in the blood-brain barrier of rat brain after cerebral ischemia. *Methods Mol. Biol.* **2011**, *762*, 333–345. [CrossRef]
94. Yang, Y.; Thompson, J.F.; Taheri, S.; Salayandia, V.M.; McAvoy, T.A.; Hill, J.W.; Estrada, E.Y.; Rosenberg, G.A. Early inhibition of MMP activity in ischemic rat brain promotes expression of tight junction proteins and angiogenesis during recovery. *J. Cereb. Blood Flow Metab.* **2013**, *33*, 1104–1114. [CrossRef] [PubMed]
95. Rom, S.; Zuluaga-Ramirez, V.; Dykstra, H.; Reichenbach, N.L.; Ramirez, S.H.; Persidsky, Y. Poly(ADP-ribose) polymerase-1 inhibition in brain endothelium protects the blood-brain barrier under physiologic and neuroinflammatory conditions. *J. Cereb. Blood Flow Metab.* **2015**, *35*, 28–36. [CrossRef] [PubMed]
96. Yang, Y.; Salayandia, V.M.; Thompson, J.F.; Yang, L.Y.; Estrada, E.Y. Attenuation of acute stroke injury in rat brain by minocycline promotes blood-brain barrier remodeling and alternative microglia/macrophage activation during recovery. *J. Neuroinflammation* **2015**, *12*, 26. [CrossRef] [PubMed]
97. Weinl, C.; Castaneda Vega, S.; Riehle, H.; Stritt, C.; Calaminus, C.; Wolburg, H.; Mauel, S.; Breithaupt, A.; Gruber, A.D.; Wasylyk, B.; et al. Endothelial depletion of murine SRF/MRTF provokes intracerebral hemorrhagic stroke. *Proc. Natl. Acad. Sci. USA* **2015**, *112*, 9914–9919. [CrossRef]
98. Wang, W.; Dentler, W.L.; Borchardt, R.T. VEGF increases BMEC monolayer permeability by affecting occludin expression and tight junction assembly. *Am. J. Physiol. Heart Circ. Physiol.* **2001**, *280*, H434–H440. [CrossRef]
99. Argaw, A.T.; Asp, L.; Zhang, J.; Navrazhina, K.; Pham, T.; Mariani, J.N.; Mahase, S.; Dutta, D.J.; Seto, J.; Kramer, E.G.; et al. Astrocyte-derived VEGF-A drives blood-brain barrier disruption in CNS inflammatory disease. *J. Clin. Investig.* **2012** *122*, 2454–2468. [CrossRef]

100. Elahy, M.; Jackaman, C.; Mamo, J.C.; Lam, V.; Dhaliwal, S.S.; Giles, C.; Nelson, D.; Takechi, R. Blood-brain barrier dysfunction developed during normal aging is associated with inflammation and loss of tight junctions but not with leukocyte recruitment. *Immun. Ageing* **2015**, *12*, 2. [CrossRef]
101. Banks, W.A.; Farr, S.A.; Morley, J.E. Permeability of the blood-brain barrier to albumin and insulin in the young and aged SAMP8 mouse. *J. Gerontol. A Biol. Sci. Med. Sci.* **2000**, *55*, B601–B606. [CrossRef] [PubMed]
102. Engelhardt, S.; Patkar, S.; Ogunshola, O.O. Cell-specific blood-brain barrier regulation in health and disease: A focus on hypoxia. *Br. J. Pharm.* **2014**, *171*, 1210–1230. [CrossRef] [PubMed]
103. Liu, W.; Sood, R.; Chen, Q.; Sakoglu, U.; Hendren, J.; Cetin, O.; Miyake, M.; Liu, K.J. Normobaric hyperoxia inhibits NADPH oxidase-mediated matrix metalloproteinase-9 induction in cerebral microvessels in experimental stroke. *J. Neurochem.* **2008**, *107*, 1196–1205. [CrossRef] [PubMed]
104. Blasig, I.E.; Bellmann, C.; Cording, J.; del Vecchio, G.; Zwanziger, D.; Huber, O.; Haseloff, R.F. Occludin Protein Family: Oxidative Stress and Reducing Conditions. *Antioxid. Redox Signal.* **2011**, *15*, 1195–1219. [CrossRef] [PubMed]
105. Bellmann, C.; Schreivogel, S.; Günther, R.; Dabrowski, S.; Schümann, M.; Wolburg, H.; Blasig, I.E. Highly conserved cysteines are involved in the oligomerization of occludin-redox dependency of the second extracellular loop. *Antioxid. Redox Signal.* **2014**, *20*, 855–867. [CrossRef] [PubMed]
106. Erickson, M.A.; Banks, W.A. Neuroimmune Axes of the Blood-Brain Barriers and Blood-Brain Interfaces: Bases for Physiological Regulation, Disease States, and Pharmacological Interventions. *Pharm. Rev.* **2018**, *70*, 278–314. [CrossRef] [PubMed]
107. Erickson, M.A.; Banks, W.A. Age-Associated Changes in the Immune System and Blood-Brain Barrier Functions. *Int. J. Mol. Sci.* **2019**, *20*, 1632. [CrossRef]
108. Ransohoff, R.M.; Engelhardt, B. The anatomical and cellular basis of immune surveillance in the central nervous system. *Nat. Rev. Immunol* **2012**, *12*, 623–635. [CrossRef]
109. Ousman, S.S.; Kubes, P. Immune surveillance in the central nervous system. *Nat. Neurosci.* **2012**, *15*, 1096–1101. [CrossRef]
110. Varatharaj, A.; Galea, I. The blood-brain barrier in systemic inflammation. *Brain Behav. Immun.* **2017**, *60*, 1–12. [CrossRef]
111. Gemechu, J.M.; Bentivoglio, M. T Cell Recruitment in the Brain during Normal Aging. *Front. Cell. Neurosci.* **2012**, *6*, 38. [CrossRef] [PubMed]
112. Xu, Y.Z.; Nygård, M.; Kristensson, K.; Bentivoglio, M. Regulation of cytokine signaling and T-cell recruitment in the aging mouse brain in response to central inflammatory challenge. *Brain Behav. Immun.* **2010**, *24*, 138–152. [CrossRef] [PubMed]
113. Stichel, C.C.; Luebbert, H. Inflammatory processes in the aging mouse brain: Participation of dendritic cells and T-cells. *Neurobiol. Aging* **2007**, *28*, 1507–1521. [CrossRef] [PubMed]
114. Ritzel, R.M.; Crapser, J.; Patel, A.R.; Verma, R.; Grenier, J.M.; Chauhan, A.; Jellison, E.R.; McCullough, L.D. Age-Associated Resident Memory CD8 T Cells in the Central Nervous System Are Primed to Potentiate Inflammation after Ischemic Brain Injury. *J. Immunol* **2016**, *196*, 3318–3330. [CrossRef]
115. Miguel-Hidalgo, J.J.; Nithuairisg, S.; Stockmeier, C.; Rajkowska, G. Distribution of ICAM-1 immunoreactivity during aging in the human orbitofrontal cortex. *Brain Behav. Immun.* **2007**, *21*, 100–111. [CrossRef]
116. Dulken, B.W.; Buckley, M.T.; Navarro Negredo, P.; Saligrama, N.; Cayrol, R.; Leeman, D.S.; George, B.M.; Boutet, S.C.; Hebestreit, K.; Pluvinage, J.V.; et al. Single-cell analysis reveals T cell infiltration in old neurogenic niches. *Nature* **2019**, *571*, 205–210. [CrossRef]
117. Suidan, G.L.; Mcdole, J.R.; Chen, Y.; Pirko, I.; Johnson, A.J. Induction of blood brain barrier tight junction protein alterations by CD8 T cells. *PLoS ONE* **2008**, *3*, e3037. [CrossRef]
118. Erickson, M.A.; Banks, W.A. Blood-brain barrier dysfunction as a cause and consequence of Alzheimer's disease. *J. Cereb. Blood Flow Metab.* **2013**, *33*, 1500–1513. [CrossRef]
119. Yamazaki, Y.; Kanekiyo, T. Blood-Brain Barrier Dysfunction and the Pathogenesis of Alzheimer's Disease. *Int. J. Mol. Sci.* **2017**, *18*, 1965. [CrossRef]
120. Marco, S.; Skaper, S.D. Amyloid beta-peptide1–42 alters tight junction protein distribution and expression in brain microvessel endothelial cells. *Neurosci. Lett.* **2006**, *401*, 219–224. [CrossRef]
121. Janota, C.; Lemere, C.A.; Brito, M.A. Dissecting the Contribution of Vascular Alterations and Aging to Alzheimer's Disease. *Mol. Neurobiol.* **2016**, *53*, 3793–3811. [CrossRef] [PubMed]

122. Wan, W.; Chen, H.; Li, Y. The potential mechanisms of Aβ-receptor for advanced glycation end-products interaction disrupting tight junctions of the blood-brain barrier in Alzheimer's disease. *Int. J. Neurosci.* **2014**, *124*, 75–81. [CrossRef] [PubMed]
123. Rosenberg, G.A. Matrix metalloproteinases and their multiple roles in neurodegenerative diseases. *Lancet Neurol.* **2009**, *8*, 205–216. [CrossRef]
124. Bourasset, F.; Ouellet, M.; Tremblay, C.; Julien, C.; Do, T.M.; Oddo, S.; LaFerla, F.; Calon, F. Reduction of the cerebrovascular volume in a transgenic mouse model of Alzheimer's disease. *Neuropharmacology* **2009**, *56*, 808–813. [CrossRef] [PubMed]
125. Carrano, A.; Hoozemans, J.J.; van der Vies, S.M.; van Horssen, J.; de Vries, H.E.; Rozemuller, A.J. Neuroinflammation and blood-brain barrier changes in capillary amyloid angiopathy. *Neurodegener. Dis.* **2012**, *10*, 329–331. [CrossRef] [PubMed]
126. Carrano, A.; Hoozemans, J.J.; van der Vies, S.M.; Rozemuller, A.J.; van Horssen, J.; de Vries, H.E. Amyloid Beta induces oxidative stress-mediated blood-brain barrier changes in capillary amyloid angiopathy. *Antioxid. Redox Signal.* **2011**, *15*, 1167–1178. [CrossRef]
127. Bruban, J.; Glotin, A.L.; Dinet, V.; Chalour, N.; Sennlaub, F.; Jonet, L.; An, N.; Faussat, A.M.; Mascarelli, F. Amyloid-beta (1–42) alters structure and function of retinal pigmented epithelial cells. *Aging Cell* **2009**, *8*, 162–177. [CrossRef]
128. Kook, S.Y.; Hong, H.S.; Moon, M.; Ha, C.M.; Chang, S.; Mook-Jung, I. $A\beta_{1\text{-}42}$-RAGE interaction disrupts tight junctions of the blood-brain barrier via Ca^{2+}-calcineurin signaling. *J. Neurosci.* **2012**, *32*, 8845–8854. [CrossRef]
129. Wan, W.; Cao, L.; Liu, L.; Zhang, C.; Kalionis, B.; Tai, X.; Li, Y.; Xia, S. Aβ (1–42) oligomer-induced leakage in an in vitro blood-brain barrier model is associated with up-regulation of RAGE and metalloproteinases, and down-regulation of tight junction scaffold proteins. *J. Neurochem.* **2015**, *134*, 382–393. [CrossRef]
130. Wan, W.B.; Cao, L.; Liu, L.M.; Kalionis, B.; Chen, C.; Tai, X.T.; Li, Y.M.; Xia, S.J. EGb761 provides a protective effect against Aβ1-42 oligomer-induced cell damage and blood-brain barrier disruption in an in vitro bEnd.3 endothelial model. *PLoS ONE* **2014**, *9*, e113126. [CrossRef]
131. Park, J.C.; Baik, S.H.; Han, S.H.; Cho, H.J.; Choi, H.; Kim, H.J.; Lee, W.; Kim, D.K.; Mook-Jung, I. Annexin A1 restores Aβ1-42-induced blood-brain barrier disruption through the inhibition of RhoA-ROCK signaling pathway. *Aging Cell* **2017**, *16*, 149–161. [CrossRef] [PubMed]
132. Mehrabadi, A.R.; Korolainen, M.A.; Odero, G.; Miller, D.W.; Kauppinen, T.M. Poly(ADP-ribose) polymerase-1 regulates microglia mediated decrease of endothelial tight junction integrity. *Neurochem. Int.* **2017**, *108*, 266–271. [CrossRef] [PubMed]
133. Chen, X.; Lan, X.; Roche, I.; Liu, R.; Geiger, J.D. Caffeine protects against MPTP-induced blood-brain barrier dysfunction in mouse striatum. *J. Neurochem.* **2008**, *107*, 1147–1157. [CrossRef] [PubMed]
134. Sarkar, S.; Chigurupati, S.; Raymick, J.; Mann, D.; Bowyer, J.F.; Schmitt, T.; Beger, R.D.; Hanig, J.P.; Schmued, L.C.; Paule, M.G. Neuroprotective effect of the chemical chaperone, trehalose in a chronic MPTP-induced Parkinson's disease mouse model. *Neurotoxicology* **2014**, *44*, 250–262. [CrossRef] [PubMed]
135. Carvey, P.M.; Zhao, C.H.; Hendey, B.; Lum, H.; Trachtenberg, J.; Desai, B.S.; Snyder, J.; Zhu, Y.G.; Ling, Z.D. 6-Hydroxydopamine-induced alterations in blood-brain barrier permeability. *Eur. J. Neurosci.* **2005**, *22*, 1158–1168. [CrossRef]
136. Huang, L.; Deng, M.; He, Y.; Lu, S.; Ma, R.; Fang, Y. β-asarone and levodopa co-administration increase striatal dopamine level in 6-hydroxydopamine induced rats by modulating P-glycoprotein and tight junction proteins at the blood-brain barrier and promoting levodopa into the brain. *Clin. Exp. Pharm. Physiol.* **2016**, *43*, 634–643. [CrossRef]
137. Kuan, W.L.; Bennett, N.; He, X.; Skepper, J.N.; Martynyuk, N.; Wijeyekoon, R.; Moghe, P.V.; Williams-Gray, C.H.; Barker, R.A. α-Synuclein pre-formed fibrils impair tight junction protein expression without affecting cerebral endothelial cell function. *Exp. Neurol.* **2016**, *285 Pt A*, 72–81. [CrossRef]
138. Pienaar, I.S.; Lee, C.H.; Elson, J.L.; McGuinness, L.; Gentleman, S.M.; Kalaria, R.N.; Dexter, D.T. Deep-brain stimulation associates with improved microvascular integrity in the subthalamic nucleus in Parkinson's disease. *Neurobiol. Dis.* **2015**, *74*, 392–405. [CrossRef]
139. Gray, M.T.; Woulfe, J.M. Striatal blood-brain barrier permeability in Parkinson's disease. *J. Cereb. Blood Flow Metab.* **2015**, *35*, 747–750. [CrossRef]
140. Drouin-Ouellet, J.; Sawiak, S.J.; Cisbani, G.; Lagacé, M.; Kuan, W.L.; Saint-Pierre, M.; Dury, R.J.; Alata, W.;

St-Amour, I.; Mason, S.L.; et al. Cerebrovascular and blood-brain barrier impairments in Huntington's disease: Potential implications for its pathophysiology. *Ann. Neurol.* **2015**, *78*, 160–177. [CrossRef]
141. Lim, R.G.; Quan, C.; Reyes-Ortiz, A.M.; Lutz, S.E.; Kedaigle, A.J.; Gipson, T.A.; Wu, J.; Vatine, G.D.; Stocksdale, J.; Casale, M.S.; et al. Huntington's Disease iPSC-Derived Brain Microvascular Endothelial Cells Reveal WNT-Mediated Angiogenic and Blood-Brain Barrier Deficits. *Cell Rep.* **2017**, *19*, 1365–1377. [CrossRef] [PubMed]
142. Greene, C.; Kealy, J.; Humphries, M.M.; Gong, Y.; Hou, J.; Hudson, N.; Cassidy, L.M.; Martiniano, R.; Shashi, V.; Hooper, S.R.; et al. Dose-dependent expression of claudin-5 is a modifying factor in schizophrenia. *Mol. Psychiatry* **2018**, *23*, 2156–2166. [CrossRef] [PubMed]
143. Takechi, R.; Galloway, S.; Pallebage-Gamarallage, M.M.; Mamo, J.C. Chylomicron amyloid-beta in the aetiology of Alzheimer's disease. *Atheroscler. Suppl.* **2008**, *9*, 19–25. [CrossRef]
144. Poduslo, J.F.; Curran, G.L.; Wengenack, T.M.; Malester, B.; Duff, K. Permeability of proteins at the blood-brain barrier in the normal adult mouse and double transgenic mouse model of Alzheimer's disease. *Neurobiol. Dis.* **2001**, *8*, 555–567. [CrossRef] [PubMed]
145. Cheng, Z.; Zhang, J.; Liu, H.; Li, Y.; Zhao, Y.; Yang, E. Central nervous system penetration for small molecule therapeutic agents does not increase in multiple sclerosis- and Alzheimer's disease-related animal models despite reported blood-brain barrier disruption. *Drug Metab. Dispos.* **2010**, *38*, 1355–1361. [CrossRef] [PubMed]
146. Mehta, D.C.; Short, J.L.; Nicolazzo, J.A. Altered brain uptake of therapeutics in a triple transgenic mouse model of Alzheimer's disease. *Pharm. Res.* **2013**, *30*, 2868–2879. [CrossRef] [PubMed]
147. Daneman, R.; Zhou, L.; Kebede, A.A.; Barres, B.A. Pericytes are required for blood-brain barrier integrity during embryogenesis. *Nature* **2010**, *468*, 562–566. [CrossRef]
148. Sagare, A.P.; Bell, R.D.; Zhao, Z.; Ma, Q.; Winkler, E.A.; Ramanathan, A.; Zlokovic, B.V. Pericyte loss influences Alzheimer-like neurodegeneration in mice. *Nat. Commun.* **2013**, *4*, 2932. [CrossRef]
149. Sengillo, J.D.; Winkler, E.A.; Walker, C.T.; Sullivan, J.S.; Johnson, M.; Zlokovic, B.V. Deficiency in mural vascular cells coincides with blood-brain barrier disruption in Alzheimer's disease. *Brain Pathol.* **2013**, *23*, 303–310. [CrossRef]
150. Tai, L.M.; Thomas, R.; Marottoli, F.M.; Koster, K.P.; Kanekiyo, T.; Morris, A.W.; Bu, G. The role of APOE in cerebrovascular dysfunction. *Acta Neuropathol.* **2016**, *131*, 709–723. [CrossRef]
151. Brandon, J.A.; Farmer, B.C.; Williams, H.C.; Johnson, L.A. Apoe and Alzheimer's Disease: Neuroimaging of Metabolic and Cerebrovascular Dysfunction. *Front. Aging Neurosci.* **2018**, *10*, 180. [CrossRef] [PubMed]
152. Bell, R.D.; Winkler, E.A.; Singh, I.; Sagare, A.P.; Deane, R.; Wu, Z.; Holtzman, D.M.; Betsholtz, C.; Armulik, A.; Sallstrom, J.; et al. Apolipoprotein E controls cerebrovascular integrity via cyclophilin A. *Nature* **2012**, *485*, 512–516. [CrossRef] [PubMed]
153. Daneman, R.; Prat, A. The blood-brain barrier. *Cold Spring Harb. Perspect. Biol.* **2015**, *7*, a020412. [CrossRef] [PubMed]
154. Zenaro, E.; Piacentino, G.; Constantin, G. The blood-brain barrier in Alzheimer's disease. *Neurobiol. Dis.* **2017**, *107*, 41–56. [CrossRef]
155. Sweeney, M.D.; Sagare, A.P.; Zlokovic, B.V. Blood-brain barrier breakdown in Alzheimer disease and other neurodegenerative disorders. *Nat. Rev. Neurol.* **2018**, *14*, 133–150. [CrossRef]
156. Greene, C.; Hanley, N.; Campbell, M. Claudin-5: Gatekeeper of neurological function. *Fluids Barriers CNS* **2019**, *16*, 3. [CrossRef]
157. Lee, H.; Pienaar, I.S. Disruption of the blood-brain barrier in Parkinson's disease: Curse or route to a cure? *Front. Biosci. (Landmark Ed.)* **2014**, *19*, 272–280. [CrossRef]
158. Pan, Y.; Nicolazzo, J.A. Impact of aging, Alzheimer's disease and Parkinson's disease on the blood-brain barrier transport of therapeutics. *Adv. Drug Deliv. Rev.* **2018**, *135*, 62–74. [CrossRef]
159. Farkas, E.; De Jong, G.I.; de Vos, R.A.; Jansen Steur, E.N.; Luiten, P.G. Pathological features of cerebral cortical capillaries are doubled in Alzheimer's disease and Parkinson's disease. *Acta Neuropathol.* **2000**, *100*, 395–402. [CrossRef]
160. Guan, J.; Pavlovic, D.; Dalkie, N.; Waldvogel, H.J.; O'Carroll, S.J.; Green, C.R.; Nicholson, L.F. Vascular degeneration in Parkinson's disease. *Brain Pathol.* **2013**, *23*, 154–164. [CrossRef]
161. Hill, K.K.; Campbell, M.C.; McNeely, M.E.; Karimi, M.; Ushe, M.; Tabbal, S.D.; Hershey, T.; Flores, H.P.; Hartlein, J.M.; Lugar, H.M.; et al. Cerebral blood flow responses to dorsal and ventral STN DBS correlate with

gait and balance responses in Parkinson's disease. *Exp. Neurol.* **2013**, *241*, 105–112. [CrossRef] [PubMed]
162. Horvath, S.; Langfelder, P.; Kwak, S.; Aaronson, J.; Rosinski, J.; Vogt, T.F.; Eszes, M.; Faull, R.L.; Curtis, M.A.; Waldvogel, H.J.; et al. Huntington's disease accelerates epigenetic aging of human brain and disrupts DNA methylation levels. *Aging (Albany Ny)* **2016**, *8*, 1485–1512. [CrossRef] [PubMed]
163. Okusaga, O.O. Accelerated aging in schizophrenia patients: The potential role of oxidative stress. *Aging Dis.* **2014**, *5*, 256–262. [CrossRef] [PubMed]

Permissions

All chapters in this book were first published by MDPI; hereby published with permission under the Creative Commo Attribution License or equivalent. Every chapter published in this book has been scrutinized by our experts. The significance has been extensively debated. The topics covered herein carry significant findings which will fuel the grow of the discipline. They may even be implemented as practical applications or may be referred to as a beginning point f another development.

The contributors of this book come from diverse backgrounds, making this book a truly international effort. This book w bring forth new frontiers with its revolutionizing research information and detailed analysis of the nascent developmen around the world.

We would like to thank all the contributing authors for lending their expertise to make the book truly unique. They ha played a crucial role in the development of this book. Without their invaluable contributions this book wouldn't have bee possible. They have made vital efforts to compile up to date information on the varied aspects of this subject to make th book a valuable addition to the collection of many professionals and students.

This book was conceptualized with the vision of imparting up-to-date information and advanced data in this field. T ensure the same, a matchless editorial board was set up. Every individual on the board went through rigorous rounds assessment to prove their worth. After which they invested a large part of their time researching and compiling the mo relevant data for our readers.

The editorial board has been involved in producing this book since its inception. They have spent rigorous hours researchir and exploring the diverse topics which have resulted in the successful publishing of this book. They have passed on the knowledge of decades through this book. To expedite this challenging task, the publisher supported the team at every ste A small team of assistant editors was also appointed to further simplify the editing procedure and attain best results f the readers.

Apart from the editorial board, the designing team has also invested a significant amount of their time in understandir the subject and creating the most relevant covers. They scrutinized every image to scout for the most suitable representatic of the subject and create an appropriate cover for the book.

The publishing team has been an ardent support to the editorial, designing and production team. Their endless efforts recruit the best for this project, has resulted in the accomplishment of this book. They are a veteran in the field of academi and their pool of knowledge is as vast as their experience in printing. Their expertise and guidance has proved useful every step. Their uncompromising quality standards have made this book an exceptional effort. Their encouragement fro time to time has been an inspiration for everyone.

The publisher and the editorial board hope that this book will prove to be a valuable piece of knowledge for researcher students, practitioners and scholars across the globe.

.ist of Contributors

ındhini Rajagopal, Flaviyan Jerome Irudayanathan and ikha Nangia
partment of Biomedical and Chemical Engineering, racuse University, Syracuse, NY 13244, USA

ına Leiz
partment of Nephrology and Intensive Care Medicine, arité-Universitätsmedizin Berlin, 12203 Berlin, Germany
lecular and Translational Kidney Research, Max-lbrück-Center for Molecular Medicine in the Helmholtz sociation (MDC), 13125 Berlin, Germany

i M. Schmidt-Ott
partment of Nephrology and Intensive Care Medicine, arité-Universitätsmedizin Berlin, 12203 Berlin, Germany
lecular and Translational Kidney Research, Max-lbrück-Center for Molecular Medicine in the Helmholtz sociation (MDC), 13125 Berlin, Germany
rlin Institute of Health (BIH), 10178 Berlin, Germany

ınnes Gonschior and Martin Lehmann
ibniz-Forschungsinstitut für Molekulare Pharmakologie MP), 13125 Berlin, Germany

lker Haucke
ibniz-Forschungsinstitut für Molekulare Pharmakologie MP), 13125 Berlin, Germany
culty of Biology, Chemistry, Pharmacy, Freie Universität rlin, 14195 Berlin, Germany

lein Plain, Wanling Pan, Deborah O'Neill, Megan Ure d Emmanuelle Cordat
partment of Physiology, The University of Alberta, monton, AB T6J 2R7, Canada

egan R. Beggs
partment of Physiology, The University of Alberta, monton, AB T6J 2R7, Canada
e Women's & Children's Health Research Institute, 11405- Avenue, Edmonton, AB T6G 1C9 Canada

aikel Farhan
e Women's & Children's Health Research Institute, 11405- Avenue, Edmonton, AB T6G 1C9 Canada
partment of Pediatrics, The University of Alberta, monton, AB T6J 2R7, Canada

Todd Alexander
partment of Physiology, The University of Alberta, monton, AB T6J 2R7, Canada
e Women's & Children's Health Research Institute, 11405- Avenue, Edmonton, AB T6G 1C9 Canada
partment of Pediatrics, The University of Alberta, monton, AB T6J 2R7, Canada

Henrik Dimke
Department of Cardiovascular and Renal Research, Institute of Molecular Medicine, University of Southern Denmark, 5230 Odense, Denmark
Department of Nephrology, Odense University Hospital, 5000 Odense, Denmark

Rocio Tapia and Sarah E. Kralicek
Department of Medicine, Division of Gastroenterology and Nutrition, Loyola University Chicago, Maywood, IL 60153, USA

Gail A. Hecht
Department of Medicine, Division of Gastroenterology and Nutrition, Loyola University Chicago, Maywood, IL 60153, USA
Department of Microbiology and Immunology, Loyola University Chicago, Maywood, IL 60153, USA
Edward Hines Jr. VA Hospital, Hines, IL 60141, USA

Thomas J. Lux, Xiawei Hu, Adel Ben-Kraiem, Jeremy Tsung-Chieh Chen and Heike L. Rittner
Department of Anaesthesiology, University Hospital of Wuerzburg, 97074 Wuerzburg, Germany

Robert Blum
Institute of Clinical Neurobiology, University Hospital of Wuerzburg, 94074 Wuerzburg, Germany

Carlos Ayala-Torres, Susanne M. Krug, Jörg D. Schulzke, Rita Rosenthal and Michael Fromm
Institute of Clinical Physiology/Nutritional Medicine, Medical Department, Division of Gastroenterology, Infectiology and Rheumatology, Charité – Universitätsmedizin Berlin, 12203 Berlin, Germany

Murat Seker and Dominik Müller
Department of Pediatric Gastroenterology, Nephrology and Metabolism, Charité – Universitätsmedizin Berlin, Charité, 13353 Berlin, Germany

Cármen Fernández-Rodríguez and Luis Alfonso Martínez-Cruz
CIC BioGUNE, Bizkaia Science and Technology Park, 801A, 48160 Derio, Spain

Matthew G. Brewer, Takeshi Yoshida, Fiona I. Kuo, Sade Fridy and Lisa A. Beck
Department of Dermatology, University of Rochester Medical Center, Rochester, NY 14642, USA

Anna De Benedetto
Department of Dermatology, University of Rochester Medical Center, Rochester, NY 14642, USA
Department of Dermatology, College of Medicine University of Florida, Gainesville, FL 32610, USA

Kana Marunaka, Mao Kobayashi, Shokoku Shu and Akira Ikari
Laboratory of Biochemistry, Department of Biopharmaceutical Sciences, Gifu Pharmaceutical University, Gifu 501-1196, Japan

Toshiyuki Matsunaga
Education Center of Green Pharmaceutical Sciences, Gifu Pharmaceutical University, Gifu 502 8585, Japan

Fábia Daniela Lobo de Sá, Eduard Butkevych, Praveen Kumar Nattramilarasu, Anja Fromm, Jörg-Dieter Schulzke and Roland Bücker
Institute of Clinical Physiology/Nutritional Medicine, Medical Department, Division of Gastroenterology, Infectiology, Rheumatology, Charité-Universitätsmedizin Berlin, 12203 Berlin, Germany

Soraya Mousavi, Markus M. Heimesaat and Stefan Bereswill
Institute of Microbiology, Infectious Diseases and Immunology, Charité-Universitätsmedizin Berlin, Campus Benjamin Franklin, 14195 Berlin, Germany

Verena Moos
Medical Department, Division of Gastroenterology, Infectiology and Rheumatology, Charité-Universitätsmedizin Berlin, 12203 Berlin, Germany

Julia C. Golz and Kerstin Stingl
German Federal Institute for Risk Assessment (BfR), Department of Biological Safety, National Reference Laboratory for Campylobacter, 12277 Berlin, Germany

Sophie Kittler and Diana Seinige
University of Veterinary Medicine Hannover, Research Center for Emerging Infections and Zoonoses, 30559 Hannover, Germany

Corinna Kehrenberg
Institute for Veterinary Food Science, Justus-Liebig-University, 35392 Giessen, Germany

Aisling Naylor, Alan Hopkins, Natalie Hudson and Matthew Campbell
Smurfit Institute of Genetics, Trinity College Dublin, Dublin 2, Ireland

Laura-Sophie Beier, Miriam Eichner and Jörg Piontek
Charité-Universitätsmedizin Berlin, Institute of Clini Physiology, Hindenburgdamm 30, 12203 Berlin, Germany

Jan Rossa, Jonas Protze, Anna Piontek and Gerd Krau
Leibniz-Forschungsinstitut für Molekulare Pharmakolo (FMP), Robert-Rössle-Straße 10, 13125 Berlin, Germany

Stephen Woodhouse and Nicole Zitzmann
Oxford Glycobiology Institute, Department of Biochemist University of Oxford, South Parks Road, Oxford OX1 3Q UK

Sophia Bergmann, Sabine Vidal-y-Sy and Johanna Brandner
Department of Dermatology and Venerology, Universi Hospital Hamburg Eppendorf, 20246 Hamburg, Germany

Holger B. Kramer
Oxford Centre for Gene Function, Department of Physiolo Anatomy and Genetics, University of Oxford, Oxford O 7DQ, UK

Laura Costea
Institute of Life Sciences, Vasile Goldiş Western Universi of Arad, 310414 Arad, Romania

Imola Wilhelm and István A. Krizbai
Institute of Life Sciences, Vasile Goldiş Western University Arad, 310414 Arad, Romania
Institute of Biophysics, Biological Research Centre, 67 Szeged, Hungary

Ádám Mészáros
Institute of Biophysics, Biological Research Centre, 67 Szeged, Hungary
Doctoral School of Biology, University of Szeged, 67 Szeged, Hungary

Hannelore Bauer
Department of Biological Sciences, University of Salzbu 5020 Salzburg, Austria

Hans-Christian Bauer and Andreas Traweger
Institute of Tendon and Bone Regeneration, Paracels Medical University—Spinal Cord Injury and Tiss Regeneration Center Salzburg, 5020 Salzburg, Austria

Attila E. Farkas
Institute of Biophysics, Biological Research Centre, 67 Szeged, Hungary
Department of Physiology, Anatomy and Neuroscien University of Szeged, 6726 Szeged, Hungary

ndex

H

I

K

L

M

N

O

www.ingramcontent.com/pod-product-compliance
Lightning Source LLC
LaVergne TN
LVHW082122200926
848484LV00008B/1991